Advanced
High Strength Steel
and Press Hardening

Advanced High Strength Steel and Press Hardening

Proceedings of the 3rd International Conference on
Advanced High Strength Steel and Press Hardening (ICHSU2016)

Xi'an, China, 25 – 27 August 2016

Editors

Yisheng Zhang
Huazhong University of Science and Technology, China

Mingtu Ma
China Automotive Engineering Research Institute, China

World Scientific

NEW JERSEY • LONDON • SINGAPORE • BEIJING • SHANGHAI • HONG KONG • TAIPEI • CHENNAI • TOKYO

Published by

World Scientific Publishing Co. Pte. Ltd.
5 Toh Tuck Link, Singapore 596224
USA office: 27 Warren Street, Suite 401-402, Hackensack, NJ 07601
UK office: 57 Shelton Street, Covent Garden, London WC2H 9HE

British Library Cataloguing-in-Publication Data
A catalogue record for this book is available from the British Library.

ADVANCED HIGH STRENGTH STEEL AND PRESS HARDENING
Proceedings of the 3rd International Conference on Advanced High Strength Steel and Press Hardening (ICHSU2016)

ISBN 978-981-3207-29-5

Preface

With the press hardening technology being under continuous development and successful industrial applications, there is an urgency to expand the research development in the technology. In this rapid growing market in advanced high strength steel and press hardening, in particularly demand from automotive industry and sustainability community to develop light-weight materials for **Body in white** or **BIW,** has motivated us to organize ICHSU2016, soon after the successful conclusion of our ICHSU2015 last year to encourage experts all over the world to get together again to exchange ideas on how to move the R&D in press hardening technology forward in the new era.

Based on this background, the 3rd International Conference on Advanced High Strength Steel and Press Hardening (ICHSU2016) is held in Xi'an, China during August 25–27, 2016. The purpose of holding ICHSU2016 is to satisfy the increasingly urgent requirement of reducing the weight of vehicle structures and increasing passenger safety. This conference arouses great interests and attentions from domestic and foreign researchers in hot stamping field, and 77 papers have been accepted, covering almost all the current topics of advanced high strength steel and press hardening technology.

The organizing committee is very grateful to the great efforts of the editors, reviewers and our publisher to publish this proceedings in such speedy manner. Our heart-felt thanks also go to all the authors, speakers and participants for their great contributions to make ICHSU2016 such a wonderful success. We shall be looking forward to meeting you again soon in the next international conference on hot stamping.

Yisheng Zhang and Mingtu Ma
Editors of the proceedings
ICHSU2016

ICHSU2016 Committees

Organized by:

- Imperial College London
- State Key Laboratory of Materials Processing and Die & Mould Technology, Huazhong University of Science and Technology
- State Key Laboratory of Vehicle NVH and Safety Technology, China Automotive Engineering Research Institute Co., Ltd.
- State Key Laboratory of Rolling and automation, Northeastern University
- National Engineering Laboratory of Clean Energy Vehicle & Power Train Systems, Tongji University
- Key Laboratory of Hubei Province for Modern Vehicle Component, Wuhan University of Technology

Co-organized by:

Materials Institution, Chinese Mechanical Engineering Society
Materials Committee of SAE-China
Shanghai Society of Automotive Engineers

Conference Organization

Chairman

Jianguo Lin, Imperial College London
Mingtu Ma, China Automotive Engineering Research Institute

Vice-Chairmen

Yisheng Zhang, Huazhong University of Science and Technology
Jianpin Lin, Tongji University
Shaodong Wang, Northeastern University
Lin Hua, Wuhan University of Technology
Xuejun Jin, Shanghai Jiaotong University
Fuh-Kuo Chen, National Taiwan University

Organizing Committee

Yisheng Zhang, Huazhong University of Science and Technology

Editorial Committee

Yisheng Zhang, Huazhong University of Science & technology
Mingtu Ma, China Automotive Engineering Research Institute

Contents

Materials & Testing I

Development of High Performance Press Hardening Steel Through Microstructure Engineering

Jian Bian[1,†], H. Z. Lu[2], W. J. Wang[2] and A. M. Guo[2]

[1]*Niobium Tech Asia, 068898 Singapore*
[2]*CITIC Metal, 100004 Beijing, China*
[†]*Email: jian.bian@niobiumtech.com*

Microstructure engineering determines predominantly the properties and applications of steel. Martensite provides the highest strength compared to any other microstructural constitutes, thus press hardening steel with martensite microstructure has been widely used in vehicles' body structure to enhance the passive safety in the event of a crash. However, for such applications there are also potential problems related to martensite microstructure, such as limited ductility (bendability), low impact toughness (low energy absorption and low fracture resistance) and low resistance to hydrogen induced cracking. Usually these properties dominate the fracture behaviour of martensitic microstructure and consequently the crash-resistance performance of press hardened components. Therefore, in order to achieve high performance, it is necessary to optimize the steel microstructure during entire metallurgical processes.

This paper introduces metallurgical solutions of microstructure engineering based on niobium metallurgy. The key point of microstructure engineering is to tailor and to fine-tune the microstructure details towards targeted applications with aims for high performance. The paper describes how to implement microstructure engineering in the press hardening steel by refining and homogenising the martensite matrix, strengthening and facilitating high angle grain boundaries in the martensite substructure, integrating a high density of homogenously distributed nano-sized NbC precipitates in the martensite matrix and finally creating a layer of soft microstructure near the surface area through controlled decarburization process during press hardening. The results and benefits of microstructure engineering to press hardening steel (22MnB5) are explained in detail.

Keywords: Martensite; Press hardening steel; Niobium metallurgy; Microstructure engineering and High performance.

1. Introduction

Press hardening steel has been increasingly used in car body structure for enhanced passive safety in the event of crash. However, the crash performance of press hardening steel under heavy impact is dependent on its fracture resistance besides strength, especially at high impact speed; lower temperature and in the worst-case scenario under hydrogen influence at the same time. Under these circumstances any premature cracking or fracture in the press hardened components will lead to uncontrolled intrusion of crash debris into the passenger cell. Therefore, from the safety point of view it is important to enhance the crash performance of conventional press hardening steel based on 22MnB5 by improving the relevant material properties, such as impact toughness, bendability and sensitivity to hydrogen embrittlement[1]. Herein the key solution is to increase the fracture resistance of martensite microstructure through microstructure engineering.

Microstructure engineering provides powerful metallurgical solutions to fine-tune the microstructural details such as microstructural heterogeneity, grain size, substructure, precipitation of micro-alloying particles and finally texture towards the targeted applications with aim for high performances. This paper demonstrates how to implement microstructure engineering based on Nb metallurgy to enable high performance through grain refinement, precipitation of nano-sized Nb precipitates in martensite matrix and finally microstructure modification in the surface area via decarburization process. The targets of microstructure engineering are to improve impact toughness, especially at lower temperature, to enhance the bendability avoiding premature cracking and to reduce the sensitivity to hydrogen embrittlement of conventional present press hardening steel.

2. Grain Refinement and Impact on High Performance Press Hardening Steel

Fracture resistance of martensite microstructure, especially at lower temperature, is very much dependent on the effective grain size. Small effective grain size provides more high angle grain boundaries to impede crack propagation, thus increases the fracture resistance. For press hardening steel the key point for grain refinement is to avoid grain coarsening during reheating process of press hardening in the entire process window or even when overheating or extended dwell time in the furnace occurs. The target is to keep the grain size of prior austenite (PAGS) as small and uniform as possible after reheating in the furnace. Microstructure engineering based on Nb metallurgy provides solutions to achieve such grain refinement due to two dominant contributions:

- Refining the microstructure of hot strip through TMCP rolling
- Grain boundary pinning by NbC precipitates during reheating process

Fig. 1 shows that through 0.05%Nb addition much smaller PAGS can be achieved under the same process conditions compared to conventional 22MnB5 after press hardening.

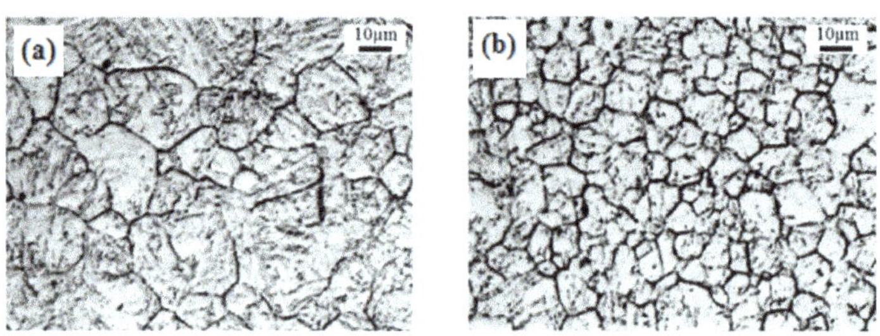

(a) 22MnB5 (b) 22MnB5+0.05%Nb

Fig. 1. Comparison of prior austenite grains between (a) 22MnB5 and (b) 22MnB5+0.05%Nb after simulated press hardening, 950 °C/6min.

- Generally conventional hot rolling process is commonly practised to produce 22MnB5. However, with Nb addition it is recommended to do TMCP rolling in order to explore the maximum effects of Nb on grain refinement. For example, rolling schedule, cooling strategy and coiling temperature should be modified accordingly. Finer microstructure in the hot strip will be inherited in the microstructure of the final product on one hand and on the other hand it will help to suppress the microstructural heterogeneity of banding structure through separate lays of ferrite and perlite which is often observed in the hot strip of produced 22MnB5. Such microstructural heterogeneity will dramatically deteriorate the bendability of press hardening steel, especially in the rolling direction, in which the banding structure prevails [1]. With decreasing PAGS the martensite microstructure will become finer as well by forming smaller packets within austenite grains. The grain boundary of prior austenite and martensite packets belongs to high angle boundary which can effectively retard crack propagation, thus increases the critical fracture stress, especially the ductile-brittle-transition temperature of martensite microstructure (DBTT). Due to the thin sheet for automotive application the stacked Charpy V-notch tests are commonly conducted for Charpy impact energy of automotive steels. Fig. 2 shows the Charpy impact energies of 1.5GPa and 1.8GPa press hardening steels as function of PAGS at room temperature. It must be pointed out that these data are from different test facilities and different sample sizes which have been reported in Bian and Mohrbacher [1] and Wang, Enloe, Singh and Horvath, [2] respectively. It is evident to see the trend that Charpy energies remain the same if PAGS is beyond 20μm, however, increase significantly if PAGS is reduced below 10μm approximately for 1.5GPa grade. It is particularly interesting to point out that the 1.8GPa grade (Δ simple) demonstrates even higher impact energy compared to 1.5GPa conventional grade (• simple) under exactly the same test conditions. The major reason for this improvement is the grain refinement through Nb micro alloying (0.06%) in the 1.8GPa grade. Normally the 1.8GPa grade should have lower impact energy than 1.5GPa grade due to higher carbon content, providing PAGS is the same. Independent investigations also confirmed that DBTT will decrease with decreasing PAGS as well[2].

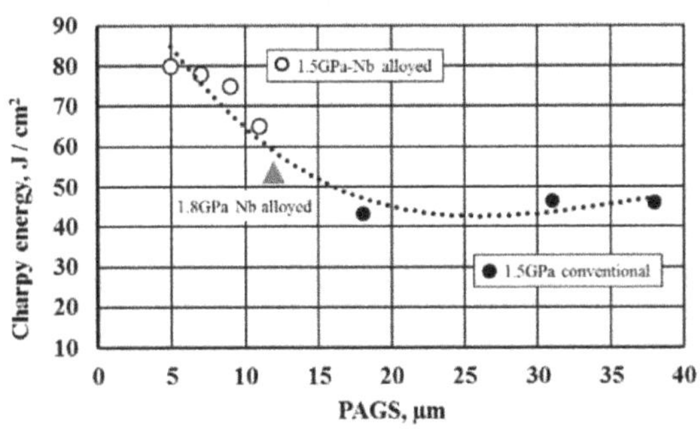

Fig. 2. Charpy impact energies as function of PAGS investigated on 1.5 and 1.8GPa press hardening steels at 25 °C [1,2].

3. Precipitation of Nano-Sized NbC and Impact on High Performance Press Hardening Steel

Another important target of microstructure engineering is to control precipitation of nano-sized NbC in press hardening steel throughout hot rolling process. It is particularly important to set the coiling temperature in the range between 630 and 650°C in order to get complete precipitation of niobium. Investigations found that the optimum precipitation of NbC can be reached in this coiling temperature range with regard to particle size, density, distribution and coherency [3]. Generally, the precipitation of NbC has following functions in the press hardening steel:

- Grain refinement through TMCP rolling
- Prevention of grain coarsening through pining effect during reheating for press hardening
- Hydrogen trapping throughout entire manufacturing processes and service life

Nano-particles of NbC precipitated during coiling process [3] can effectively pin the grain boundaries of austenite and retard the grain coarsening, which takes place to conventional press hardening steel 22MnB5 unavoidably, especially with increasing temperature during reheating. Due to pining effect the grain growth rate of Nb alloyed grade is 3 times lower than conventional 1.5GPa grade, calculated according to the regression equations listed in Fig. 3. The stabile grain structure of Nb alloyed steel, especially at higher reheating temperate and extended dwell time in the furnace makes it possible to enlarge the process window for press hardening, as shown in Fig. 4. The white area in Fig. 4 represents the process window for connectional 22MnB5 with point A standing for minimum process conditions (lowest temperature and shortest furnace time for austenitization and homogenization), point B standing for the normal process conditions and point C standing for the maximum process conditions (highest temperature and longest furnace time). Independent investigations found that austenite grain size becomes critically large (about 35μm) if reheating conditions are beyond point C so that components made from such steel might crack in the validation tests [2]. In the common practice heated material will be scraped in such cases in order to avoid component failures. However, providing pining effect through NbC the process window can be enlarged significantly to higher temperature (point D) and longer furnace dwell time (point E) without producing abnormal austenite grains (Fig. 3), thus press hardening process becomes much more robust due to Nb addition. Another important functionality of nano-sized NbC precipitates in the final microstructure is to serve as hydrogen traps and to reduce the sensitivity to hydrogen embrittlement [4].

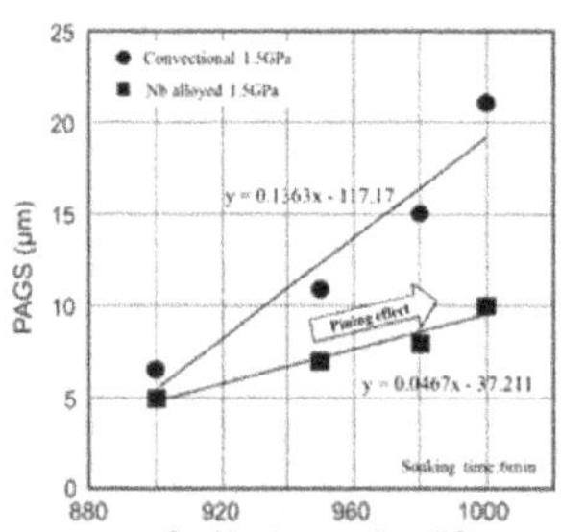

Fig. 3. PAGS of conventional and Nb alloyed 1.5GPa PHS as function of soaking temperature.

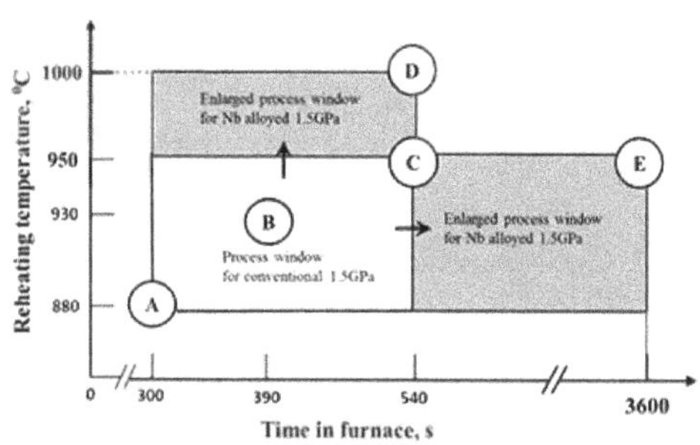

Fig. 4. Process window of conventional and Nb alloyed 1.5GPa in relation to soaking temperature and time.

4. Microstructure Modification in the Surface Area and Impact on High Performance Press Hardening Steel

Microstructure modification in the surface area belongs to standard technology for many steel grades and for different applications. For example, through surface carburization or decarburization martensite or ferrite will exclusively form in the surface area which provides high wear resistance in case of carburization and high formability in case of decarburization. The combination of hard surface with soft core or wise verse will create unique properties and applications. In practice of press hardening by using uncoated material surface decarburization is often observed after press hardening due to uncontrolled furnace atmosphere with air penetration. The decarburization depth is usually around 20 to 50µm. De Cooman et al. found that the surface decarburization, generally regarded as a negative side effect due to slight loss of strength, is beneficial to improve bendability of PHS [5]. Apart from strength and elongation bendability is the most important and currently the only quantitatively specified material property for 1.5GPa grade by many OEMs. Especially, for high strength PHS with 1.8-2.0GPa UTS all crash relevant material properties such as impact toughness, bendability and hydrogen embrittlement become very critical due to higher carbon content and the soft ferrite in the surface area through decarburization can be a powerful enabler to boost these properties. Therefore, it is important to develop decarburized press hardening steel, especially for high strength PHS with 1.8-2.0GPa UTS. The decarburized press hardening steel could bring following potential benefits:

- Improvement of impact toughness (to be verified)
- Improvement of bendability (experimentally confirmed)
- Improvement of hydrogen embrittlement behaviour (promising results from preliminary investigations)
- Improvement of weldability (to be investigated)

Basically there are two possibilities to conduct decarburization process for press hardening steel, either in the reheating furnace for uncoated material or in the annealing / coating process for coated material. Production experiences verified that technologically it is feasible to achieve 20-50µm deep surface decarburization in the conventional reheating furnaces of any kind under conventional process conditions by just adjusting the dew point in the protective atmospheres. Due to surface decarburization ferrite or ferrite / perlite will form in the surface area after die quenching, as shown in Fig 5. By the bending test the outer bend will be under the highest tensile stress and ferrite has much better formability than martensite and can impede crack propagation, thus improve bendability. The principle of improving bendability through surface decarburization is explained in Fig. 6. Fig 7 shows a press hardened B-pillar with approximately 50µm deep surface decarburization. The bending tests on the samples from B-pillar demonstrate around 80 degree bending angle uniformly by the sample thickness of 1.8mm. In this particular case the minimum bending angle for 1.8mm is 55 degree required by OEMs and the improvement of bending angle due to the surface decarburization is more than 60% (unpublished paper). However, the decarburization technology for press hardening steel needs to be further developed for mass production. Nb micro-alloying could be beneficial to facilitate decarburization process through grain refinement (grain boundary diffusion of carbon) and to suppress grain coarsening in decarburized ferrite zone.

Fig. 5. Optical micrograph showing surface decarburization after heat treatment.

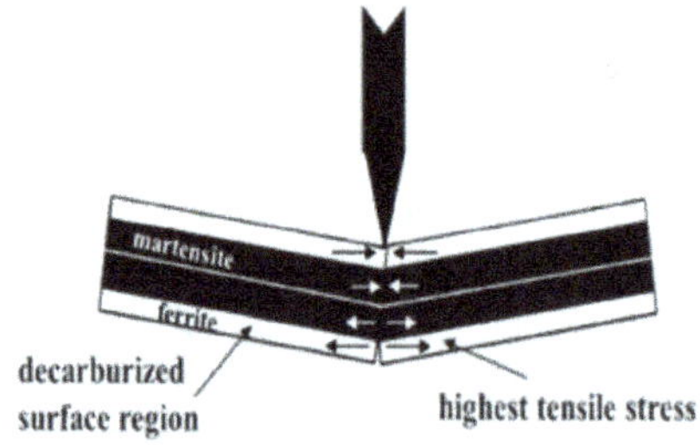

Fig. 6. Influence of surface decarburization on bending behaviour of steel sheet.

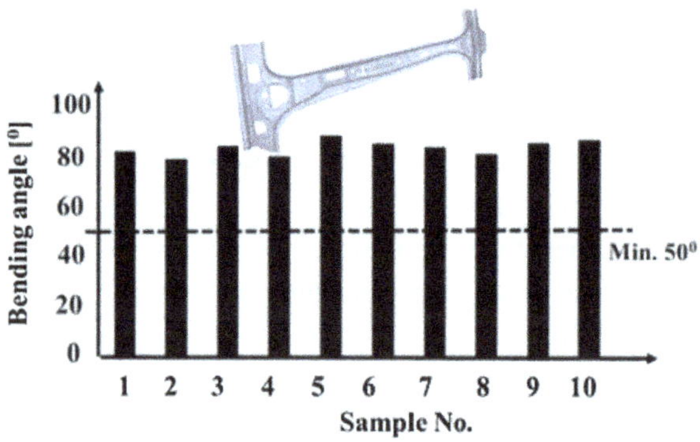

Fig. 7. Bending angle of samples taken from decarburized B-pillar with 50μm decarburized zone, according to VDA238-100(thickness 1.8mm).

5. Discussion

Nb metallurgy plays a key role in microstructure engineering of press hardening steel by providing two fundamental metallurgical effects of gain refinement and precipitation of nano-sized NbC particles in the steel matrix. For micro-alloyed steels grain refinement can be achieved most effectively through Nb micro-alloying plus TMCP rolling while precipitation of NbC can be controlled during cooling and coiling process after hot rolling. The optimum precipitation of Nb in the aspect of particle size, density, distribution and morphology can help to preserve the TMCP rolled fine-grained microstructure throughout annealing and press hardening process to the final product by pining grain boundary and suppressing grain coarsening. Fined-grained martensite microstructure contains smaller packets. The packet boundary belongs to high angle boundary which can effectively retard crack propagation, thus increase the critical fracture stress, especially at lower temperature. Consequently, fined-grained martensitic steel has higher impact energy and lower DBTT which is generally required for the crash relevant structure parts. Grain refinement will also improve the microstructural heterogeneity during phase transformation by providing more nucleation sites for the new phases, thus avoiding banding structure consisting of separate layers of ferrite and perlite which is often observed in the hot strip of produced 22MnB5. Such microstructural heterogeneity will dramatically deteriorate impact behaviour and bendability of press hardening steel. The nano-sized Nb precipitates in the martensite matrix has another important mission to serve as hydrogen traps for higher resistance to hydrogen embrittlement caused by hydrogen penetrated into press hardening steel at any production stage or during the service life. Therefore, high performance press hardening steel should have fine-grained martensite microstructure with finely and homogeneously dispersed NbC particles. Consequently, intensive research has been focused on 22MnB5 with Nb addition for the last few years. In order to boost the performance of conventional press hardening steel (22MnB5) the optimum Nb content was found to be around 0.05% [1]. From the production side it is important to pay attention to following aspects:

- to dissolve Nb completely during slab reheating
- to reduce TiN precipitation at high temperature as much as possible
- to reduce large inclusions and content of S, P and H

TiN precipitated at high temperature will trigger co-precipitation of Nb with Ti and Nb co-precipitated at high temperature will not make notable contributions to microstructure engineering. Besides, large precipitates like TiN usually serve as stress concentrators, thus deteriorate impact toughness and bendability. It is advisable to limit N and Ti addition or even to take Ti out completely. In this case N will be stabilized by Al. For the same reasons it is important to limit S and P content and especially, hydrogen content in the press hardening steel. P and H tend to segregate on the grain boundary of prior austenite causing brittle behaviour of intergranular fracture. Finally, microstructure engineering based on Nb metallurgy is a powerful enabler to make high performance press hardening steel which is characterized by higher impact toughness against cracking, higher bending angle for energy absorption, higher resistance to hydrogen induced cracking to ensure component functionality, more weldable due to surface decarburization.

6. Conclusion

Crash performance of press hardening steel under heavy impact is dependent on its fracture resistance, especially at high impact speed; lower temperature and in the worst-case scenario under hydrogen influence at the same time. Under these circumstances any premature cracking or fracture in the press hardened components will lead to uncontrolled intrusion of crash front into the passenger cell. From material side the fracture resistance of press hardening steel is controlled by the microstructural features of martensite after die quenching. This paper demonstrated that the high performance steels should have following microstructural features:

- Fine-grained microstructure (smaller PAGS and finer martensite substructure)

- Homogeneously dispersed nano-sized NbC in the steel matrix
- Soft microstructure near the surface region through surface decarburization

The results shown in this paper verified clearly that such microstructural features can be achieved by microstructure engineering based on Nb micro-alloying which provides a powerful enabler to enhance the performance of conventional press hardening steel based on 22MnB5. The major benefits of microstructure engineering to press hardening steel are summarized as follows:

- higher impact energy, especially at lower temperature
- higher bending angle according to 3-point-bending test
- higher resistance to hydrogen embrittlement.

It can be expected that such improvements on press hardening steel will consequently enhance the component functionality and vehicle safety during the service life.

References

1. J. Bian and H. Mohrbacher, in Proc. of AIST International Symposium on New Developments in Advanced High Strength Sheet Steels, (Colorado, USA, 2013).
2. J. Wang, C. Enloe, J. Singh and C. Horvath, Effect of Prior Austenite Grain Size on Impact Toughness of Press Hardened Steel, SAE Int. J. Mater. Manf. **9**, 488 (2016).
3. J. Bian, H. Mohrbacher, H. Lu and W. Wang, Development of Press Hardening Steel with High Resistance to Hydrogen Embrittlement, in HSLA Steels 2015, Microalloying 2015 & Offshore Engineering Steels 2015: Conference Proceedings, (John Wiley & Sons, Inc., 2015).
4. H. Lu, S. Zhang, B. Jian, H. Mohrbacher and A. Guo, Solutions for Hydrogen-Induced Delayed Fracture in Hot Stamping, Adv. Mater. Res. **1063**, 32 (2015).
5. W.S. Choi and B.C. De Cooman, Characterization of the Bendability of Press-Hardened 22MnB5 Steel, Steel Res. Int. **5**, 824 (2014).

Low Density High Strength and High Toughness Steel with Ferrite and Martensite Laminated Structure

W. Q. Cao[1,a,†], H. F. Xu[1,2,b], M. D. Zhang[1,2,c], C. X. Huang[3,d], H. Dong[1,e] and Y. Q. Weng[1,f]

[1]*Special Steel department of Central Iron and Steel Research Institute (CISRI), Beijing 100081, P. R. China*

[2]*School of Materials Science and Engineering, Huazhong University of Science and Technology, Wuhan 430074, P. R. China*

[3]*School of Materials Science and Engineering, Sichuan University, Chengdu 610065, P. R. China*

E-mail: [a,†]caowenquan@nercast.com, [b]xuhaifeng228@163.com, [c]15901022010@139.com, [d]chxhuang@scu.edu.cn, [e]donghan@nercast.com, [f]wengyuqing@ nercast.com

The impact toughness of high strength steel decreases with increasing strength in general and thus leads to extensive research to improve the toughness of high strength steel. In this study a series high strength low carbon and low alloy steels alloyed with manganese and aluminum were designed, hot deformed in the ferrite and austenite dual phase region and air cooling down to room temperature. The microstructure observation indicates that the hot deformation introduces a fibered ferrite phase into the martensite matrix in the bar samples or laminated ferrite phase into the martensite matrix in the plate samples. Charpy V-notch toughness measurements at room temperature reveal that the designed steels are remarkably enhanced high up to 150-250Jin bar steel and 200-450J in plate steel with tensile strength varied from 1.0-1.5GPa, both of which are nearly 10 times higher than that of the conventional low alloy steels with similar strength level. This finding gives us another way to produce high strength steel with ultrahigh impact toughness by simple alloying design and hot deformation in the ferrite and austenite dual phase region.

Keywords: Ultrahigh strength steel; Impact toughness; Area reduction; Ferrite fiber or laminate; Interface delaminating.

1. Introduction

Toughness is an indication of the capacity of a steel to absorb energy and is dependent on strength as well as ductility and the critical impact toughness is strongly concerned as one of the screening parameters of steel properties in many structural steel applications to ensure the safety [1,2], such as the Pipe line steel, Bridge steel, Car steel, and Aeronautic steel, et.al. With today's situation of severe environmental pollution and natural resource limitation, steels with much higher strength are applied in many industries for weight lightening and safety improvement. However, the impact toughness of low alloy steel normally decreases with tensile strength increasing [3-5].

The dependence of the V-notch impact toughness at room temperature are plotted in Fig.1(a) as a function of tensile strength of different steels from various literatures, such as the precipitated hardening steel, maraging steel, pipeline steel, nano-bainite steel and other bar steels. For the low alloy steel as revealed in Fig.1 (a), the impact toughness decreases from ~300-400J at strength level of 0.5GPa (pipeline steel), to 50-100J at 1.0 GPa (medium carbon steel) and to only 10-40J at 1.5GPa (High carbon steel) [6-8]. However for high alloyed steel, such as the high alloyed precipitated hardening steel, austenite steel and nanobainite steel, the impact energy of about 200J at strength level of 1.5GPa and about 50J at 2.0GPa could be obtained, which is several times higher than that of low alloy steels [9-11]. The underlined toughening mechanism is summarized in Fig.1 (b) that the toughness value of these low alloy steels is mainly determined by the area reduction of the tensile specimens. This dependence (even though there are some deviation due to some unknown reasons) of impact toughness on the area reduction (Z) could be easily related to the similar local deformation capacity both in tensile experiment and in the Charpy impact toughness measurement. Future development of the high toughness to ultrahigh strength needs to be done as indicated by the arrow in Fig.1(a) and may be realized by the trend as indicated by the arrow in Fig.1(b). In order to get this target, the dependence of toughness on the strength of low alloy steels has to be altered, i.e., to find a way to break up the relationship between toughness and local deformation capacity as shown in Fig.1(b), which demonstrates that the impact toughness is controlled by the local deformation capacity in constitutive materials.

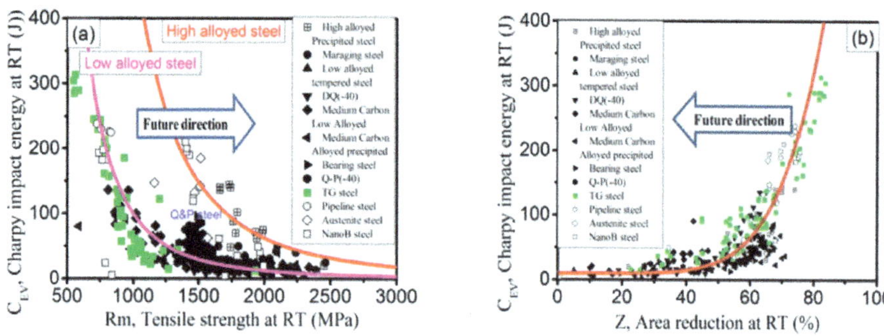

Fig. 1. The dependence of V-notch toughness on the tensile strength and ductility of the steels (a) Impact toughness varied with tensile strength Rm, (b) Impact toughness varied with area reduction A of the tensile specimens.

Apart from the local deformation capacity controlling impact toughness as indicated above, remarkably enhanced impact toughness has been revealed in the composite materials with laminated structure [12-18]. Composite laminated materials consist of different constitutive materials and are alternately separated by discrete interfaces. The alternatively distributed interface are capable of arresting propagating cracks and is responsible of the high impact and fracture resistance of the composites, much better than the constitutive material components separately [19-22]. It was proposed that this remarkably enhanced impact toughness is resulted from the interface delamination initiated from the laminate interface and thus blunt the crack tip when the crack front intersects with the interface of the laminates [23-24]. For example, when the aluminum of 5xxx and 1xxx were hot rolled into a laminated structure with 19 layers, the impact toughness was increased with 18times increment. In addition, when low carbon steel was hot rolled together with ultrahigh carbon steel, a laminated steel composite would developed and gives very high impact toughness and very low ductile to brittle transition temperature comparing with the single materials [16]. Even though the laminated composites can improve the impact toughness significantly, their conventional production procedure is very complex, which increases the cost significantly and restricts their production rate in large quantity [25, 26].

In this research, a series of low carbon and low alloy steels alloyed with Mn and Al were designed and fabricated by conventional melting, casting and hot rolling. Both the martensite and ferrite laminated microstructure and the remarkably enhanced impact toughness were examined and demonstrated in this research. The underlined formation mechanism of the laminated microstructure and the toughness-enhancing mechanism of this new designed low carbon and low alloy steel will be discussed tentatively in this paper.

2. Experimental

The chemical composition of FeMnAlC steels with C of 0.05-0.2%, Mn of 5% and Al of 3-4% are designed, where Mn is added to obtain the air hardening capacity, Al is to widen the dual phase region and to decrease the density ~5%, and hot deformation is applied to develop the laminated microstructure. These steels were prepared by high frequency induction furnace in a vacuum atmosphere and casted into ingots with 50Kg. The ingots were homogenized at 1250°C with two hours, forged at 1200 °C-850°C into plates and bars. The final dimension of the forged plates is 40mm in thickness and 200mm in width. The dimension of bars is ϕ16mm in diameter. Then some forged plates hot rolled into thin plates with thickness of 12mm after heating to 1150°C (70% thickness reduction) and then there were are cooled to room temperature. Both the forged bars and the hot rolled thin plates were annealed at 200°C with 2 hours as the materials for microstructure and property examination in this research.

The microstructure of the bars and the plates were examined by optical microscopy (OM), scanning electron microscopy (SEM), electron back scattered diffraction (EBSD) and transmission electron microscopy (TEM). The volume fractions of the phases were measured according to the color difference in the optical micrograph and calculated by using the Thermo-Calc software with TCFE 7 database provided by the CISRI-TCS Joint Open Laboratory. The dog-bone shaped tensile specimens with gauge length of 25 mm and diameter of 5 mm were cut with their axial directions paralleling to the longitudinal direction. Tensile tests were conducted in a universal testing machine with a strain rate of 10^{-3} s^{-1} at room temperature. The micro-hardness of different phases was measured in the transverse plane to identify the property of the laminated ferrite phase and the martensite phase. Standard specimens for Charpy V-notch impact measurement at different temperature were cut from the thin plates and bars to examine the toughness and its anisotropy.

3. Results

3.1. *Microstructure observation of impact samples*

The microstructures of the hot rolled steel with chemical composition of 0.05C5Mn3Al, 0.10C5Mn3Al and 0.15C5Mn3Al are examined by EBSD and the results are given in Fig.2. It can be seen that a laminated structure developed after hot rolling with ferrite and martensite alternatively arranged along the thickness direction. The white phase is the ferrite and grey phase is the martensite phase. In addition, the volume fraction and the thickness of the ferrite lamina decreases with increasing of carbon from 0.05% to 0.2% at 5%Mn and 3%Al. Based on the phase diagram of xC5Mn3Al, the microstructure is composed with ferrite phase and the austenite and both of them would be flattened during to the hot rolling. Thus in this dual phase laminated microstructure, the ferrite laminated comes from the already existed flattened ferrite grains, but the martensite lamina is a phase transformation product from the hot rolled flatted austenite grains.

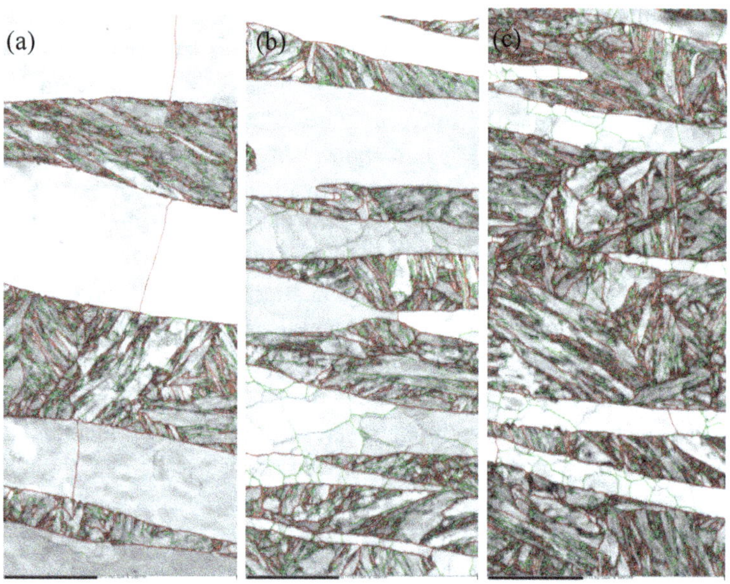

Fig. 2. Laminated microstructure of the hot rolled xC5Mn3Al examined by EBSD (a) 0.05C5Mn3Al, (b) 0.1C5Mn3Al and (c) 0.15C5Mn3Al, in which the high angle boundary with misorientation no less than 15degrs is revealed by red color and the low angle boundary is shown by green color.

3.2. *Mechanical properties and impact toughness of studied FeMnAlC steel*

The Engineering stress-strain curves of studied steel bars were measured by tensile test and were calculated and were given in Fig.3 (a) and the impact toughness of these bars were measured at different temperature and the results were given in Fig.3 (b). It can be seen that for the xC5Mn3Al steel both yield strength (Rp0.2) and tensile strength (Rm) increase with increasing of the carbon content, but their total elongation decreases with increasing of carbon content. In addition, both the high upper shelf energy of the impact toughness (USE) and the low ductile to brittle transformation temperature (DBTT) can be observed in Fig.3 (b). The USE is about 150-260J for the forged bars, which varied with carbon content and is about 10times higher than that of the convention martensite steel with same strength level as shown in Fig.1(a). The DBTT for our studied bars is about -75℃, which do not change with carbon content and are significant lower than that of the conventional martensite steel. This implies that the introduction of fibered ferrite into the martensite matrix would increase the USE remarkably and decrease the DBTT significantly.

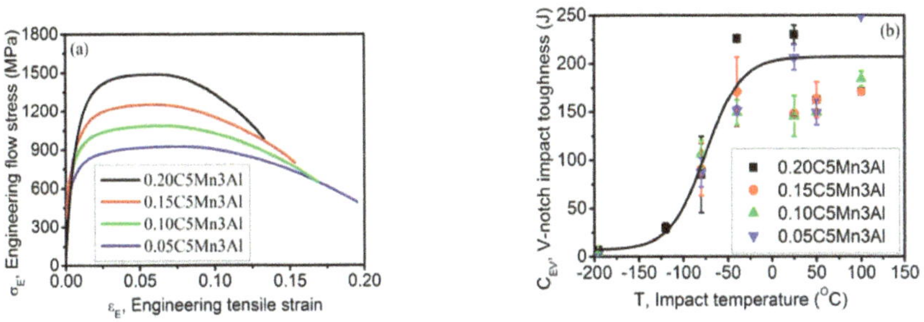

Fig. 3. Mechanical property and impact toughness of the forged bars of xC5Mn3Al steels (a) Stress-strain curves at room temperature and (b)V-notch Charpy impact toughness measured at different temperature.

The mechanical properties and Charpy impact toughness at room temperature were measured and the results were given in Table 1, in which the impact toughness of the plates were measured in two different specimen sampling, one is the notch parallel to the normal direction of the rolling plane (named as E_{CV-NT}) and the other is along the transverse direction (named as E_{CV-TN}). It can be seen that the strength increases but ductility decreases with increasing of carbon content. The toughness measured in the rolling plane is varied from 174 to 406J (the measured deviation was given in the bracket), which is 5-10 times higher than that measured in the transverse plane.

Table 1. Room mechanical properties and the impact toughness of hot rolled plates with 70% and 40% thickness reduction.

Steels/Strain	Rm (MPa)	Rp0.2 (MPa)	A5 (%)	Z (%)	E_{CV-NT} (J)	E_{CV-TN} (J)
0.05C5Mn3Al/70%	806	517	22.0	69.5	360(40)	119.9(10)
0.10C5Mn3Al/70%	984	644	18.3	66	362(20)	97.3(10)
0.15C5Mn3Al/70%	1199	785	14.8	54	406(30)	79.5(18)
0.20C5Mn4Al/70%	1468	700	12.8	23	174(23)	23.5(3)

4. Discussions

4.1. *Development of the laminated dual phase microstructure of FeMnAlC steels*

Based on the laminated microstructure observation as shown in Fig.2, the microstructure evolution during hot deformation process could be divided into three steps, the first is the formation of very coarse grained and granular shaped $\alpha+\gamma$ microstructure developed in the austenization by the portioning of C, Mn and Al elements at temperature of about 1200℃, the second step is the microstructure from the very large sized and gradually shaped grained microstructure to laminated α+γ microstructure by the pancaking of the grains of both a and g grains during hot deformation process. And finally is the phase transformation of the laminated austenite phase to laminated martensite phase and finally developed into a α+M laminated dual phase steel. Thus the microstructure evolution of our designed dual phase FeMnAlC steels are mainly controlled by elements portioning controlled grain growth in austenization process, the pancaking of the coarse grains to laminated microstructure and the phase transformation from austenite to martensite, which is schematically illustrated in Fig.4.

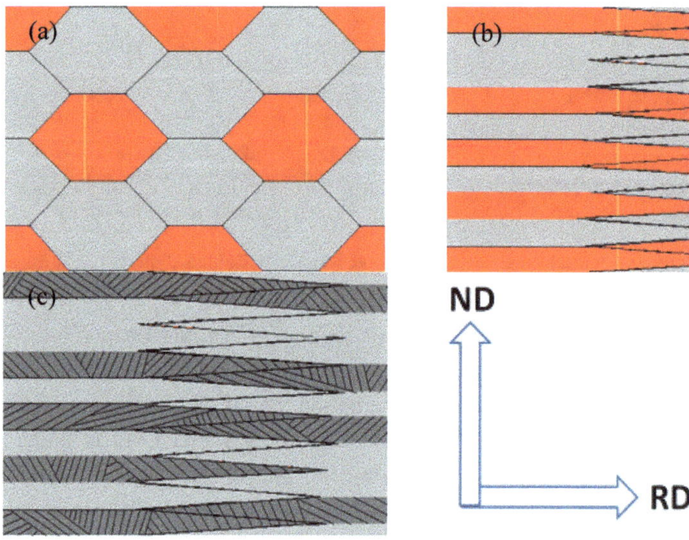

Fig. 4. The microstructure evolution during hot rolling process from coarse and granular grained dual phase microstructure in (a), to a α+γ dual phase laminated microstructure after 70% thickness reduction at high temperature in (b) and finally to a a+M dual phase laminated microstructure in (c).

4.2. *Comparison of the impact toughness between our studied steel and the conventional high strength steel*

The high strengthened dual-phase steels with coarse grain structure usually assume very low toughness, especially under the low temperature range, which is only about not enough to meet the requirement of the damage tolerance for the application. Over the past two decades, the dual-phase steel consisting with ferrite phase and martensite phase The dual-phase steels consisting with ferrite and martensite normally assume a relative low impact toughness of ~10-30J [27,28]. However, our designed dual phase steels with laminated microstructure assume toughness of ~150-400J as shown in Fig. 3 and Table 1. For the hot forged bar steels, the toughness is ~150-260J at room temperature with DBTT about -75℃, both of these values are much superior than that of the conventional steels. For the hot rolled plates, the toughness is about 174-406J at room temperature, indicating extrahigh toughness if the V-notch is aligned in the rolling plane. However, the toughness of the hot rolled plates in the transverse plane is only about 20-120J at room temperature as given in Table 1, which is much lower than that in the rolling plane but still much higher than that of the conventional dual phase steels. Furthermore, the elongation and toughness as shown in Table 1 is also much higher than that of the conventional dual phase steels. The big difference between the tensile strength and the yield strength implies a very high work hardening rate in the initial deformation process, which is essential for the formation capacity of the dual phase steels.

5. Conclusion

The mechanical properties, the V-north type impact toughness and the microstructure of the new developed steel with composition FeMnAlC were examined. The conclusions could be given as follows.

1) The laminated ferrite/martensite duplex microstructure has been well developed by suitable chemical design with FeMnAlC and hot deformation in the α+γ or δ+γ region, which could be explained by the panking of the grains of different phases with very high stability to suppress the granularity during dynamic recrystallization process.
2) The extra-high impact toughness of ~150-450J has been found in the high temperature forged bars and hot rolled steel plates with yield strength, tensile strength and the total elongation of the new developed steel is about 0.7-1.0GPa, 0.8-1.5GPa and 15-20%, indicating a very good combination of strength and ductility. Furthermore, the ductile to brittle transition temperature is reduced about 100°C, implying significantly low ductile to brittle transition temperature of the ultrahigh strength steel.
3) Future work should be done to understand the dependence of toughness on the type of microstructure between granular structured steel and laminate structured steel, which may provide us a promising way to fabricate high strength and high toughness steel and other alloys.

Acknowledgments

This research was supported by both National Natural Science Foundation of China (NSFC) No. 51371057 and National Basic Research Program of China (973 program) No. 2010CB630803.

References

1. R.O. Ritchie, The conflicts between strength and toughness, *Nature Materials* **10**, 817 (2011).
2. Y. Kimura, T. Inoue, F. Yin and K. Tsuzaki, Inverse Temperature Dependence of Toughness in an Ultrafine Grain-Structure Steel, *Science* **320**, 1057 (2008).
3. Cottrell, Int. J. Pressure Vessels piping, 64, 171 (1995).
4. V.F. Zakey, E.R. Parker, G. Thomas and J.W. Morris, *Mat. Sci. Eng. A* **16**, 201 (1974).

5. I.W. Morris, Jr., Z. Guo, C.R. Krenn and Y.H. Kim, The Limits of Strength and Toughness in Steel, *ISIJ Inter.*, 41, 599 (2001).
6. T.V. Philip, T. J. McCaffrey, *in Metals Handbook* (ASM International, Novelty, OH, ed. 10, 1990), vol. 1, pp. 430.
7. Y. Tomita, Development of fracture toughness of ultrahigh strength, medium carbon, low alloy steels for aerospace applications, *Int. Mater. Rev.* **45**, 27 (2000).
8. Y. Tomita, Low-Temperature improvement of mechanical properties of aisi 4340 steel through high-temperature thermomechanical treatment, *Metall. Trans. A.* 22A, 1093 (1991).
9. J.L. Maloney and W.M. Garrison, The effect of sulfide type on the fracture behavior of HY180 steel, *Acta Mater.* **53**, 533 (2005).
10. L.E. Iorio and W.M. Garrison Jr., The effects of titanium additions on AF1410 ultra-high-strength steel, *Metall. Mater. Trans. A.* **37**, 1165 (2006).
11. G. J. Spaeder, Impact transition behavior of high-purity l8Ni maraging steel, *Metall. Trans.* **1**, 2011 (1970).
12. C.M. Cepeda-Jimenez, J.M. Garcia-Infanta, M. Pozuelo, O.A. Ruano and F. Carreno, Impact toughness improvement of high-strength aluminium alloy by intrinsic and extrinsic fracture mechanisms via hot roll bonding, *Scripta Mater.* **61**, 40 (2009).
13. B.A. Movchan and F.D. Lemkey, Mechanical properties of fine-crystalline two-phase materials, *Mater. Sci. Eng. A.* **224**,136 (1997).
14. F. Carreno, J. Chao, M. Pozuelo and O. A. Ruano, Microstructure and fracture properties of an ultrahigh carbon steel–mild steel laminated composite, *Scripta Mater.* **48**, 1135 (2003).
15. C.M. Cepeda-Jiménez, M. Pozuelo, J.M. García-Infanta, O.A. Ruano and F. Carreño, Influence of the alumina thickness at the interfaces on the fracture mechanisms of aluminium multilayer composites, *Mater. Sci. Eng. A.* **496**, 133 (2008).
16. M.Z. Quadir, A. Wolz, M. Hoffman and M.Ferry, Influence of processing parameters on the bond toughness of roll-bonded aluminium strip, *Scripta Mater.* **58**, 959 (2008).
17. A.Y. Chen, D.F. Li, J.B. Zhang, H.W. Song and J. Lu, Make nanostructured metal exceptionally tough by introducing non-localized fracture behaviors, *Scripta Mater.* **59**, 579 (2008).
18. C.M. Cepeda-Jiménez, M. Pozuelo, O.A. Ruano and F. Carreño, Influence of the thermomechanical processing on the fracture mechanisms of high strength aluminium/pure aluminium multilayer laminate materials, *Mater. Sci.& Eng. A.* **490**, 3197 (2008).
19. D.W. Kum, T. Oyama, J. Wadsworth and O.D. Sherby, The impact properties of laminated composites containing ultrahigh carbon (UHC) steels, *J. Mech. Phys. Solids.* **31**,173 (1983).
20. S. Lee, J. Wadsworth and O.D. Sherby, Impact properties of a laminated composite based on ultrahigh carbon steel and Hadfield manganese steel, *Res. Mechanica.* **31**, 233 (1990).

21. D.W. Kum, T. Oyama, O.A. Ruano and O.D. Sherby, Development of ferrous laminated composites with unique microstructures by control of carbon diffusion, *Metall. Trans. A* **17**, 1517 (1986).
22. D.R. Lesuer, C.K. Syn, O.D. Sherby, J. Wadsworth, J.J. Lewandowski and W.H. Hunt, Mechanical behaviour of laminated metal composites, *Int. Mater. Rev.* **41**, 169 (1996).
23. C.K. Syn, D.R. Lesuer and O.D. Sherby, Enhancing tensile ductility of a particulate-reinforced aluminum metal matrix composite by lamination with Mg-9%Li alloy *Mater. Sci Eng A* **206**, 201 (1996).
24. C.K. Syn, D.R. Lesuer, J. Wolfenstine and O.D. Sherby, Layer thickness effect on ductile tensile fracture, *Metall. Trans. A*. **24**, 1647 (1993).
25. M.S. Joo, D.W. Suh, J.H. Bae and H.K.D.H. Bhadeshia, Role of delamination and crystallography on anisotropy of Charpy toughness in API-X80 steel, *Mater. Sci. Eng. A.* **546**, 314 (2012).
26. F. Carreno, J. Chao, M. Pozuelo and O.A. Ruano, Microstructure and fracture properties of an ultrahigh carbon steel-mild steel laminated composite, *Scripta Mater.* **48**, 1135 (2003).
27. S.S.M. Tavares, P.D. Pedroza, J.R. Teodósio and T. Gurova, Mechanical properties of a quenched and tempered dual phase steel, *Scripta Mater.* **40**, 887 (1999).
28. Y. J. Chao, J. D. Ward and R. G. Sands, Charpy impact energy, fracture toughness and ductile-brittle transition temperature of dual-phase 590 steel, *Mater. Des.* **28**, 551 (2007).

Study on the Mechanical Property Control of Hot Stamped Parts Based on the Grain Size

F. F. Li[1], J. P. Lin[1,2,†], Y. Lin[1] and Y. J. Yang[1]
[1]*School of Mechanical Engineering, Tongji University, Shanghai, 201804, P.R. China*
[2]*National Engineering Laboratory of Clean Energy Vehicle & Powertrain System, Shanghai, 201804, P.R. China*
[†]*E-mail: jplin58@tongji.edu.cn*

The mechanical properties of boron steel 22MnB5 are determined by its microstructure, and the microstructural features mainly include phase fraction and grain size. The property control of hot stamped parts based on the grain size was studied in the following experiments. The effect of heating temperature and dwell time on grain size of boron steel was analyzed so that the heating parameters could be determined according to the required grain size. The effect of grain size on the strength and elongation properties was researched as well in order to allow property control by adjusting the grain size.

Keywords: Hot stamping; Boron steel 22MnB5; Mechanical property control; Grain size.

1. Introduction

The mechanical property control of hot stamped parts is key to control the crash performance of safety parts. We draw a particular attention to the significance of the parts' geometries and techniques to control them, and form theories accordingly. The mechanical properties of hot stamped parts are determined by the phase fraction, grain size, and distribution of microstructure at whole part range. This paper studies the property control of hot stamped parts with respect to grain size.

2. Experiment

In order to investigate the mechanical properties of boron steel with different grain sizes, and the effect of heating parameters (heating temperature and dwell time) on the grain size, the heat treatment experiment with heating temperatures of 850 and 950 °C and dwell times of 5, 20 and 30 min was designed. The grain size of austenite cannot be measured directly as it exists at elevated temperatures, so the grain size of martensite after water quenching is used to characterize the grain size.

The thickness of boron steel is 1.8 mm. The specimen was machined according to "GB/T 228-2002 standard test method for tensile testing of metallic materials at room temperature". The tensile tests were conducted on a ZWICK/ROELL Z050 universal testing machine with an equivalent strain rate of 0.001/s. In terms of metallography analysis, the specimens were mounted in epoxy resin, ground, and polished to a mirror finish by using 500, 1200, and 2000 grit SiC paper, and followed by 3- and 1-μm diamond pastes. A 4 % nitric acid alcohol solution was used to reveal the microstructure.

3. Effect of Heating Temperature and Dwell Time on the Grain Size of Boron Steel

The microstructure of boron steel heated to 850°C is shown in Fig. 1. The average grain intercept method introduced by the "GBT6394-2002 Determination method of average grain size of metal" is used to calculate the average grain size of boron steel. The average grain size is 9.5, 13.6 and 15.7μm for the dwell time of 5, 20 and 30 min, respectively. For the dwell time of 5 min, the austenitizing of boron steel has begun but not completed, there still exists some undissolved ferrite. Ferrite can hinder the growth of austenite grain, and makes the austenite grain relatively small. With the extension of dwell time, ferrite dissolves gradually, and austenite grain grows larger, but the growing trend is not very obvious.

(a) (b) (c)

Fig. 1. Microstructures with dwell times of (a) 5, (b) 20, and (c) 30 °C/s at the heating temperature of 850°C.

The microstructure of boron steel heated to 950°C is shown in Fig. 2. The average grain size is 13.7, 17.1 and 19.2 μm for the dwell time of 5, 20 and 30 min, respectively. Full martensite can be obtained when boron steel was heated to 950 °C and then quenched by water. However, microstructural morphology varies with dwell time. With the increasing of heating temperature, the grain size distribution becomes more uniform, but the average grain size becomes larger.

(a) (b) (c)

Fig. 2. Microstructures with dwell times of (a) 5, (b) 20, and (c) 30 °C/s at the heating temperature of 950 °C.

When austenite has just formed completely, the newly formed austenite grains contact with each other. At this moment, the austenite grains are not uniform due to that earlier nucleated grains grow larger. Meantime, the grains grow larger continuously during the process of heating or dwell because of high energy generated by bending at the grain boundary. The driving force for grain growth is the free energy of grain boundary (interfacial energy). Boundaries tend to move toward center of curvature which cause that large grains grow larger and small grains shrink and even disappear. Thus, the number of grains per unit volume decreases continuously [1].

The empirical formula of normal grain growth under isothermal condition can be expressed as [2]:

$$\bar{D} = at^b \tag{1}$$

where, $\bar{D}$ is average grain diameter (μm), t is dwell time (min), and a and b are constant relating to material and temperature.

Taking the logarithm on both sides of Eq. (1): $ln\,\bar{D} = lna + blnt$, where b=0.228. By inserting heating temperature into a, it can be described as: $a = 0.021 * T - 10.926$. Therefore, the empirical formula of grain growth can be rewritten as:

$$\bar{D} = (0.021 * T - 10.926) * t^{0.228} \tag{2}$$

where, T is heating temperature (°C).

In order to verify the accuracy of the Eq. (2), the experimental grain sizes with dwell time of 3 min measured by Hou et al. [3] were cited. The comparison shown in Fig. 3 emphasizes the accuracy of empirical formula of grain growth based on the heating temperature and dwell time.

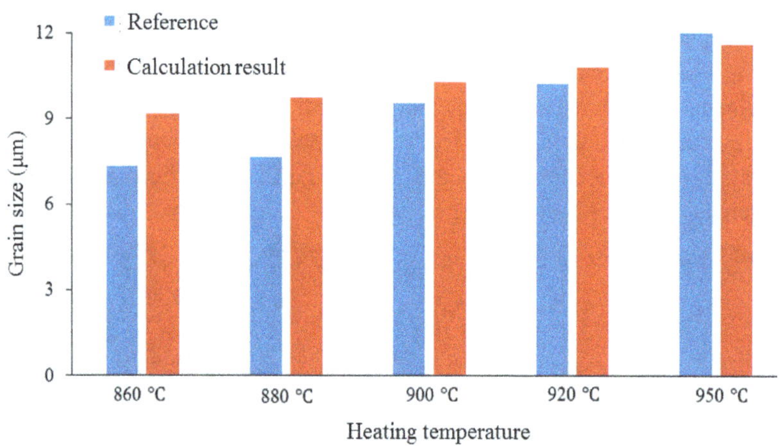

Fig. 3. Comparison of referenced [3] and calculated grain sizes.

4. Effect of Grain Size on the Strength and Elongation of Boron Steel

The grain sizes and corresponding material properties with different heat treatment parameters are shown in Table 1. The yield strength, tensile strength and elongation of boron steel all decrease with increasing grain size.

The relationship between grain size and strength of boron steel obeys the classic Hall-Petch equation [4]:

$$\sigma = 696 + 3086 \cdot d^{-1/2} \quad (3)$$

And the relationship between grain size and tensile strength of boron steel can be expressed as:

$$\sigma = 386 + 2723 \cdot d^{-1/2} \quad (4)$$

Table 1. Grain sizes and corresponding material properties of boron steel.

Grain size (μm)	Yield strength (MPa)	Tensile strength (MPa)	Elongation (%)
19.2	992	1355	3.64
17.1	1080	1475	3.44
13.7	1196	1594	3.76
15.7	1173	1579	3.78
13.6	1191	1600	4.34
9.5	1206	1630	5.72

The calculated tensile and yield strengths are in good agreement with those obtained by experiment or cited from reference, as shown in Fig. 4.

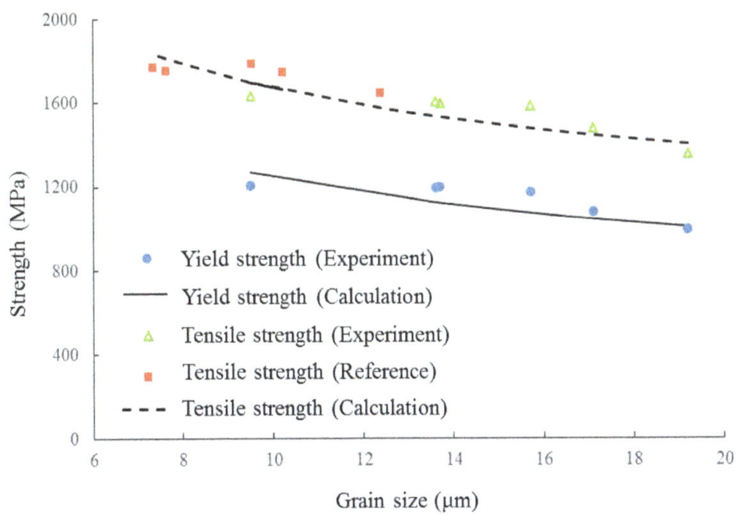

Fig. 4. The relationship between grain size and strength.

Smaller austenite grains mean higher boundary density or higher surface-to-volume ratio, which can provide more nucleation sites for martensitic transformation. And more nucleation sites make the formed particles smaller. In addition, large grain size leads to crack more easily based on Griffith's theory [6]. For small grains, the amount of precipitation phase per unit area is relatively low and the resistance to dislocation motion is relatively small due to the large grain boundary area. The motion and plastic deformation in grain interiors may trigger new dislocations in adjacent grains, so that the material exhibits good ductility. On the contrary, for coarse grains, with the rapid reduction of grain boundaries, the density of precipitation phase on grain boundaries increases significantly so that barriers to dislocation motion are produced. The coordinated deformation of grains is restricted and plasticity decreases significantly, so that the material with high strength exhibits brittleness [7]. Based on the relationship research among heating parameters, grain size and mechanical properties, boron steel with grain size of 10 μm has good comprehensive mechanical properties.

5. Conclusion

Based on the experimental and theoretical analysis, the relationship among mechanical properties, grain size and heating parameters of boron steel was studied. The conclusions drawn are as follows:

1) The grain size of austenite increases with increasing heating temperature and dwell time. The empirical formula of grain growth based on the heating parameters was established, so that the control method of grain size based on the heating parameters was thus obtained.
2) The yield strength and tensile strength of boron steel decrease with increasing grain size, and the Hall-Petch equations were established to describe this relationship. The elongation decreases with increasing grain size as well, the reason was analyzed from the aspect of dislocation motion. Therefore, the control method of mechanical properties based on the grain size was thus obtained.

Acknowledgment

The authors sincerely thank the financial supports from National Natural Science Foundation of China (No. 51375346).

References

1. Z. Q. Fu, Research of experiment and numerical simulation of recrystallization process of B1500HS boron steel, Master thesis, Shandong University (2012).
2. G. L. Hu and X. W. Xie, in *The heat treatment of steel (principle and Technology)*, (Northwestern Poly. Univ. Press, 2012).
3. H. M. Hou, L. Ying, X. F. Wu, Y. S. Wang, L. X. Zhao and P. Hu, Effect of heating temperature on microstructure and austenite grain of 22MnB5, *Forg. Equip. Manuf. Technol.* **89**, 47 (2012).
4. X. D. Li, X. P. Ma, S. V. Subramanian, C. J. Shang and R. D. K. Misra, Influence of prior austenite grain size on martensite–austenite constituent and toughness in the heat affected zone of 700MPa high strength line pipe steel, *Mater. Sci. Eng. A* **141**, 616 (2014).

5. Y. He, G. Y. Su, W. S. Qu, F. Y. Kong, L. He and K. Yang, Grain size and its effect on tensile property of ultra-purified 18Ni maraging steel, *Acta Metall. Sin.* **53**, 38 (2002).
6. C. F. Wang, M. Q. Wang, J. Shi, W. J. Hui and H. Dong, Microstructural characterization and its effect on strength of low carbon martensitic steel, *Iron and Steel* **57**, 42 (2007).
7. M. C. Zhao, F. X. Yin, T. Hanamura, K. Nagai and A. Atrens, Relationship between yield strength and grain size for a bimodal structural ultrafine-grained ferrite/cementite steel, *Scripta Mater.* **57**, 857 (2007).

Improved Scale Adhesion of Uncoated Hot Stamped Steel

L. Levander
TTE R&D PH Research, Gestamp
Luleå, Sweden
E-mail: llevander@se.gestamp.com

The yearly production of hot stamped automotive parts is increasing due to several advantages over the conventional cold forming processes, such as allowing the manufacture of components of much higher strength and of more complex geometries. Hot stamping of uncoated boron steel results in the formation of a loose oxide on the steel surface. This oxide has to be subsequently removed by shot blasting for satisfactory painting. In this paper, a method to avoid shot blasting is presented. The method is based on a chemical pre-treatment allowing anchoring of the formed oxide to the steel surface during hot stamping. Characterization by Scanning Electron Microscopy (SEM) as well as results from adhesion tests and a cyclic corrosion test of the produced material indicate that the scale adhesion is unaffected by subsequent process steps (cleaning, phosphating, and painting). The results also indicate that the corrosion properties of the hot stamped material are directly affected by scale adhesion; improved adhesion results in improved corrosion resistance. Furthermore, the chemical pre-treatment allows more rapid heating than non-modified material.

Keywords: Scale adhesion; Hot stamping; Boron steel; Corrosion resistance.

1. Introduction

The yearly production of hot stamped automotive parts is increasing due to several advantages over the conventional cold forming processes, such as allowing manufacture of components of much higher strength and of more complex geometries. Today, hot stamping of uncoated 22MnB5 boron steel results in formation of non-adherent oxides on the steel surface, and requires shot blasting in order to remove the named oxide for satisfactory painting and welding. Research has been focused on preventing oxidation [1-3], while strategies to control the oxidation in hot stamping and to understand what factors influence the adhesion of the produced scale have received very little attention.

In this paper, a new strategy to eliminate shot blasting is presented, including a chemical pre-treatment of the uncoated 22MnB5 boron steel followed by controlled oxidation, resulting in excellent scale adhesion. Since hot stamped components are subsequently assembled, cleaned, phosphated and e-coated by the car manufacturer, test material has also been subjected to such processing and the scale adhesion has been evaluated after each step. Finally, e-coated material has been subjected to cyclic corrosion testing in order to investigate how the adhesion of the remaining scale affects the corrosion resistance.

2. Experimental

2.1. *Material*

The material used in this study was 1.4 mm thick cold rolled, uncoated 22MnB5 steel from SSAB, both untreated as well as subsequently chemically pre-treated. For adhesion and corrosion testing, one blank of hot stamped reference material was shot blasted at Gestamp. For the heating trial, 1, 4 mm thick Usibor 1500P AS150 from ArcelorMittal (AlSi-coated 22MnB5) was also used as a reference.

2.2. *Hot stamping and heating measurements*

Hot stamping was performed in the research facility at Gestamp, Luleå, in the hot stamping line. Blanks of 390x490 mm were heated in a controlled atmosphere and thereafter quenched between flat dies. Heating measurements were also performed in the same research facility, but in a small lab furnace kept at 900°C, using an air atmosphere. The time elapsed for the samples to reach 850°C was registered. At this temperature the austenitic transformation was expected to be completed.

2.3. *Adhesion testing*

To qualitatively evaluate the scale adhesion of the produced material, a pressure-sensitive tape method, modified from ISO8502-3, was chosen. An adhesive tape was pressed to the surface, and then removed with the loose oxide adhering to it. The tape was attached to a white background for evaluation. The more oxide that adhered to the tape and was removed from the surface, the worse was the scale adhesion.

2.4. *Scale characterization by Scanning Electron Microscopy (SEM)*

In order to obtain intact areas of the scales, the sample was cut with a saw from the opposite side of interest, through about 90% of the thickness, and dipped in liquid nitrogen. Then the sample was clamped in a vice and fractured by a hammer with a single hit. The cross-sections were thereafter observed with FE-SEM (Merlin, Zeiss).

2.5. *Subsequent processing and corrosion testing*

In order to examine how the scale of hot stamped material and its adhesion to the substrate is affected by the subsequent treatments, samples of 75x100 mm were laser cut from a large sheet and subjected to cleaning and Zn-phosphating by EBP European body panels in Olofström, and finally tape tested. Some samples were also e-coated by EBP, and were then evaluated with respect to dry paint adhesion; a cross-hatch cut was made through the paint layer and thereafter a tape was attached to the scribed surface.

In order to evaluate how the remaining scale affects the corrosion properties, three e-coated panels of each material were exposed to a VDA 233-102 accelerated corrosion test. Before the exposure a scribe with a width of 1 mm was made through the paint layer down to the base material. The evaluation was made according to Renault standard 02-00-005 and Volvo standard VCS 1021, 29.

3. Results and Discussion

3.1. *Hot stamping and heating measurements*

The temperature of the pre-treated material rose more rapidly than for the reference uncoated material, reaching 850°C at around 96 s compared to 116 s for the reference. The AlSi-coated material was heated significantly slower, and reached 850°C at about 200 s.

After hot stamping of the reference material there was a big amount of scale left in the tool (see Fig. 1), and there was excessive flaking of the oxide from the hot stamped blank. The corresponding tape was completely black. However, no dust was seen in the tool after hot stamping of the pre-treated blank, and practically no oxide could be seen on the tape after the tape test.

3.2. *Scale characterization by Scanning Electron Microscopy (SEM)*

For the untreated reference, the produced oxide came off in big flakes, as seen in Fig. 2a. For the pre-treated material, SEM observations indicate that introduced cavities in the steel surface anchored the formed oxide to the steel, preventing flaking of the oxide (Fig. 2b).

3.3. *Subsequent processing and corrosion testing*

The results from the adhesion tests of hot stamped, Zn-phosphated, as well as e-coated material are found in Table 1. Photographs of panels exposed to the cyclic corrosion test are also found in the same table.

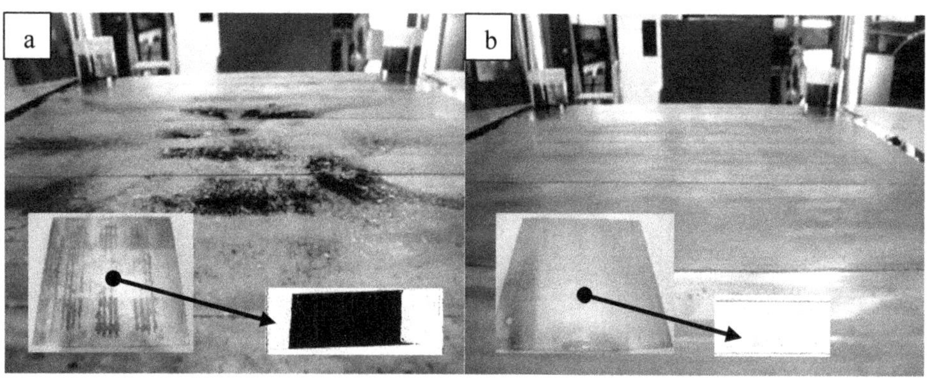

Fig. 1. Hot stamping of a) uncoated reference material, as well as b) pre-treated material. The images show the tool surface after hot stamping and the amount of oxide that has flaked off from the blanks, as well as the produced blanks and the resulting tape tests.

The poor scale adhesion of the hot stamped reference material was not improved by either phosphating or e-coating. SEM-observations of e-coated reference material (see

Fig. 2c) indicated that the scale is still present after subsequent processing, and that the e-coating adheres rather well to the oxide, but the adhesion of the entire system is very poor due to the oxide-metal decohesion. After cyclic weathering severe blistering of the coating was seen, which is not surprising since water and chloride can easily reach the steel substrate underneath the coating, promoting corrosion.

For the pre-treated material, adhesion testing indicated that the scale adhesion was still excellent after phosphating as well as after e-coating. Imaging by SEM (Fig. 2d) revealed that, close to the fracture (from sample preparation), oxide was attached to both the steel substrate and the coating. After the corrosion test, no blistering was seen and the delamination from the scribe was comparable to that found on shot blasted reference material. Further investigation is however necessary in order to draw more conclusions about the final corrosion properties.

4. Conclusion

The effect of the pre-treatment on scale adhesion was very clear; for the untreated reference, spallation of the scale was evident. However, for the pre-treated material no visible spalling was seen and SEM observations indicated that introduced cavities act as anchoring points, preventing flaking of the oxide.

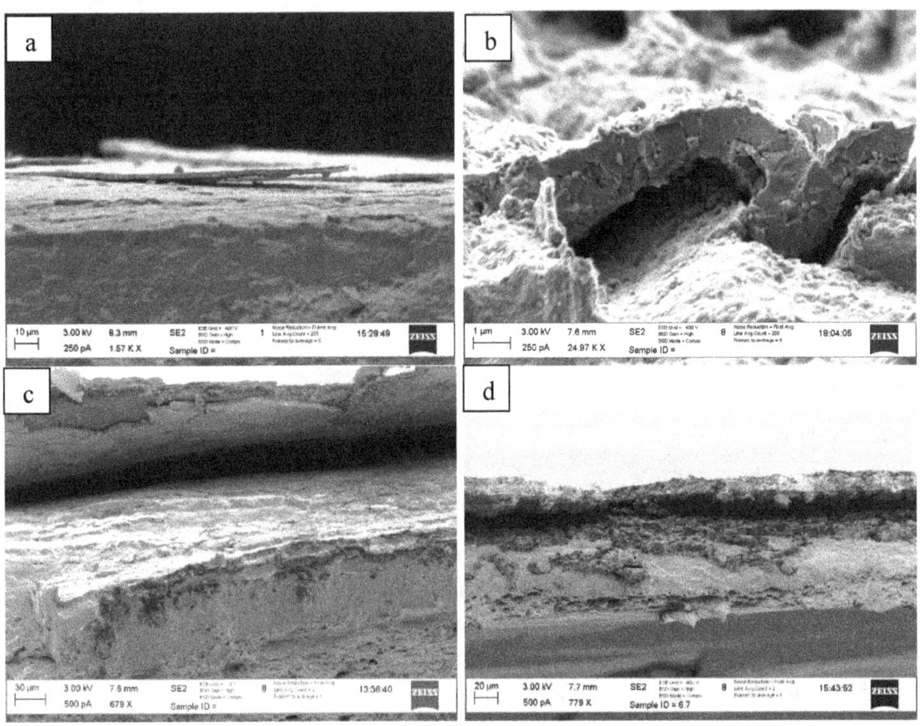

Fig. 2. SEM-images of fractured cross-sections of a) hot stamped reference material and b) hot stamped pre-treated material, as well as c) e-coated hot stamped reference material, and d) e-coated hot stamped pre-treated material.

Tape tests indicated that the scale adhesion of the materials is more or less unaffected by the phosphating step as well as the e-coating process.

The results from the corrosion tests also indicated that the initial scale adhesion of the hot stamped material is directly affecting the corrosion resistance – a poor initial scale adhesion lead to severe blistering of the e-coating when subjected to a corrosive environment, while such blistering was not seen on the samples with good initial scale adhesion.

As a positive side effect, the use of the pre-treatment also leads to shortening of the furnace dwell time, as compared to the conventional process.

Table 1. Adhesion as well as corrosion results for reference, shot-blasted, and pre-treated material, after press-hardening, after Zn-phosphating, as well as after e-coating.

Material	After hot stamping	After Zn- phosphating	After e-coating	
	Tape Test	Tape test	Dry Paint Adhesion	Corrosion test (VDA)
Reference	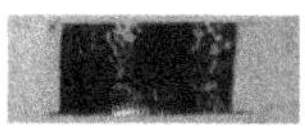			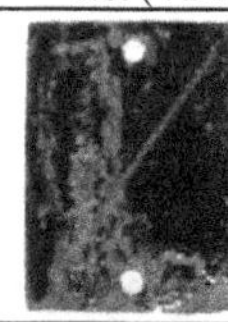
Shot blasted reference			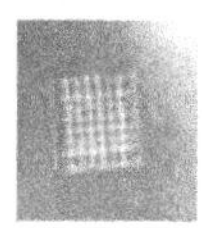	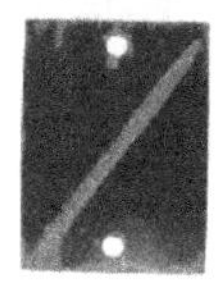
Pre-treated			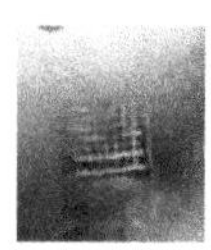	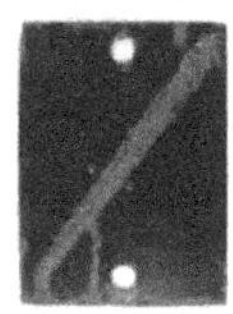

References

1. K. Mori and D. Ito, Prevention of oxidation in hot stamping of quenchable steel sheet by oxidation preventive oil, *CIRP Ann. Manuf. Techn* **58**, 267 (2009).
2. C. Choi, J. Cho, Y.J. Park and J.S. Kim, Process window study of heat resistant nano-coated steel, in *18th Int. Corrosion Congress*, (Perth, Australia, 2011).
3. M. Yekehtaz, C. Klesen, S. Benfer, W. Bleck and W. Fürbeth, Effect of nano-particulate sol–gel coatings on the oxidation resistance of high-strength steel alloys during the press-hardening process, *Mater. Corros* **63**, 940 (2012).

Application Research on Antioxidant Hot Formed PHS1500A Steel

H. L. Liu†, Y. Hu, H. G. Gao, Y. Chen, J. Huang and G. W. Jiang

Products Institute, Benxi Steel Plates Co., Ltd.,
Benxi, P.R. China
†E-mail: daliang_lhl@hotmail.com

Most of the hot formed steel used is the traditional bare cold-rolled plate, and a thermal oxide scale is obtained due to the high temperature oxidation during heat treatment. This seriously affects the service life of the mold. Indeed, the production process has to be shut down regularly for cleaning, which has become a bottleneck constraint in industrial production. Besides, having lots of oxide scale shedding also leads to an unhealthy working environment. In the previous study, a special surface modification technology had been used in the laboratory for the traditional bare cold-rolled plate of PHS1500, in order to improve the surface quality of steel plate during the heating and pressing process. The results show that the surface oxide scale consisted mainly of Fe_3O_4 instead of an FeO layer. The problem of thermal oxide scale has therefore been solved. In this study, we focus on the application of the product in the industry.

Keywords: Hot formed steel; PHS1500A; Surface modification.

1. Introduction

Hot press forming with quenching is an alternative technology to produce ultra-high strength steels with a tensile strength above 1500MPa for automotive bodies. Just like PHS1500 steel, it has a good forming performance, high forming strength and obviously effects that improve crash worthiness and reduce vehicle weight, which encourages increasingly wide applications in white body [1-4].

However, most of the PHS1500 steel used is the traditional bare cold-rolled plate, and a thermal oxide scale is obtained due to the high temperature oxidation during heat treatment. Shot blasting is then needed after preparing the parts. As we know, too much oxide scales shedding lead to a bad working environment [5, 6].

This paper introduces a new anti-oxide scale off hot press forming steels has been developed, named PHS1500A, which uses a special surface modification technology without any additional coating and alter the material properties to improve the hot forming steel surface quality during the heating and pressing processes. This leads consequently to the suppression of the oxide scale shedding in the hot stamping process.

2. Experiment and Preparation

2.1. *Theoretical analysis*

It is widely acknowledged that oxidized steel generally includes surface oxide scale, consisting of an outermost hematite layer (Fe_2O_3), an intermediate magnetite (Fe_3O_4), and a square iron ore layer (FeO) in contact with the steel [7]. Figure 1 presents the surface pattern of the part of the traditional bare cold-rolled plate. It has become apparent that, the oxide scale is easily fall off. Besides, the oxide scale has the characteristics of layered structure. It has been analyzed by XRD, it shows that oxide scale layer structure mainly Fe_3O_4, FeO and Fe_2O_3. The oxide scale is very easy to fall off.

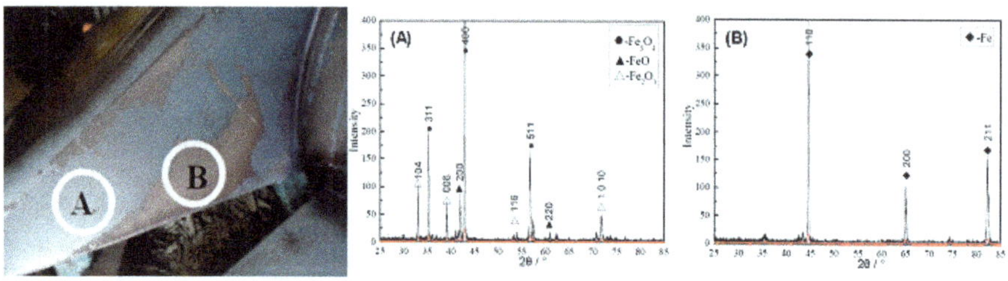

(a) PHS1500. (b) XRD for "A". (c) XRD for "B".

Fig. 1. Surface appearance and structure of the traditional stamping parts.

Also several studies reported that the effects of the initial scale structure, which is prone to change based on temperature and atmospheric factors. Such as, at temperature above 560°C in a non-oxidizing atmosphere the scale consists of only FeO layer. However, the scale with FeO layer shows low adhesion, besides the FeO is unstable below 560°C, it transforms to a mixture of Fe_3O_4 and Fe during cooling process. This leads to the oxide scale, which consistent of mixed of FeO and Fe_3O_4 or Fe_2O_3, exfoliate easily.

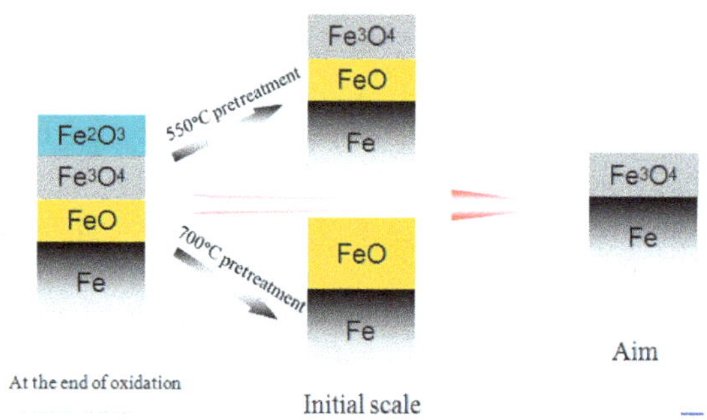

Fig. 2. PHS1500A schematic illustration.

To this end, if an appropriate method is adopted, such the Fe_3O_4 dense structure is formed directly on the surface of part, the phenomenon of oxide scale shedding can be effectively improved and resolved, as shown in Fig.2. A special method has been used and anti-oxidation hot press steel PHS1500A is developed in Benxi Iron and Steel Group. Aim to avoid the FeO substructure layer, the pre-Fe_3O_4 dense surface oxide layer has been generated, so that the FeO substructure layer cannot be produced during heating, Fe_3O_4 closely integrated with the steel matrix, thus avoiding the shedding of oxide scale.

2.2. *Sample preparation*

In the present study, the chemical composition of both PHS1500A and the traditional PHS1500 steel is the classical 22MnB5 components, as shown in table 1. The only different is the surface. In this study a special surface modification technology has been used, as shown in Figure 3, the PHS1500A steel has formed a layer of dense Fe_3O_4 before stamping.

Table 1. Chemical composition of the PHS1500 steel, wt%.

C	Si	Mn	Cr	Als	Ti	B
0.23	0.16	1.4	0.20	0.06	0.03	0.002

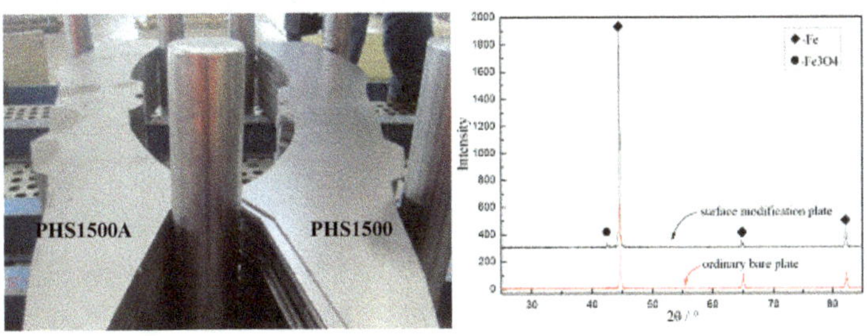

Fig. 3. PHS1500A and PHS1500 surface structure.

3. Result Analysis

3.1. *Surface oxide scale structure*

Stamping process used for the production of one mold with two parts, the experiments clearly show that the traditional PHS1500 stamping process a large number of oxide scale scattered, but the PHS1500A stamping process is almost no oxide scale off. There is also a big difference between the part surfaces after stamping. As PHS1500A obtained the gray uniform oxide scale, and non-shedding phenomenon, but PHS1500 surface shown random fall off the facial features.

Both PHS1500 and PHS1500A two kinds of hot forming steel parts surface quality and surface rust layer structure has been studied, as shown in Figure 4. The surface rust layer structure of PHS1500 parts is still a large number of FeO exist. However, the surface of PHS1500A is almost no FeO diffraction peaks has been detected, and the morphology is significantly dense rust layer, as shown in Figure 5.

3.2. *Mechanical properties*

Because the chemical composition of PHS1500A steel has not been changed, the final mechanical properties of stamping parts with ordinary PHS1500 are no difference also. The actual test result shows that the tensile strength still reached 1490MPa ~ 1525MPa, and the elongation ≥7%, it has been meet the requirements.

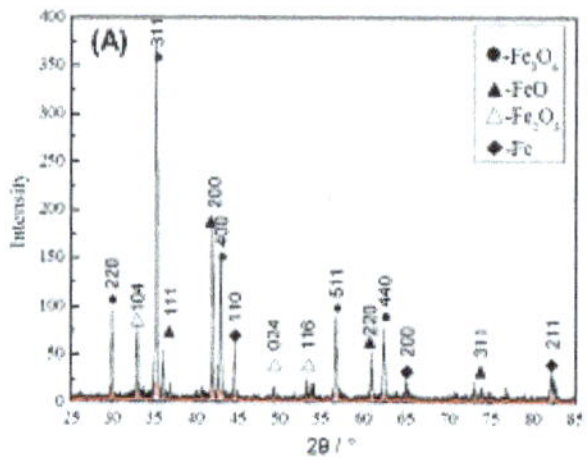

(a) XRD for PHS1500

(b) PHS1500

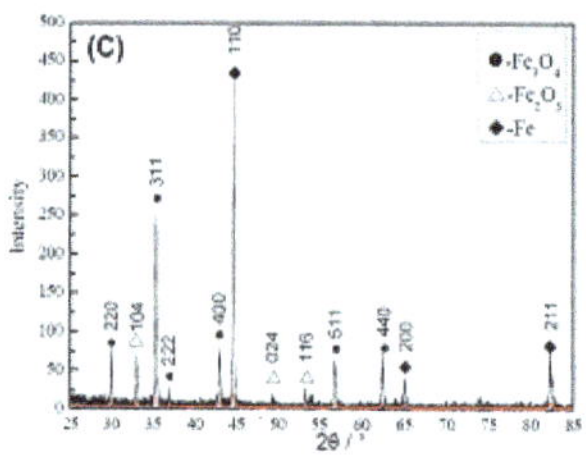

(c) XRD for PHS1500A

(d) PHS1500A

Fig. 4. Surface structure of PHS1500A and PHS1500.

(a) PHS1500

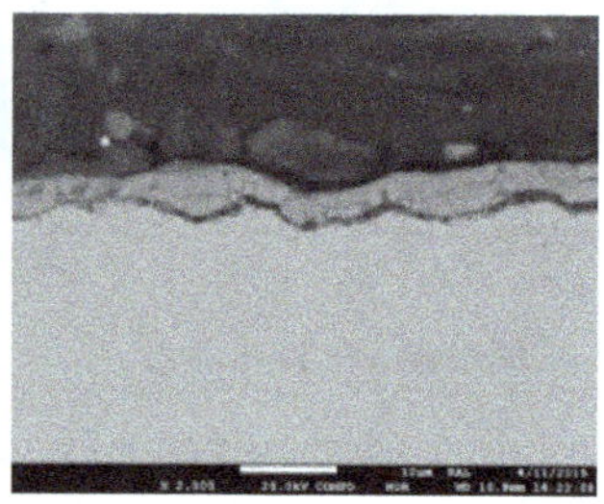

(b) PHS1500A

Fig. 5. PHS1500A and PHS1500 parts rust appearance.

4. Conclusion

The traditional bare cold-rolled plate PHS1500 after heating includes surface oxide scale, consisting of Fe_3O_4, Fe_2O_3, and FeO. The oxide scale has the characteristics of layered structure, and it is inhomogeneous deciduous.

However, the PHS1500A steel used the special surface modification technology, and the pre-Fe_3O_4 dense surface oxide layer has been generated, so that the FeO substructure layer cannot be produced during heating, Fe_3O_4 closely integrated with the steel matrix, thus avoiding the shedding of oxide scale.

Nevertheless, the chemical composition of PHS1500A steel is same as the traditional bare cold-rolled plate PHS1500. And the properties and microstructures suggest that the mechanical properties are consistent with each other.

Acknowledgement

The authors like to thank the Batch of Scientific and Technological Projects (2015105004) of Liaoning Province for the support of the project.

References

1. M.T. Ma, H.L. Yi, H.Z. Lu and X.M. Wan, On the lightweighting of automobile, *Eng. Sci.* **11**, 24 (2009).
2. W.J. Liu, An introduction to advanced hot-formed steel for automobile, *Acta Metall. Sin.* **27**, 373 (2014).
3. D.W. Fan, H.S. Kim and B.C.D. Cooman, A review of the physical metallurgy related to the hot press forming of advanced high strength steel, *Steel Res. Int.* **80**, 241 (2009).
4. M.G. Lee, S.J. Kim, H.N. Han and W.C. Jeong, Application of hot press forming process to manufacture an automotive part and its finite element analysis considering phase transformation plasticity, *Int. J. Mech. Sci.* **51**, 888 (2009).
5. R. Endo, T. Yagi, M. Ueda and M. Susa, Thermal Diffusivity Measurement of Oxide Scale Formed on Steel during Hot-rolling Process, *ISIJ Int.* **54**, 2084 (2014).
6. R. Matsumoto, Y. Osumi and H. Utsunomiya, Reduction of friction of steel covered with oxide scale in hot forging, *J. Mater. Process. Technol.* **214**, 651 (2014).
7. H. Tanei and Y. Kondo, Effects of Initial Scale Structure on Transformation Behavior of Wüstite, *ISIJ Int.* **52**, 105 (2012).

Research Status of Hydrogen-Induced Delayed Cracking in Martensitic Ultra-High Strength Steel

L. F. Song[1,2,†], Y. Zhao[1,2] and M. T. Ma[1,2]

[1]*China Automotive Engineering Research Institute CO., LTD.*
Automotive Lightweight Engineering Research Center, Chongqing, 401122, China
[2]*Chongqing Automotive Lightweight Engineering Research Center*
Chongqing 401122, China
[†]*E-mail: songleifeng@caeri.com.cn*

Ultra-high strength steel (UHSS) has now been widely used in automobiles to form anti-intrusion components due to its strength, cost advantages and lightweight, leading to energy saving and responding to the emission reduction demand for automotive industry. Steel strength in automobiles is increasingly important, making it the focus of current research. Steel ductility and hydrogen-induced delayed cracking (HIDC) resistance are crucial and therefore need further comprehensive studies. In this paper, based on original research results, the test methods and evaluation standards of HIDC susceptibility were compared and analyzed in detail in order to find an acceptable and universal test and evaluation method, and help with the application of UHSS in the automobile sector.

Keywords: Hydrogen-induced delayed cracking; Ultra-high strength steel; Constant load; Hydrogen content; Electrochemical hydrogen pre-charging.

1. Introduction

In recent years, these topics including passenger safety, lightweight technologies, energy saving and emission reduction have been widely concerned in global automotive industry, HSS and UHSS were accepted by auto manufacturer as the main lightweight materials due to its' high strength and low cost. In order to find the further potential of automotive lightweight and safety, 1800MPa hot stamping steel has been developed in Mazda motor [1].

Due to the ultra-high strength, the UHSS simultaneously exhibited limited ductility and formability, Because of the increasing use of MS UHSS, it is important to understand their performance under service conditions. In particular, the issue of hydrogen-induced delayed cracking of these steels under service condition was investigated [2].

The HIDC susceptibility of martensitic UHSS mainly depended on steel strength, hydrogen content and loading stress, previous study showed that HIDC can result in the decrease of yield stress or a loss of ductility. But until now, even one method were not applied as industry agreed standard, combination with our present research work, some critical effect factors on HIDC were proposed and discussed in detail in this paper, expected to found a universal test and evaluation method for HIDC [3].

2. Key Factors of HIDC

Previous study showed that the steel strength, loading stress and hydrogen content were the main factors which influence the HIDC susceptibility of UHSS, also the critical fracture stress and hydrogen content were expected to be obtained for HIDC susceptibility evaluation [4].

2.1. *Steel strength*

High strength martensitic steels exhibited a high HIDC sensitivity after quenching and tempering treatment, and the sensitivity increased with higher steel strength. Significant HIDC fracture would occur when the steel strength exceed 1300Mpa [5].

2.2. *Loading stress*

Loading stress was an important factor for HIDC susceptibility evaluation. In each of these typical applications, under normal service, the martensitic UHSS is subjected either to zero stress, or a constant low stress. However, there are high loading rates and high stresses under crash situations.

2.3. *Hydrogen content*

In terms of HIDC, previous research mainly focused on the strength and ductility reduction of UHSS under special environment, such as H_2S saturated solution. In recent years, the critical hydrogen content was proposed to build a relationship between hydrogen content and delayed cracking stress.

2.4. *Test condition*

About HIDC sensitivity test, H_2S saturated solution, 0.1M HCl, 3.5%NaCl and 0.1M NaOH solutions were always used to simulate the corrosion in auto service in previous research, present work focus on electrochemical hydrogen charging to change the diffusible hydrogen content of steels during loading process or before.

3. Test Methods of HIDC

HIDC of UHSS was widely concerned by researchers, especially the steel manufacturer. It is an ideal situation of HIDC to carry out components test in actual service environment, but the test period was long. Various accelerated test methods were proposed by different researchers, which could be divided into these types as follow:

(1) Constant stress or strain: critical fracture stress and fracture time,
(2) SSRT test: fracture stress and plastic parameter,
(3) Quasi-static tensile test: strength and plastic parameter,
(4) Hydrogen charging: high pressure hydrogen charging, solution charging and electrochemical charging,

(5) Hydrogen permeation test: the effective hydrogen coefficient, D_{ap}, and the total hydrogen concentration on the input side of the specimen, C_0.

3.1. *Constant load test*

In this test, smooth and notched specimens were stretched under constant stress (strain) carried out by mechanical apparatus, simultaneously, these specimens were exposed in air or in special solutions. The mechanical apparatus were shown in Fig.1 [6].

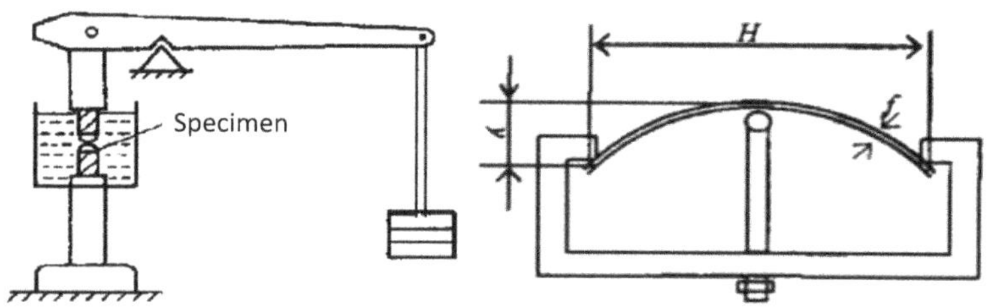

Fig. 1. Mechanical apparatus of constant loading test, (a) constant stress stretching, (b) constant stress bending.

The three points bending test was conducted by Liu (2007) [7], the maximum stress loaded at the top of the specimen was calculated by equation as follow:

$$\sigma_{max} = \frac{6Ety}{H^2} \tag{1}$$

Where, E-material elastic modulus (MPa); t-thickness (mm); y-maximum deflection (mm); H-the distance between supporting points. The result showed that the fracture time was prolonged under high temperature annealing (350°C), the HIDC sensitivity of 1500MPa UHSS was decreased with elevated annealing temperature.

Constant stress stretching test of 22MnB5 with notched specimens was investigated by Zhang (2013) [8]. The WOL corrosion inhibitor was used in this test. The critical stress σ_c could be calculated as follow:

$$\sigma_c = \frac{\sigma_f+\sigma_n}{2} \tag{2}$$

Where, the σ_f-the minimum stress within set time (t_c=100h); σ_n-the maximum stress within set time. While σ_N could be conducted by tensile test in air; σ_c/σ_N was used in this paper to evaluate the sensitivity of HIDC. The hydrogen content was also analyzed after stretching in inhibitor under critical stress for 100h. The test result showed that 200°C tempering treatment can significantly improve the HIDC resistance of hot stamping steel with maximum notched critical fracture stress and time, and the hydrogen content was obviously increased after stretching for 100h under critical stress.

3.2. *Slow strain rate test (SSRT)*

The slow strain rate test was widely used by researchers as accelerate test of HIDC, the tensile strength and elongation contrastive analysis before and after hydrogen charging were conducted to evaluate the sensitivity of HIDC.

SSRT of different grades hot stamping steel was carried out by Zhang (2013) [9] with using WDML-300kN type slow tensile testing machine. 0.04mm/min tensile speed was applied in this test, Electrochemical test was conducted for sample hydrogen pre-charging.

The rate of notched tensile strength decrease (RNS) was used to evaluate the susceptibility of HIDC, the result showed that value of RNS decreased with lower carbon content and elevated tempering temperature, as shown in Fig.2.

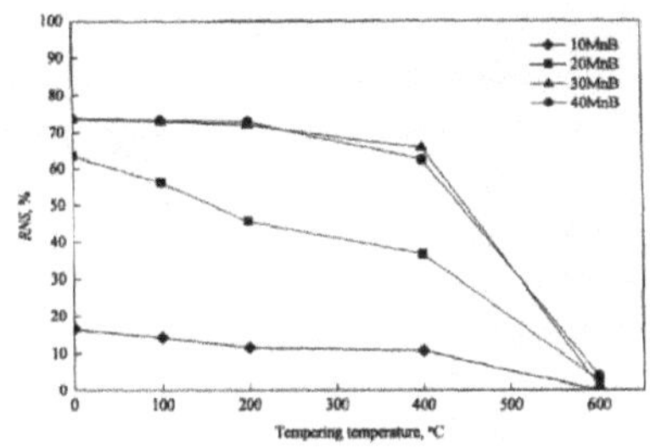

Fig. 2. Value of RNS with different carbon content and tempering temperature.

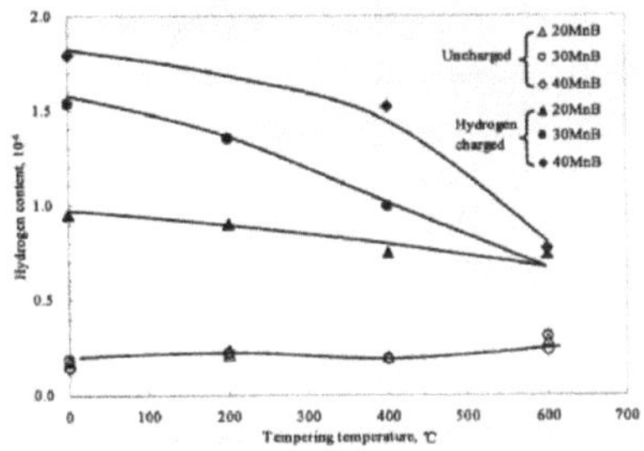

Fig. 3. The whole hydrogen content analysis for sample before and after pre-charging.

Hydrogen content analysis result (Fig.3) showed that the diffusible hydrogen content significantly increased after electrochemical pre-charging, and the hydrogen content decreased with higher temperature.

3.3. *Quasi-static tensile test*

Quasi-static tensile test was more and more conducted in present research combination with hydrogen charging process, the strength and elongation before and after hydrogen charging were tested to evaluate the susceptibility of HIDC.

Quasi-static tensile test of AHSS were studied by Venezuela, Zhou, Liu, Zhang and Atrens [10], and Kolk [11], on a servo-hydraulic loading machine equipped with a high pressure gas test chamber. The material was tested either under helium or hydrogen atmosphere.

To evaluate the effect of hydrogen embrittlement, the change in ductility was considered according to equation:

$$S = (A_{bHe} - A_{bH2})/A_{bHe} \tag{3}$$

Where A_{bHe} is the breaking strain under helium atmosphere and A_{bH2} is the breaking strain under hydrogen atmosphere. The result was shown in Table 1. A larger value of HE-index showed a higher susceptibility of hydrogen embrittlement.

Table 1. Mechanical properties of materials tested under H and He atmospheres.

Steel grade	Yield strength, MPa		Tensile strength, MPa		Elongation, %		HE-index,%
	He	H	He	H	He	H	
DP900	1048	1047	1282	1265	9	5	47
CP600	680	654	807	810	15	12	19
CP950	1061	1054	1274	1257	10	6	38
TRIP500	542	533	883	798	30	10	66

3.4. *Hydrogen charging*

There are several types of hydrogen charging methods, such as high pressure hydrogen, solution and electrochemical charging. During the loading process, hydrogen charging could be continuous charging and pre-charging [12], in which a method of Hydrogen absorption in aqueous solution at free corrosion potential was proposed. For hydrogen charging by corrosion in acid, HCl solutions or HCl with CH_3COOH/CH_3COONa buffered solutions were often used.

5% HCl solution immerse test of hot stamping steel was conducted by author, and the test result showed that the elongation of test plate significantly reduced after immersing over 144h due to the crack and corrosion holes emerged in the sample (Fig.4).

The British Standard also proposed the Hydrogen absorption in high pressure hydrogen gas. Cathodic hydrogen charging method proposed by British Standard indicated that the hydrogen is forced to diffuse into the specimens by the cathodic charging method. The charging solution should be prepared and the chemical compositions of the solutions are listed in Table 2.

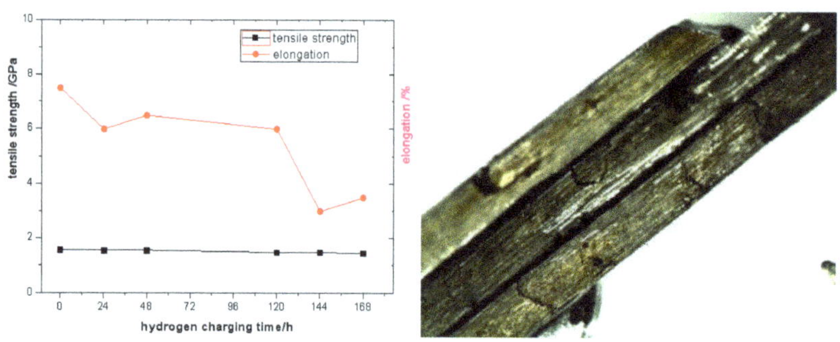

Fig. 4. Influence of 5% HCl charging on strength and elongation/microstructure.

Table 2. Chemical composition of the solutions for hydrogen charging.

Charging solution	Element	Content g/L	Mark
Solution1	NaCl	30	Large amount of hydrogen
	NH4SCN	3	
Solution2	NaOH	4	Small amount of hydrogen

After hydrogen pre-charging, Cd plating should be conducted to prevent the release of hydrogen from the specimens during constant loading test. The elements for Cd plating solution and the amounts of each element are listed in Table 3.

Table 3. Elements for Cd and Zn plating solutions and the amounts of each element.

Solution	Element	Type	Amount
Cd plating solution	$Cd(BF_4)_2$	Solution[(C)=50%]	427g
	NH_4BF_4	solute	48g
	H_3BO_4	solute	21.6g
	H_2O(distilled)	Solvent	460ml
	PH	-	3 to 5

The specimens were cathodically charged using a solution of 6% H_2SO_4 and 0.5g/L As_2O_3 as a hydrogen-recombination poison were used in this test by reference [11].

3.5. *Hydrogen permeation test*

The ASTM 148 [13] has proposed a method of hydrogen permeation test, which was widely applied by researchers and equipment manufacturers [14].

The apparent diffuse coefficient D_{ap} and diffusible hydrogen concentration C_0 could be obtained by hydrogen permeation test, the permeation apparatus was double-cell structure, as shown in Fig.5.

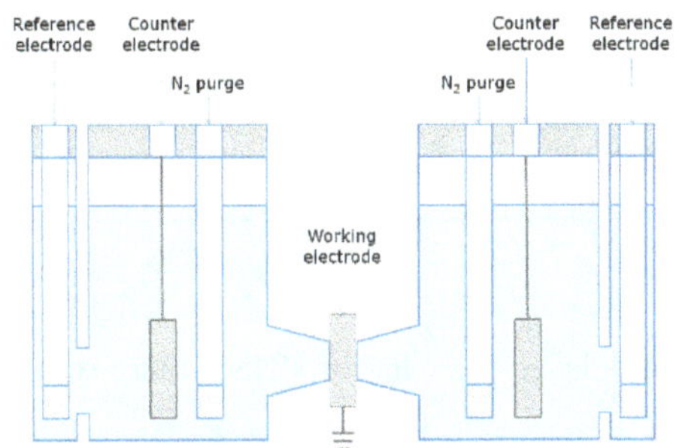

Fig. 5. Test principle of hydrogen permeation.

The coefficient D_{ap} and C_0 can be calculated by using the equation [15]:

$$D_{ap} = \frac{L^2}{6t_{0.63}} \tag{4}$$

$$C_0 = 1.31 \times 10^{-6} \frac{I_\infty L}{D_{ap} S} \tag{5}$$

Where L is the specimen thickness, I_∞ is saturated current density, S is the area of specimen through which diffusion occurs and $t_{0.63}$ is the time-lag in seconds for 0.63 saturation.

Hydrogen permeation test of Nb-added hot stamping steel under taken by Zhang, Huang, Sun, Liao and Lu [15]. The result demonstrated that the diffusion coefficient of hydrogen in the Nb containing steel is lower than that of non-micro-alloying steel, and the hydrogen concentration C_0 shows an opposite trend, which were shown in Fig.6 and Table 4.

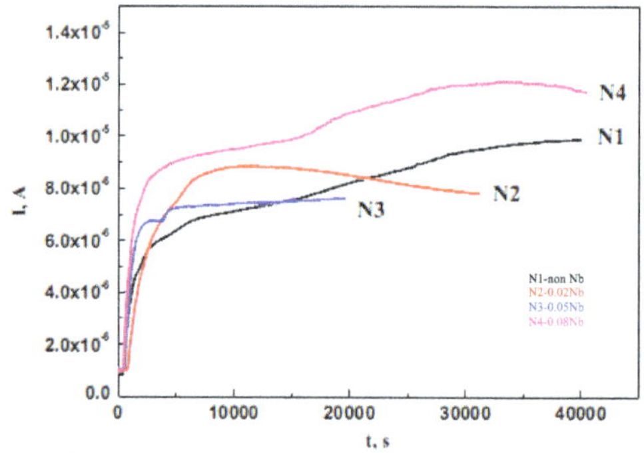

Fig. 6. Hydrogen permeation curves of test steels.

Table 4. Apparent diffusion coefficient and concentration of hydrogen for test steels.

I=1(mA/cm^2)	N1/non Nb	N2/0.02Nb	N3/0.05Nb	N4/0.08Nb
$D_{ap}(10^{-7}cm^2/s)$	10.16	7.97	2.94	5.96
$C_0(10^{-6})$	0.399	0.798	2.065	1.130

4. Conclusion

About hydrogen-induced delayed cracking of UHSS, there was no universal test method in automotive industry.

Previous study has showed that no evident change in hydrogen concentration during solution immersing charging, electrochemical cathode charging was the trend of charging high content of hydrogen concentration with adding hydrogen- recombination poison. Most of the test methods presently did not introduced the high strain rate test to evaluate susceptibility of HIDC. Otherwise, some previous research found that the a higher content of hydrogen concentration was obtained by using graphite counter electrode during the electrochemical hydrogen charging process than using Pt counter electrode also the presence of stress or strain in the steel increases the hydrogen concentration.

Based on the above-mentioned information, there are several suggestions for hydrogen-induced delayed cracking of ultra-high strength martensitic steel:

(1) Present research indicated that there was minimal influence of hydrogen charging in dilute solution (HCl, NaOH and NaCl, etc,) at free corrosion potential on hydrogen concentration in sample.

(2) It is feasible to increase the hydrogen content of UHSS by electrochemical charging with increasingly negative applied potential and the type of electrolyte and counter electrode.

(3) Previous study showed that some percent of hydrogen would release from hydrogen pre-charged samples at a short time, electroplating of Cd or Zn should be considerably carried out to prevent the release of hydrogen from specimens. but substantial strain rate test of hydrogen pre-charged specimens could be conducted within a short time.

(4) It is insecure to conduct loading test under high pressure hydrogen gas environment and itis not recommended due to the danger of experiment.

(5) There was mature equipment of hydrogen permeation test, which can be carried out independently as one of the methods for evaluation of HIDC sensitivity.

(6) Presently, constant stress test with electrochemical hydrogen continuous charging would be accepted combination with hydrogen content analysis for evaluation of HIDC sensitivity. Further research should focus on quasi-static tensile test or high strain rate stretching and impact test with electrochemical hydrogen pre-charging and hydrogen content analysis.

Acknowledgment

Work partially funded by Major state basic research development program of China (No. 2016YFB0101605).

References

1. O. Bouaziz, H. Zurob and M. Huang, Driving Force and Logic of Development of Advanced High Strength Steels for Automotive Applications, *Steel Research Int.* **84**, 937 (2013).
2. S. Lynch, Hydrogen embrittlement phenomena and mechanisms, *Corros. Rev.* **30**, 105 (2012).
3. J. Q. Wang, A. Atrens, D. R. Cousens, P. M. Kelly, C. Nockolds and S. Bulcock, Measurement of grain boundary composition for X52 pipeline steel, *Acta Mater.* **46**, 5677 (1998).
4. J. A. Ronevich, J. G. Speer and D. K. Matlock, Hydrogen embrittlement of commercially produced advanced high strength steels, *SAE Int. J. Mater. Manuf.* **3**, 255 (2010).
5. J. Venezuela, Q. Liu, M. Zhang, Q. Zhou and A. Atrens, The influence of hydrogen on the mechanical and fracture properties of some martensitic advanced high strength steels studied using the linearly increasing stress test, *Corros. Sci.* **99**, 98 (2015).
6. SEP: 1970-2011, Test of the resistance of Advanced High Strength Steels (AHSS) for automotive applications against production related hydrogen induced brittle fracture (2011).
7. J. Y. Liu, Study on delayed fracture resistance of High strength automotive steel, Phd thesis, Shanghai jiaotong University (2007).
8. Y. J. Zhang, C. Zhou, W. J. Hui, S. L. Chen and H. Dong, hydrogen induced delayed fracture behavior of a low carbon Mn-B ultra-high strength steel sheet after hot stamping, *Acta Metallurgica Sinica.* **10**, 1153 (2013).
9. Y. J. Zhang, Study on hydrogen induced delayed fracture behavior of ultra-high strength steel sheets, Phd thesis, Iron and steel research institute (2013).
10. J. Venezuela, Q. Zhou, Q. Liu, M. Zhang and A. Atrens, Influence of hydrogen on the mechanical and fracture properties of some martensitic AHSS, *Corros. Sci.* **5**, 40 (2016).

11. M. L. O. Kolk, Hydrogen Embrittlement in AHSS Limits Use in Lightweight Body, *Adv. Mater. Process* **169**, 22 (2011).
12. BS ISO 16573:2015 Steel — Measurement method for the evaluation of hydrogen embrittlement resistance of high strength steels, BSI Standards Publication (2015).
13. ASTM G148-97, Standard Practice for Evaluation of Hydrogen Uptake, Permeation, and Transport in Metals by an Electrochemical Technique (2011).
14. Autolab Application Note COR07, Hydrogen permeation experiments with PGSTAT302F (2012).
15. S. Zhang, Y. Huang, B. Sun, Q. Liao and H. Lu, Effect of Nb on hydrogen-induced delayed fracture in high strength hot stamping steels, *Mater. Sci. Eng. A* **626**, 136 (2015).

Crack Propagation Research of Aluminum-Silicon Coated Boron Steel Under Plane Stress State

Y. X. Liu and Y. L. Wang†

State Key Laboratory of Material Processing and Die and Mould Technology, Huazhong University of Science and Technology, Wuhan 430074, China

†*E-mail: wangyilin@mail.hust.edu.cn*

In order to explore the crack propagation of aluminum-silicon coated boron steel, biaxial tensile tests under complex loads were performed.The biaxial tensile device is installed in the AG-100kN machine at high-temperature, which provides plane-stress conditions. Strain rates of 0.1, $1s^{-1}$ were selected as the forming process parameters. Real data obtained from the biaxial tensile tests produced stress-strain curves of the coated boron steels, which could then be used to simulate the crack propagation path more accurately. When the fatigue crack kept growing in tensile tests, the path could be observed by Photron FASTCAM Viewer, which is in good correspondence with the result of the tests. Fracture surfaces of the coated steels were inspected under SEM. This paper reports the significance to study the microscopic mechanism of failure and the great importance in practical application of aluminum-silicon coated boron steel in the forming process.

Keywords: Boron steel; Al-Si coating; Crack; Biaxial tensile.

1. Introduction

With the rapid development of automobile industry, lightweight car bodies become the focus of automobile research. The application of high strength steel sheets on automobiles has been increasingly popularized. Among them is the high strength steel obtained by reducing thickness or section size of parts, which has obvious advantages in realizing the lightweight quality as well as meeting the requirements of improving security [1-2] at the same time.

The Al-Si coated boron steel is commonly used in hot stamping, which can prevent the boron steel from oxidation and decarburization. However, coating cracks and failure may occur due to the brittleness and thermal expansion mismatch of the inter-metals under stress deformation, which may increase the corrosion resistance of steel sheet [3]. In general service conditions, materials always inevitably undergo complicated loads. It will aggravate failure and fracture of the material. Biaxial tensile testing could provide complex plane-stress condition. Therefore, many researchers adopted biaxial tensile testing to study the mechanical behavior under plane stress. In order to explore the complex deformation behavior of aluminum-silicon coated boron steel, the stress-strain relations under bidirectional loading path are investigated with theoretical, experimental and finite element methods (FEM).

2. Experimental Procedure

2.1. *Materials*

The tested boron steel has a thickness of 1.8 mm. The chemical composition of the substrate boron steel is shown in Table 1. The coating was hot-dipped on the substrate in a molten bath containing Al-10 wt. % Si.

Table 1. Chemical composition of the steel substrate in wt. %.

C	Si	Mn	Cr	Al	Ti	B	Ni	Nb	V	N
0.22	0.23	1.2	0.16	0.04	0.03	0.002	0.012	0.013	0.004	0.005

A set of biaxial tensile tests were performed to evaluate the effect of strain rate on the cracking behavior of the Al-Si coating. Dimension properties of the tensile test specimen are shown in Fig. 1.

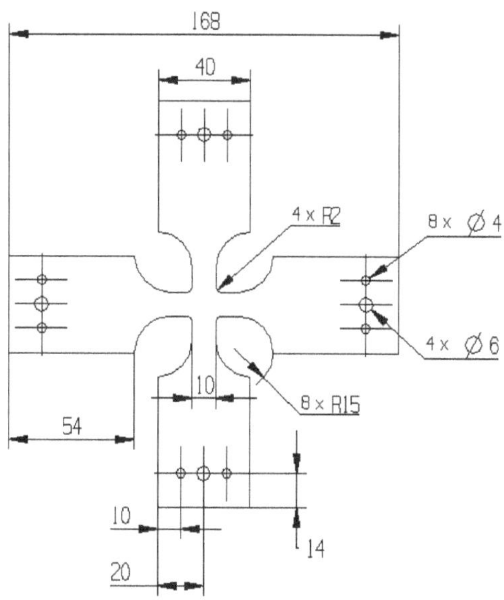

Fig. 1. Dimension properties of the tensile test specimen (mm).

2.2. *Biaxial tensile testing*

The experiments were conducted by the biaxial tensile testing device which was controlled by AG-100kN material high-temperature testing machine, as shown in Fig. 2. Material testing machine compressed the biaxial tensile clamp in the vertical direction, made the samples to stretch uniformly at the same speed on the X (rolling direction) and Y (perpendicular to the rolling direction) directions, until the crack was observed in the center of specimen crack or the sample broken. Strain rates of 0.1, 1 s^{-1} were selected as the forming process parameters. The whole fracture process was recorded by the high-speed cameras (Photron FASTCAM Viewer).

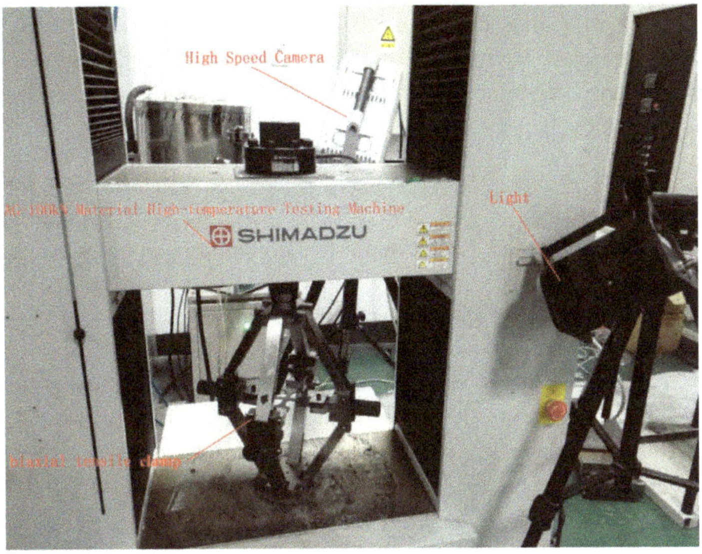

Fig. 2. Testing equipment.

During the test, the datas were collected on the X and Y directions, which can be used to calculate the true stress-strain curve, as shown in Fig. 3.

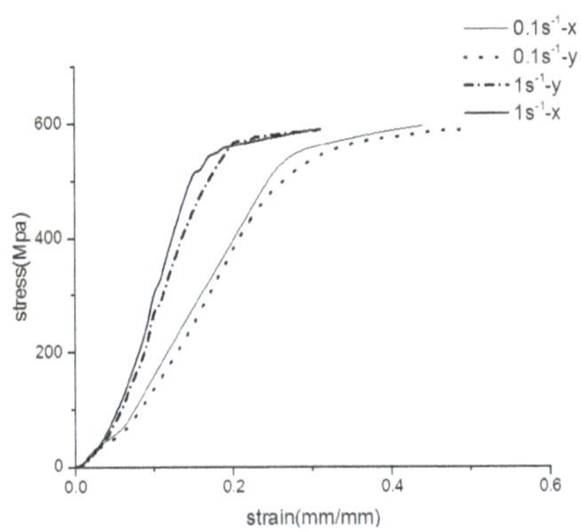

Fig. 3. True stress-strain curve of biaxial tensile.

The relationship between the stress and strain of sheet metal is the basis of the study of its mechanical behavior, and most of the materials in the process of forming are at the bidirectional loading stage, so the study of stress-strain of the biaxial tensile will be essential for technology, die design and finite element simulation. Then, biaxial tensile stress-strain relationship will be deduced, which can be compared with the above test results and verify the precision of different yield criterion.

By Drucker postulate

$$d\varepsilon_{ij}^{p} = d\lambda \frac{\partial \varphi}{\partial \sigma_{ij}} \tag{1}$$

where $d\varepsilon^p_{ij}$ is the plastic strain increment; $d\lambda$ is Nonnegative proportion coefficient; ϕ is yield criterion expression. By type (1), the plastic strain increment can be expressed by stress under different yield criterion [2]. Set up the loading ratio of two directions under the bidirectional loading paths is 1: k, so

$$\sigma_y = k\sigma_x \tag{2}$$

According to the principle of work of plastic deformation per unit volume,

$$\sigma_x d\varepsilon_x^p + \sigma_y d\varepsilon_y^p = \bar{\sigma} d\bar{\varepsilon}^p \tag{3}$$

σ_x and σ_y are respectively the stress of x and y directions in the process of the biaxial tensile; $d\varepsilon^p{}_x$ and $d\varepsilon^p{}_y$ are respectively the plastic strain increment of two directions; σ is equivalent stress; $d\varepsilon^p$ is equivalent plastic strain increment. By type (3),

$$d\bar{\varepsilon}^p = \frac{\sigma_x d\varepsilon_x^p + \sigma_y d\varepsilon_y^p}{\bar{\sigma}} \tag{4}$$

In plane principal stress state, $\sigma_3 = 0$, Because this article is the biaxial tensile tests of cruciform specimen, so two orthogonal the stretching direction of the cruciform specimen are the principal stress direction, thus Mises yield criterion can be written as the following common form

$$\phi = \sigma_x^2 - \sigma_x\sigma_y + \sigma_y^2 = \bar{\sigma}^2 \tag{5}$$

According to Eq. (2) and Eq. (5),

$$\begin{cases} \sigma_x = \dfrac{\bar{\sigma}}{\sqrt{k^2 - k + 1}} \\[2ex] \sigma_y = \dfrac{k\bar{\sigma}}{\sqrt{k^2 - k + 1}} \end{cases} \tag{6}$$

According to Eq. (1), Eq. (2) and Eq. (5),

$$\begin{cases} d\varepsilon_x^p = d\lambda \dfrac{\partial\varphi}{\partial\sigma_x} = d\lambda(2\sigma_x - \sigma_y) \\[2ex] d\varepsilon_y^p = d\lambda \dfrac{\partial\varphi}{\partial\sigma_y} = d\lambda(2\sigma_y - \sigma_x) \end{cases} \tag{7}$$

In this paper, the two main directions of stress and loading rate (k) are constant, thus the Eq. (7) subsequently infers

$$\frac{\varepsilon_x^p}{\varepsilon_y^p} = \frac{d\varepsilon_x^p}{d\varepsilon_y^p} = \frac{k-2}{1-2k} \tag{8}$$

According to Eq. (6), for the same load ratio k (constant), σ/σ_x is constant ($\sigma/\sigma_x=\sqrt{(k^2-k+1)}$). Therefore, by Eq. (2), and Eq. (4) for integrals

$$\varepsilon_x^p + k\varepsilon_y^p = \frac{\bar{\sigma}}{\sigma_x}\bar{\varepsilon}^p \tag{9}$$

According to Eq. (7), Eqs. (8) and (9),

$$\begin{cases} \varepsilon_x^p = \dfrac{(2-k)\bar{\varepsilon}^p}{2\sqrt{k^2-k+1}} \\ \varepsilon_y^p = \dfrac{(2k-1)\bar{\varepsilon}^p}{2\sqrt{k^2-k+1}} \end{cases} \tag{10}$$

On the premise of a given relationship between equivalent stress and equivalent plastic strain, the plastic strain of two directions can be obtained by Eq. (10). On the basis of the equivalent plastic strain and the known relationship between equivalent stress and equivalent plastic strain, equivalent stress can be obtained, therefore, the plastic stress of two directions can be obtained by Eq. (8). According to Eq. (8) and Eq. (10), based on the Mises yield criterion, the true strain in x and y directions are as follow.

$$\begin{cases} \varepsilon_x = \dfrac{(2-k)\bar{\varepsilon}^p}{2\sqrt{k^2-k+1}} + \dfrac{1}{E}\dfrac{\bar{\sigma}}{\sqrt{k^2-k+1}}(1-\mu k) \\ \varepsilon_y = \dfrac{(2k-1)\bar{\varepsilon}^p}{2\sqrt{k^2-k+1}} + \dfrac{1}{E}\dfrac{\bar{\sigma}}{\sqrt{k^2-k+1}}(k-\mu) \end{cases} \tag{11}$$

Form Fig. 4. Mises yield criterion can more accurately describe stress-strain relations of the aluminum-silicon coated boron steel under the plane stress condition.

2.3. *Finite element simulation of biaxial tensile*

In the biaxial tensile tests, the force of the cruciform specimen is complex, which is difficult to analysis. Generally, the method of finite element simulation analysis can be adopted. The dimension properties of the cruciform specimen was as shown in Fig. 1. Import the model into Abaqus software. Using ABAQUS to simulate the fracture in the biaxial tensile tests, the plastic stress and strain curve and parameters such as elastic modulus and poisson's ratio can adopt the values above. The specimen was applied the same displacement of 2 mm at the end the blessing. Boundary condition of simulation and experiment are consistent. The results were as shown in Fig. 5.

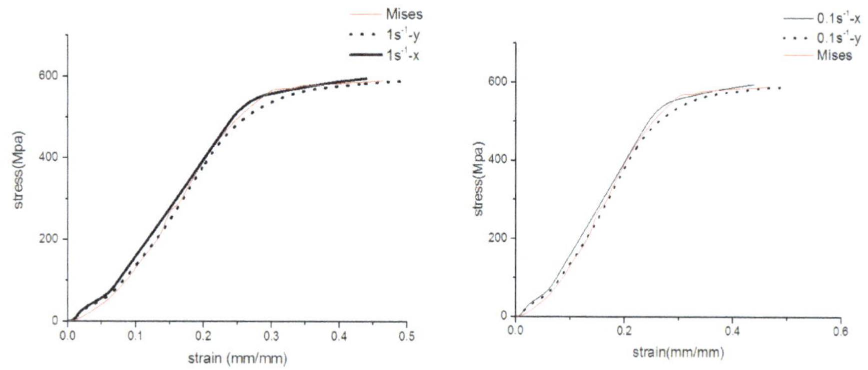

(a) $\dot{\varepsilon}=1s^{-1}$ (b) $\dot{\varepsilon}=0.1s^{-1}$

Fig. 4. Theoretical and true stress-strain curve of biaxial tensile.

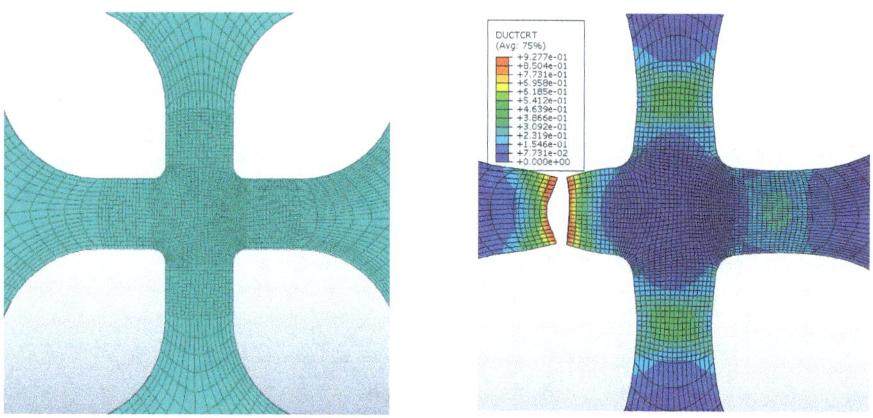

(a) Finite element meshing. (b) Fracture.

Fig. 5. Simulation results.

3. Results and Discussion

3.1. *Observation and discussion of in-situ tensile process*

Without thinning processes in the central, the crack of specimen was expanded in the arm, and the cross-section shape of the specimen was maintained. Images taken by high-speed video are as shown in Fig. 6. The path of the crack propagation can be seen clearly.

Comparing the results of finite element simulation with the biaxial tensile test, the width of arm was from 10 mm to 8.8 mm, the simulation result was 8.5 mm, the error of which was 3.4%; the distance from the fracture to the center was 14.0mm, the simulation result was 13.1, the error of which was 6.4%. Therefore, experiment results are consistent with simulation's (Fig. 7), which can verify the feasibility and the accuracy of FEM to predict the deformation of aluminum-silicon coated boron steel in biaxial tensile test.

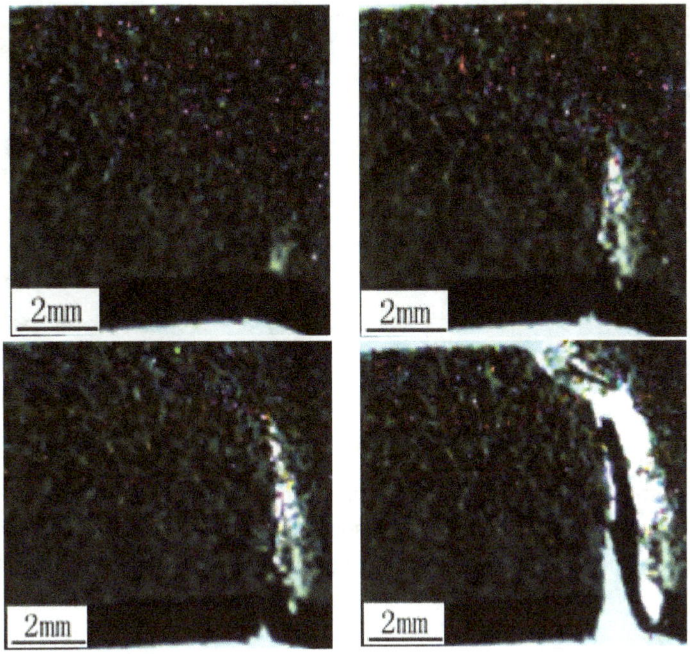

Fig. 6. Crack propagation ($\dot{\varepsilon}$ =0.1s^{-1}).

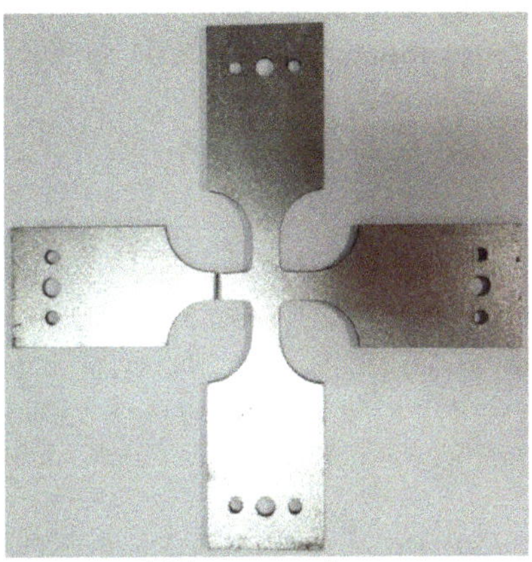

(a) Fracture specimen.

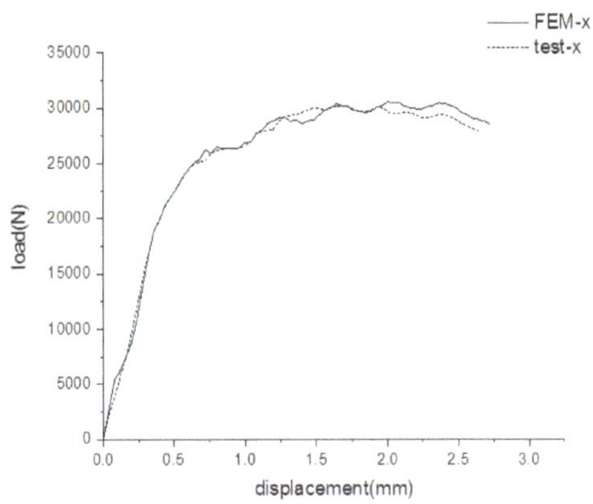

(b) Displacement-load diagram.

Fig. 7. Compare experiment results with simulation's ($\dot{\varepsilon}$ =0.1s^{-1}).

3.2. *Observation and discussion of in-situ tensile fracture*

Micro-structure and failure morphology of the coating layer were inspected by means of scanning electron microscopy (SEM, FEI/QUANYA 200). Ultra-depth optical microscopy (KEYENCE VHX-1000C) was used to observe the crack morphology of the coating. The typical SEM fracture morphology of in-situ tensile samples are as shown in

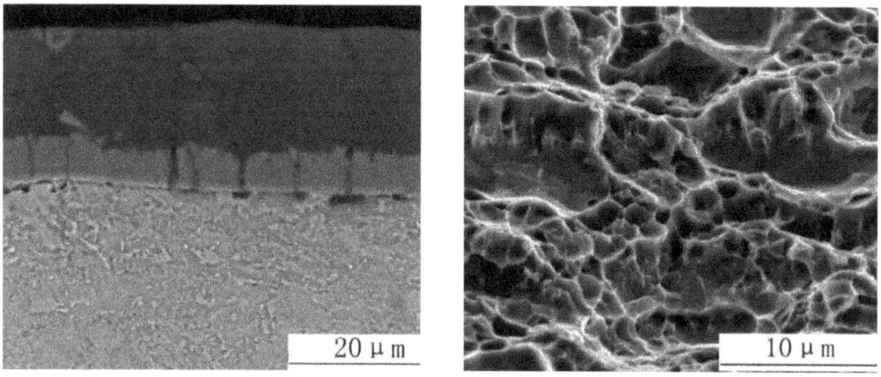

(a) Micro-structure of the coating layer. (b) The failure morphology of the fracture specimen.

Fig. 8. Fracture surfaces of in-situ tensile specimens ($\dot{\varepsilon}$ =0.1/s).

Fig. 8.The Micro-structure of the coating layer on the specimen is shown in Fig. 8 (a). And the failure morphology of the fracture specimen ($\dot{\varepsilon}$ =0.1/s) are shown in Fig. 8 (b).

The fractograph in Fig. 8 (b) exhibits typical brittle fracture pattern of the coating. Therefore, the Al-Si coated boron steel could be regarded as an elastic film attached to an elasticperfectly plastic substrate. Under plane-stress condition, the direction of boron steel's crack propagation is parallel to the direction of shear stress. The fatigue crack's initiation is near round corner, and the crack extends along the direction perpendicular to the principal stress' direction, which forms a fracture. The actual direction of crack extension and the principal stress is substantially perpendicular, which is in good correspondence with the rule of stress.

4. Conclusion

(1) The stress-strain curve of aluminum-silicon coated boron steel has been obtained from the tests conducted by the biaxial tensile testing device under plane strain state.

(2) Using ABAQUS to simulate the fracture in the biaxial tensile tests, the experiment and simulation results are consistent, which can verify the feasibility and the accuracy of FEM based on Mises yield criterion to the predict the deformation of aluminum-silicon coated boron steel in biaxial tensile test.

(3) The path of the crack propagation can be seen clearly by High-speed video. The actual direction of crack extension and the principal stress is substantially perpendicular, which is in good correspondence with the rule of stress.

(4) In this paper, the results of the study provide important help for the study of the crack propagation law of aluminum-silicon coated boron steel under the plane stress state, which are of great importance in practical application of aluminum-silicon coated boron steel in forming process.

Acknowledgments

This research work is financially supported by the National Natural Science Foundation of China (grant no. U1564203) and National Natural Science Foundation of China (grant no. 51405171).

References

1. H. Karbasian and A. E. Tekkaya, A review on hot stamping, *J. Mater. Process Tech.* **210**, 2103 (2010).
2. Z. X. Gui, Y. S. Zhang, H. Q. Li and M. T. Ma, Hot Stamping and Blank Designing for a Vehicle Bumper Using Ultra High Strength Steel (UHSS), *Adv. Mater. Res.* **690-693**, 2240 (2013).
3. C. Wang, Y. S. Zhang, X. W. Tian, B. Zhu and J. Li, Thermal contact conductance estimation and experimental validation in hot stamping process, *Sci. China Technol. Sc.* **55**, 1852 (2012).

A Study on the Impact Resistance of LWB Hot Stamped B Pillar

D. Zhang[1,†], B. Yang[1], R. Ge[2] and Z. H. Chang[1]

[1]*Dongfeng Motor Corporation Technical Center, Wuhan, China*

[2]*R&D Center of Wuhan Iron & Steel Corporation, Wuhan, China*

[†]*E-mail: zhangding@dfmc.com.cn*

During side impact, hot stamped Laser Welded Blank (LWB) B-pillar improves the living space of the driver and passenger's head. There are already some vehicles applied with hot stamped LWB B-pillar, of boron steel as the upper and HSLA as the lower part. The test of hot stamped LWB B-pillar was only focused on the microstructure and hardness of laser weld after heat treatment. There was no generally acknowledged test to evaluate the quality of laser weld after heat treatment under dynamic load. In this paper, the drop test is used to evaluate the quality of laser weld under dynamic load in lateral direction.

Keywords: Laser Welded Blank; Hot stamping; Variable strength B-pillar; Drop test.

1. Introduction

Due to the demand for reducing vehicle weight, improving safety and crashworthiness qualities, the need to manufacture automobile structural components by UHSS is desirable [1]. In the last few years, the use of hot stamping boron steel for the body-in-white structure has grown tremendously, such as A/B-pillar reinforcements, door beams, roof rails, and other structural members. If a vehicle is involved in a collision or gets turned upside down, these hot stamped components help to prevent intrusion, thus preserve the integrity of the passenger compartment. However, it was observed in some actual crash tests that the energy-absorbing capacity of part was sacrificed for increased tensile strength because of less deformation. There is a potential to improve the energy absorbing capability of these hot formed components by introducing regions of reduced strength and increased ductility, referred to here as "tailored properties". There are currently several methods of achieving these tailored properties, including: variable furnace heating zones, tailor welded blanks, advanced tool materials with reduced thermal conductivity, in-die heating, and post heat-treatment [1]. This new flexibility allows automotive designers to "tune" performance of a part based upon specific needs. For example, some center pillar reinforcements have incorporated lower strength materials that are laser welded to the PHS to allow deformation at the bottom of the pillar. These designs enable energy absorption where it does not adversely affect the occupant while maintaining high strength in the upper pillar section to resist deformation and provide superior occupant protection.

In this paper, laser welded blank by boron steel and HSLA was applied in hot stamping process in line with traditional boron steel design. Through the side impact simulation, analyzed how to improve the energy absorbing capacity by the hot stamped LWB B-pillar. Meanwhile, the toughness of weld area after heat treatment would be test by drop tower crash test.

2. LWB Hot Stamped B-pillar Reinforcement

In this paper, the laser welded material used for experiment and simulation was HC420LA of 1.5 mm thickness. The location of laser weld is shown as "red line" in Figure 1. The laser welded blanks was austenitized in a furnace for 5 min at 930 °C, then quenched in water by a cooling rate around 50 °C /s. The mechanical properties of steel after quenching change concerning to its chemical elements. Consequently, the strength of HSLA and HF steel has different changes, see Table 1. The "transition zone" is approximately 0.3mm wide and the hardness of LWB distribution along the transition zone was as shown in Figure 2.

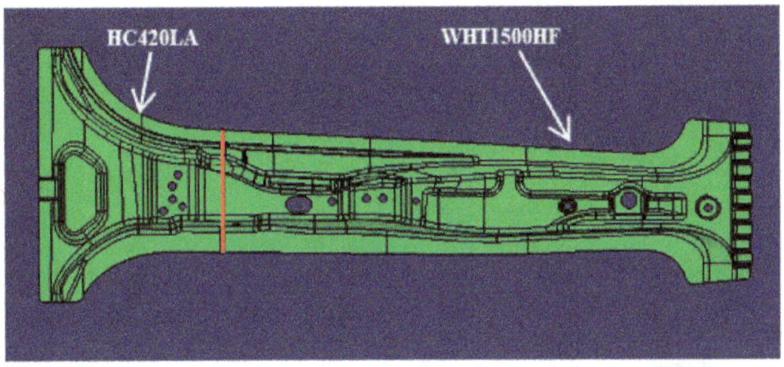

Fig. 1. Tailor welded blank of B-pillar reinforcement.

Table 1. Mechanical properties of LWB.

Material	Steel type	Thickness /mm	YS (N/mm^2)		TS (N/mm^2)	
			Before quenching	After quenching	Before quenching	After quenching
HC420LA	HSLA	1.5	424	464	544	615
WHT1500HF	HF	1.5	306	1108	511	1566

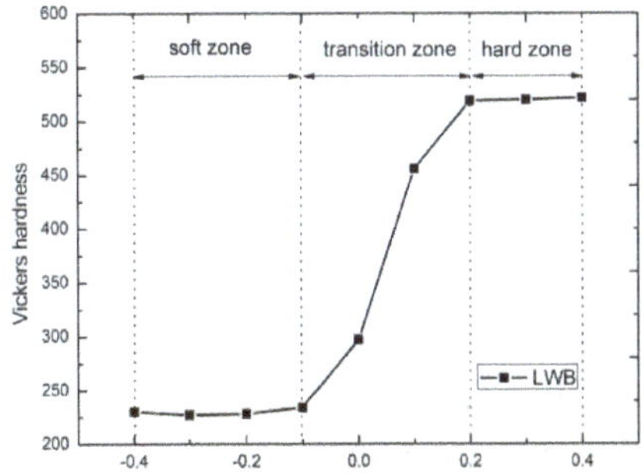

Fig. 2. Vickers micro-hardness distribution along the transition zone.

2.1. *Side impact simulation*

In side impact simulation, the B-pillar reinforcement model was derived from a certain DongFeng vehicle. The CAE model set up for the side impact according to C-NCAP Regulations.

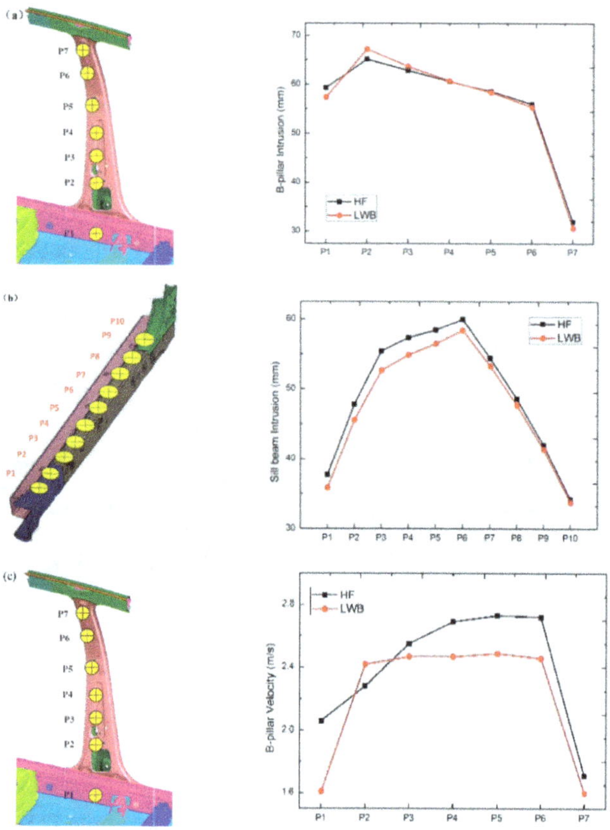

Fig. 3. Side impact simulation results: (a) Intrusion of B-pillar reinforcement; (b) Intrusion of sill beam; (c) Intruding velocity of B-pillar reinforcement.

Table 2. Living space for passengers.

	HF	LWB	IIHS standard
Living space/mm	335	333	≥125

7 key points from B-pillar were taken for quantitative analysis in Figure 3(a). LWB scheme basically has lower intrusion than HF scheme, except Point 2 and Point 3 which located in soft zone. There were 10 key points from sill beam taken for quantitative analysis in Figure 3(b). LWB scheme obviously has lower intrusion than HF scheme. In conclusion, B-pillar had a bending deformation at soft zone. Therefore, the distance between B-pillar to longitudinal centerline of driver's seat [2] of LWB was 2mm less than HF, (see Table 2). Meanwhile, the intruding velocity of Point 2 on LWB hot stamped B-pillar was higher than HF, see Figure 3(c). In summary, through the deforming process,

the lower part of LWB hot stamped B-pillar absorbed more energy than HF, and LWB almost had a slight wider living space and an obvious slower intruding velocity for passengers' head, which improved the security for passenger's head during side crash.

3. Drop Tower Crash Test

In a vehicle side crash, B-pillar is subjected to bending loads [3]. Therefore, a drop tower crash test has been designed to examine the crash behavior of laser weld after heat treatment with several different impact energies. From the side impact simulation in section 2, LWB would absorb more energy at the laser weld than HF at the same position of B-pillar, see Table 3.

Table 3. Energy absorbed in laser weld position.

	HF	LWB
Energy(J)	406.7	505.9

3.1. *Procedures*

In the drop tower test, a punch is dropped from a height at least 1m to achieve an impact energy. The laser weld blank was hot formed into a U-shaped specimen with a height of 75mm and a width of 90mm, the schematic diagram as shown in Figure 4. In a drop, the falling weight aimed at the laser weld of U-shaped specimen. Through adjusting the mass weight mass and falling height, the maximum impact energy can be up to 1500 J. The initial drop energy was chosen to be 450 J.

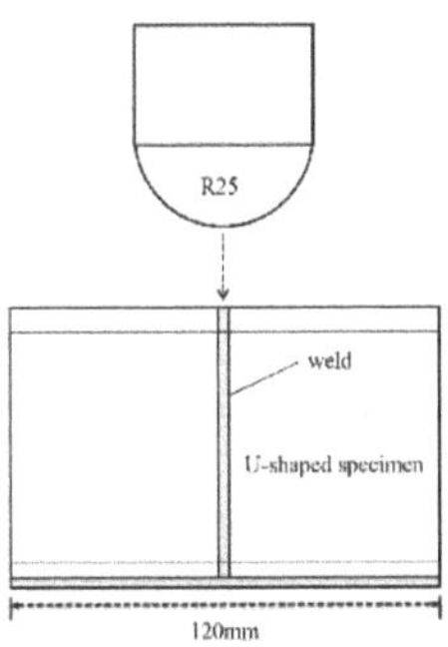

Fig. 4. Schematic diagram of drop tower crash test.

3.2. *Results*

After crashing test, the U-shaped specimen had a bending deformation, as shown in Figure 5. Due to the punch aimed at the laser weld, the area nearby the laser weld sunken down and the laser weld of U-shaped specimen undertook a tensile stress.

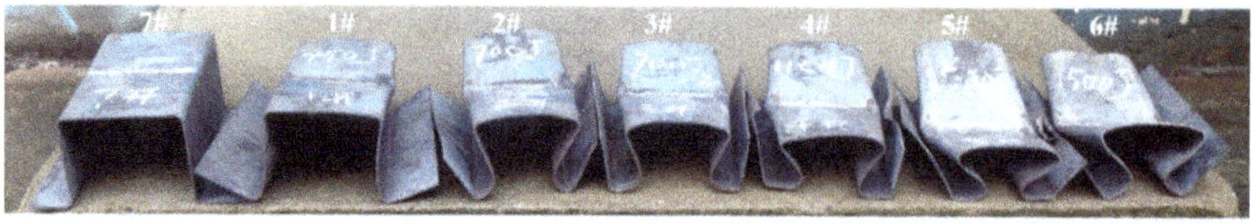

Fig. 5. U-shaped specimen after crash testing: 1 # - 450 J, 2# - 700 J, 3 # - 700 J, 4 # - 1100 J, 5 # - 1500 J, 6 # - 1500 J, 7 # - 0 J.

The laser weld on the edge of U-shaped specimen had pulled widened, specimen 5# is shown in Figure 6. Compared to specimen 7 #, with the increasing impact energy, the height of U-shaped specimen decreased, meaning the more impact energy loaded on laser weld. In this study, even though the maximum impact energy of 1500 J, there was no crack on laser weld, shown in Figure 6.

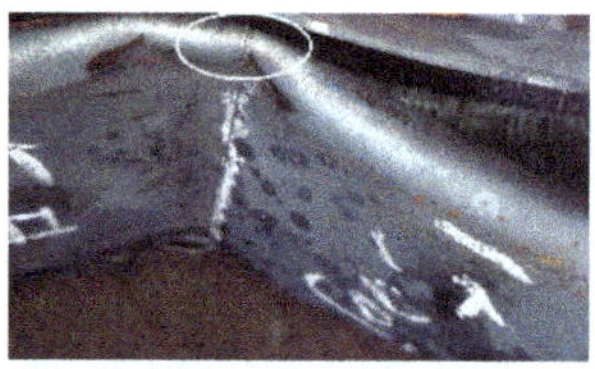

Fig. 6. 5 # specimen after crash testing.

4. Conclusion

Due to the lower strength of laser welded HC420LA, the lower part of LWB hot stamped B-pillar absorbed more energy. Compared to HF B-pillar reinforcement, the LWB hot stamped B-pillar reinforcement had better performance in side impact with a slight wider living space and an obvious slower intruding velocity for occupants, which improved the security for passenger's head during side crash. In this study, a specific drop tower crash test was introduced to examine the crash behavior of laser weld after heat treatment. The results showed that the laser weld with heat treatment was strong enough to resist a crash energy, which 2 times more than undertook in C-NCAP.

References

1. H. Karbasian and A. E. Tekkaya, A review on hot stamping, *J. Mater. Process. Tech.* **210**, 2103 (2010).
2. Side Impact Crashworthiness Evaluation Test Report, Insurance Institute for Highway Safety, 2015.
3. T. M. Link and C. M. Jensen, Three-Point Bending Crash Performance of Advanced High Strength Steels, *SAE Technical Paper*, No. 2009-01-0797 (2009).

Research on the Deformation Behavior of Aluminum Alloy Sheets Under Various Temperatures

J. Zhou[a], J. P. Zhang[b], H. L. Yan[c] and M. T. Ma[d]

China Automotive Engineering Research Institute CO., LTD, Chongqing 401122, China

E-mail: [a]zhoujia@caeri.com.cn, [b]zhangjunping@caeri.com.cn, [c]yanhuanli@caeri.com.cn, [d]mamingtu@caeri.com.cn

This paper aimed at investigating the deformation behavior of non heat treatable aluminum alloys of the 5052 H32 and heat treatable aluminum alloys of the 6016 T4P after six months natural aging. The deformation behavior and mechanical properties of 5052 H32 and 6016 T4P aluminum alloy sheets after six months natural aging under different temperature conditions were studied, the deformation potential of the two aluminum alloy at room and elevated temperature were investigated. The results indicated that the 6016 aluminum alloy sheet exhibit better mechanical properties at room temperature, but the 5052 H32 aluminum alloy sheet performs poor formability. 5052 H32 aluminum alloy sheet shows super plastic phenomenon at elevated temperature, the super plastic deformation behavior were analyzed by SEM and microstructure.

Keywords: Deformation behavior; Microstructure; Aluminum alloy sheet.

1. Introduction

In recent years, because of the possession of light mass, high strength, easy recycling and high corrosion resistance of aluminum alloys, which have been attracting elevating attention, they have been showing a remarkable increase in the applications in automotive industry and other engineering areas as structural materials [1-5]. However, the widespread applications of wide aluminum alloy sheets are restricted because of their various and complex process technologies, and higher cost than steel sheets [6-10].

Some of aluminum alloys will exploit the phenomenon of superplasticity under severe deformation conditions [11-13]. Currently, there is increased commercial interest in superplasticity, coming from both the automotive and aerospace industries.

Due to the low resistance of these materials to plastic flow, the materials are formed into complex shapes using gas pressure with minimal energy expenditure in the superplasticity forming process, which reduces tooling costs since only one single-surface tool is required. The equipment from the complexity of parts that can be formed in a single operation, which leads to reduced part counts and a reduced need for fasteners and connectors, decreasing the weight of the final product. The superplasticity forming is a slow process due to the low strain-rate range in which superplasticity normally. The decreased forming times can be extremely beneficial through microstructural refinement. Superplastic 2xxx, 5xxx aluminum alloys by static or dynamic recrystallization, few studies have been made of the grain refinement of 6xxx alloys for superplasticity [14].

[a] Corresponding author.

This paper presents the main achievements of a research project aimed at investigating the applicability of the hot forming technology to non heat treatable aluminium alloys of the 5052 H32 and 6016 T4P aluminum alloy sheets after six months natural aging, and the superplasticity phenomenon of aluminum alloys under hot forming condition. State of the art knowledge is provided in terms of material response to deformation and formability limits.

2. Materials and Experiments

The material utilized in the tests are the heat-treatable aluminum alloy 6016 T4P aluminum alloy sheets after six months natural aging supplied as 1.2 mm thick blanks of 600 mm × 600 mm and non heat-treatable aluminum alloy 5052 H32 supplied as 2.2 mm thick blanks of 800 mm × 800 mm. The alloys chemical compositions are reported in Table 1.

Table 1. Chemical compositions of the 5052 H32 alloy and 6016 T4P aluminum alloy sheets after six months natural aging (wt%).

	Si	Cu	Mg	Zn	Mn	Cr	Fe	Ti	Al
5052	≤0.25	≤0.1	2.2-2.8	≤0.1	≤0.1	0.15-0.35	≤0.4	-	Balance
6016	1.0-1.5	≤0.2	0.25-0.6	≤0.2	≤0.2	≤0.1	≤0.5	≤0.15	Balance

This section presents the testing procedures adopted for the laboratory tests and the obtained results in terms of mechanical properties and formability at room and elevated temperatures.

3. Tensile Tests at Room and Elevated Temperatures

The uni-axial tensile tests at room and elevated temperatures were carried out on a SANS CMT5305 universal tester machine. The length and width with size of 6mm at the parallel segment, the samples were obtained along the rolling direction of the sheet, two holes were processed the ends of the specimen with 8mm diameter for holding in the test processing. The dimensions of tensile sample show in Fig. 1. A furnace was used to heat the sample to the testing temperature. The tensile sample for elevated temperature was heated to the testing temperature, held at this temperature for 90 s, and then strained till fracture. The tensile speed for room and elevated temperatures was 3mm/min.

The results of the tensile tests are presented in Fig. 2, showing respectively the material stress, strain at fracture.

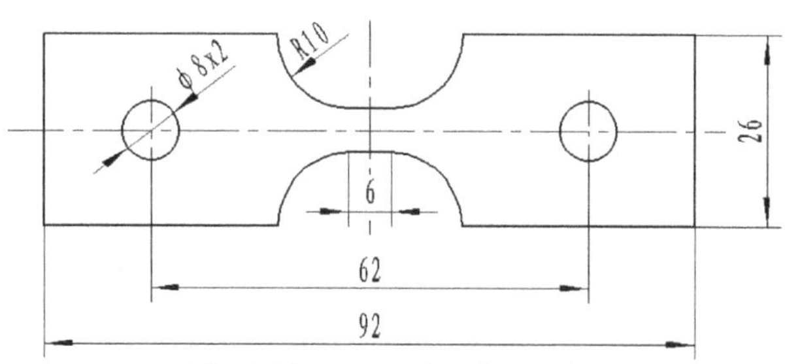

Fig. 1. Dimensions of tensile sample.

4. Results and Discussion

Table 2 presents strain-stress data of the 6016 T4P aluminum alloy sheets after six months natural aging and 5052 H32 aluminum alloy sheet at room temperature and 300°C. It can be noted that the 6016 T4P aluminum alloy sheets after six months natural aging perform better mechanical properties than the 5050-H32 aluminum alloy sheet at room temperature, it has higher tensile strength and yield strength. When the temperature rose to 300°C, the deformation resistance of the aluminum alloy sheet was significantly decreased, the magnitude of decreasing was more than 100MPa. The plastic deformation properties of these two kinds of aluminum alloy sheets undergone significant changes at elevated temperatures. The elongation of 6016 T4P aluminum alloy sheets after six months natural aging occurs slightly increased at 300°C, while the elongation 5052 H32 aluminum alloy sheet perform significantly increasing (see Fig. 2), it can be noted that the elongation of 5052 H32 aluminum alloy is more than 100%, which shows superplastic phenomenon (In general, if the elongation of material is more than 100%, it can be called superplasticity). Under this deformation condition, the average elongation for 5052 H32 aluminum alloy was 103% at 300°C. The Stress-strain plots for these tensile tests of 5052 H32 aluminum alloy are shown in Fig. 2.

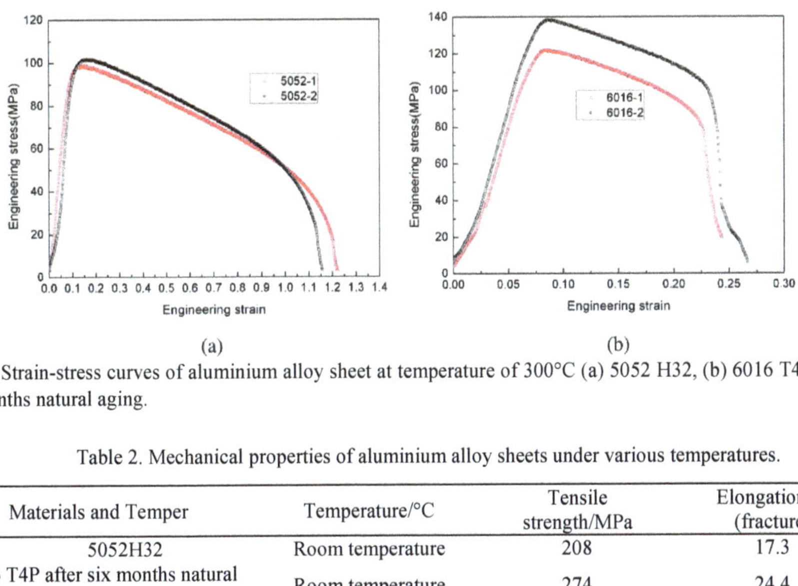

Fig. 2. Strain-stress curves of aluminium alloy sheet at temperature of 300°C (a) 5052 H32, (b) 6016 T4P after six months natural aging.

Table 2. Mechanical properties of aluminium alloy sheets under various temperatures.

Materials and Temper	Temperature/°C	Tensile strength/MPa	Elongation/% (fracture)
5052H32	Room temperature	208	17.3
6016 T4P after six months natural aging	Room temperature	274	24.4
5052 H32-1	300	103	121.3
5052 H32-2	300	100	115.2
6016 T4P after six months natural aging-1	300	121	24.2
6016 T4P after six months natural aging-2	300	138	26.6

To analyze the deformation behaviors of 5052 and 6016 aluminum alloy, the microstructures of them under various states were obtained, which include the microstructures of original, 300°C with Soaking time 30 min, Air cooling and 300°C, Soaking time 30 min, Water quenching. The microstructures show in Fig. 4 (transverse direction), and the average grain sizes of them are listed in Table 3.

The original microstructure of 5052 H32 aluminum alloy obtained in the sheet form is also shown in Fig. 3(a) and Table 3 for comparison. 6016 T4P after six months natural aging was found to have an average grain size of 50μm, which is coarse than 5052 H32 aluminum alloy in original condition (average grain size of 30μm). A significant difference in the grain size in the two alloys is evident. Clearly, 6016 T4P after six months natural aging has larger grain size than 5052 H32.

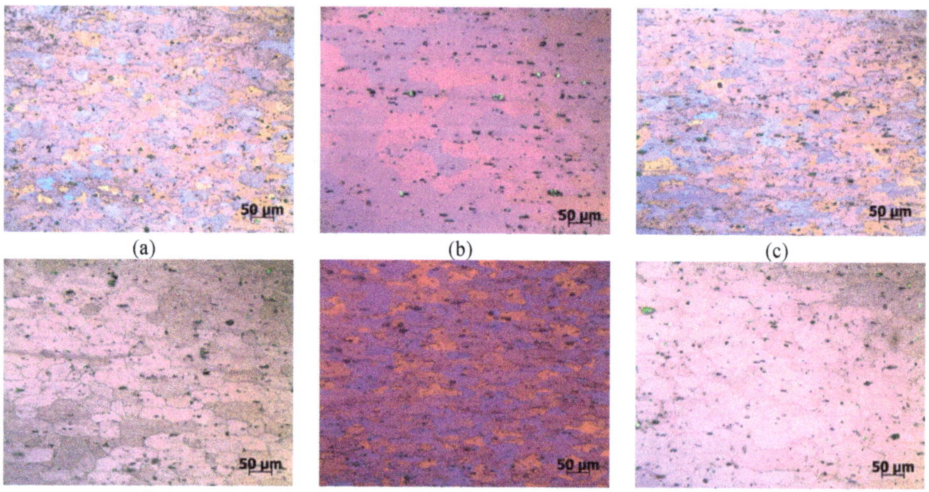

Fig. 3. Microstructures of aluminium alloy under various states (a) 5052 H32, (b) 6016 T4P after six months natural aging, (c) 5052 H32, 300°C, Soaking time 30 min, Air cooling, (d) 6016 T4P after six months natural aging, 300°C, Soaking time 30 min, Air cooling, (e) 5052 H32, 300°C, Soaking time 30 min, Water quenching, (f) 6016 T4P after six months natural aging, 300°C, Soaking time 30 min, Water quenching.

Table 3. Average grain size of aluminium alloy under various states.

State of aluminium alloy	Average grain size/μm
5052 H32	30
6016 T4P after six months natural aging	50
5052 H32, 300°C, Soaking time 30 min, , Air cooling	26
6016 T4P after six months natural aging, 300°C, Soaking time 30 min, Air cooling	43
5052 H32, 300°C, Soaking time 30 min, Water quenching	26
6016 T4P after six months natural aging, 300°C, Soaking time 30 min, Water quenching	41

An average grain size of 50μm and the equiaxed grain structure were observed in the as-received material of 6016 T4P after six months natural aging as shown as Fig. 3(b) and Table 3. It can be noted that the average grain sizes have slight decrease after soaking time 30 min under temperature of 300°C, which are still large than 40 μm as shown in Fig. 3(d), (f) and Table 3. The original microstructure of 5052 H32 aluminum alloy has finer grain size of 30μm, it has a slight decrease after 300°C with Soaking time 30 min with a reduction of 4μm.

Fig. 4 shows the fracture morphologies of tensile fracture for 5052 H32 and 6016 T4P after six months natural aging. Figs. 4(a), (b) show any dimples on the fracture surface. Several coarse dimples and many small dimples were captured over the fracture surface in fibre zone [15]. The fractograph presents a ductile fracture mode. Several deep dimples were formed during the tension process, which is consistent with the high material ductility. Several small, shallow dimples can be observed in fibre zone in Figs. 4(c), (d). The density of dimples decreased and their shallowness is consistent with the low material ductility.

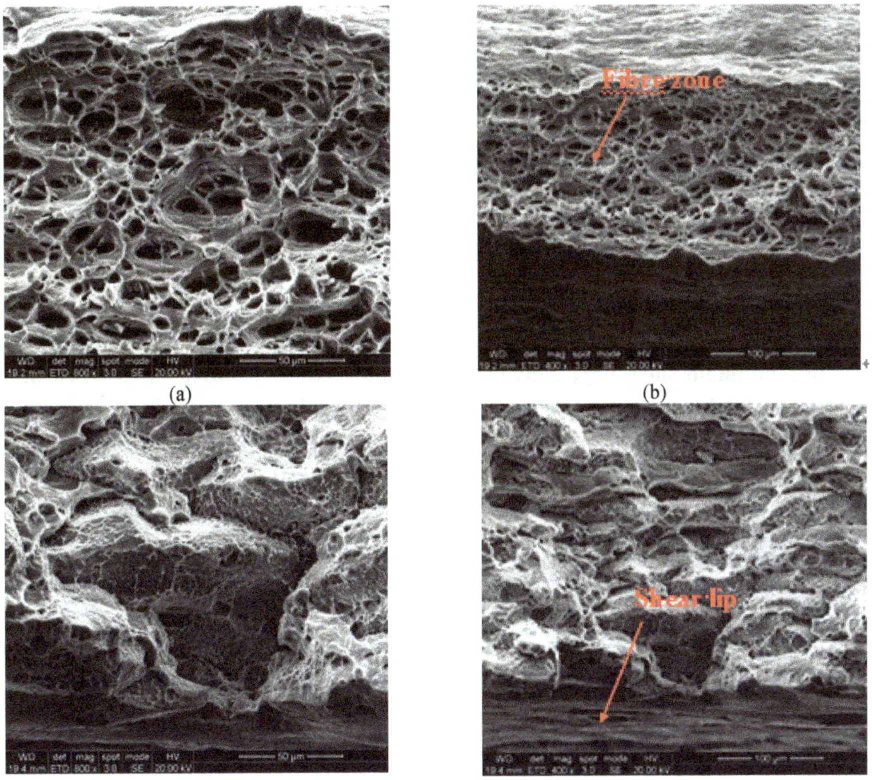

(a) (b) (c) (d)

Fig. 4. Fracture morphologies of tensile fracture for aluminum alloy under 300 °C (a) 5052 H32, 800X resolutions, (b) 5052 H32, 400X resolutions, (c) 6016 T4P after six months natural aging, 800X, resolutions, (d) 6016 T4P after six months natural aging, 400X, resolutions.

5. Conclusion

The deformation behavior and mechanical properties of 5052 H32 and 6016 T4P after six months natural aging aluminum alloy sheets under different temperature conditions were studied. When the temperature rose to 300°C, the deformation resistance of the two aluminum alloy sheets was significantly decreased, the magnitude of decreasing was more than 100MPa. The plastic deformation properties of them had undergone significant changes at elevated temperatures. The elongation of 5052 H32 aluminum alloy is more than 100%, which shows superplastic phenomenon. The Microstructures and SEM fracture morphologies were obtained to analyzed this phenomenon. The results indicate that 5052 H32 has finer grain size and presents a ductile fracture mode. Several deep dimples were formed during the tension process, which is consistent with the high material ductility. 6016 T4P after six months natural aging aluminum alloy sheet has several small, shallow dimples, and the density of dimples decreased and their shallowness is consistent with the low material ductility.

Acknowledgments

This work was funded by Chongqing Basic and Frontier Research Program for Distinguished Young Scientists Project (cstc2013jcyjjq60001) and Chongqing Training Program for Science and Technology Talents (cstc2013kjrc-qnrc60003). Supports from major industrial technology working on research and development in key Technologies for high formability aluminum automotive sheets, and Chongqing Automotive Collaborative Innovation Center contribution to the project are herewith acknowledged.

References

1. J. Zhou, F. Wang and X. M. Wan, Optimal Design and Experimental Investigations of Aluminum sheet for Lightweight of Car Hood, *Materials Today: Proceedings* **2**, 5029 (2015).
2. J. Zhou, X.M. Wan and Y. Li, Advanced aluminum products and manufacturing technologies applied on vehicles presented at the EuroCarBody conference, *Materials Today: Proceedings* **2**, 5015 (2015).
3. Loorentz and Y. G. Ko, Effect of differential speed rolling strain on microstructure and mechanical properties of nanostructured 5052 Al, *J. Alloy. Compd.* **586**, 205 (2014).
4. K.J. Kim, H.T. Jeong, K.S. Shin and C.W. Kim, Texture evolution of rolled AA5052 alloy sheets after annealing, *J. Mater. Process Tech.* **187-188**, 578 (2007).
5. S.L. Xia, M. Ma, J.X. Zhang, W.X. Wang and W.C. Liu, Effect of heating rate on the microstructure, texture and tensile properties of continuous cast AA 5083 aluminum alloy, *Mater. Sci. Eng. A* **609**, 168 (2014).
6. M.P. Liu, T.H. Jiang, J. Wang, Q. Liu and Z.J. Wu, Aging behavior and mechanical properties of 6013 aluminum alloy processed by severe plastic deformation, *Transactions of Nonferrous Metals Society of China* **24**, 3858 (2014).
7. H.Y. Song, Y.S. Kim and W.J. Nam, Mechanical properties of ultrafine grained 5052 Al alloy produced by accumulative roll-bonding and cryogenic rolling, *Metals and Materials International* **12**, 7 (2006).
8. H. Yuan, J. Li, X.Y. Kong, C.C. Yui, Q.X. Yang and W.C. Liu, Strain hardening and orientation hardening/softening in cold rolled AA 5052 aluminum alloy, *Mater. Lett.* **62**, 4085 (2008).

9. M.T. Ma, X.Y. Bi, J.H. You and H.Z. Lu, Research progress of property and its application of aluminium alloy auto sheet, *Materials for mechanical engineering* **34**, 1 (2010).
10. M.T. Ma, H.Y. Yang, E.M. Wu and Q.S. Jin, Research progress of the cause for orange peel of aluminum alloy sheet during tensile deformation, *Eng. Sci.* **16**, 6 (2014).
11. T.G. Langdon, Experimental observations in superplasticity, Superplastic Forming of Structural Alloys, *The Metallurgical Society of AIME*, 27 (1982).
12. J. Pilling and N. Ridley, Cavitation in aluminum alloys during superplastic flow, in In *Proc. Symp. on Superplasticity in Aerospace I*, (Phoenix, AZ, 1988).
13. C.C. Bampton, J. Wadsworth and A.K. Ghosh, Superplastic aluminum alloys, *Treatise on Materials Science and Technology* **31**, 189 (1989).
14. L.P. Troeger and E.A. Starke Jr, Microstructural and mechanical characterization of a superplastic 6xxx aluminum alloy, *Mater. Sci. Eng.* A **277**, 102 (2000).
15. M.T. Ma, H.Y Yang, E.M. Wu and Q.S. Jin, Research progress of the cause for orange peel of aluminum alloy sheet during tensile deformation, *Eng. Sci.* **16**, 4 (2014).

Effects of Solution Temperature and Time on Microstructure and Properties of a 6061 Aluminum Alloy

J. Ren, Y. Liu, Y. M. Jin, B. Zhu and Y. S. Zhang†

State Key Laboratory of Materials Processing and Die and Mould Technology, Huazhong University of Science & Technology, Wuhan, 430074, China

†E-mail: zhangys@mail.hust.edu.cn

Solution treatment is an important step in aluminum alloy hot stamping; and solution time and temperature are important parameters in solution treatment. Therefore, the effects of solution temperature and time on microstructure and mechanical properties were investigated by optimizing them respectively; using the process of solution treatment, water quenching, and artificial aging (200°C for 3 h) on a 6061 aluminum alloy sheet. The strength and hardness test results show comprehensive mechanical performance and production efficiency. After solution treatment at 560°C for 10 min, water quenching and artificial aging at 200°C for 3 h, the 6061 aluminum alloy sheet can obtain a preferable microstructure and properties. This was done under the condition of meeting the same strength and other properties, greatly reducing the aluminum thermoforming production cycle and improving productivity.

Keywords: Aluminum alloy hot forming; Solution temperature; Solution time; Microstructure; Artificial aging.

1. Introduction

With the increase of car ownership, the energy consumption and emissions of pressure increasing, energy conservation and emissions reduction is urgently needed. Research shows that weight loss of 10 % per car, fuel consumption can be reduced by 6 % - 8 % [1-3], therefore, the car lightweight is the most direct and most effective energy conservation and emissions reduction measures. The density of aluminum is only one-third of steel, aluminum alloy weight loss is more effective in the same conditions of mechanical properties, but there are some problems which aluminum alloy at room temperature exits forming low plasticity, a big rebound and the total elongation small et al. In order to solve these problems, aluminum alloy hot forming technology arises at the historic moment.

Aluminum alloy hot forming, generally by the solution treatment, quenching, and aging of three stages. The long-playing of solution treatment and aging have become a main problem for aluminum alloy hot forming in the actual production. D. F. Zhang etc.[4] consider that 535°C solution heat 4 h and then 180°C ageing for 3 h, the 6061 aluminum alloy sheet can obtain preferable microstructure and properties; S. L. Li etc. [5] believe that solution treatment at 560°C for 40 min, water quenching and artificial aging at 565°C for 6h to obtain preferable microstructure and properties. Under the premise to meet the required performance, it takes so long time that affect productivity and cannot be applied to the actual production line. Hence, this paper researches the stage of solution treatment, under the condition of meet the same strength and other properties, by appropriately increasing the solution temperature to shorten solution time.

In this paper, we control the solution temperature and solution time, respectively, using the process of solution treatment + water quenching + artificial aging (200°C * 3 h) on 6061 aluminum alloy sheet (The process shown in Fig. 1). Comprehensive mechanical performance and production efficiency, after solution treatment at 560°C for 10 min, water quenching and artificial aging at 200°C for 3 h, the 6061 aluminum alloy sheet can obtain preferable microstructure and properties. Under the condition of meet the same strength and other properties, greatly reducing the aluminum hot forming production cycle and improving productivity.

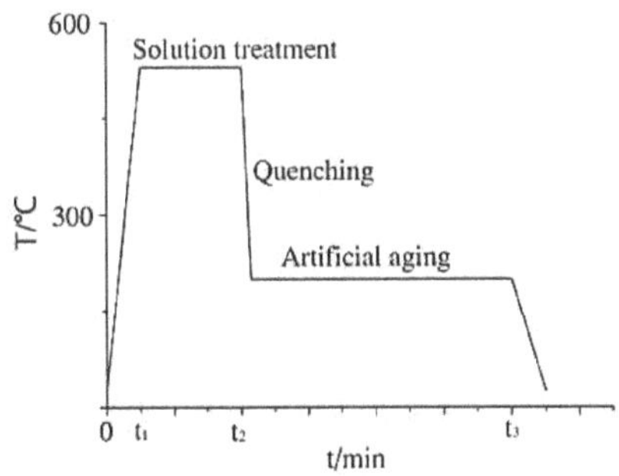

Fig. 1. Aluminum alloy hot forming process.

2. Experimental

Type 6061 aluminum alloy is age-hardening alloy, the parameters of solution treatment has great influence on microstructure and properties. 6061 aluminum components used in the experiment shown in Table 1, in T6 state, the hardness is 128.75 HV, tensile strength is 339.70 MPa. Tensile test specimens used in a thickness of 1.5mm, a shape as shown in Fig. 2.

Table 1. Chemical composition in 6061aluminium alloys (wt, %).

Si	Cu	Fe	Mg	Zn	Ni	Mn	Ti	Cr	Al
0.676	1.495	0.4227	1.1036	0.114	0.0058	0.078	0.0273	0.1647	Bal.

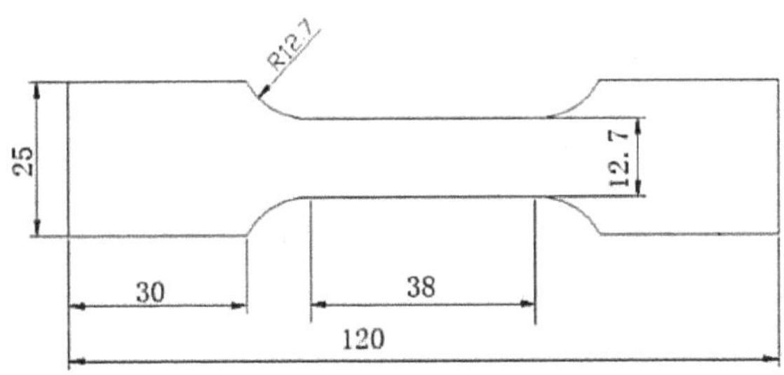

Fig. 2. Schematic of the tensile test samples; unit: mm.

Experimental process:

1) The temperature is set to 500-590°C (interval 10°C), put the specimen into the solution furnace heat insulation for 30 min and quickly water quenching, then put into the aging furnace 200°C aging for 3 h. The sample was subjected to hardness and strength testing and elongation measuring, to give proper solution temperature from the results.
2) Using the optimal solution obtained from the experiment (1) was subjected to solution heat treatment at different times, respectively for 10 min, 15 min, 20 min, 25 min and 40 min. Then, the specimen was subjected to hardness and strength testing and elongation measuring, to give proper solution time from the results.

3. Results and Discussions

3.1. *Effect of solution temperature and time respectively on mechanical properties*

The mechanical properties of 6061 aluminum sheet heat different temperature and artificial aging at 200°C for 3 h is shown in Fig. 3. With the increase of temperature, tensile strength and hardness of the alloy are rapidly rising, strength and hardness increases slowly after 560°C. Comprehensive consideration of various performance indicators, the most appropriate of 6061 aluminum alloy solution temperature is 560°C.

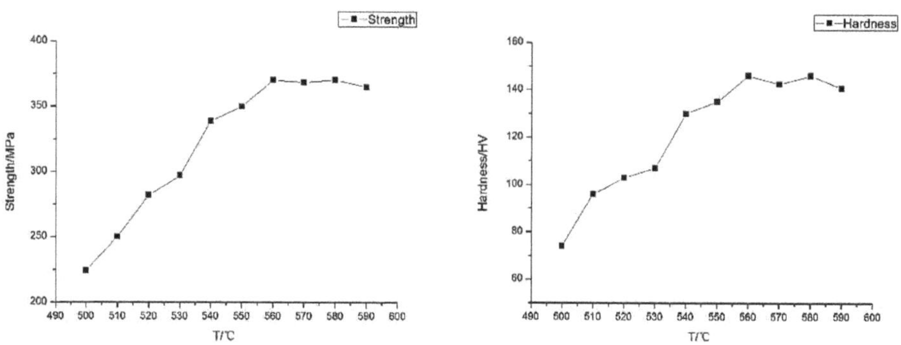

(a) (b)

Fig. 3. Effect of solution temperature on properties of the tested alloy (a) Tensile properties (b) Hardness.

Experiment with the appropriate solution treatment temperature of 560°C for different time and after artificial aging at 200°C for 3 h, mechanical properties as shown in Fig. 4. As the growth of the time, tensile strength and hardness of the alloy increase is not obvious or even present downward trend. Comprehensive consideration of various performance indicators, the 6061 aluminum alloy suitable solution time is 10 min.

In summary, the optimum parameter of the solid solution of 6061 aluminum alloy is 560°C * 10 min.

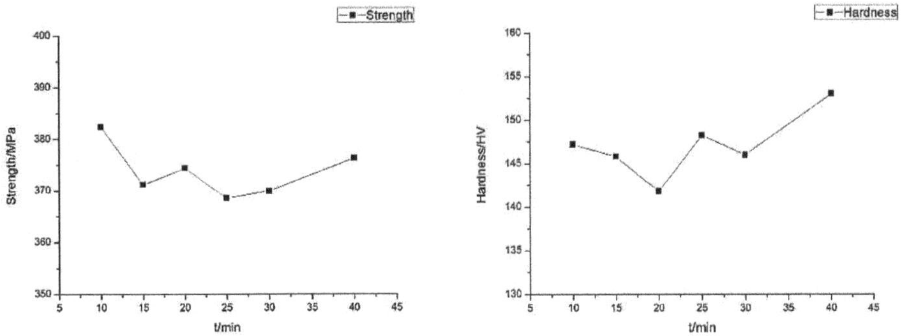

(a) (b)

Fig. 4. Effect of solution time on properties of the tested alloy (a) Tensile properties (b) Hardness.

3.2. *Effect of solution temperature and time respectively on microstructure*

The microstructure of 6061 aluminum alloy sheet after solution treatment at different temperatures for 30 min and artificial aging is shown in Fig. 5. Aluminum is mainly composed of α (Al) and various alloy elements which can dissolve into aluminum alloy (such as Mg, Si, Cu, etc.), of which 6061 aluminum alloy by α (Al) + β (Mg_2Si) two-phase eutectic composition, when the alloy is heated to the eutectic temperature, β phase is dissolved into the matrix phase form a single phase solid solution α (solution process as shown in Fig. 5). The actual microstructure process of solution is as shown in Fig. 6. It shows that the black part is Mg_2Si, grey part is the Al substrate. As the solution temperature increases, change in the microstructure of the alloy sheet is very clear.

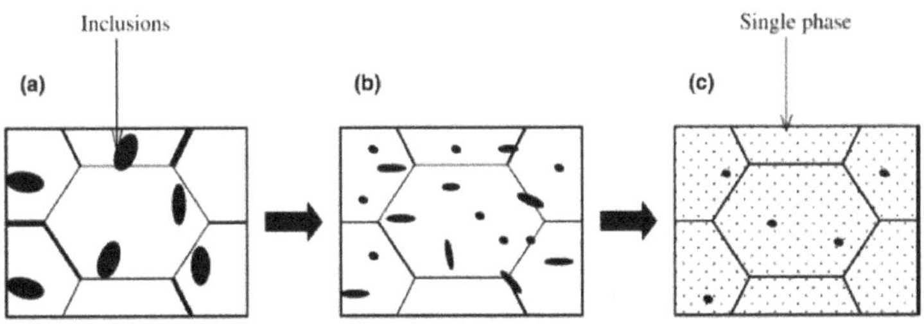

Fig. 5. Schematic diagram showing the microstructure evolution for the material [6]: (a) at the beginning of SHT; (b) intermediate SHT; (c) close to finishing SHT.

When the solution temperature is 500°C, the intra-crystalline and grain boundary residual many big Mg_2Si phase particles (black part of the Fig. 6), namely β phase dissolve is not sufficient; When 520°C solution, grain has been grown, the residual phase in instarcrytalline and grain boundary has been reduced; When 540°C solution, grain continue to grow up, strengthening phase in the alloy particle decreased significantly; When 560°C solution, the residual phase further reducing, grain further growth, strengthened alloy phase substantially fully dissolved into the matrix; When 590°C solid solution, local grain boundaries compound melting widened and there are compound melting ball, suggesting that the alloy over burn phenomenon occurred at this temperature.

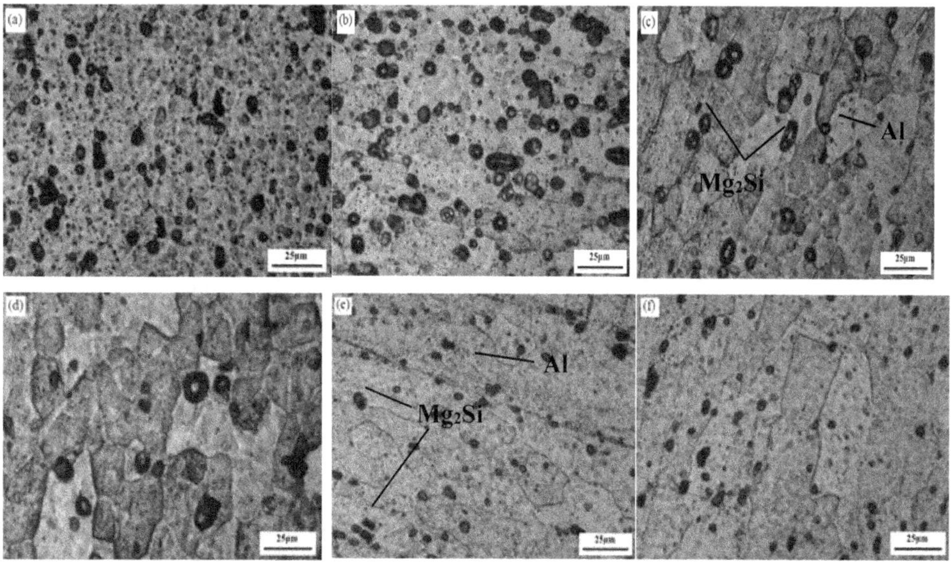

Fig. 6. Effect of solution temperature on microstructure of the alloy solution-treated for 30 min: (a) 500°C (b) 520°C (c) 540°C (d) 550 °C (e) 560 °C (f) 590°C.

The microstructure of 6061 aluminum alloy sheet after solution treatment at different time at 560°C and artificial aging is shown in Fig. 7. When 560°C solution, the strengthening phase has been fully dissolved into the matrix, so the microstructure changed little at different solution time as the solution time increases.

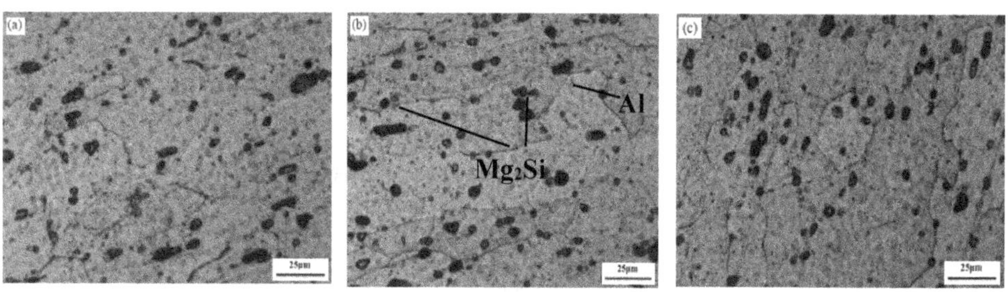

Fig. 7. Effect of solution time on microstructure of the alloy solution-treated at 560°C: (a) 10 min (b) 20 min (c) 40 min.

The supersaturated solid solution obtained after solution treatment and water quenching, at the time through the aging process, Mg, Si atoms gathered in the plane of the aluminum matrix to form a solute atom-rich region, the so-called GP zone. Since the GP zone and the substrate is formed to maintain a coherent relationship between atoms on the boundary of the parent phase and GP total area. In order to adapt to the arrangement in the form of two different atoms at same time in the vicinity of the boundary will produce coherent elastic strain, crystal lattice distortions lead to dislocation motion is blocked, resulting in increased hardness of the alloy. With Mg, Si atoms further enriched, will gradually lose its coherent relationship with the substrate, the substrate is formed with partial coherent phase of β (having a hexagonal crystal structure of the shaft) and the elastic strain of the surrounding matrix will have the reduced barriers to dislocation motion is decreased, the hardness decreases. In the latter part of aging β' phase completely lost coherent relationship with the matrix, into stable β (Mg_2Si) phase, after aging strengthen, the mechanical properties such as strength and hardness are improved.

The greater the degree of supersaturation solid solution, the greater the driving force off the solvent, dispersed precipitates more fully, the greater the strengthen effect. On the premise of not happened overburn phenomenon, the higher the solution temperature, the greater the concentration of alloying elements in solution, the solubility of a supersaturated solid solution after quenching was greater, and aluminum alloy sheet obtain higher strength after aging.

4. Conclusion

1) From the mechanical properties, with the increase of solution temperature, the tensile strength of 500°C to 560°C alloy hardness increases rapidly, the tensile strength increased by 146 MPa, the hardness increased by 72 HV; but there have no obvious increase for tensile strength and hardness when the solution temperature over than 560°C.
2) From the microstructure of view, with the increase of temperature, grain growth, the second phase Mg_2Si constantly dissolved, dissolve substantially complete when the temperature heated to 560°C, and the grain changed little when the temperature continues rising; 560°C solution at different times, microstructure changed little for the elements dissolved rapidly at 560°C.
3) Comprehensive mechanical performance and production efficiency, the 6061 aluminum alloy sheet can obtain preferable microstructure and properties after solution treatment at 560°C for 10 min, water quenching and artificial aging. The alloy by increasing the solution temperature reduces the solution time.

Acknowledgments

This research work is financially supported by the National Natural Science Foundation of China (grant no. U1564203) and National Natural Science Foundation of China (grant no. 51405171).

References

1. M. T. Ma, H. L. Yi, H. Z. Lu and X. M. Wan, On the light weighting of automobile, *Engineering Science* **11**, 20 (2009).
2. C. Joseph and K. Benedy, Light metals in automotive applications, *Light Metal Age* **10**, 34 (2000).

3. H. Yang, L. Li, Q. Wang, W. P. Chen, Z. M. Liu and W. J. Qi, Research on the development of advanced forming for lightweight alloy materials area, *Chinese Journal of Mechanical Engineering* **46**, 31 (2010).
4. D. F. Zhang, G. Tan, L. Liu, Y. Yang, X. W. Chen and Y. Chen, Effects of heat treatments on the microstructure and mechanical properties of a 6061 aluminum alloy, *Heat Treatment of Metals* **40**, 184 (2015).
5. S. L. Li, Z. Q. Huang, F. L. Jiang, W. P. Chen, Z. M. Liu and W. J. Qi, Effect of solution temperature on microstructure and property of a 6061 aluminum alloy, *Transactions of Materials and Heat Treatment* **34**, 31 (2013).
6. R. P. Garrett, J. Lin and T. A. Dean, An investigation of the effects of solution heat treatment on mechanical properties for AA 6xxx alloys: experimentation and modelling, *International Journal of Plasticity* **21**, 1640 (2005).

Effects of Different Quenching Conditions on the Mechanical Properties of 6061 Aluminum Alloy in Hot Stamping

Y. Liu, J. Ren, Y. M. Jin, Y. S. Zhang and B. Zhu[†]
State Key Laboratory of Materials Processing and Die & Mould Technology
Huazhong University of Science and Technology
Wuhan, 430074, P.R. China
[†]E-mail:zhubin26@hust.edu.cn

Due to the demand for reduction in weight and energy saving, the automation in the manufacturing of aluminum alloy sheets has been gathering pace. The effects of different holding pressures and holding time on the mechanical properties of 6061-T6 aluminum alloy during quenching by cold dies were investigated. The mechanical properties of 6061-T6 aluminum alloy quenched by cold upper die and hot lower die with different temperature and holding time at a certain pressure were also studied to explore the feasibility of pre-ageing with hot die in hot stamping to reduce the time of aging. The results showed that higher holding pressure and a holding time of 6-10 s were recommended while quenching by cold dies, so that higher UTS and hardness could be obtained. With the increase of hot die's temperature from 100°C to 200°C and holding time, the UTS and hardness increased, and higher UTS and hardness than the specimen quenched by water can be obtained. As a result, we demonstrated that the feasibility and possibility of pre-ageing with hot die to reduce the aging time.

Keywords: 6061 aluminum alloys; Hot stamping; Holding pressure; Holding time; Pre-ageing.

1. Introduction

With the demand for weight loss and energy saving, the need to manufacture automobile structure components from high performance sheet metal is apparent. The mass of a car's body structure can be reduced by over 40% by using high performance sheet metal such as aluminum alloys [1]. In order to improve the formability, overcome the spring back of the part and avoid thermal distortion of formed part during heat treatment, hot forming-quenching integrated process (so-called hot stamping of aluminum) has been developed recently for heat-treatable aluminum alloy forming, which was proposed firstly by Lin et al. [2]. Quenching and aging are important procedures during hot stamping for heat-treatable aluminum alloy and suitable process need to be developed so the effects of different quenching conditions on the mechanical properties of aluminum alloy should be focused on. O. K. Abubakre et al. gave a study on the effects of different quenching media on the mechanical properties of 6061aluminum alloy [3]. Literatures [4-5] made efforts on quench sensitivity of aluminum alloy and showed that the cooling rate during quenching should be enhanced as high as possible in the quenching sensitive temperature range to suppress the heterogeneous precipitation to get optimal mechanical properties. In N. Li's work [6], a cold die was utilized to quench. Few researchers concentrated on the quenching conditions in die during hot stamping process. As we know, aging takes too much time and one reason may be that it takes considerable time to heat the aluminum alloy to a certain temperature especially the temperature is relatively low. As a result, the effects of different holding pressures and holding time on the mechanical properties of 6061-T6 aluminum alloy during quenching by cold dies were investigated and the mechanical properties of 6061-T6 aluminum alloy quenched by cold upper die and hot lower die with different temperature and holding time at a certain pressure were also studied to explore the feasibility of pre-ageing with hot die in hot stamping to reduce the time of aging.

2. Experiment

Main chemical compositions of the as-received 6061-T6 aluminum alloy used for the investigation are shown in Table 1. The original Vickers hardness was 112.3 HV and the ultimate tensile strength (UTS) was 337 MPa. Specimens were cut to dimensions of approximately 260 mm×170 mm×1.6 mm. Flat dies with heating control system and a servo press was used in the investigation. The servo press can adjust the holding pressure and holding time while the flat dies can be heated to different temperatures so different quenching conditions were brought in. Heat treatments were conducted using laboratory chamber furnaces.

Table 1. Main chemical compositions of the as-received 6061-T6 aluminum alloy [wt. %].

Chemical elements	Mg	Si	Mn	Cu	Fe	Cr	Ni	Zn	Ti	Al
Contents	1.1036	0.676	0.078	0.1495	0.4227	0.1647	0.0058	0.21	0.0273	Rem

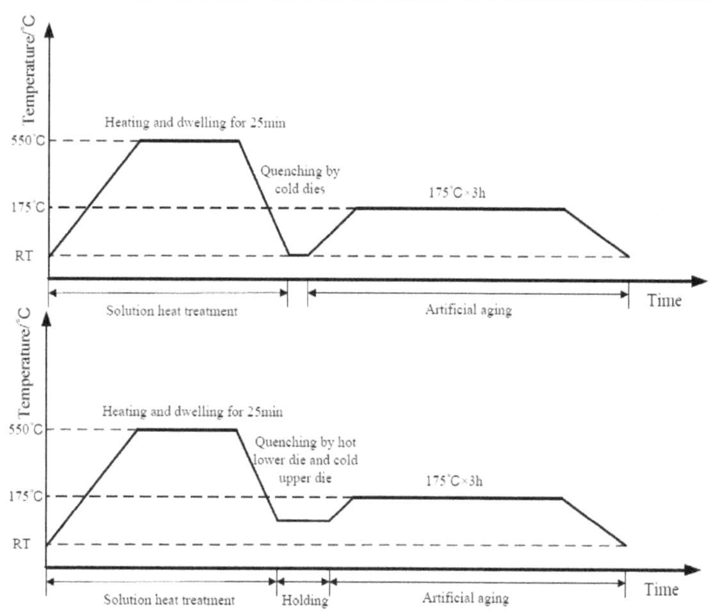

Fig. 1. The designed procedure of heat treatments for the HFQ and artificial aging process.

The designed procedure of heat treatments for the HFQ and artificial aging process is schematically illustrated in Figure 1 and specimens with different quenching conditions are shown in Table 2. For the investigation of quenching conditions, three groups of tests were carried out: (A) Specimens were quenched by cold flat dies with different holding pressures and holding time; (B) Specimens were quenched by cold upper die and hot lower die with different temperature and holding time at a certain pressure of 20MPa; (C) Specimens were water quenched and its specimen number was C. All specimens were heated to SHT temperature 550°C and soak for 25 min, and then they were aged under the condition of 175°C × 3 hours. A K-type thermocouple was implanted into the specimens and the temperature history during quenching process was recorded. Vickers hardness (HV) of specimens was tested with 9.8 Kg loading force and tensile tests were also carried out.

Table 2. Specimen No. and it's quenching condition.

Specimen No.	Holding pressure/MPa	Holding time/s	Temperature of lower die/°C
A1	5	5	RT
A2	5	10	RT
A3	10	5	RT
A4	10	10	RT
A5	15	5	RT
A6	15	10	RT
A7	20	5	RT
A8	20	10	RT
B1	20	5	100
B2	20	10	100
B3	20	5	150
B4	20	10	150
B5	20	5	200
B6	20	10	200

3. Results and Discussion

3.1. *Hardness and tensile tests*

Figure 2 shows the UTS and hardness of specimens. Fig. 2(a) shows the UTS and hardness of specimens quenched by cold flat dies with different holding pressures and holding time while Fig. 2(b) shows the UTS and hardness of specimens quenched by cold upper die and hot lower die with different temperature and holding time at a certain pressure of 20MPa. As shown in Fig. 2(a), generally the UTS and hardness of specimens with a holding time of 10 s are lager than that of specimens with a holding time of 5 s when they have the same holding pressure and we got higher UTS and hardness when the specimen have higher holding pressure and longer holding time. Of course we don't need much holding time with cold dies and 10 s is enough for the consideration of production efficiency. As for Fig. 2(b), it has a similar conclusion that the UTS and hardness of specimens with a holding time of 10 s are lager than that of specimens with a holding time of 5 s when specimens have the same holding temperature. Otherwise some specimens have higher UTS and hardness than specimen quenched by water and almost all specimens have mechanical properties close to that of specimen quenched by water.

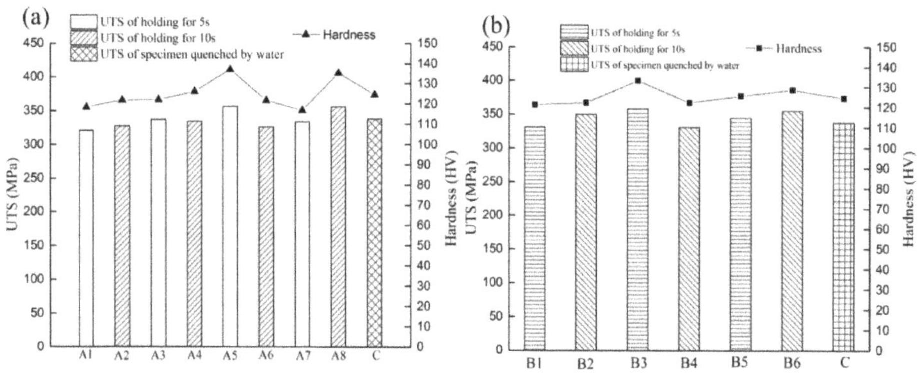

Fig. 2. (a) UTS and hardness of specimens quenched by cold flat dies with different holding pressures and holding time; (b) UTS and hardness of specimens quenched by cold upper die and hot lower die with different temperature and holding time at a certain pressure of 20 MPa.

3.2. *Cooling curve and TTP diagram*

Figure 3 shows the cooling curves of specimens A8, B2, B4 and B6 in Time Temperature Precipitation (TTP) diagram of the 6061 aluminum alloy [7]. The cooling curves are solid line and the dotted lines are extending lines or reference lines. In fact, when the cooling curve reaches to the end, they dropped to the room temperature because the specimens were not on the hot die anymore. Extending lines mean the phase transformation if the certain temperature was kept. In the HFQ process, the precipitates (in any phase) are dissolved and grow to the required size (β'' phase) during the ageing process. A typical ageing process for 6061 involves heating and dwelling the material, so as to provide sufficient energy to allow precipitates to be evenly distributed throughout the material. On a microstructural level, this allows the material to pass through several stages, in particular: α (SSSS) $\rightarrow$co-clusters $\rightarrow$GP zones (I) $\rightarrow$GP zones (II) /β'', which increases the strength of the material. However, ageing for a longer time will pass the material through the $\beta'' \rightarrow \beta' \rightarrow \beta$ phases which would induce over-age and reduction in strength [8]. For specimens A8, B2 and B4, their temperatures are below GP zone solvus and it generates fine dispersed high fraction of GP zones; for specimen B6, because its temperature dropped to the room temperature and its real cooling curve still passed through the GP zones. As a result, when they were heated to 175°C and dwell for 3 hours (about 10000 s), the phase transformation of GP $\rightarrow$ β'' took place and the required β'' phase generated, leading to a high UTS and hardness. As we can see from the TTP diagram, high aging temperature and rapid heating during aging process takes less time to get the required phase transformation. Combined with the mechanical properties of specimens quenched by cold upper die and hot lower die, there exists the feasibility of pre-ageing with hot die to reduce the aging time, making hot stamping of aluminum alloy with high efficiency.

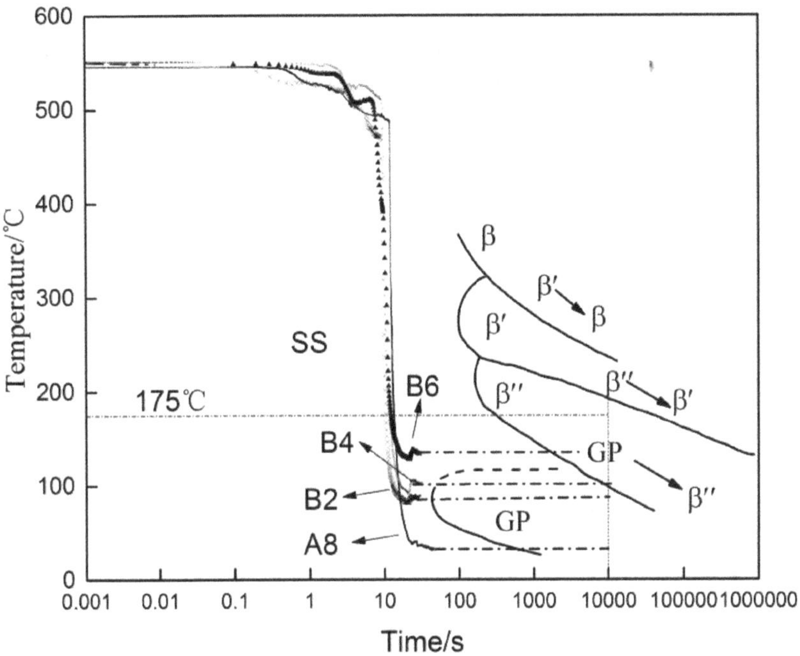

Fig. 3. Cooling curves of specimens A8, B2, B4 and B6in TTP diagram of the 6061 aluminum alloy.

4. Conclusion

The effects of different holding pressures and holding time on the mechanical properties of 6061-T6 aluminum alloy during quenching by cold dies were investigated and the mechanical properties of 6061-T6 aluminum alloy quenched by cold upper die and hot lower die with different temperature and holding time at a certain pressure were also studied to explore the feasibility of pre-ageing with hot die in hot stamping to reduce the time of aging. The results showed that higher holding pressure and a holding time of 6-10 s were recommended while quenching by cold dies so higher UTS and hardness can be obtained; with the increase of hot die's temperature from 100°C to 200°C and holding time, the UTS and hardness increased and higher UTS and hardness than specimen quenched by water can be obtained. As a consequence, there exists the feasibility of pre-ageing with hot die to reduce the aging time, making hot stamping of aluminum alloy with high efficiency.

Acknowledgments

This research work is funded by the National Natural Science Foundation of China (grant No. U1564203) and National Natural Science Foundation of China (grant no. 51405171).

References

1. E. Fakir, L. L. Wang, D. Balint, J. P. Dear, J. G. Lin and T. A. Dean, Numerical study of the solution heat treatment, forming, and in-die quenching (HFQ) process on AA5754, *Int. Mach. Tool. Manu.***87**, 39 (2014).
2. R. P. Garrett, J. Lin and T.A. Dean, Solution heat treatment and cold die quenching in forming AA6xxx sheet components: feasibility study, *Adv. Mater. Res.* **673**, 6 (2005).
3. K. Abubakre, U. P. Mamaki and R. A Muriana, Investigation of the Quenching Properties of Selected Media on 6061 Aluminum Alloy, *J. Miner. Mater. Charact. Eng.* **8**, 303 (2009).
4. S. C. Bergsma, M. E. Kassner, X. Li and R. S. Rosen, The Quench Sensitivity of Hot Extruded 6061-T6 and 6069-T6 Aluminum Alloys, in *Proc. 3rd International Conference on Processing and Manufacturing Advanced Materials*, (Las Vegas, American, 2000).
5. H. Y. LI, C. T. Zeng, M. S. Han, J. J. Liu and X. C. Lu, Time–temperature–property curves for quench sensitivity of 6063 aluminum alloy, *T. Nonferr. Metal. SOC.* **23**, 38 (2013).
6. N. Li, M. S. Mohamed, J. Cai, J. Lin, D. Balint, and T. A. Dean, Experimental and Numerical Studies on the Formability of Materials in Hot Stamping and Cold Die Quenching Processes, in *AIP Conference Proceedings* **1353**, 1555 (2011).
7. D. Dafir, R. Saadani and A. Jamil, Analysis of Thermoelectric Power TTP Diagrams of Aluminum Alloy 6061 Produced by Two Different Techniques, *Int. J. Res. Rev. Mech. Des. Simul.* **2**, 86 (2012).
8. N. Li, J. H. Zheng, C. Zhang, K. L. Zheng, J. G. Lin and T. A. Dean, Investigation on Fast and Energy-Efficient Heat Treatments of AA6082, in *HFQ processes for Automotive Applications* (*MATEC Web of Conferences*. **21**, 05015 (2015)).

Microstructure and Mechanical Properties of Aluminum 7050 Alloy Sheet in Hot Stamping-Quenching Integrated Process with Cold Dies

Y. Chen[1], Y. Liu[2], C. W. Tian[1], M. H. Cai[1], H. Ding[1,†] and Y. S. Zhang[2]

[1]*School of Materials Science and Engineering, Northeastern University, Shenyang, 110819, P. R. China*

[2]*State Key Laboratory of Material Processing and Die & Mould Technology, Huazhong University of Science and Technology, Wuhan, 430074, P. R. China*

†*E-mail: hding2013@163.com*

In this study, cold dies were proposed to use in hot stamping–quenching integrated process which aimed to improve the formability and avoid thermal distortion for a heat-treatable Aluminum 7050 alloy during hot-stamping. During this hot stamping–quenching integrated process, influences of initial base metal temper and heat treatment schedule after hot stamping on the microstructure and mechanical properties were investigated systematically. Two initial base metal tempers were applied in AA7050 alloy, over-aged T7 and natural-aged T4. The temperature of hot stamping was 450°C, meanwhile, the strain rate was $1\times10^{-1}s^{-1}$. Both traditional and novel heat treatment schedules after hot-stamping were used in order to discuss the effects of heat treatment processing parameters on the microstructure and mechanical properties. The microstructure was analyzed by optical microscope (OM) and transmission electron microscope (TEM), and the mechanical properties were measured by Vickers hardness and tensile tests. The correlation between the schedules, microstructure and mechanical properties of the Al alloys were discussed as well.

Keywords: Hot stamping; Aluminum alloy; Base metal temper; Heat treatment.

1. Introduction

Aluminum alloys are widely used to meet the need for lightweight structural metals in automotive and aerospace in industries because of their benefits such as good corrosion resistance, high strength-to-weight ratio and advantageous structural performance [1, 2]. However, further application is limited by their poor formability and high spring-back at room temperature. The tricky problem can be solved by forming at elevated temperature (namely, hot forming). Deformation resistance decreases and formability increases with forming temperature increasing. Meanwhile, spring-back would be reduced significantly while forming at high temperature [3].

Although hot forming can increase formability and reduce spring-back, it may lead to softening (recovery and recrystallization) and destroy the desirable microstructure. The following heat treatment is required to restore it and improve its strength, which often results in thermal distortion of the formed parts during quenching. To solve the above problems, a hot forming-quenching integrated process called Solution heat treatment, Forming and cold-die quenching (HFQ), for producing high strength and high precision Al-alloy sheet parts, has been developed by Lin [3]. Heated sheet after solution heat treatment was formed within cold-dies, whose keys are combining the quenching and hot forming in one operation. Cooling within cold-dies after hot forming can reduce the effect of thermal stress on shape of formed parts, especially those with complex and non-closed sections. However, artificial aging heat treatment is always needed after HFQ in order to improve the mechanical properties of the formed sheets. The traditional schedule of artificial aging is long (measured in hours) and not suitable for industry, therefore, novel schedule of artificial aging needs to be proposed to decrease the processing period. In this study, both a traditional and a novel heat treatment schedule after hot stamping were used in order to discuss the effects of heat treatment processing parameters.

AA7050 is a precipitation-hardened Al-Zn-Mg-Cu aluminum alloy [4]. It has been widely applied to produce aircraft parts due to its light weight and high strength. Over the years, different aging tempers, such as T7 and T4, have been developed for this kind of alloys. Each aging temper provides the alloy with different performances. A high tensile strength can be obtained by over aged T7 temper and the formability after natural aging T4 temper is excellent. Because of the different performances of AA7050 under T7 and T4 tempers, the hot stamping behavior on this alloy can differ. Hence it becomes necessary to conduct a research on the effect of the initial base metal temper on the microstructure and the mechanical properties of hot-stamped AA7050. In the present work, influences of initial base metal temper and heat treatment schedule after hot stamping on the microstructure and mechanical properties of AA7050 alloy were investigated systematically.

2. Experimental

Rolled AA7050 aluminum alloy sheets of 2 mm were applied in this study, and the nominal chemical compositions of this alloy are listed in Table 1. Two heat treatments (T7 and T4) were adopted for AA7050 plates before hot-stamping. Over aged T7-temper involved a combination of solution treatment at 480 °C for 80 min followed by water quenching and subsequent artificial aging at 165 °C for 12 h [5]. Natural aged T4-temper was conducted by solution treatment at 480 °C, soaking for 80 min followed by air cooling and aging at ambient temperature for 20 days. The plates were processed by HFQ and the detailed process schedule is shown in Fig. 1.

Table 1. Nominal chemical compositions (wt.%) of the studied AA7050 aluminum alloys.

Alloy	Mg	Zn	Cu	Mn	Fe	Cr	Al
AA7050	2.22	5.82	2.14	<0.3	<0.1	0.1-0.25	Bal

Heated sheet after solution heat treatment (480 °C) was transferred and located to the cold-dies quickly, hereafter stamped into required shape (U-shape in this work), then held for 5s within cold-dies to reduce the temperature rapidly to the ambient temperature. Finally, two kind of artificial aging was applied on the hot-stamped sheets in order to discuss the effects of heat treatment processing parameters. The traditional artificial aging was conducted at 165 °C for 6 h while the novel one was at 190 °C for only 30 min. The designated of samples produced with different process parameters in brief form was listed in Table 2.

Table 2. Experimental process parameters combination.

Samples	Designation in brief	Hot-stamping	Artificial aging
AA7050-T4 before Hot-stamping	T4-BM	No	
AA7050-T7 before Hot-stamping	T7-BM	No	
AA7050-T4 after Hot-stamping	T4-SH-T	Yes	Traditional schedule
	T4-SH-N	Yes	Novel schedule
AA7050-T7 after Hot-stamping	T7-SH-T	Yes	Traditional schedule
	T7-SH-N	Yes	Novel schedule

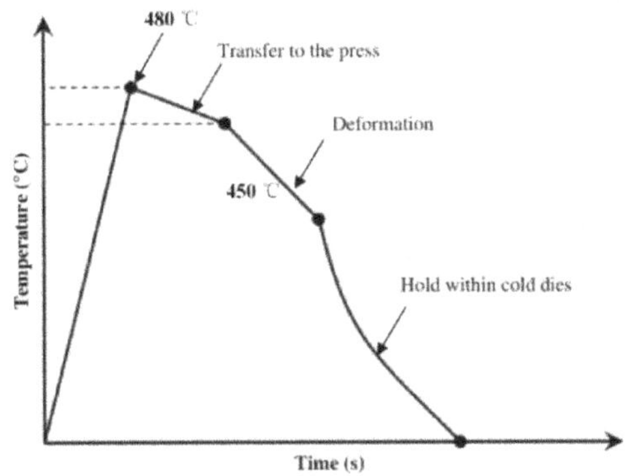

Fig. 1. A schematic diagram shows the thermal history for hot-stamping of present work.

Microstructure of the hot-stamped sample was characterized using an Olympus DSX-500 optical microscopy (OM) and transmission electron microscopy (TEM). The samples for OM were ground and polished and then etched using Keller's reagent (190 ml water, 2 ml hydrofluoric acid, 3 ml hydrochloric acid and 5 ml nitric acid). Thin foils for TEM observation were prepared by jet electro-polishing with a solution of 70% methanol and 30% nitric acid at -30 °C and 19 V. The TEM study was performed on a Tecnai G220 TEM. The mechanical properties of the hot-stamped samples were evaluated by Vickers hardness and tensile tests. The Vickers hardness measurement was carried out under a load of 50 g for 10 s. Room-temperature tensile tests were conducted on a universal testing machine with strain rate of 10^{-3} s^{-1}. The tensile specimens were cut with a gauge length of 20 mm and width of 4 mm, and the thickness of the cross-section is 1.6 mm.

3. Results and Discussion

The microstructure of the base metal under different heat treatment conditions before hot-stamping are shown in Fig. 2. The difference in the morphology of grains is insignificant between T7-BM and T4-BM, where all the microstructure primarily consists of elongated grains due to the rolling (Fig. 2a and b). Differently, the morphology of the precipitates in the base metal under different tempers is much more significant. Under T7 temper (Fig. 2c), a high density of particles on the order of 10-20 nm disperse uniformly throughout the matrix. The selected area diffraction pattern (SADP) indicates the fine particles are the strengthening precipitates, η (η')-$MgZn_2$ phase. The fine η (η') forms during artificial aging. It was reported that the η'-phase was described as a slightly strained version of the η-phase [6]. Therefore, the two phases of η and η' will not be distinguished in the present study. Unlike the T7 temper, only a few coarse Fe-Mn particles are observed in the matrix under T4 temper, as shown in Fig. 2d. With a T4 heat treatment, the strengthening precipitates firstly dissolve during solution heat treatment and then the Guinier-Preston (GP) zones precipitate at room temperature [7-9], however, GP zones are hard to discern here due to their fine features and coherent relationship to the matrix.

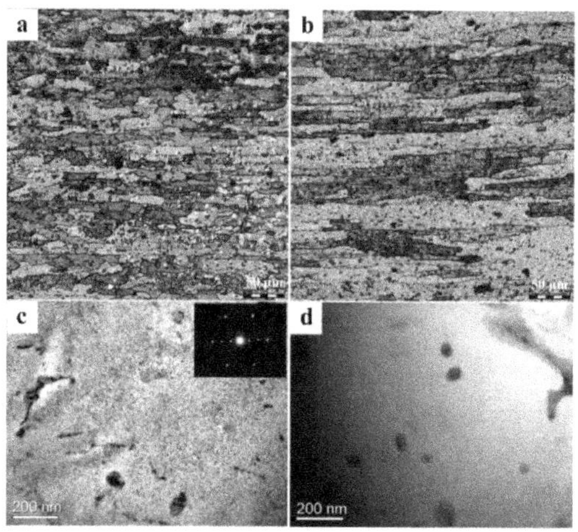

Fig. 2. The microstructure before hot-stamping of T4-BM and T7-BM samples analyzed by OM (a)-(b) and TEM (c)-(d).

After hot-stamping with artificial aging, the morphology of grains keeps similar, while the distribution of the strengthening precipitates varies obviously. As shown in Fig. 3, a great number of fine η (η') reformed during the artificial aging regardless of the initial base metal temper, and the amount of the strengthening precipitates is much higher than that in the base metal before hot-stamping. It seems that the effect of the initial base metal temper on evolution of the strengthening precipitates during hot forming is negative, where the distribution and morphology of the precipitates are similar among all hot-stamped samples. It indicates that a 30 min solution heat treatment before hot-stamping is long enough for the dissolution of the precipitates, decreasing the difference in microstructure between T4 and T7 temper. Besides, similar to the traditional artificial aging schedule, the novel artificial aging schedules also induce the reformation of precipitates, which is mainly because the reforming kinetics of the precipitates get higher when higher artificial aging temperature used, and enough amount of precipitates can be reformed within a relatively short period.

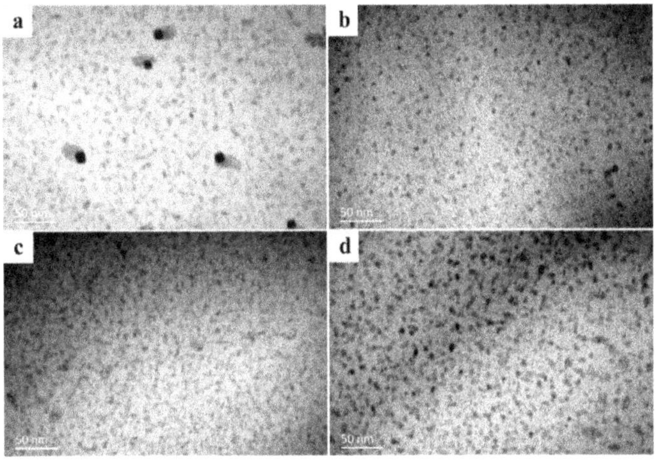

Fig. 3. TEM micrographs of the strengthening precipitates in the T4-SH (a) -T, (b) -N and T7-SH (c) -T, (d) –N.

The hardness profiles measured are presented in Fig. 4, it can be seen that the hardness of T4-BM sample is lower than that of T7-BM, which is due to existence of the fine η (η′) in T7 sample. The hardness of the hot-stamped samples all increases compared with that of the base metal. For precipitation-hardened aluminum alloys [10], fine and great amount of precipitates is beneficial to strengthen the materials. And the increment in the hardness of all hot-stamped samples is mainly attributed to these reformed strengthening precipitates during artificial aging. Furthermore, it can be seen that the hardness of hot-stamped samples using novel artificial aging schedule is a bit lower than that using traditional artificial aging schedule.

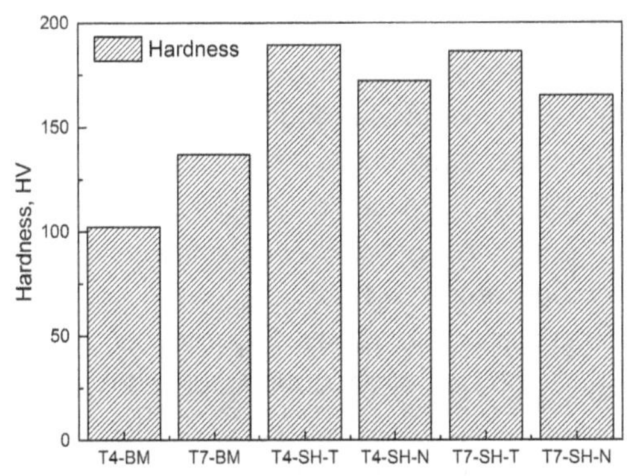

Fig. 4. Hardness distribution of AA7050 using different processing parameters.

The results of the tensile tests are shown in Fig. 5, and the tensile properties are consistent with the results of hardness. The tensile strength of all samples gets increased after hot-stamping with artificial aging. Although the mechanical properties using novel schedule are a bit lower than that using traditional schedule, the ultimate tensile properties can be also achieve to the 500 MPa level.

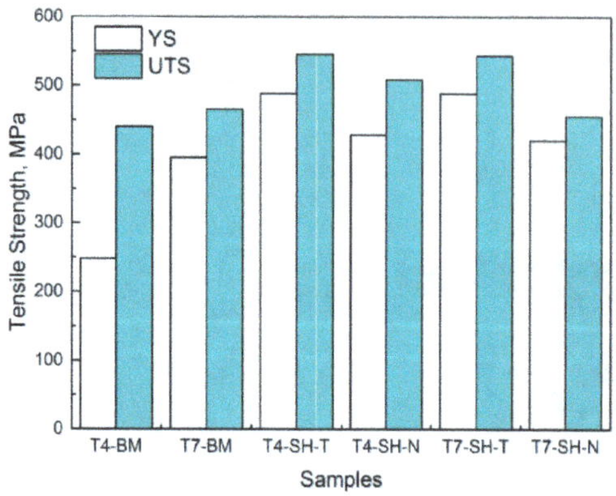

Fig. 5. Tensile strength of AA7050 using different processing parameters.

4. Conclusion

In summary, a 30 min solution heat treatment was long enough for the dissolution of the strengthening precipitates regardless of the initial base metal tempers, which weakened the effect of initial base metal temper on the microstructure and mechanical properties of AA7050 during hot-stamping. A novel artificial aging schedule which used higher temperature with shorter heating period was proposed, and the mechanical properties using the novel artificial aging schedule was a bit lower than those using traditional schedule, however, the ultimate tensile strength using novel schedule could also attain nearly 500 MPa.

References

1. X. Fan, Z. He, S. Yuan and P. Lin, Investigation on strengthening of 6A02 aluminum alloy sheet in hot forming-quenching integrated process with warm forming-dies, *Mater. Sci. Eng. A* **587**, 221 (2013).
2. Y. Chen, H. Ding, J. Li, Z. Cai, J. Zhao and W. Yang, Influence of multi-pass friction stir processing on the microstructure and mechanical properties of Al-5083 alloy, *Mater. Sci. Eng. A* **650**, 281 (2016).
3. X. Fan, Z. He, S. Yuan and K. Zheng, Experimental investigation on hot forming-quenching integrated process of 6A02 aluminum alloy sheet, *Mater. Sci. Eng. A* **587**, 154 (2013).
4. R. Brown, W. Tang and A.P. Reynold, Multi-pass friction stir welding in alloy 7050-T7451: Effects on weld response variables and on weld properties, *Mater. Sci. Eng. A* **115**, 513 (2009).

5. Y. Chen, H. Ding, Z. Cai, J. Zhao and J. Li, Effect of initial base metal temper on microstructure and mechanical properties of friction stir processed Al-7B04 alloy, *Mater. Sci. Eng. A* **650**, 396 (2016).
6. J.Q. Su, T.W. Nelson, R. Mishra and M. Mahoney, Microstructural investigation of friction stir welded 7050-T651 aluminum, *Acta. Mater.* **51**, 713 (2003).
7. K. Stiller, P.J. Warren, V. Hansen, J. Angenete and J. Gonnes, Investigation of precipitation in an Al-Zn-Mg alloy after two-step ageing treatment at 100 and 150 °C, *Mater. Sci. Eng. A* **270**, 55 (1999).
8. X.J. Jiang, J. Tafto, B. Noble, B. Holme and G. Waterloo, Differential scanning calorimetry and eletron diffraction investigation on low-temperature aging in Al-Zn-Mg alloys, *Metall. Mater. Trans. A* **339**, 31 (2000).
9. C.B. Fuller, M.W. Mahoney, M. Calabrese and L. Micona, Evolution of microstructure and mechanical properties in naturally aged 7050 and 7075 Al friction stir welds, *Mater. Sci. Eng. A* **527**, 2233 (2010).
10. Y. Chen, H. Ding, J. Li, J. Zhao, M. Fu and X. Li, Effect of welding heat input and post-welded heat treatment on hardness of stir zone for friction stir-welded 2024-T3 aluminum alloy, *T. Nonferr. Metal. Soc.* **25**, 2524 (2015).

Research of the Mechanical Performance After Galvanization of Innovative Multilayer Electric Power Fitting

Y. M. Jin, Y. P. Zheng, Z. J. Zhu and Y. S. Zhang†

State Key Laboratory of Materials Processing and Die and Mould Technology, Huazhong University of Science & Technology, Wuhan 430074, China

†E-mail: zhangys@mail.hust.edu.cn

Hot stamping ultra-high strength steel not only plays an important role in automotive field, but also has potential applications in the transmission line fittings industry. Combined with hot stamping technology, an innovative multilayer electric power fitting has been proposed. This paper offers a presentation of the structure of this multilayer electric power fitting, and then carries out the forming experiments of the multilayer stack structure on the independent development hot stamping mold. Considering the actual application status, mechanical performance after galvanization of the innovative power fitting was also investigated.

Keywords: Hot stamping; Electrical power fitting; Multilayer; Galvanization.

1. Introduction

Transmission line fittings are important parts which related to the safe operation of overhead transmission line. Among the many power fitting products, suspending and connection fittings are widely used. However, the traditional fitting products are mainly produced by casting or machining a whole piece of thick plate, which have the disadvantage of heavy weight and much metal consumption. So the demand for developing new material, improving the tensile strength and yield strength of the material, and reducing weight in powering fittings has become increasingly prominent. As a result, the traditional Clevis-Clevis body is replaced by multilayer stack of ultra-high strength boron steel sheet, which produced by hot stamping to achieve the purpose of realization of lightweight electrical fittings [1].

Based on the typical electric connection fitting Clevis-Clevis, a new method for reduction of weight in electrical power fittings has been proposed under the premise of ensuring strength [2, 3]. Presentation of the structure of this multilayer electric power fitting was made, and then the forming experiments of the multilayer stack structure were carried out on the independent development hot stamping mold. Considering the actual application status, mechanical performance after galvanization of the innovative power fitting was also investigated.

2. Structure Presentation

Clevis-Clevis is one kind of connection fittings used to change directions in overhead transmission line. Traditional large-tonnage level Clevis-Clevis, as shown in Fig. 1 is usually manufactured by casting or forging process, with a wall thickness of up to about 15-20mm and a nominal weight of 7.0Kg. Although the mechanical strength meets the application requirement, the component is thick and heavy, with low material utilization. It will increase the weight load of transmission line tower due to its large quantity use.

In Fig. 2, it is the improved structure of a new type Clevis-Clevis with high-performance boron steel. The fitting body is composed of multilayer U-shaped stack steel sheets of ultra-high strength boron steel manufactured by hot bending process. The thickness of each layer is 1.8 mm, spot welding, riveting and other separate or composite connections are applied among the layers to achieve a rigid and stable connection. The new type Clevis-Clevis of multilayer stacked structure has better improved mechanical property as well as light weight (with a nominal weight about 2.7Kg), attribute to the characteristics of hardened boron steel with high yield and tensile strength.

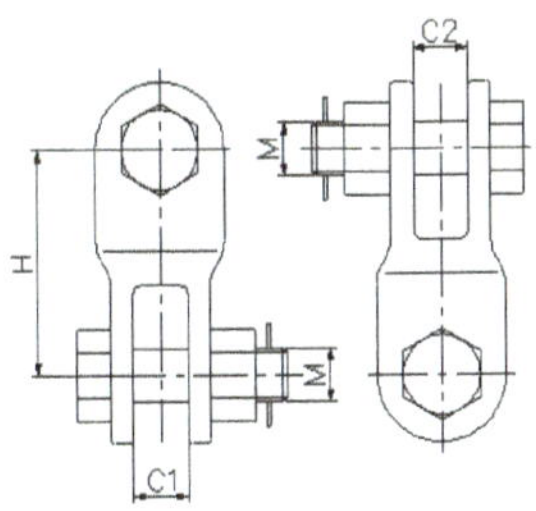

Fig. 1. The traditional Clevis-Clevis.

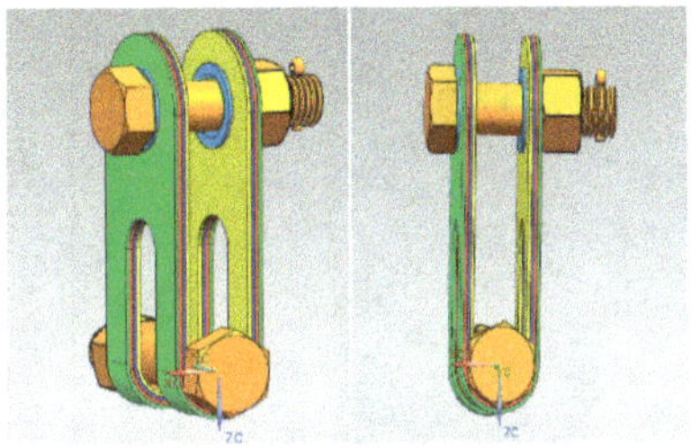

Fig. 2. Multilayer stack structure of UHSS.

3. Multilayer Stack Sheets Forming Experiments

Fig. 3 is the multilayer stack hot bending mold and equipment, including heating equipment, the servo mechanical press and the mold. The 1.8mm high strength boron steel sheet with the size of 310 mm×70 mm is used during the experiment, the chemical composition of the steel can be found in Table 1. The double plates will be heated to 930°C firstly and then placed in the mold after 5-6 minutes holding, with the servo mechanical press stamping with 1820KN and holding for 20s, the products are taken out [4].

We choose the hardness analysis method to testing the mechanical properties of the part, and three areas are selected to get the hardness specimens: A is lateral wall area, B represents transition region; and C is the bottom zone. Fig. 4 shows that in spite of different area, the hardness of part is concentrated in 500HV, we can draw the conclusion that each layer of multilayer stack plates has realized almost full martensite phase, and the mechanical properties of different positions are close to homogeneous.

Table 1. Chemical composition of the 22MnB5 substrate in wt.%.

C	Si	Mn	P	S	Cr	Ni	Mo	Ti	B
0.23	0.17	0.95	0.01	0.002	0.24	1.50	0.21	0.023	0.019

Fig. 3. Multilayer stack hot bending mould and equipment.

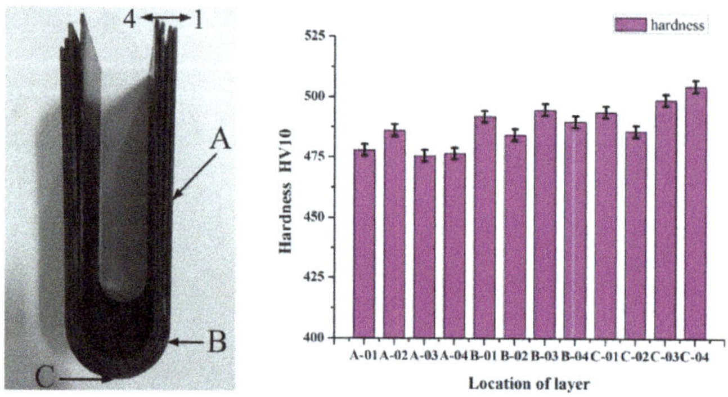

Fig. 4. Hardness results of the hot stamping multilayer stacks.

4. Galvanization Experiments and Product Performance Test

4.1. *Galvanization experiments*

The surface of the Clevis-Clevis must be galvanized during the actual application status, because the naked steel structures in the air will rust soon, leaving the life of power fitting decreased greatly due to the corroding. At present the main method of corrosion prevention is hot galvanization, the specific process is shown in Fig. 5.

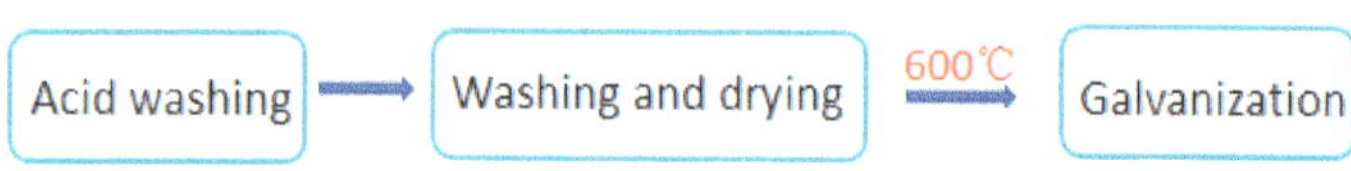

Fig. 5. Specific process for hot galvanization.

For large tonnage level Clevis-Clevis, when it ruptures under overload condition, we hope the failure mode is ductile fracture rather than brittle fracture. In this case, the catastrophic accident can be avoided even the part has a local fracture, which requires the part has a good plasticity and toughness. The strength of the high strength steel laminate decreases and the elongation increases due to the tempering effect of galvanization, which improving the overall performance of the part. Therefore, the further research for mechanical properties of high strength steel after galvanization is conducted.

Firstly, the tensile specimens were cut down from the side wall on the area A in Fig. 4 in each plate, the size of the specimen is shown in Fig. 6. Then tensile experiments were conducted on SHIMADZU AG -100KN material properties testing machine. Tensile strength properties and elongation data are collected during the process.

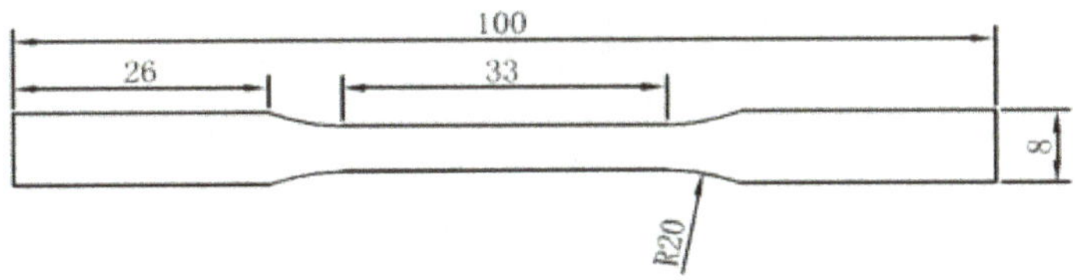

Fig. 6. Size of the specimen.

Fig. 7 has shown tensile strength and elongation change of the samples which are cut down from same position of the plates 1, 2, 3 and 4 before and after the hot-dip galvanizing. As can be seen from the figure 7(a), the intensity distribution between the high strength steel laminates is uniform before galvanizing, and the average tensile strength is 1568MPa. However, after hot-dip galvanizing, because of the softening caused by the tempering, the average tensile strength decreased to 992MPa, about 63% of the initial tensile strength. From the perspective of elongation (figure 7(b)), tempering during hot-dip galvanizing makes the average elongation increasing from 7.43% to 9.88%. In a word, the mechanical properties of high strength steel parts had a reduction after galvanization, however, the process improved plasticity properties, which made the parts have a better overall performance.

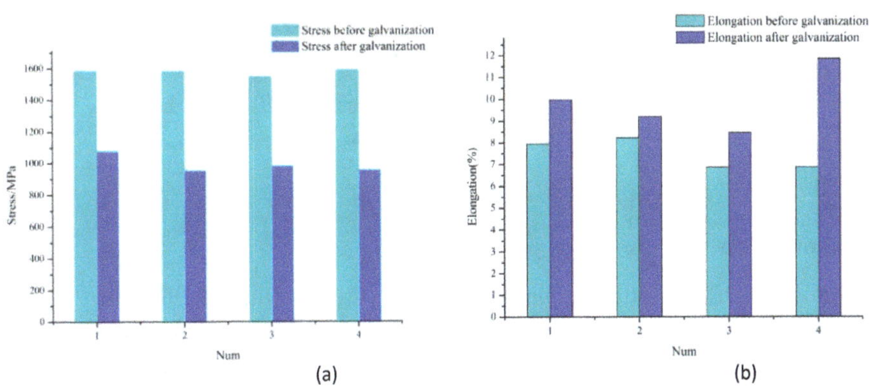

Fig. 7. Mechanical properties comparison before and after galvanization (a) tensile strength (b) Elongation.

4.2. *Product performance test*

In order to test whether the mechanical properties of the innovative multilayer electrical power fitting after galvanization can meet requirements of national standards, which qualify power fitting for using when it is able to withstand 1.2 times nominal load without damage, we need to test the mechanical properties of the new type Clevis-Clevis under 240KN load. The test was conducted in an electrical power fittings company in south of China, and the nominal load of the test equipment is 1000KN. The Product performance test process is shown in Fig. 8.

Fig. 8. Product performance test process.

Finally, test results showed that when the load reached 240KN even 280KN, this new type of Clevis-Clevis had no fracture failure, and showed excellent mechanical properties. As shown in Fig. 9.

Fig. 9. Product performance test result.

5. Conclusion

1) An innovative multilayer electrical power fitting structure combined with hot stamping technology was proposed, which has better improved mechanical property as well as light weight attribute to the characteristics of hardened boron steel with high yield and tensile strength.
2) Multilayer stack sheets forming experiments were carried out, we have gained the multilayer stack part with homogeneous mechanical properties on the independent development hot bending mold.
3) Considering the actual application status, mechanical performance after galvanization of the innovative power fitting was also investigated, the results show that: the mechanical properties of high strength steel parts had a reduction after

galvanization, however, the process improved plasticity properties, which made the parts have a better overall performance.

References

1. M.L. Zhou, L. Wang, Z.J. Wang and Y.S. Zhang, Analysis of the design and mechanical performance about hot stamping power fitting products, *Adv. Mater. Res.* **1063**, 272 (2014).
2. C. Nan and T. Li, Development of High Strength Hardware, *Electric Power Construction* 48 (2001).
3. J. Dong, Electrical fittings Manual, *China Electric Power Press*, (Beijing, 2001).
4. Y.M. Jin, Z.J. Wang and Y.S. Zhang, Investigate of the Mechanical and Heat Transfer Performance of High Strength Steel Hot Stamping Power Fitting Products, in *advanced high strength steel and press hardening: proceedings of the 2nd international conference (ICHSU2015)*, (Changsha, China, 2015).

Continuous Cooling Transformation of 1700 MPa Grade Hot Stamping Steel

W. Y. Liu[†], K. H. Hu, J. P. Zheng and R. D. Han

Research and Development Center, WISCO,
Wuhan, 430080, P.R.China
[†]*E-mail: liuwenyan2006@aliyun.com*

Experiments were carried out upon the continuous cooling transformation (CCT) of 1700 MPa grade hot stamping steel by Formastor-F dilatometer. Microstructures and Vickers hardness were analyzed and a CCT diagram was drawn. The CCT diagram shows that when the cooling rate is more than 5°C /s, martensite is obtained and Vickers hardness is 529~546 HV5, which displays that the test steel has a good hardenability. Bainite is gotten with the cooling rate below 5°C /s. Pearlite appears when the cooling rate is lower than 2°C /s and ferrite begins to form while the cooling rate is under 1°C /s. The CCT diagram of 1700 MPa grade hot stamping steel has been right shift compared with the 1500 MPa grade due to the addition of 0.03% carbon and 0.3% manganese which increase hardenability of steels and also have good effect to improve martensite hardening capacity and then increase the strength of steel. The 1700 MPa grade steel has the broader cooling rate range from 10°C /s 50°C /s to obtain a full martensite.

Keywords: 1700 MPa; Hot stamping steel; Martensite; Vickers hardness.

1. Introduction

To meet automotive lightweight and enhance crash safety of vehicles, high and ultra-high strength steels with thinner gauge are drastically prone to be used in automotive industry. In the last decade, the development of hot stamping steel and hot stamping process has made the application come true. The heated blanks are immediately quenched in dies during stamping and martensite transformation from austenite is obtained to increase the strength of steel drastically. Hot stamping process has many advantages such as low forming force, small spring back and shape distortion, better formability, and moderate tool wear in comparison with the traditional forming processes at room temperature for high strength steels which are easy to cause forming cracks, bad spring back and difficult stamping especially for complex parts. Therefore hot stamping steels, as super high strength steels in auto-body parts, are widely applied in the reinforcing parts of automobile anti collision such as door beam, bumper beam, side beam, stiffener of A and B pillars. Hot stamping steel usually exhibits a ferritic-pearlitic microstructure before quenching and should possess better hardenability by having a certain amount of carbon in combination with alloying elements such as boron, manganese, chromium which can improve the hardenability of steel rapidly [1-5].

In the last decade, studies on the development and characteristics of hot stamping materials and process, the numerical simulation, the effect of hot stamping parameters on the mechanical properties and microstructures of hot stamping steel have been published [6-9].

WISCO has developed 1300 MPa, 1500 MPa and 1700 MPa grade hot stamping steel in recent years. This paper mainly studied the continuous cooling transformation (CCT) of 1700 MPa grade hot stamping steel, that is to say, the characteristics of microstructures and hardness under various cooling rates. The CCT diagram is practical as well as being of assistance to the hot forming engineers.

2. Materials and Experimental Methods

The steel used in this study was molten by converter, hot rolled, and then cold rolled to 1.5 mm thin sheet. The tensile strength of the tested steel is between 500 and 600 MPa before quenching. The chemical composition (in weight percent) is given in Table 1. The specimens with the dimensions of 10 mm in length and 3 mm in diameter had to be machined oriented in rolling direction from the plates after rough rolling for the guaranteed dimensions.

Table 1. Chemical composition of test steel (wt. %).

C	Si	Mn	Al	Cr	B
≤0.2	≤0.3	≤1.6	≤0.050	≤0.3	≤0.003

The dilatometric experiments, which were used to determine the CCT diagram and critical points of Ac_1, Ac_3 for hot stamping steel, were carried out on a full automatic dilatometer, Formastor-F, of which the principle is based on induction heating. The specimens were heated to 900°C for austenization, holding for 10 min, then cooled to room temperature at the cooling rate ranging from 0.1 to 50°C /s. Transformation temperature for the corresponding microstructural products can often be obtained by locating the temperature at which the dilation versus temperature curves start to deviate from linearity. The CCT diagram can then be constructed by plotting the temperature versus time. After the dilatometric tests, the microstructures were etched by the solution of alcohol with nitric acid (4% in volume percent) and then examined using Olympus GX71 light microscope. The hardness tests were accomplished by Type FV-700 Vickers-Hardness Tester with a 5 kg load. An average hardness value was calculated by taking three measurements along the same direction in each specimen for hardness testing.

3. Results and Discussions

According to the thermal expansion curves obtained from the dilatometric experiments, the corresponding microstructures and Vickers hardness of the steel and phase transformation theory, the transformation temperature at different cooling rates are confirmed and the CCT diagram of hot stamping steel is drawn. The CCT diagram is illustrated in Fig. 1.

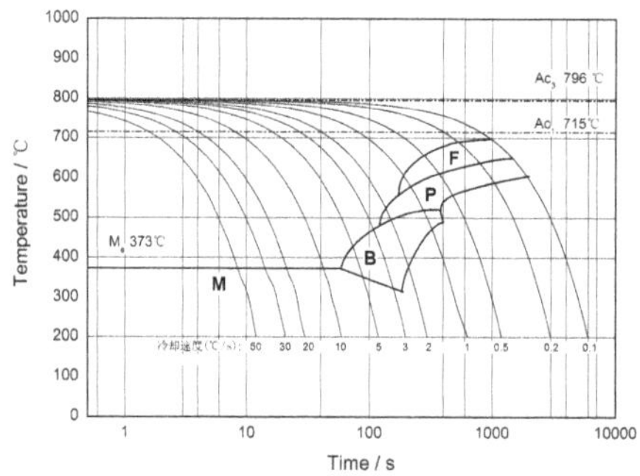

Fig. 1. CCT diagram.

Phase transformation temperatures, microstructure type and Vickers hardness are shown in Table 2. The critical points of Ac_1 and Ac_3 which express the initial and final temperature of the austenite phase transformation are obtained from the dilatometric curve by continuous heating rate of 200°C per hour. The testing results show that Ac_1 and Ac_3 are 715°C and 796°C respectively. Microstructure of as-received base material and optical micrographs at the representative cooling rates from 0.1°C/s to 50°C/s are shown in Fig. 2.

Table 2. Phase transformation temperature, microstructure type and Vickers hardness.

Cooling rates, °C/s	Initial temperature, °C	Final temperature, °C	Microstructure type	Average Vickers hardness value, HV5
0.1	696	602	P+F	179
0.2	688	589	P+ F	199
0.5	660	557	P+ F	205
1	603	472	B+P+F	243
2	525	377	B+P	287
3	474	-	B+M	342
5	428	-	M+B	464
10	373	-	M	529
20	373	-	M	533
30	373	-	M	533
50	373	-	M	546

As shown in Table 2 and Fig.1~2, the microstructure of the as-received base material is mainly pearlite and ferrite. Martensite starts to appear when the cooling rate is 3°C/s. When the cooling rate is greater than 5°C/s, a full martensite can be obtained and the Vickers hardness values are from 529 to 546 HV5. Bainite is obtained with the cooling rate below 5°C/s. The slower the cooling rates become the less martensite forms and the more bainite forms. Pearlite starts to be observed when the cooling rate is lower than 2°C/s and ferrite begins to form while the cooling rate is under 1°C/s. The microstructures are predominantly pearlite and ferrite when the cooling rates are in the range of 0.1 to 0.5°C/s and the average values of Vickers hardness are between 179 and 205 HV5. The CCT diagram indicates that under the condition of the desired cooling rates (10~50°C/s), a full martensite has been obtained and the Vickers hardness values are from 529 to 546 HV5.

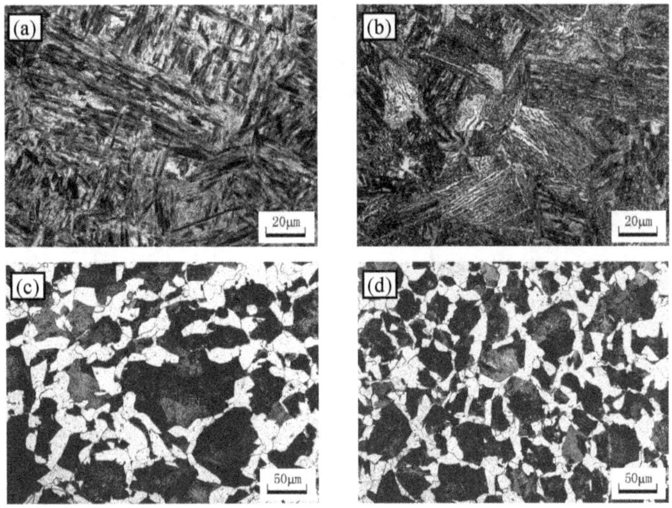

Fig. 2. Optical micrographs at the representative cooling rates and as-received base material (a) 30°C /s, (b) 3°C /s, (c) 0.1°C /s, (d) as-received base material.

In comparison with the CCT diagram of 1500 MPa grade hot stamping steel [10], that one of 1700 MPa grade has been right shift, that is to say, the phase transformation of ferrite, pearlite and bainite have delayed, and the transformation temperature of ferrite is inclined to decrease. These two strength grade hot stamping steels both contain boron, manganese, and chromium which can all stabilize austenite and increase the hardenability of steel. When the cooling rates are between 20°C /s and 50°C /s, these two strength steels can both obtain a full martensite but different martensite hardening capacity. The average Vickers hardness values of 1500 MPa grade are 422~464 HV5 [10], those of 1700 MPa grade are 533~546 HV5. The 1700 MPa grade hot stamping steel obtains a full martensite under the broader cooling rate range from 10°C /s 50°C /s. On the basis of the chemical compositions of 1500 MPa grade steel, the 1700 MPa grade steel increases 0.03% carbon since the content of carbon has to be controlled no more than 0.2% (in weight) to fulfill a good weldability. With the increase of carbon mass fraction, the probability of ferrite nucleation is reduced and the diffusion rates of atoms in austenite slow down. The precipitation rate of ferrite becomes slow. So Fig.1 exhibits the rule of right shift of ferrite's transformation and decrease of ferrite's transformation temperature. Transformation of Pearlite and bainite also delays but transformation temperatures change little. To guarantee the desired strength, the content of manganese needs to be enhanced by 0.3%. Manganese is one of the most principal elements to get good hardenability. So the addition of 0.03% carbon and 0.3% manganese has good effect to improve martensite hardening capacity and then increase the strength of steel.

4. Conclusion

The CCT diagram of 1700 MPa grade hot stamping steel shows that when the cooling rate is more than 5°C/s, a full martensite is obtained, which indicates that the test steel has a good hardenability. Bainite is obtained with the cooling rate below 5°C/s. Pearlite appears when the cooling rate is lower than 2°C/s and ferrite begins to form under 1°C/s. The microstructures are predominantly pearlite and ferrite with the cooling rates at the range of 0.1 to 0.5°C/s. The 1700 MPa grade steel has the broader cooling rate range from 10°C/s 50°C/s to obtain a full martensite and the Vickers hardness values are from 529 to 546 HV.

In comparison with the CCT diagram of 1500 MPa grade hot stamping steel, that one of 1700 MPa grade has been right shift. The addition of 0.03% carbon and 0.3% manganese stabilizes austenite and increase hardenability of steels and also has good effect to improve martensite hardening capacity and then increase the strength of steel.

References

1. M.T. Ma, H.L. Yi and H.Z. Lu, On the lightweighting of automobile, *Eng. Sci.* **11**, 20 (2009).
2. D.W. Fan, H.S. Kim and B.C.D. Cooman, A review of the physical metallurgy related to the hot press forming of advanced high strength steel, *Materials Technology: Steel Research Int.* **80**, 241 (2009).
3. X.Q. He, Y. Wang, H.L. Wei and Z Zhou, The mechanical performance and microstructure of 1500 MPa hot rolled steel for hot forming, in *The 2nd International Conference on Advanced High Strength Steel and Press Hardening*, (Changsha, China, 2015).
4. Y.K. Gao, D. Gao and H.Y. Yu, Analysis on high-strength steel hot forming technology for automobile, *Automobile Technology* **8**, 56 (2010).
5. K.H. Hu and R.D. Han, A novel approach to developing hot stamping parts with tailored properties, in *The 2nd International Conference on Advanced High Strength Steel and Press Hardening*, (Changsha, China, 2015).
6. R. Gi, Y.J. Bi and H.L. Wei, Bendability improvement by optimizing the amount of micro alloy in boron steel, in *The 2nd International Conference on Advanced High Strength Steel and Press Hardening*, (Changsha, China, 2015).
7. H. Wang, R. Ge and S. Y. Zhou, An introduction to thermal pressing forming & its development, *J. of Wuhan Engineering Institute* **26**, 52 (2014).
8. J. Zheng, B.Y. Wang, M. D. Huang and D Cui, Effect of hot stamping parameters on the mechanical properties and microstructure of cold-rolled 22MnB5 Steel Strips, *International J. of Minerals, Metal. Mater.* **21**, 544 (2014).
9. H.L. Yi and J.J. Dong, Hot press forming process and HPF Steel, *Shandong Metallurgy* **31**, 17 (2009).
10. W.Y. Liu, K.H. Hu and G.L. Yuan, Continuous cooling transformation of 1500 MPa grade hot stamping steel, in *The 2nd International Conference on Advanced High Strength Steel and Press Hardening*, (Changsha, China, 2015).

Materials & Testing II

The Response Characteristics of High Strength Steel for Automotive Under High Strain Rates

M. T. Ma[1,†], Y. Zhao[1], J. Zhou[1] and G. Y. Li[2]

[1]*China Automotive Engineering Research Institute CO., LTD, Chongqing 401122, China*

[2]*Central Iron & Steel Research Institute, Beijing 100081, China*

[†]*E-mail:mingtuma@126.com*

The response characteristics of high strength steel for automotive under high strain rates were comprehensively reviewed, the relationship of vehicle lightweight with improvement of vehicle safety and the application of high strength steel were discussed, and the demand of response characteristics of materials for crash safety simulation was proposed. The developments of high-speed tensile testing equipment, test methods, data processing and constitutive equations were further discussed, and finally presented the test results, response characteristics, the related constitutive equation and typical application of various types of automotive high strength steel under high strain rates. Perfection of tests for response characteristics of high-strength steel under high strain rates and deep understanding of its deformation mechanisms will contribute to the development and applications of high-strength steel.

Keywords: High strength steel; Strain rate; Mechanical properties; Constitutive equation.

1. Introduction

After 2000, China's automobile industry develops rapidly [1], and also the steady development of the world automobile industry [2]. In 2015, the total auto output in the world is more than 90 million, auto ownership is more than 1.1 billion, China's auto output is more than 24 million, auto ownership is 170 million. In 2014, global oil consumption of is more than 4.2 billion tons, car consumption takes up of 60%, thus CO_2 emissions in atmospheric is 84 million tons.

The data released on May 6, 2015 show that global CO_2 content in atmospheric exceeded 400ppm first time, the energy conservation for automotive industry is imperative. Automotive lightweight is one of the most direct and effective means for energy conservation. To deal with the third industrial revolution, the electric, intelligent and lightweight will be taken as the development direction for the automotive industry.

In 2011, on the European Car Body Engineering Strategy Forum, Fabbrica Italiana Automobili Torino (F.I.A.T) proposed impact of various factors on fuel consumption [3], such as the power system performance increases 10%, fuel consumption can be reduced to about 10%, the rolling resistance decreases 10%, fuel consumption can be reduced to about 1.5%, air resistance decreases 10%, fuel consumption can be reduced to about 2.7%, vehicle weigh reduces 10%, fuel consumption can be reduced to about 3.5% to 6%. In those four kinds of energy conservation, the effect of lightweight ranks in second place. The European Aluminium Association adopted a large number of experiments indicate that [4], every vehicle weight reduces 10%, savings fuel of 6-8%, and the corresponding reduction in emissions has a similar value. Considering the battery energy density limits for electric automotive, lightweight is more important to new energy vehicles, but it must remain vehicle's safety. The safety regulations become more strict around world, such as in the United States, in order to ensure the car rollover safety, the top crush strength increases from 2.5 times of the vehicle weight to 4.2 times.

An important means of optimization design for automotive NVH and safety is the application of computer simulation, which can reduce a lot of trial workload, thereby reduce the development cost of vehicle. However, it must have response characteristics of the material under high strain rate for computer simulations. In recent years, people are interested in testing and application of response characteristics under high strain rates for automotive high strength steel. The present paper will take a comprehensive review on the test results, response characteristics, the related constitutive equation and typical application of various types of automotive high strength steel under high strain rates.

2. High Speed Tensile Testing Equipment, Test Methods

The testing of material strain rate sensitivity, which started at the dynamic fracture strength test of track steel [5], at the same time, Thomas Young proposed elastic wave theory to describe the dynamic fracture strength [6]. In 1872, J. Hopkinson completed the first dynamic test demonstration, and realized the dynamic loading for metal wire [7, 8]. A split Hopkinson pressure bar (split Hopkinson pressure bar-SHPB) was invented in 1945 [9]. 1970s, ZWICK and other companies began development and production of high-speed tensile testing machine [10].

Currently, the testing equipment of material properties under high strain rate including pendulum test machine, drop test machine, the flywheel test machine, propelled compression testing machine, hydraulic tensile testing machine, split Hopkinson pressure bar. The split Hopkinson pressure bar and hydraulic servo test equipment have a broad application. When performing high-speed tensile, load measured by load package, or force sensors measured by posted strain gauge on a fixture or sample. Strain measurements contain laser extensometer, Doppler laser, displacement sensors, strain gages and strain field optical measuring. Gauge dimension of high-speed tensile specimen usually taken 10 × 20mm [11, 12]. Nominal strain rate of the specimen has relationship with gauge length, stretching displacement and velocity, as shown in Fig. 1. How to reduce the load fluctuations in the high-speed tensile test processing is an important technology, related work has been done in Xiao[13], Xia, Zhu, Wang and Zhou [14]. The original data collection of high speed tensile and smooth of stress-strain curve are very important. The smoothing method for measured stress-strain curves include average method of the stress fluctuations upper limit and lower limit, Savitzky-Golay method (polynomial fitting), Fourier method, LS-DYNA simulation, Hollomon equation fitting, the average method and polynomial fitting have better application. Besides, it can effectively reduce the dynamic load shock by adjusting a reasonable match between load measuring system holding portion of mass and stiffness [15].

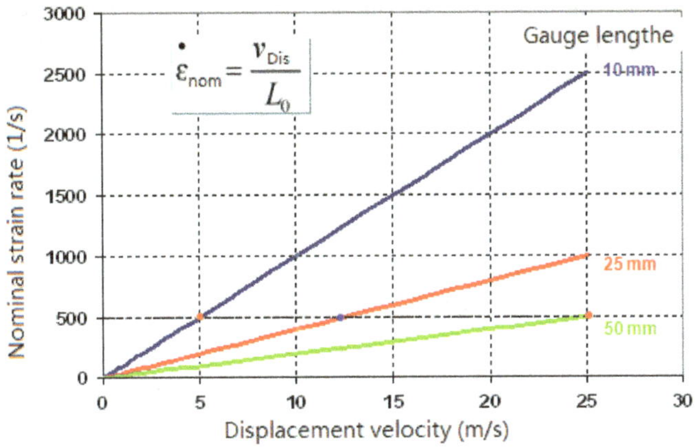

Fig. 1. The relation of nominal strain rate with gauge length and displacement speed during tension.

3. Constitutive Equation for High Speed Tensile Testing Equipment, Test Methods

When vehicle crashed, different component of the different place or different place of the same component will be subjected to different collision speeds or different strain rates [16]. The number strain rate of in high-speed tensile tests is limited, the deformation and strain rate at any point in the computer simulation will be obtained.

To provide all kinds of material stress-strain curves under any strain rates for computer simulation, it is necessary to establish constitutive equations under high speed tensile. Currently, used constitutive equation including: Johnson-Cook equation [17], modified Johnson-Cook equation [18], Cowper-Symonds equation [19], Zerilli-Armstrong equation [20], modified Zerilli-Armstrong equation [21], dual power function model [22], Arrhenius equation [23, 24]. Johnson-Cook equation, modified Johnson-Cook equation are fitting empirical equations based on the experimental data, which contain strain hardening item, strain rate hardening item and temperature impact item. To steel materials, it has high heat capacity, high melting point, as stamping parts and safety parts applied at room temperature, which is less affected by temperature, so the temperature item can be neglected. Cowper-Symonds equation, an exponential equation, is also a mathematical empirical equation based on work hardening and strain rate hardening, the flow stresses are divergent calculated using the equation under various strain rates, it is adopted to describing flow stress curves of the deformation for bullet. In this equation, there are many ways to deal with the static alignment of flow stress, the first one is to use Hollomon equation, the second one is to take into account the plastic hardening modulus approach, the last one is to use secant modulus approach. Zerilli-Armstrong equation is a materials dynamics constitutive model based on dislocation dynamics, which consider the effects of tensor orientation factor, dislocation density, average velocity of dislocation, grain size, strain hardening and temperature, and proposes descriptive model of rheological properties for FCC and BCC metals.The modified Zerilli-Armstrong equation is mainly used to describe the rheological properties of materials under high temperatures.

The dual power function model is mainly used for constitutive model which taking into account the rate of hardening and strain hardening at the same time. Arrhenius equation is adopted to describe the relationship between flow stress, strain rate and temperature of materials under high temperatures, it also considers the plastic deformation activation energy at the same time. Currently, Johnson-Cook equation, modified Johnson-Cook equation are usually adopted to deal with high speed tensile data.

There are many factors to effect the dispersion of high speed tensile test data which contain test equipment, test specimen gauge size, smoothing method measurement method, acquisition and experimental curves strain test data, test fixture design, test environment and the selected constitutive equation.

4. Response Characteristics of High Strength Steel Under High Strain Rates

In recent years, with the development of automotive lightweight technology, high-strength steel and advanced high strength steel for vehicle develop rapidly. Currently, high strength steel for automotive includes high strength steels, the first generation of advanced high strength steel, with product of strength and ductility of 20000MPa%. The second generation of advanced high strength steel, with product of strength and ductility of 60000 MPa%;. The third generation of advanced high strength steel, with product of strength and ductility of 30000-40000 MPa%.

Yan and Xu [25] have studied the response characteristics of high speed tensile testing for ordinary high-strength steel and first generation of advanced high strength steel in detail, and obtained the relevant parameters for Johnson-Cook equation. The result indicated that the elastic modulus amount of the studied steel does not change with changing of strain rates, but the yield strength, tensile strength increases with increasing of strain rate. With increasing of quasi-static strength, the effect of increasing the flow stress caused by increasing of strain rate decreased, but while the strain rate increases by one magnitude, the increasing value of the tensile strength of steel in each category is 22MPa. Effect of strain rate on the uniform elongation and total elongation is relatively complex, not simple relationship between the increase or decrease, in a certain range of strain rate, and sometimes increase and sometimes decrease. Different results from different levels of strength dual-phase steel, flow curves at different strain rates tested are not identical, some parallel, some convergence, there are divergent [26,27]. Usually at low strain (e.g. 15%), most of the flow curves are parallel or slightly divergent nature, but at high strain (e.g. greater than 20%), there is phenomenon of convergence.

Bullet-proof steel based on martensite matrix, the results of high speed tensile testing indicate that B-class and C-class ballistic plate has obvious strain rate sensitivity, while the fracture was observed under different strain rates, the dimples show shallow and thinner with increasing of strain rates, which is consistent with the strain rate sensitivity of steel. With the obtained measured data, Johnson-Cook equation for Bullet-proof steel and Cowper-Symonds equation for bullet can be built, the computer simulation of target practice and shooting test are in good agreement [28, 29].

A series of studies have studied the effect of strain rate on the second generation of high-strength steel TWIP steel's mechanical properties and microstructure evolution [30-38]. Xiong [33] have studied the strain rate on five TWIP steels (No. 1-5) dynamic tensile properties, the basic ingredient is 0.007-0.024% of C, Mn contents are 14.3%, 18.8%, 25.4%, 28.1% and 32.3%, respectively. Al content are 2.29%, 2.31%, 2.28%, 2.25% and 3.79%, respectively. Si contents are 2.60%, 2.65%, 2.55%, 2.55% and 2.30%, respectively. Its quasi-static and dynamic mechanical properties are shown in Table 1, it can be noted that the yield strength, tensile strength of this type of steel increase with increasing of strain rate, only No. 1's elongation and uniform elongation increased with increasing strain rate, the other four kinds of steel decrease with increasing of strain rate, due to the different manganese content of steel, different austenite content, which have different effect on dislocation cross slip, slip and thermal activation. The final results of the strength properties is the combined effects of strain hardening, strain rate hardening and softening of the adiabatic temperature rise at high strain rate.

Table 1. The mechanical properties of quasi-static tension and dynamic tension.

Materials	Strain Rate /s^{-1}	Yield Strength $R_{0.2}$/MPa	Tensile Strenght R_m/MPa	Uniform Elongation A_{gt}/%	Elongation A_t/%	Product of Strength and Ductility R_m*A_t (MPa.%)
	3.3e-02	283	1005.3	22.6	24.9	25032
N0.1#	760	471	1183	25	40.5	47912
	1380	516	1302	31.7	50.2	65881
	3.3e-02	272	699.4	58.3	62.9	43992
N0.2#	850	436	905	34.5	48.5	43893
	1480	490	1075	33.9	57.1	61383
	3.3e-02	275	631.6	61.1	68.5	43265
N0.3#	870	458	791	36.5	58	45878
	1490	511	810	36.3	60.3	48843
	3.3e-02	255	596.4	60.2	66.7	39780
N0.4#	830	454	815	34	56.5	46048
	1560	474	825	35	60.4	49830
	3.3e-02	277	578	59.9	66	38148
N0.5#	850	438	736	30	51.5	37904
	1540	458	795	32	56.8	45156

Tang, Wu and Zan [35] have studied the microstructures evolution and deformation behaviors of low C (content 0.03%) 18Mn-3Al-3Si and 21Mn-3Al-3Si of TRIP/TWIP effect Symbiosis high speed steel during high-speed tensile with strain rate of 1000s^{-1}, the composition before and after tensile show in Table 2. It can be noted that the changes is more complex after high speed tensile deformation and the changes of microstructures are closely relate to the steel layer fault energy. Wu, Tang and Li [38] have similar results on microstructures evolution of 18Mn-3Al-3Si TRIP/TWIP steel under quasi-static and high strain rate.

Table 2. Microstructure composition of 18Mn-3Al-3Si and 21Mn-3Al-3Si of before and after tensile deformation.

Steel	Before tensile			After tensile			
	γ	α_F	ε	γ	αF	ε	α'
18Mn-3Al-3Si	67.8	23.6	8.6	51.6	23.6	6.9	17.9
21Mn-3Al-3Si	76.1	12.7	11.2	65.3	12.7	10.8	8.2

Q & P and Q & PT steels are TRIP steels with martensitic matrix, which are now the industrialization of the third generation of steel. In 2003, Professor Speer proposed quenching allocation process technology, referred to as Q & P process [39, 40]. Q & PT steel is steel with quenching, tempering distribution process, taking into account the precipitation of alloy carbides precipitation strengthening which can be further improved strength [41, 42], the strength level of this type of steel is similar to nano-bainitic steel developed by Bhadeshia et al. [43-45]. Liu, Wang and Liu [46] studied the effect of strain rate on Q & PT steel, the results show that both yield and tensile strength increase with increasing of strain rate, the change elongation with various strain rates is more complex, strain rate from 10 s^{-1} to 100 s^{-1}, a substantial increase in elongation, and then with increasing of strain rate, the elongation decreases, which is closely related to the reduction of residual austenite. The microstructure observation near fracture showed that with increasing strain rate, the microstructures retained reduced austenite and unconverted residual austenite is significantly stretched. Comparison of fiber and shear fracture lip zone for static and dynamic tensile samples, the quasi-static tensile fracture has deeper dimples, and more tearing ridge, The dynamic tensile fracture has lighter dimples, and torn crack narrowed part, this change is consistent with mechanical properties. Wang[47] studied the deformation mechanism of dynamic mechanical properties, mechanical properties under different strain rates are shown in Table 3.

Table 3. The mechanical properties of Q-P-T steel and Q&T steel at various stain rates.

Sample	Strain Rate/s^{-1}	Tensile Strength /MPa	Total Elongation/%	Product of Strength and Ductility /MPa%)
Q&PT Steel	10^{-4}	1322	16.9	22342
	10^{1}	1355	17.2	23306
	10^{3}	1620	18.8	30456
Q&T Steel	10^{-4}	1438	12.1	17400
	10^{1}	1485	11.7	17375
	10^{3}	1820	10.5	19110

It can be noted that Product of Strength and Ductility of Q & PT steel is much higher than the Q & T steel under various strain rates, and the fracture morphology show that fracture ductile of Q & PT steel is significantly higher than the Q & T steel. With the increasing of strain rate, the strength and ductility Q & PT steel increase, which is more suitable for collision member in safety applications for automotive.

5. Conclusion

Automotive lightweight, safety technology, computer simulation technology, which have the need to test the materials response characteristics under high speed tensile.

There are many kind of high speed tensile testing equipment, which are suitable for the different modes of deformation. Hydraulic servo testing machine and Hopkinson Bar are more commonly used. Test equipment, the size of the sample, strain measurement, collection and data smoothing method and fitting the constitutive equation have an impact on the high speed tensile test results. Adopting two methods to measure each, it will increase test data reliability and comparability.

There are two kind of constitutive equation to describe the response characteristics of metal materials under high strain rate, one is the experience of the constitutive model, the other is the physical nature of the constitutive model based on the material deformation. Johnson Cook model, Cowper-Symonds equation, dual power function model are more commonly used.

High strength steel and first generation, second generation and third generation advanced high strength steel have obviously strain rate sensitivity, ie, with the rising of strain rate, the strength increased significantly, but the change total elongation and the uniform elongation for part of steels are more complex. TRIP and TWIP steel steel, due to the presence of austenitic, the strain rate sensitivity exhibit more complex relationships, The final results of the strength properties is the combined effects of strain hardening, strain rate hardening and softening of the adiabatic temperature rise at high strain rate. Perfection of tests for response characteristics of high-strength steel under high strain rates, and deep understanding of its deformation mechanisms will contribute to the development and applications of high-strength steel.

References

1. M.T. Ma, H.L. Yi and H.Z. Lu, On the lightweight of automobile, *Eng. Sci.* **11**, 20 (2009).
2. Y. Liu, Society of Automotive Engineers of China, China Automobile lightweight technology innovation strategic alliance, China FAW Group Corporation R&D Center, China Automotive Lightweight Development-Strategy and Path, *Beijing Institute of Technology Press*, 2 (2015).
3. F. Antonio, Strategies in car body engineering 2011, (Bad Neuheim, Germany, 2011).
4. M. Bertram, K. Buxmann, R. Chase, P. Furrer, B. Gilmont, J. Hannagan and G. Rebitzer, Improving sustainability in the transport sector through weight reduction and the application of aluminum, in *The Symposium on the International Aluminum Conference of Application in Automobile Industry. Improving Sustainability in the transport sector*, (Dalian, China, 2007).
5. W.K. Hatt and E. Marburg, Preliminary report on the present state of knowledge concerning impact tests, *Ohio, USA: Process American Society Testing Materials* **1**, 27 (1899).
6. T.A. Young, Course of Lectures on Natural Philosophy and the Mechanical Arts, *London, England: Joseph Johnson* **1**, 144 (1807).

7. J. Hopkinson, Further experiments on the rupture of iron wire, *Process Manchester Literature Philosophy Society* **11**, 119 (1872).
8. J. Hopkinson, On the rupture of an iron wire by a blow, *Process Manchester Literature Philosophy Society* **11**, 40 (1872).
9. H. Kolsky, An investigation of the mechanical properties of materials at very high rate of loading, *Proc. Phys. Soc. Lond.* **62**, 676 (1949).
10. S. Maschinen, High-speed Tension Machines (1977).
11. SEP1230, The Determination of the Mechanical Properties of Sheet Metal at High Strain Rates in High-Speed Tensile Tests.
12. ISO26203-2, Metallic materials-Tensile testing method at high strain rates-Part 2: Servo-hydraulic and other test systems.
13. X.R. Xiao, Dynamic tensile testing of plastic materials, *Polymer Testing* **27**, 164 (2008).
14. Y. Xia, Z. Zhu, K. Wang and Q. Zhou, Design and verification of a strain gauge based load sensor for medium-speed dynamic tests with a hydraulic test machine, *International Journal of Impact Engineering* **88**, 139 (2016).
15. B. Yin, Research on dynamics testing and simulation methods of material for development of automobile collision safety performance, in *High speed tensile testing technology for Lightweight and Safety of Automotive Forum*, (Chongqing, China, 2015).
16. A. Koprivc, Testing solutions for the automotive industry, in *22th international forum for materials testing*, 102.
17. G.R. Johnson and W.H. Cook, A constitutive model and data for metals subjected to large strains, high strain rate and high temperatures, in *proceedings of the seventh international symposium on ballistics*, (Hague, Netherlands, 1983).
18. K. Xu, C. Wong, B. Yan and H. Zhu, A high strain rate constitutive model for high strength steels, *SAE Technical Paper* (2003).
19. G.R. Cowper and P.S. Symonds, Strain hardening and strain rate effects in the impact loading of cantilever beams, *Brown University. Applied Mathematics report* 28 (1958).
20. F.J. Zerilli and R.W. Amstrong, Dislocation mechanica-based constitutive relations for material dynamic calculations, *Journal Application Physic* **61**, 1861 (1987).
21. D. Samantaray, S. Mandal, U. Borah, A.K. Bhaduri and P.V. Sivaprasad, A thermo-viscoplastic constitutive model to predict elevated-temperature flow behavior in a titanium-modified austenitic stainless steel, *Mater. Sci. Eng. A* **526**, 1 (2009).
22. J.R. Klepaczko, A Practical Stress-strain-strain-rate temperature constitutive relation of the powerform, *Journal of Mechanical Working Technology* **15**, 143 (1987).
23. J. Jonas, C. Sellars and W. Tegart, Strength and structure under hot working condition, *International Metal Reviews* **130**, 1 (1969).
24. H. Shi, A.J. Mclaren, C.M. Sellars, R. Shahani and R. Bolingbroke, Constitutive equation for high temperature flow stress of aluminum alloys, *Mater. Sci. Eng.* 1997, **13**, 210 (1997).

25. B.Yan and K. Xu, High steain rate behavior of advanced high strength steels for automotive applications, in *44th MWSP (metal working and steel processing) conference*, 493 (2002).
26. J.P. Zhang, G. Fang, Q.S. Jin and M.T. Ma, Mechanical properties and microstructure of DP steel sheets under dynamic loads, in *Proceedings of The 2nd international conference on advanced high strength steel and press hardening*, (Changsha, China, 2015).
27. C.L.T.D. POSCO, Automotive steel data book, *Copylight by POSCO* 47 (2008).
28. M.T. Ma, G. Chen and Y.H. Ma, The Development and Application Research of Light Weight Heat Treated B-grade Bullet Proof Steel, *Engineering* **12**, 2 (2014).
29. M.T. Ma, G. Chen, Y.H. Ma, Z.G. Li and Y. Feng, The Development and Application Research of Light Weight Heat Treated C-grade Bullet Proof Steel, *Adv. Mater. Res.* **1063**, 21 (2015).
30. J.P. Zhang, X.F. Duan, Z.M. Shi and F.S. Han, Effects of Temperature and Strain Rate on Tensile Properties of TWIP Steel, *Materials for Mechanical Engineering* **39**, 4 (2015).
31. X.M. Qing, L.Q. Chen and W. Deng, Effects of strain rate on mechanical properties of Fe-23Mn-2Al-0.2C TRIP steel, *Chinese Journal of Materials Research* **25**, 278 (2011).
32. O. Grassel, L.Kruger and G. Frommeyer, High strength Fe-Mn-(Al,Si) TRIP/TWIP steels development-properties-application, *Int. J. Plasticity* **16**, 1391 (2000).
33. R.G. Xiong, Research on strain behavior of TWIP steel under various strain rates, Master Thesis, Shanghai University (2008).
34. R.E. Reed-Hill, The inhomogeneity of plastic deformation, *Ohio: ASM, Metals Park* 285 (1971).
35. Z.Y. Tang, Z.Q. Wu and N. Zan, Microstructure evolution and deformation behavior of TRIP/TWIP steel with high manganese content, *Acta Metall. Sin.* **47**, 1426 (2011).
36. R.E. Schramm and R.P. Reed, Stacking fault energies of seven commercial austenitic stainless steels, *Metallurgy Transaction* **6**, 1345 (1975).
37. N.I. Noskva and V.A. Pavlov, Stacking fault in nickel solid solution, *Physics Metal and Metallography* **14**, 86 (1962).
38. Z.Q. Wu, Z.Y. Tang and H.Y. Li, Effect of strain rate on microstructure evolution and mechanical behavior of a low C high Mn TRIP/TWIP steel, *Acta Metall. Sin.* **48**, 593 (2012).
39. J.G. Speer, D.K. Matlock, B.C. De Cooman and J.G. Schroth, Carbon partitioning into austenite after martensite transformation, *Acta Mater.* **51**, 2611 (2003).
40. D.K. Matlock, V.E. Bräutigam and J.G. Speer, Application of the quenching and partitioning（Q&P）process to a medium-carbon, high-Si microalloyed bar steel, *Mater. Sci. Forum* **426-432**, 1089 (2013).

41. X.D. Wang, N. Zhong, Y.H. Rong and T.Y. Hsu, Novel ultrahigh-strength nanolath martensitic steel by quenching-partitioning-tempering process, *J. Mater. Res.* **24**, 260 (2009).
42. X.D. Wang, Z.H. Guo and Y.H. Rong, Mechanism exploration of an ultrahigh strength steel by quenching-partitioning-tempering process, *Mater. Sci. Eng. A* **529**, 35 (2011).
43. F.G. Caballero, H.K.D.H. Bhadeshia, K.J.A. Mawella, D.G. Jones and P. Brown, Very strong low temperature bainite, *Mater. Sci. Technol.* **18**, 279 (2002).
44. H.K.D.H. Bhadeshia, High performance bainitic steels, *Mater. Sci. Forum* **500-501**, 63 (2005).
45. C. Garcia-Mateo, F.G. Caballero and H.K.D.H. Bhadeshia, Mechanical properties of low-temperature bainite, *Mater. Sci. Forum* **500-501**, 495 (2005).
46. C. Liu, L. Wang and Y. Liu, Effect of Strain Rates on Tensile Deformation Behavior of Quenching and Partitioning Steel, *Special Steel Technology* **18**, 18 (2012).
47. Y. Wang, Plasticized mechanism and dynamic mechanical properties of advanced high-strength Q-P-T steel, PhD thesis, Shanghai Jiao Tong University (2012).

Test-Piece and Test-Rig Designs to Obtain High Accuracy Experimental Data for Boron Steel at Hot Stamping Conditions

M. Ganapathy, N. Li[†] and J. G. Lin
Mechanics of Materials, Department of Mechanical Engineering, Imperial College London, London, SW7 2AZ, UK
[†]E-mail: n.li09@imperial.ac.uk

M. Abspoel and D. Bhattacharjee
Research and Development, Tata Steel
IJmuiden, The Netherlands

Hot stamping is a new manufacturing process that benefits in meeting the most important requirements and regulations like weight reduction, fuel efficiency and safety performance of automobiles. The finite element analysis acts as a tool to investigate the feasibility of manufacturing a part with hot stamping process. The accuracy of finite element analysis depends on the precision of the material model. Visco-plastic damage constitutive equations require close-fitted material constants, which are obtained from thermo-mechanical tests using a Gleeble simulator for various strain rates and temperatures. Due to resistance heating principle, a normal sample is not homogeneously heated along the gauge length, which can result in different phases along the gauge length while attempting to heat the steel sample to above austenite temperature. The stress-strain data obtained at this condition is not accurate enough to precisely simulate the hot stamping process. In order to obtain uniform thermo-mechanical properties of full austenite, a new self-resistance heating test rig and technique to achieve accurate stress-strain data at hot stamping conditions is proposed in this study.

Keywords: Hot Stamping; Thermo-mechanical test; Boron steel; Electrical-resistance heating; Temperature fields.

1. Introduction

In a hot stamping process, boron steel is first heated to a temperature of ~ 900 °C and soaked for a few minutes to get homogeneous temperature. Subsequently it is air cooled during transportation from furnace to forming press and deformed in a stamping tool followed by immediate quenching. Initial material is in ferrite/pearlite microstructure; during soaking, the material is in austenite condition; after deformation and fast cooling the final product mostly consists of a martensitic phase. The stress-strain behavior of boron steel at hot stamping conditions is necessary for forming simulations, which is experimentally replicated with the help of a Gleeble thermo-mechanical test. Gleeble has various features to simulate the actual process in laboratory, but it has limitations in high temperature tensile test. A large temperature gradient could exist along the gauge of a uniaxial sample in a typical Gleeble test due to electrical resistance heating and a water cooled copper jaw, as shown Figure 1. This diminishes the quality of the hot tensile tests, because a non-uniform temperature field would cause non-uniform deformation along the gauge length. If large temperature gradient exists during heating of boron steel above 900 °C, microstructures obtained could be a combination of various phases in the sample, like austenite in the center of the sample and ferrite phase near the grips [7]. If the temperature dependent material constants of visco-plastic constitutive equations are determined from this kind of experimental data, it could lead to wrong predictions and interpretations in finite element simulations. Numerous researchers [1-7] have investigated the consequence of non-uniform temperature fields within sample gauge section by electrical resistance heating on Gleeble. However, most of the studies focused on to the optimization of sample geometries and few researchers have discussed about the cause of thermal gradients either through experimental studies or finite element modeling. An overview of temperature distributions along uniaxial samples found in literatures are shown in Figure 2.

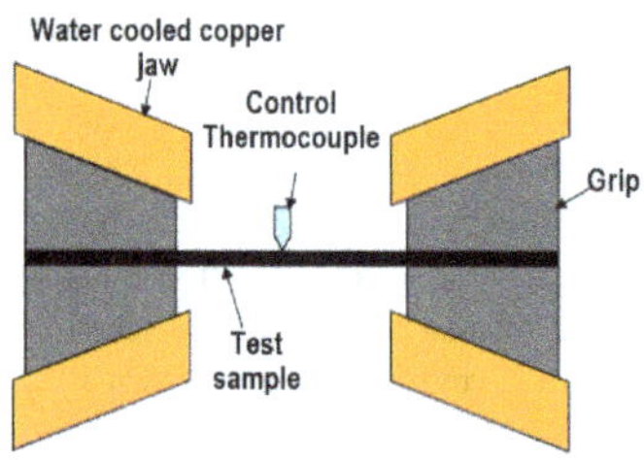

Fig. 1. Working principle of Gleeble thermo-mechanical test.

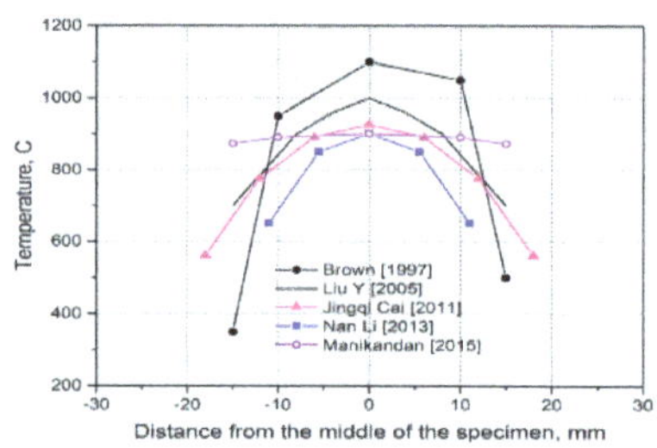

Fig. 2. Temperature profile from various test results.

Abspoel, Neelis and van Liempt, [7] have proposed a new sample geometry [Figure 3c] and test procedure to achieve uniform temperature distribution and consequently homogeneous deformation for 30 mm gauge length. Manikandan et al., [5] further analyzed the "van Liempt" sample and concluded that the "van Liempt" sample design is able to achieve a more uniform temperature over the 30mm gauge length, and thus obtain a more uniform strain distribution over the sample gauge length for the material deforming at hot stamping test conditions. However, the design increases the complexity of tests. In addition, due to limited space available in the Gleeble, the ratio of gauge length to width is less than 5, which does not meet the ISO 6892-2:2011 standard [8]. Therefore, a study was started to achieve a homogeneous temperature profile within a sample that fits the standard. This resulted in a redesign of the grips to achieve less heat loss towards the gripping system.

2. Development of New Test Rig Design Concept

Schematics represented in Figures 4 (a) to (d) illustrate the basic principles of electrical resistance heating by using conventional standard copper and stainless grips, "van Liempt" Sample [7], and new grips proposed in this study.

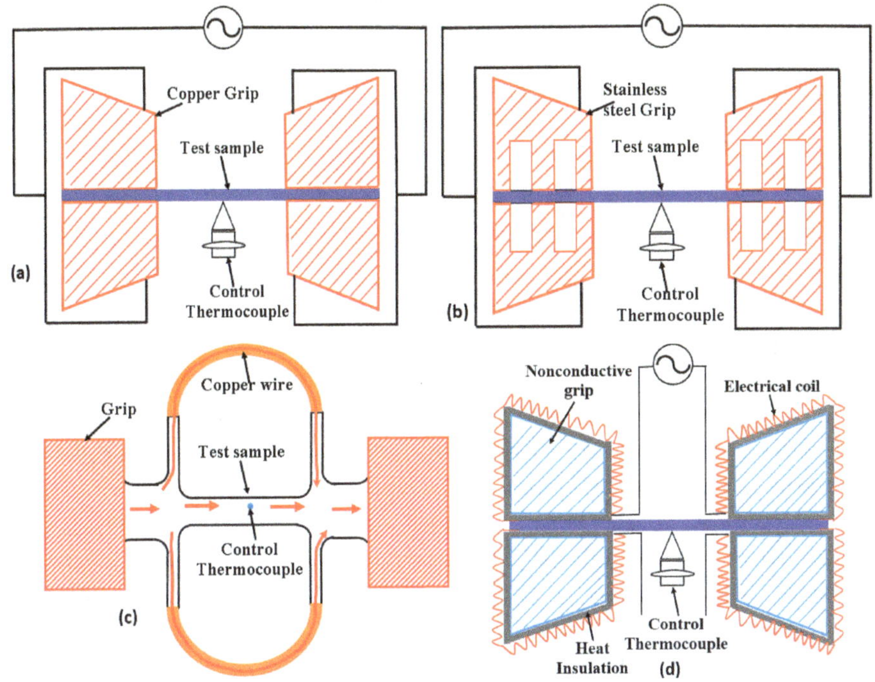

Fig. 3. Gleeble heating concepts (a) standard copper grips,(b) standard stainless steel grips, (c) "Van Liempt" test sample, (d) New grips.

Figure 3 (a) shows the resistance heating principle of a testing system adopting standard copper grips, in that metallic grips clamp a sheet sample and alternative current directly passes through grips to the sample. The sample gets heated up due to joules effect; at the same time, the heat transfers from the sample to the grips and water cooled copper jaws which tightly hold the grips, resulting in temperature gradient along the gauge of sample. Figure 3 (b) shows the resistance heating principle with a set of standard stainless steel grips with reduced contact area adopted to decrease the heat lost from sample so as to reduce the thermal gradient. Figure 3 (c) shows the resistance heating principle with the "van Liempt" sample [7]. The objective of the "van Liempt" sample geometry was to achieve a uniform temperature distribution along the gauge length by creating a buffer to the heat loss to the water cooled grips with a supplementary region between the gauge section and the shoulders, so that the gauge section is isolated from the grips and equal current density throughout the sample was achieved by copper wire. The experimental and sample design details for the "van Liempt" sample can be obtained from Ganapathy, et al [5] and Abspoel et al [7]. Figure 3 (d) shows the resistance heating principle of a new proposed grip design. In this design, two concepts were incorporated for the purpose of achieving uniform temperature distribution along the gauge length: (1) Excess electrical energy is supplied to generate more heat energy through additional length of electrical wind route arrangement in the current flow path. (2) Reduced heat loss to the water cooled copper jaws from the sample by separating the sample from solid grips with thermal insulation layers. Standard dog-bone shape samples can be used in this design.

3. Experimental Results and Discussion

Figure 4 shows the temperature distributions along the gauge length of the sample during soaking at 900 °C (Solid lines) and onset of deformation at 700 °C (Dash lines) when a sample is clamped by 3 different types of grips as introduced in Section 2. A dog-bone sample with 10mm width and 80mm parallel length was designed for this study. It has been observed that there is a large temperature gradient along the gauge section in both standard copper and stainless steel grips. But in case of the "van Liempt" sample the temperature deviation during soaking is within ~25 °C and onset of deformation is within ~5 °C, which is almost uniform for the entire 30 mm gauge length. In the case of using newly proposed grips, the maximum temperature variations are 90 °C during soaking and 10 °C at onset of deformation for 80 mm gauge length. With 40mm gauge length (5 times of gauge width), the temperature variations are 10 °C during soaking and no deviation at onset of deformation, which can be reasonably taken as uniform. In standard grips the gauge section was highly affected by the water-cooled copper jaws and electrical resistance variation with respect to sample geometry. In the "van Liempt" sample design, the gauge section is isolated from the grip by a supplementary region, so the temperature gradient within the gauge length is negligible. In the new grips design, heat loss to grip was reduced by heat insulation layer between grip and sample. Also, the influence of the geometry of samples in a dog-bone shape on thermal gradient under resistance heating was eliminated by providing excess electrical energy through additional electrical resistance heating coil arrangement.

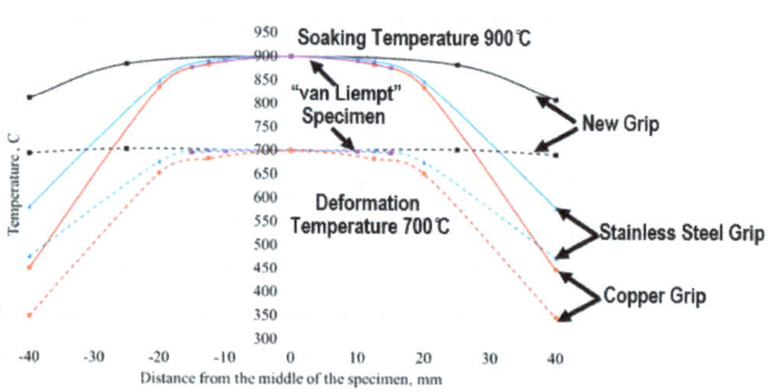

Fig. 4. Comparison of temperature distributions along gauge length by using different types of grips.

4. Conclusion

In order to achieve a homogeneous temperature field along the gauge length (no less than 5 times of gauge width) during austenitization and deformation of boron steel at hot stamping conditions, an innovative resistance heating grips design has been developed and fabricated in this study. The experimental results confirm that the new set of grips is able to obtain a homogeneous temperature distribution along the gauge length of test sample, so as to obtain a more uniform micro structural distribution during heating, and more uniform strain distribution under deformation in the critical areas of the sample.

Acknowledgment

This work was funded by Tata Steel UK. The authors would like to thank Tata Steel for financial support on this project.

References

1. S. G. R. Brown, J. D. James and J. A. Spittle, "A 3D numerical model of the temperature-time characteristics of samples tested on a Gleeble thermomechanical simulator.", *Model. Simul. Mater. Sc.* **5**, 539 (1997).
2. Y. Liu, J. Lin, D. Farrugia and T.Dean, A numerical and experimental study of cavitation in a hot tensile axisymmetric testpiece, *J. Strain Analysis* **40**, 571 (2005).
3. J. Q. Cai, Modelling of phase transformation in hot stamping of boron steel, Diss. Imperial College London, 2011.
4. N. Li, Fundamentals of materials modelling for hot stamping of UHSS panels with graded properties, 2013.
5. M. Ganapathy, N. Li, J. G. Lin, M. Abspoel, H. Guido and D. Bhattacharjee, Analysis of new Gleeble tensile sample design for hot stamping application, *Matec Web of Conferences* **21**, 1 (2015).
6. N. A. Lindeman, a technique for applying direct resistance heating current to a specific location in a sample under test while substantially reducing thermal gradients in the sample gauge length, US Patent, No. 7363822 (17 AUG, 2006).
7. M. Abspoel, B. Neelis and M. P. Liempt, Exploring process and material parameters for hot forming, in *7th Forming Technology Forum*, (University of Twente, the Netherlands, 2014).
8. ISO 6892-2:2011, Metallic materials – Tensile testing – Part 2: Method of test at elevated temperature, International Standard, 2011.

The Corrosion Properties of Zinc Coated Boron Steel for the Direct Press Hardening Process

M. Jönsson[a] and D. Berglund[b]

Gestamp,

Luleå, 971 25 Luleå, Sweden

E-mail: [a]*MJönsson@se.gestamp.com,* [b]*DBerglund@se.gestamp.com*

Due to continuously higher demands from different organizations and severe legislation on passive automotive safety and the effort to reduce vehicle emissions, the use of high- and ultra-high strength components in both car body and closures have increased drastically during the last two decades. An increased interest has been directed towards the use of press hardened steel (sometimes referred to as hot stamped steel) in underbody parts. However, the lower part of the body in white (BIW) can be highly affected by corrosion due to the harsh road environment. Consequently, enhanced corrosion protection properties are desired. In order to meet the higher demands for corrosion protection of press harden steel, zinc coated material for the direct press hardening process will be more frequently used in future car bodies.

In the present study the corrosion properties of zinc coated material for the direct press hardening process have been investigated and compared to references i.e. Aluminized 22MnB5, bare press-hardened 22MnB5 and standard galvanized steel for cold forming. The results show that zinc coated material has an electrochemical potential significantly below the potential of the base material, the boron steel. Hence, it gives a cathodic corrosion protection of the substrate. However the cathodic protection is influenced by the dwell time (total time in furnace) in the press hardening process. The heating dwell time governs the amount and type of Fe–Zn-intermetallic phases formed. Consequently the cosmetic corrosion behaviour of the zinc coated material is connected to the heating dwell time in the press hardening process. The results show that an optimum corrosion resistance is obtained after short dwell time. This can probably be attributed to a combined effect of the galvanic corrosion protection and the thickness of the layer.

Keywords: Press hardening; Ultra-high strength steels; Zinc; GA; GI; Corrosion; Process.

1. Introduction

Today, both uncoated as well as coated 22MnB5 material is used in the press hardening industry. Mainly two types of coatings are used today; AlSi and Zinc. Of the two, the AlSi coating is most commonly used. The AlSi coating prevents oxidation (scaling) and surface decarburization. The AlSi layer also gives a degree of corrosion protection. However it has been argued that an AlSi coating cannot be considered a sacrificial coating for 22MnB5 substrate [1], due to the low potential difference between the AlSi surface coating and the 22MnB5 substrate. Instead the AlSi layer provides a barrier corrosion protection.

Zinc coated steel sheets have been used in the automotive industry for a long time and the positive corrosion properties of Zn coated steel are well documented [2]. In the press hardening industry Zinc coated material is today used in the indirect press

[a] Corresponding author.

hardening process and not in the direct press hardening process. The main difference between the indirect and the direct process is the forming temperature. In the indirect process, the major forming step is done in cold condition before the heat treatment. In the direct forming process the forming is made in hot condition. Up until now the forming of zinc coated steel in hot condition has been limited by the creation of surface defects, such as liquid metal embrittlement (LME) and micro cracks. However using new processes and tool technology together with newly developed steel, with a slightly different chemical composition compared to the commercially available 22MnB5, the forming can be performed at a lower temperature. Consequently the problem with LME and micro cracks can be overcome [3]. Today zinc-coated material for the direct press hardening process is close to be commercialized. In the present investigation the corrosion properties of zinc coated steel for the direct press hardening have been evaluated and the results have been compared to commercially available coatings. A special focus has been given to the influence of the heating dwell time on the corrosion properties.

2. Experimental

2.1. *Material and process*

The direct press hardening process have earlier been described in the literature [3]. In the press hardening process, it is important to have as short heating- and soaking time as possible to obtain a short process cycle time or to reduce the length of the short furnace. The total time in furnace (heating time plus soaking time) is referred to as the *dwell time*. The maximum allowed dwell time is important for process robustness because it allows longer time for process interruption without scrapping material in the furnace. In the present work, the zinc coated material was heated using three different dwell times i.e. short (S), medium (M) and long (L). These dwell times and temperatures were chosen in accordance with internal recommendations. Due to company confidentiality, these times and temperatures are not revealed in the study. The press hardening experiments in the present work have been conducted at Gestamps fully automated research line located in Luleå, Sweden. In the present work, both galvanized material (GI) and galvannealed (10% Fe on the surface) (GA) material have been evaluated. In addition to the zinc coated material for the direct process reference panels of uncoated 22MnB5 (uncoated), AlSi-coated 22MnB5 (AlSi), and cold-formed zinc coated steel (Z100) have been included in the study. A summary of the material included in the study can be seen in Table 1 and Table 2.

Table 1. Zn-coated materials included in the corrosion test, The type and weight (g/m2) of the coating is specified together with the dwell time. GI-NHT- Galvanized not heat treated (dwell time= 0s). GI-S, M, L-Galvanized with short, medium or long dwell time. GA-NHT-Galvanized not heat treated (dwell time=0s). GA-S, M, L-Galvanized with short, medium or long dwell time.

Coating	GI-140 g/m^2				GA-140 g/m^2			
Name	GI-NHT	GI-S	GI-M	GI-L	GA-NHT	GA-S	GA-M	GA-L
Dwell time	-	Short	Medium	Long	-	Short	Medium	Long

Table 2. Reference materials/coatings included in the corrosion test. The type and weight (g/m2) of the coating is specified. GI-100 Galvanized cold formed steel (dwell time= 0s).

Coating	GI-100 g/m^2	AlSi-150	-
Name	Z100	AlSi	Bare 22MnB5

2.2. *Investigation of corrosion properties*

The corrosion panels were exposed to the VDA 233-102 accelerated corrosion test. The duration of the test is 6 weeks. The 1 week cycle consists of alternating salt spray (1% NaCl), wetting, freezing and ambient (or drying) phases. In this study the corrosion resistance has been evaluated on painted panels where the corrosion has been quantified by measuring the delamination of paint from a scribe. So called cosmetic corrosion.

The galvanized panels have also been investigated electrochemically. A solution of 100g/l ZnSO4*7H2O+200g/l NaCl was used as electrolyte. The temperature was controlled to 25°C. Further, a current of 11.7mA/cm2 was applied on the samples. The electrolyte and the applied current were chosen to be the same as those used by Autengruber et al. [4]. After the electrochemical test the depth of the dissolved area was measured using a micrometer. The measured depth was divided with the exposure time in each measurement. Consequently, the X-axis in the Galvanostatic measurement presents a depth indication instead of a time factor.

2.3. *Characterization of the coating layer*

Cross section of the un-exposed and exposed samples were investigated with both light optical microscopy (LOM), model Olympus BX60M, as well as scanning electron microscope (SEM) model Hitachi S-3400N equipped with EDS model QUANTAX system and a Bruker SDD X-flash Detector 5010. The surfaces of samples with different dwell time were investigated using X-ray diffraction (XRD) model PANalytical Empyrean X-ray diffractometer equipped with a PixCel3D detector and a graphite monochromator. CuKa1 radiation with l = 1.540598 Å at 45 kV and 40 mA was used. 2θ was varied in the range 10-80° at a scanning speed of 0.026°/s. The XRD spectra were analysed with the software Highscore Plus using the ICDD-JCDPS PDF-2 database.

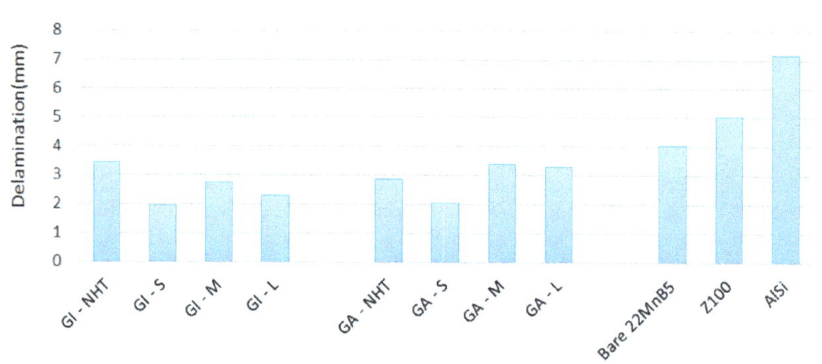

Fig. 1. The delamination from the scribe on panels exposed to 6 weeks of VDA 233-102 accelerated corrosion test.

3. Results

3.1. *Results from corrosion investigations*

The results from the accelerated corrosion test are shown in Fig. 1.

The GI and GA-coated material for the direct press hardening process show very good corrosion protection properties, better than the references i.e. bare 22MnB5, AlSi and Z100. Further, the panels with the short and medium dwell time show the best results, with low delamination from the scribe. The panels without heat treatment show a poorer cosmetic corrosion result. This includes the GI and GA panels without a heat treatment (NHT-GI, NHT-GA) as well as the cold formed Z100 material which show a poorer cosmetic corrosion resistance compared to the GI and GA material. Figure 2 shows galvanostatic measurement at a current density of 11.7 mA/cm^2, of the galvanized (GI) boron steel.

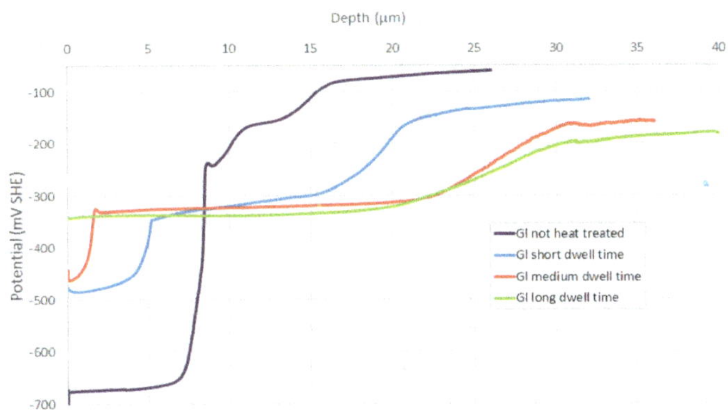

Fig. 2. Galvanostatic measurement of GI pre-coated steel with different dwell time. Electrolyte: 100g/l ZnSO4*7H2O+200g/l NaCl, T= 25°C i= 11.7mA/cm^2.

As can be seen in the figure the GI material with no heat treatment (GI-NHT) has an initial potential of approximately -680 mV at the surface. At a distance of 15-20 μm the potential increases to around -50mV. It should be remembered that the depth given on X-axis is an approximate value. For the short dwell time material (GI-S) the potential is higher at the surface, with an initial potential of approximately -470 mV. At a distance of 5μm the potential increases towards -320 mV and at a depth of approximately 20μm the potential increase again to a level of -120 mV. For the medium dwell time (GI-M) the potential on the surface is around -450 mV. However the "depth" of this potential is only a few μm. For the long dwell time (GI-L) the initial potential at the surface is even higher than for the GI-M samples (-340 mV). At a depth of 20 μm for GI-S samples and at a depth of 30 μm for the GI-M and GI-L samples, respectively, it reaches the potential of the base material, the press harden boron steel. The potential of the base-material the press hardened steel should diverge against the same value. However, as seen Fig. 2, the final potential of the base material seems to be dependent of the heating dwell time. The results are a bit unexpected and no explanation can be provided for this behaviour. The same tendency was reported by Autengruber [4].

3.2. *Characterization of the coating layer*

Scanning electron microscopy images of cross sections of the galvanized (GI) samples with different dwell times can be seen in Fig. 3 and Fig. 4.

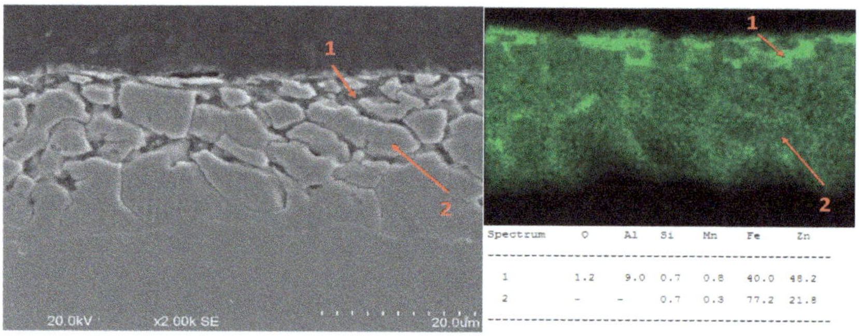

Spectrum	O	Al	Si	Mn	Fe	Zn
1	1.2	9.0	0.7	0.8	40.0	48.2
2	-	-	0.7	0.3	77.2	21.8

Fig. 3. Left picture: SEM image of cross section of the galvanised (GI) boron steel, short dwell time. Right picture, EDS element mapping of the Zn content and EDS point analysis of the element composition in point 1 and 2.

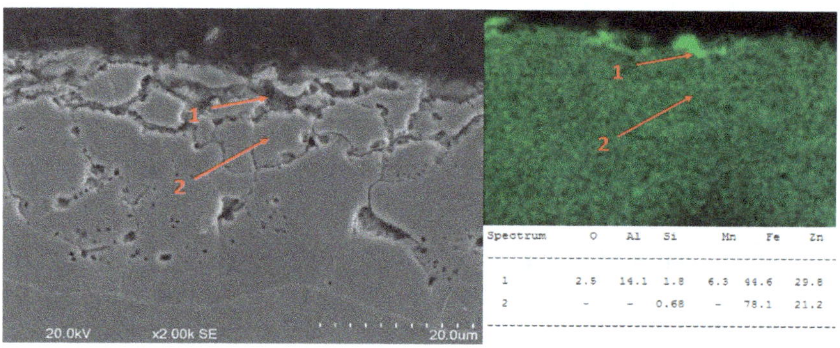

Spectrum	O	Al	Si	Mn	Fe	Zn
1	2.5	14.1	1.8	6.3	44.6	29.8
2	-	-	0.68	-	78.1	21.2

Fig. 4. Left picture: SEM image of cross section of the galvanised (GI) boron steel, long dwell time. Right picture: EDS element mapping of the Zn content. EDS point analysis of the element composition in point 1 and 2.

On the left hand side, the electron image can be viewed and to the right hand side, the Zn content in EDS mapping mode is shown together with point analysis. The analysed points are marked in the SEM image and the corresponding EDS mapping image. Fig. 3 shows the galvanised GI-S sample. The EDS mapping image shows a phase with high content of zinc (Point 1). According to the point analysis the phase contains a zinc content of approximately 50 wt.%. This higher zinc phase is primarily located near the surface and between the grains. It should also be noted that there is a high amount of Al in point 1. Fig. 4 shows the cross section after long dwell time. The EDS mapping shows that only a small area with high zinc content can be found. XRD investigations of the surfaces of galvanized samples with different dwell times can be seen in Fig. 5.

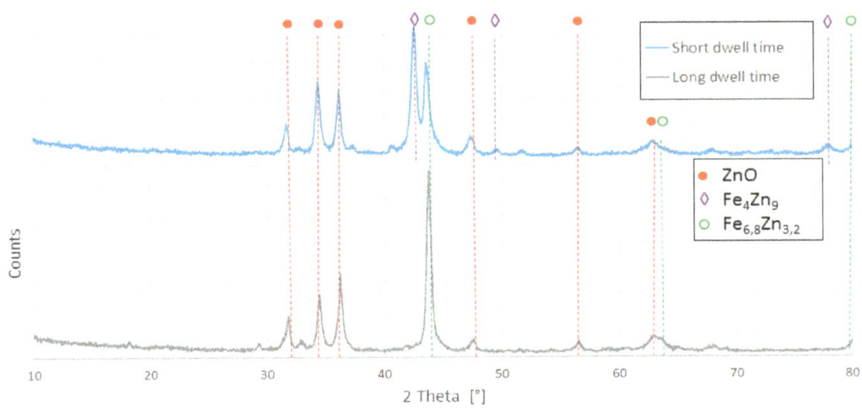

Fig. 5. XRD spectra of Galvanized boron steel after short and long dwell time.

Using the software Highscore Plus, three main phase components can be identified i.e. ZnO, Fe_4Zn_9 and $Fe_{6,8}Zn_{3,2}$. After short dwell time large peaks that can be assigned to ZnO and to the binary phase Fe_4Zn_9, peaks that can be assigned to $Fe_{6,8}Zn_{3,2}$ are also observed. However the intensity of these peaks are lower compared to the intensity of those found for the Fe_4Zn_9 phase. After a long dwell time the peaks assigned to the phase Fe_4Zn_9 have disappeared. Only the peak attributed to $Fe_{6,8}Zn_{3,2}$ and to ZnO can be seen in the spectra.

4. Discussion

The zinc coated hot stamped steel show a better corrosion behaviour compared to the references i.e. AlSi coated material, Z100 and bare 22MnB5. The results from this investigation supports the results obtained by Dosdat et al [5]. Also the appearance of the corrosion attack is very similar to the attacks seen by Dosdat et al. It is a bit surprising to see that the bare 22MnB5 show a comparably good corrosion behaviour in this test, with lesser delamination from the scribe compared to the references, i.e. AlSi coated material and Z100. However, the same order of cosmetic corrosion were reported by Dosdat et al. [5].

For both the GI and GA material the short heat treatment shows the best cosmetic corrosion behaviour compared to samples with longer dwell time but also better than the samples not heat treated at all. The results obtained are supported by the results reported by Autengruber et al. [4]. The explanation for this optimum corrosion behaviour after short dwell time can be found in the microstructure of the coatings. The SEM images (Fig. 3 and Fig. 4) show that a phase with a higher zinc content is located close to the surface of the coating, primarily located towards the grain boundaries. According to the point analysis the phase contains a high zinc content (around 50%). The XRD show the presence of three main phase components in the GI material, identified as ZnO, Fe_4Zn_9 and $Fe_{6,8}Zn_{3,2}$. After long dwell time the peaks for the Fe_4Zn_9 phase, with a high zinc content, disappeared. Leaving, beside the ZnO, the $Fe_{6,8}Zn_{3,2}$ phase, with a lower zinc content.

Comparing the potentials obtained in the electrochemical potential measurements with the results obtained in the SEM and the XRD investigation. The potential values can be assigned to different intermetallic phases. The potential -470 mV, seen for both GI short dwell time and medium dwell time, can be attributed to the Fe_4Zn_9 phase. The phase $Fe_{6,8}Zn_{3,2}$ can most likely be attributed to a potential of around -320 mV. The potential values of the different phase are similar to the values reported by Autengruber et al. [4]. Both of the Fe–Zn-phases i.e. the Fe_4Zn_9 phase and the $Fe_{6,8}Zn_{3,2}$ phase, are significantly below the potential of the base material, the boron steel. Hence, indicating that both of these phases present in the coating of press hardened galvanized steel sheet offers cathodic corrosion protection for the substrate. The optimum corrosion resistance obtained after short dwell time can thus probably be attributed to a combined effect of the galvanic corrosion protection, the amount and type of Fe–Zn-intermetallic phases formed.

5. Conclusion

The results in the present study show that zinc coated material for the direct press hardening process show better cosmetic corrosion properties compared to the references used i.e. Aluminized 22MnB5, bare press-hardened 22MnB5 and standard galvanized steel for cold forming. The zinc coated material has an electrochemical potential significantly below the potential of the base material, the boron steel, thus, offering a cathodic corrosion protection for the substrate. However the cathodic protection is influenced by the heating dwell time in the press hardening process. The heating dwell time governs the amount and type of Fe–Zn-intermetallic phases formed. Consequently the cosmetic corrosion behaviour of the zinc coated material is connected to the heating dwell time in the press hardening process.

References

1. C. Allély, L. Dosdat, O. Clauzeau, K. Ogle and P. Volovitch, Anticorrosion mechanisms of aluminized steel for hot stamping, *Surf. Coatings Technol.* **238**, 188 (2014)
2. A. R. Marder, The metallurgy of zinc-coated steel, *Prog. Mater. Sci.* **45**, 191 (2000).
3. K. Isaksson, M. Jönsson and D. Berglund, The direct press hardening process for Zn-coated ultra-high strength steels, in *ICHSU 2015*, (Changsha, China, 2015).

4. R. Autengruber, G. Luckeneder and A. W. Hassel, Corrosion of press-hardened galvanized steel, *Corros. Sci.* **63**, 12 (2012).
5. L. Dosdat, J. Petitjean, T. Vietoris and O. Clauzeau, Corrosion resistance of different metallic coatings on press-hardened steels for automotive, *Steel Res. Int.* **82**, 726 (2011).

Morphology Alternation for Iron-Zinc Coating After Hot Stamping

F. Fang[†], Y. J. Bi, G. W. Feng, R. Ge and Y. Chen
Research and development center, Wuhan iron and steel Co.
Wuhan, Hubei, China
[†]*E-mail: honeyfreda@163.com*

The morphology of a hot press forming steel galvannealed by HDPS (hot dipping process simulator) and hot stamped with different re-austenization temperatures in laboratory was studied in the paper. The results showed that, coating thickness may have a great effect on morphology transformation after hot stamping. Two typical thicknesses of 23 and 42 μm galvannealed samples were chosen, with coating morphology of delta phase and zeta phase (only thicker coating) transformed to Zn-rich gamma phase (Γ-Fe_4Zn_9) and Fe-rich alpha phase (α-Fe (Zn)) after hot stamping. The coating morphologies of three areas of the omega-shaped samples were analyzed, mainly composed of Γ-Fe_4Zn_9, α-Fe (Zn) and ZnO, with none of the microcracks penetrated to the matrix. The corrosion property can be guaranteed by Zn content in the two phases (Γ-Fe_4Zn_9 and α-Fe (Zn)) in which both were greater than 30%. The external oxides were also observed and the alloying element Si promoted Mn_2SiO_4 formation along grain boundaries.

Keywords: Iron-zinc intermetallics; Microcrack; Hot stamping.

1. Introduction

Hot Stamping of boron alloyed steels gained more and more importance over the last years. This is reflected in the increasing number of hot stamped parts in modern car bodies and consequently in an increasing number of corresponding production lines [1]. In the mean time, the requirement profile for surface coatings has become ever more demanding. As well as protection against scale and formability both in the direct and indirect forming processes, it is also necessary to ensure suitability for different further processing techniques typically used in the automotive industry, such as resistance spot welding and painting. However, corrosion protection in particular has become increasingly important. The aluminum-silicon coating only provides protection against corrosion due to its barrier effect. Active corrosion protection is primarily guaranteed by zinc components in the coating [2-5].

Considering the process chain of hot stamping, there are four major research fields: Materials and coating, heating technologies, new tool-concepts, tool materials and process routes and following operations [6]. Also, when the sheet is drawn out of the bath, delicately controlled high pressure air knives blow specific amounts of the liquid zinc from steel substrate [7]. Verifying the actual amount of zinc deposited on the steel is critical for performance and profitability. If the coating is too thin, it will not provide the necessary corrosion protection; if it is too thick, the galvanize producer is losing money by over-coating with more zinc than required. Thus, galvannealed samples with different thicknesses were used to conduct series of heating and die cooling processes in this research, to study the effect of coating thickness on formation of micro cracks and alternation of zinc-iron phases after hot stamping.

2. Experimental Procedure

2.1. *Coating*

The hot stamping steel used in the paper was industrially produced with a regular cold-rolled mill finish to thickness of 1.5mm. The material chemistry was 22MnB5 with a small amount of Cr and Ti to achieve the desirable mechanical properties.

The galvannealed samples were conducted on Iwatani-Surtec HDPS simulator. The coupons comprised 110mm × 220mm with the long axis parallel to both the rolling and dipping directions, cleaned and dried per standard practice [8]. The samples were preheated to 700°C for 3 min under protective atmosphere (15%H2-85%N2) with dew point of -30°C, hot dipped in zinc bath with 0.15%Al solved. The air knife parameters and dipping time were changed to achieve different coating thicknesses. The galvannealing process was stabilized to eliminate its effect on later hot stamping process.

2.2. *Hot stamping test*

Hot stamping experiment was carried out on a hot forming pilot line with a set of U-channel dies. Two galvannealed samples were laser welded and heated around 900°C for 5 minutes in a chamber furnace without protective atmosphere. The samples were manually transferred to the U-channel die for hot stamping to form omega-shape samples. The time before die closing was 8 s, and pressing temperature was below 600°C.The temperature curves of the chamber furnace was described in detail by Yunjie Bi [9].

Specimens were laser cut from three areas of the omega-shaped samples: wall side, bottom side and bended corner. The coating morphology and phase analysis were performed on SEM (fei Quanta 400) equipped with Oxford EDAX system on the cross section of the specimens. Coating thickness before and after hot stamping was measured by Glow Discharge Optical Emissivity Spectroscopy (GDOES) using a Horiba Jobin-Yvon Profiler, the results were averaged by three points on each side.

3. Results and Discussions

The sample coating thicknesses after galvannealing with different air knife parameters and dipping times were measured by GDOES, two typical thicknesses of 23 and 42 μm were chosen for later hot stamping process, respectively. The initial galvannealed coating predominantly consisted of a uniform delta (δ) phase ($FeZn_7$). With increasing coating thickness, a little amount of zeta (ζ) phase ($FeZn_{13}$) forms on the surface, which indicated insufficient alloying treatment due to the thick coating. After reheating in the chamber furnace for reaustenization, the coating morphology changed and became thicker. In differing to other samples, the three areas of samples were described as W1/C1/B1 (thinner coats) and W2/C2/B2 (thicker coats), in which W-wall side, C-bended corner, B-bottom side.

3.1. *Wall side*

SEM analysis revealed that the Zn coating after hot stamping generally contained two distinctive phases, i.e. the Zn-rich gamma phase (Γ-Fe_4Zn_9) and Fe-rich alpha phase (α-Fe (Zn)), and no microcrack appeared in this area. Surface oxidation occurred during annealing, leading to a surface covered with ZnO. There was no significant evaporation of Zn from the coating layer during the annealing process. Coatings contained greater than 30 (at%) Zn in α-Fe(Zn) which verified cathodic protection potential through corrosion testing and electrochemical work. The results conformed to the recent findings of Y. Kim and J.R. McDermid [10]. Γ-Fe_4Zn_9 dissapeared in the thinner coating as shown in Fig 2a, the coating was complete α-Fe(Zn) with some ZnO on the surface, and three phases were all observed in Fig 2b. Coating weight seems to be an essential variable in defining a process window, with wide variations in final microstructure seen as a function of coating weight.

The results analysed above showed that, although the wall side received most deformation from the die, the coating morphology seemed good enough that distributed homogenously and continuously, with no penetrated microcrack. In addition, the corrosion property can be guaranteed by Zn content in the two phases (Γ-Fe_4Zn_9 and α-Fe (Zn)) that both were greater than 30%.

Fig. 1. Coating morphology after hot stamp (a) W1, (b) W2.

3.2. *Bended corner*

The micro cracks appeared in the coating area but none of which penetrated into the matrix in bended corner. The outer side morphology of the bended corner was displayed in Fig. 2. There is a little amount of Γ-Fe_4Zn_9 remained in the thinner coating and ZnO on coating surface.

As the matrix was 22MnB5, there were alloying elements of Mn, Si, and B in the steel. The external oxides displayed that the alloying element Si promoted Mn_2SiO_4 formation along grain boundaries. Vahid A. Lashgari [11] had mentioned that oxidation mode changes from external oxidation to internal oxidation with increasing dew point (from -45 to +10 °C). Oxides are formed along grain boundaries and the oxide precipitates are not uniformly distributed. The results in the paper were conformed to the above research.

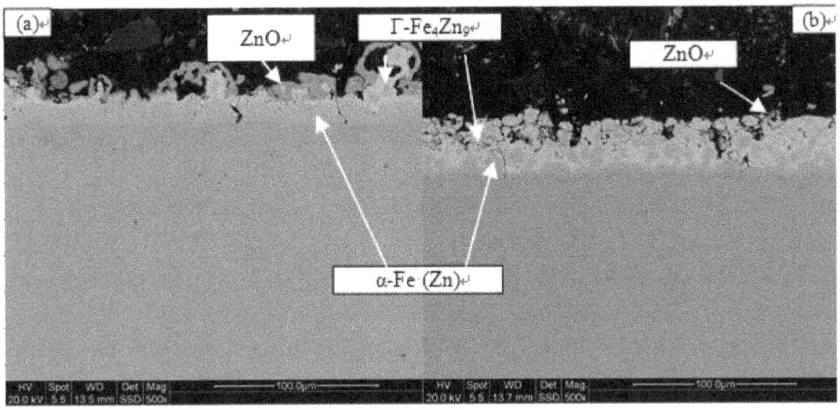

Fig. 2. Coating morphology after hot stamp (a) C1, (b) C2.

3.3. *Bottom side*

The bottom side of the sample received no deformation from the U-channel die, the pressing effect was the same to planar die. The coating thicknesses were 32 and 48 μm, respectively. The coating became thicker and under-continuous after hot stamping. Three phases were all exited in thinner and thicker coatings, as shown in Fig. 3. Microcracks were observed but no penetration to the matrix.

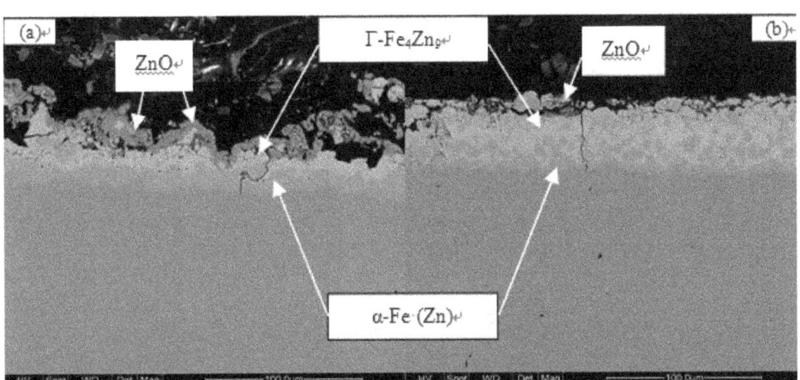

Fig. 3. Coating morphology after hot stamp (a) B1, (b) B2.

4. Conclusion

Two typical thicknesses of 23 and 42 μm galvannealed samples were chosen for hot stamping process, the coating morphology of delta phase and zeta phase (only thicker coating) transformed to Zn-rich gamma phase (Γ-Fe_4Zn_9) and Fe-rich alpha phase (α-Fe (Zn)). The coating morphologies of three areas of the omega-shaped samples were mainly composed of Γ-Fe_4Zn_9, α-Fe (Zn) and ZnO, with microcracks none of which penetrated to the matrix. The corrosion property can be guaranteed by Zn content in the two phases (Γ-Fe_4Zn_9 and α-Fe (Zn)) that both were greater than 30%.

References

1. R. Kolleck and R. Veit, Current and Future Trends in the field of Hot Stamping of Car Body Parts, *GALVATECH* 40 (2011).
2. M. Köyerl, P. Kuhn and J. Kondratiuk, New Coating for hot press forming, *GALVATECH* 451 (2011).
3. R. Autengruber, G. Luckeneder and A.W. Hassel, Corrosion of press-hardednd galvanized steel, *Corr. Sci.* **63**, 12 (2012).
4. K. Thomas, S. Harald and L. Gerald, Zinc coated press-hardening steel-challengers and solutions, *Proceedings hot sheet metal forming of high-performance steel* 345 (2015).
5. Y.G. Kim, I.J. Kim, J.S. Kim, Y.I. Chung and D.Y. Choi, Evaluation of Surface Crack in Resistance Spot Welds of Zn-Coated Steel, *Mater. Trans.* **55**, 171 (2014).
6. K. Ralf and V. Robert, Current and Future Trends in the field of Hot Stamping of Car Body Parts, *GALVATECH* 40 (2011).
7. B. Christopher, H. Bernd and R. Zhang, Measurement of Zinc/Magnesium Coatings Utilizing Online XRF Sensors, *GALVATECH* 323 (2013).
8. E.M. Bellhouse and J.R. McDermid, Selective Oxidation and Reactive Wetting during Hot-Dip Galvanizing of a 1.0% Al - 0.5% Si TRIP-Assisted Steel, *Metall. Mater. Trans. A* **43**, 2426 (2012).
9. Y.J. Bi, G.W. Feng and F. Fang, The development of the coated hot forming steels in *WISCO. Proc. of CHS2 5th International Conf. on Hot Sheet Metal Forming of High-Performance Steel*, (Toronto, Canada, 2015).
10. Y. Kim, J.R. McDermid, J. Kish and M.J. Worswick, ZCO-70: Nature of Coating Development and Substrate Microcracking during Hot Press Forming, *GAP meeting* (2015).
11. V.A. Lashgari, W.C. Mao and W.G. Sloof, Oxidation Phenomena in Advanced High Strength Steels, IZA project: ZCO-62, *GAP meeting* (2015).

The Research on Al-Si Coating for Hot Forming Steel

S. L. Zhou[a], L. Cui[b], Y. G. Liu[c], W. Zhang[d], H. Zhan[e], J. C. Jin[f] and L. Chen[g]
Technology Center of Ma'anshan Iron and Steel Company, Ma'anshan, P.R. China
E-mail: [a]superzsl11@163.com, [b]cuilei1219@163.com, [c]lyghotmail@hotmail.com, [d]zwniat@126.com, [e]cquzhanhua2000@yahoo.com.cn, [f]jinjiachun1988@163.com, [g]chenlekuaile@163.com

Hot forming technology can not only improve the impact resistance and safety performance of the car, but also reduce its weight, so it has become a popular research topic in China and abroad. Hot forming steel is divided into uncoated hot forming steel and coated hot forming steel. Hot stamping parts of coated hot forming steel do not need shot blasting treatment and have better corrosion resistance compared with uncoated hot forming steel, so coated hot forming steel has been used more widely. Al-Si coating is the most mature coating for hot forming steel, and the Zn coating hot forming steel is still in the research stage because of the narrow process window. This paper focuses on Masteel Al-Si coating hot forming steel, whose coating composition, matrix microstructure and mechanical properties before and after hot stamping is studied. The research shows that Masteel Al-Si coating hot forming steel has good performance, and it can meet the requirements of mainstream car manufacturers.

Keywords: Impact resistance; Safety; Hot forming steel; Al-Si coating; Corrosion resistance.

1. Introduction

With the increasingly strict requirements of automobile industry emission regulations, the lightweight of the automotive industry has become the main theme for the moment [1]. Hot forming technology meets the market demand, it can reduce the weight of the body material under the premise of ensuring the safety of the car body structure [2, 3]. Since 2000, the demand for hot forming parts has increased rapidly. The production of hot forming parts reaches 107 million in 2007, the global hot forming production line reaches 110 in 2009. It is estimated that the world's demand for hot stamping parts will be more than 600 million in 2015, the corresponding amount of hot forming steel will be nearly 2 million tons.

Hot forming steel is divided into uncoated hot forming steel and coated hot forming steel. Hot stamping parts of coated hot forming steel do not need shot blasting treatment and have better corrosion resistance compared with uncoated hot forming steel, so coated hot forming steel has been used more widely. Application of Al-Si coating hot forming steel is the most mature, and the Zn coating hot forming steel is still in the research stage because of the narrow process window [4,5]. This article mainly carries on the comparative analysis on the material properties of Masteel Al-Si coating hot forming steel before and after hot stamping.

[a] Corresponding author.

2. Experimental Materials and Process

The material is 22MnB5 of Masteel, whose chemical composition is shown in table 1, and its thickness is 1.2mm.

Table 1. Element content of 22MnB5(wt%).

C	Si	Mo	P	S	Al	B
0.23	0.27	0.002	0.021	0.001	0.045	0.003

The experiment of hot dip Al-Si is by aluminum silicon production line, and the target thickness of coating is 25μm.

The process parameters of hot stamping are: heating temperature: 930°C, holding time: 3 min, pressing rate 600mm/s, cooling rate>50°C /s.

3. Results and Discussion

3.1. *Coating composition*

Figure 1 for the Al-Si coating section morphology, table 2 for the coating composition analysis results. From figure 1, the Al-Si coating is mainly composed of two parts, near the side of the substrate coating C and the upper part of the coating B, and the thickness of C is about 5μm while the thickness of B is about 20μm. Using EDS to analyze its composition, coating C is mainly ternary alloy phase Fe2SiAl7, coating B is Al-Si phase, and there is very thin Fe2Al5 and FeAl3 phase between the substrate and coating.

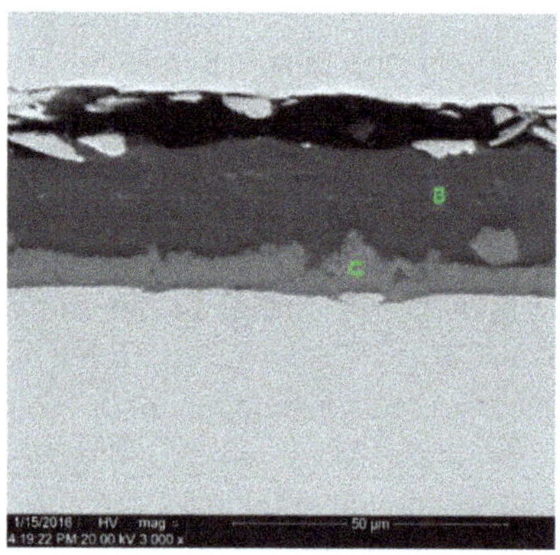

Fig. 1. Al-Si coating section morphology.

Table 2. Al-Si coating composition (wt%).

Element	Fe	Al	Si
B	-	81.70	18.30
C	31.04	57.11	11.85

Figure 2 is the section morphology of the coating after hot stamping, table 3 is the composition of the coating after hot stamping. From figure 2, the Al-Si coating becomes thicker after hot stamping, and its thickness is about 35μm. Because of the diffusion of Fe in the substrate to the coating, the coating becomes thicker, and it makes Fe content of the coating increase. The coating is composed of Fe-Al binary alloy phase and Fe-Al-Si ternary alloy phase, and each phase from the upper part of the coating to the substrate is $FeAl_2$、Fe_2SiAl_2、$FeAl_2$、Fe_2SiAl_2、Fe_3Al.

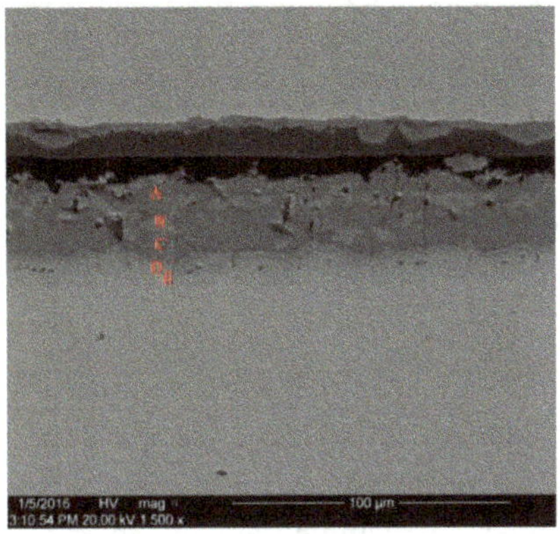

Fig. 2. The coating section morphology after hot stamping.

Table 3. The composition of the coating after hot stamping (wt%).

Element	Fe	Al	Si
A	44.30	54.65	1.05
B	63.10	29.42	07.48
C	44.83	54.22	0.95
D	61.88	26.67	11.45
E	78.05	16.05	5.91

3.2. *Matrix microstructure*

The matrix microstructure of Al-Si coating before and after hot forming is shown in figure 3, the microstructure analysis method refers to GB/T13298-1991. The matrix microstructure of Al-Si coating is ferrite, pearlite and dispersed carbide particles before hot stamping; The matrix microstructure of Al-Si coating changes into tempered martensite after hot forming. Because the steel plate is heated to 930°C and holding the temperature for a period of time, the matrix changes into austenite structure, then rapid cooling after hot stamping (cooling rate: > 50°C /s), and the austenite changes into martensite. And the material temperature after cooling is about 200 degrees, so it causes self-tempering, and then the martensite changes into tempered martensite.

The test of matrix hardness after hot stamping refers to GB/4340.1-2009, the test results are shown in table 4.

Table 4. The matrix hardness after hot stamping.

Test	1	2	3	4	5	Average
HV30	458	464	470	468	467	465

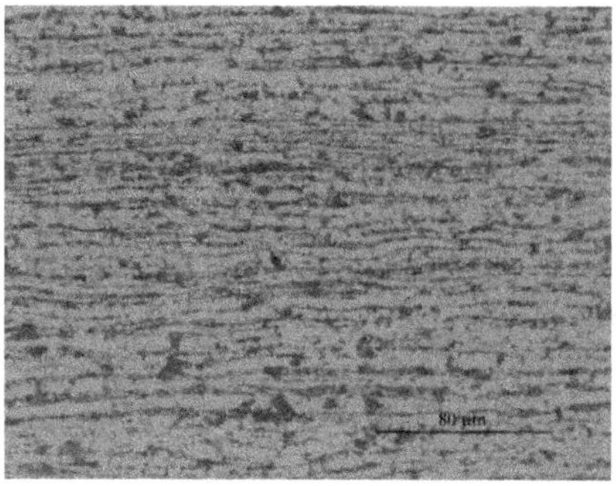

(a) Before hot forming.

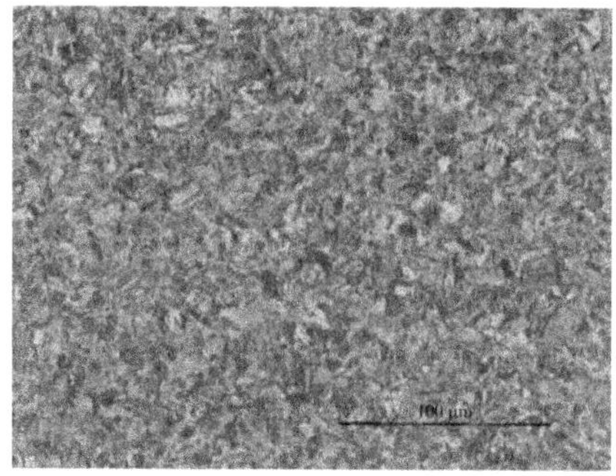

(b) After hot forming.

Fig. 3. The matrix microstructure.

3.3. *Mechanical properties*

Due to the formation of martensite after hot stamping, the strength of the material increases but the elongation decreases. Mechanical properties of Al-Si coating hot forming steel before and after hot stamping are shown in table 5, table 6, referring to GB/T228.1-2010, and the stress-strain curve is shown in figure 4. So the strength of Al-Si coating hot forming steel after hot stamping is much higher than the strength before hot stamping, Rm from an average 587Mpa to an average 1503Mpa. But the elongation of fracture after hot stamping decreases sharply, which decreases from 23.91 to 6.48.

Table 5. Mechanical properties before hot stamping.

Test	$Rp_{0.2}$/Mpa	R_m/Mpa	A_{80}/%
1	434	577	22.99
2	453	596	25.76
3	448	587	22.99
Average	445	587	23.91

Table 6. Mechanical properties after hot stamping.

Test	$Rp_{0.2}$/Mpa	R_m/Mpa	A_{50}/%
1	987	1465	6.48
2	1033	1499	6.42
3	1038	1517	6.36
4	1056	1530	6.66
Average	1029	1503	6.48

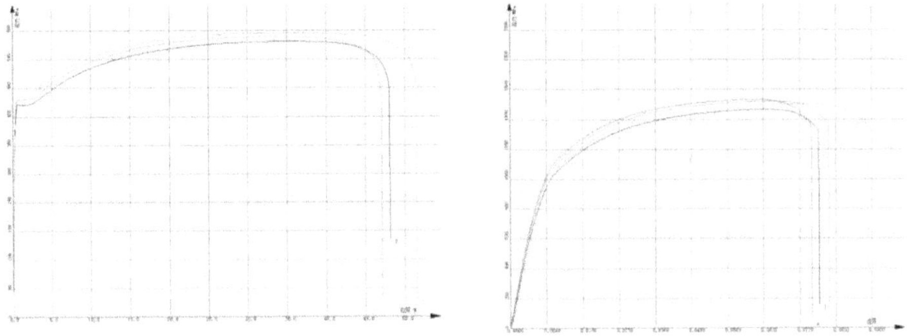

(a) Before hot stamping. (b) After hot stamping.

Fig. 4. Stress-strain curve.

4. Conclusion

1) Al-Si coating is mainly composed of two parts before hot forming, the upper part is Al-Si phase, the other part near the side of the substrate is Fe_2SiAl_7. After hot forming, because of the diffusion of Fe in the substrate to the coating, the coating becomes thicker, and the coating is composed of Fe-Al binary alloy phase and Fe-Al-Si ternary alloy phase.
2) The matrix microstructure of Al-Si coating is ferrite, pearlite and dispersed carbide particles before hot stamping, which changes into tempered martensite whose microhardness is high after hot forming.
3) After hot forming, the average tensile strength of Masteel Al-Si coating hot forming steel is 1503Mpa, the average fracture elongation is 6.48. It shows that Masteel Al-Si coating hot forming steel has good mechanical performance, and it can meet the requirements of mainstream car manufacturers.

References

1. S. Huang, Y. Zhao and C. He, Shear fracture of advanced high strength steels, *J. Iron and Steel Research, International* **21**, 938 (2014).
2. D.W. Fan and B.C. D. Cooman, State of the Knowledge on Coating Systems for Hot Stamped Parts, *Steel research international* **83**, 412 (2012).
3. H. Karbasian and A.E. Tekkaya, A review on hot stamping, *J. Mater. Process. Tech.* **210**, 2103 (2010).
4. Ing-form galvanized UHSS to body-in-white application. Genava: Associazione Italiana di Metallurgia, Galvatech'11(2011).
5. P.X. Deligne and B.H. Gross, On the exceptional series, and its descendants, *C. R. Math. Acad. Sci. Paris* **335**, 877 (2002).
6. C.W. Lee, D.W. Fan, I.R. Sohn, S.J. Lee, and D. Cooman, Liquid-metal-induced embrittlement of Zn-coated hot stamping steel, *Metal. Mater.Trans A* **43**, 5122 (2012).

Microstructure Evolution Characterization of the Third Generation Automotive Steel and its Influence on the Mechanical Properties

J. Y. Ma[†], X.D. Liu, Z. F. Wang and Z. J. Deng

Research and Development Center of Wuhan Iron and Steel (Group) CORP., Wuhan 430080, China

[†]E-mail: ma_jia_yan@163.com

Microstructure evolution characteristics of the third generation automotive steel in hot rolling, cover annealing, cold rolling and simulation continuous annealing procedure conditions were observed by optical microscope (OM) and transmission electron microscope (TEM), and the relationship between microstructure and properties was analyzed. The results showed that a large number of short rod-like with lager size and granular carbides existed in the cover annealing plate, while in continuous annealing plate there were only a small spherical carbides; The average grain sizes of continuous annealing plate are smaller than that of cover annealing plate. Above results indicated that solid solution strengthening and fine grain strengthening effects in continuous annealing plate significantly higher than that in cover annealing plate. Although the tensile strength of the continuous annealing plate is 200MPa higher than that of the cover annealing plate, the elongation and strong plastic product are lower than that of the cover annealing plate. Tempering again at lower temperature may be a good way to improve the strength plastic product of continuous annealing plate.

Keywords: The third generation automotive steel; Microstructure evolution characteristics; Properties.

1. Introduction

The performance including strength and plasticity of automotive steel is put forward increased demands in order to accommodate the dual requirements of lightweight and safety. China Gangyan Technology Group in the laboratory successfully developed out the third generation automotive steel with the strong plastic product of 30GPa%. The results of the previous studies show that when the microstructure of the third generation automotive steel contains martensite matrix and about 30% residual austenite, the product of tensile strength and elongation can arrive at 30GPa%~40GPa%. Therefore, the third generation automobile steel has good comprehensive properties and application prospect [1-3]. Currently, Wuhan Iron and Steel is committed to the development of the third generation automotive steel with the content of Mn about 5%~10%.

The production process of the third generation automotive steel contains hot rolling, cover annealing, cold rolling, and continuous annealing. The organization of each process has a certain influence on it of the posterior procedure, so it is necessary to study detailed the microstructure evolution characterizationin each process, which is helpful to guide process improvement and improve product performance.

This paper reports a detailed microstructure evolution characterization of the third generation automotive steel with 5Mn in different production processes conditions by a variety of detection and analysis means. The influence of the microstructure on the performance is discussed.

2. Test Material and Experimental Method

The main chemical components of the test materials are shown in Table 1. The Mn content is 5%. The main function of Mn is to enlarge the austenite region and raise the stability of austenite.

Table 1. Chemical composition of the test steels(wt%).

C	Mn	P	S	N	Als
0.1	5	<0.005	<0.0017	<0.0054	0.1

Test materials are hot-rolled sample 1#, cover annealing sample 2#, cold rolling sample 3# and simulation continuous annealing sample 4#. Main production process is: vacuum induction smelting, forging into the slab, hot rolled 3.8 mm, coiling, cover annealing at 650°C, cold-rolled to thin plate of 2 mm, simulation continuous annealing with heating temperatures at 675°C holding time in 30min and cooling rate at 15 °C /s.

Metallographic samples were etched with 4% nitric acid alcohol solution after grounding and polishing. The microstructures were observed by optical microscope (OM) and transmission electron microscope (TEM). TEM thin specimens were prepared using the double jet electrolytic method. Then the fine microstructure and precipitates were observed on JEM 2100F field emission TEM. The micro-area composition can be achieved by INCA EDS.

3. Experimental Results and Analysis

3.1. *Mechanical properties*

The mechanical properties of samples 2# and 4# are shown in Table 2. R_{eL} and R_m denote as lower yield strength and tensile strength, respectively, while A represents elongation. The results show that the tensile strength of sample 4# is 200 MPa higher than that of sample 2#, while the elongation of sample 2# is 13.5% higher than that of sample 4#, and the strong plastic product of sample 2# is 5.4GPa % bigger than that of sample 4#.

Table 2. Mechanical properties of test steels.

Sample No	R_{eL} / MPa	R_m / MPa	A_{50mm} / %	$R_m \times A$ /(GPa%)
2#	607	739	35.5	25.6
4#	631	939	22	20.2

3.2. *Microstructure observation results*

Figure 1 shows the microstructure of manganese steel in four different processes condition by OM. The microstructures of 1# mainly consist of bainite and a small amount of M/A islands with serious composition segregation, as shown in Fig. 1 (a, b). The microstructure of sample 2# is mainly bainite, while that of 3# are deformation organization. The microstructures of sample 4# mainly consists of bainite and martensite. Due to the limited resolution of optical metallographic microscope, the fine organization of samples needs to be further confirmed by TEM.

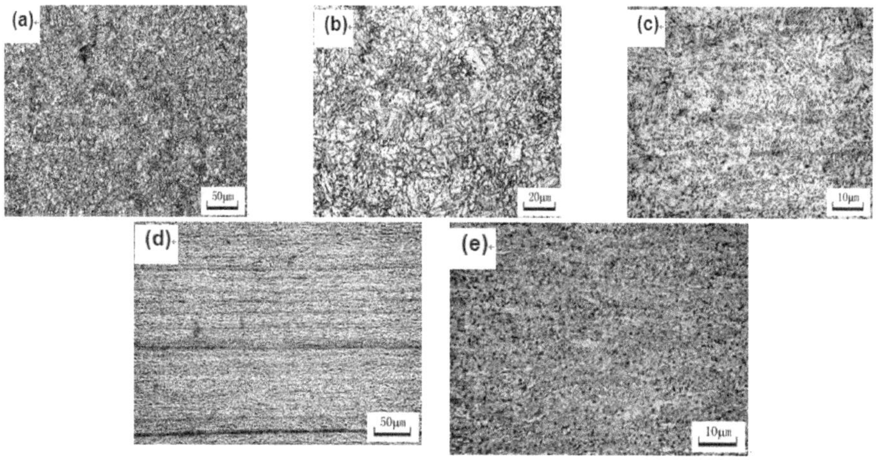

Fig. 1. Microstructures by OM of medium magnesium steels in different processes (a,b) 1#, (c) 2#, (d) 3#, (e)4#.

The microstructures of medium magnesium steel in four different processes conditions by TEM are shown as in Figure 2. For the sample 1#, the microstructures mainly consist of lath bainite, a small amount of M/A islands distributing along the bainite lath boundaries and a lot of acicular or granular carbides precipitates within lath, as shown in Fig. 2(a~c). The dislocation density is higher in sample 1#. As shown in Fig. 2 (d~g), the microstructures of sample 2# are made of lath bainite with bigger size than sample 1#, superfine crystalline ferrite, a small amount of lath martensite and austenite. The ferrite grain size is mainly from 540 nm to 2μm. A large number of stick precipitates of 70~580 nm and granular carbide precipitates of 30~70 nm exist in lath bainite matrix. As shown in Fig. 2(h, I), for sample 3#, the microstructure is made of deformation organization, lots of granular carbide precipitation phase and a small amount of needle carbides. After slipping and deformation, a large number of cellular substructures are formed. The dislocation density is lower intracellular than that in cell wall with tangled dislocations. For sample 4#, the microstructure mainly contains bainite, ultrafine crystalline ferrite, matensite and austenite. There are small quantity of spherical carbides of 80~270 nm existing in the sample 4#, as shown in Fig. 2 (j, k).

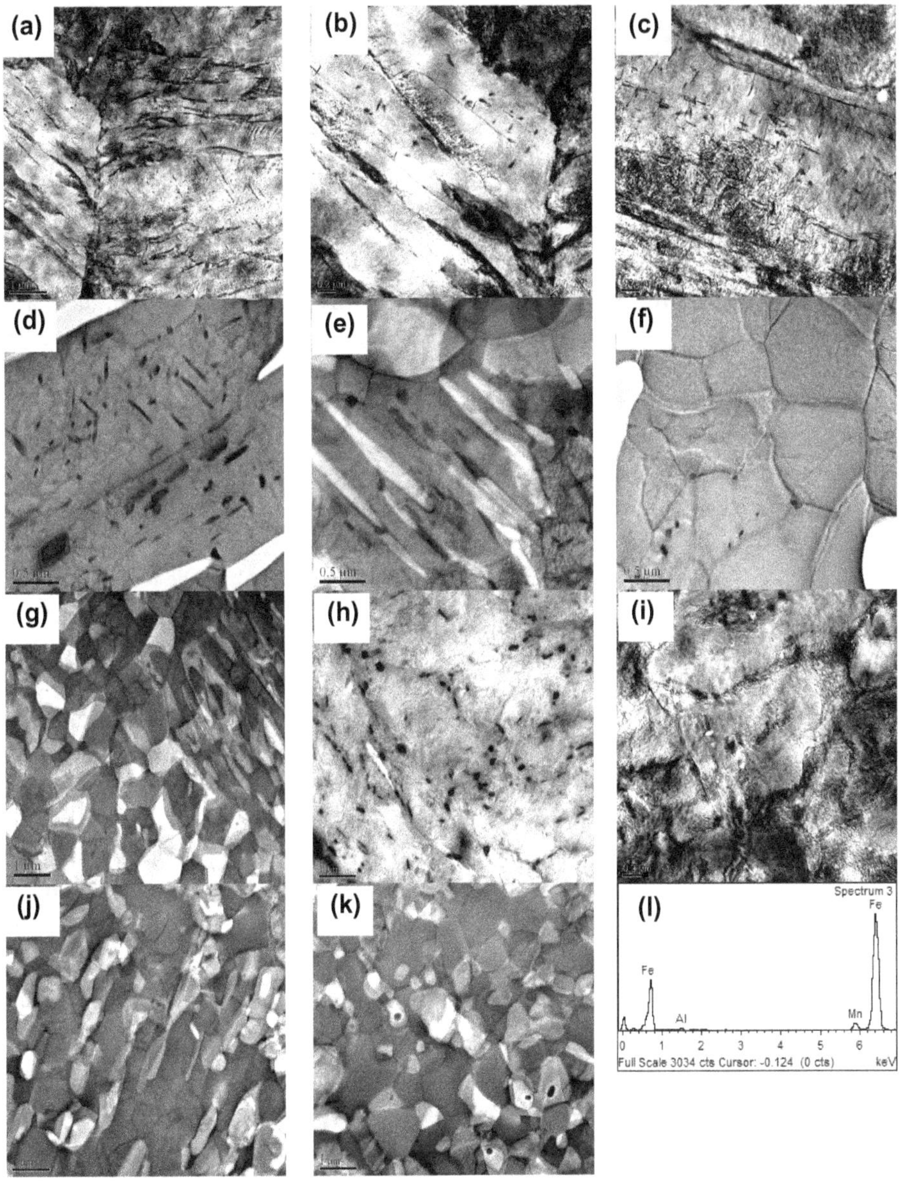

Fig. 2. Microstructures by TEM of medium magnesium steels in different processes (a~c) 1#, (d~g) 2#,(h, i) 3#, (j, k) 4#, (l) EDS of carbide.

3.3. *Result analysis and discussions*

Because of the higher Mn content and more severe composition segregation in steel, direct cold rolling of hot rolled sample is likely to cause craze. Therefore, the cover annealing was done at 650°C in two phase region before cold rolling, which can not only make composition uniform and grains fine, but also improve the mechanical performance and reduce the residual stress.

Generally speaking, in the process of producing the third generation automobile steel, the strong plastic product of the continuous annealing plate should be better than that of the cover annealing plate, and it should be at least 30GPa%. In fact, according to the above results, the performance of the continuous annealing plate has not yet arrived at the requirements of the third generation steel, and even its strong plastic product is just 20.2GPa% lower than the cover annealing plate. In the following, we analyzed the possibility of reasons from the aspects of microstructure. For the sample 2#, there are a large number of stick and granular carbide whose size get bigger than carbides in sample 1# and the bainite bath get wider. Above results shows that the carbide has not completely solid solution and most carbide grew up in the process of annealing at 650°C, so precipitation strengthening effect is not obvious. Meanwhile, because of the slower cooling speed, there were some carbides precipitations which may result that solid solution strengthening weakens. Although the strength is low, the plastic is much better. After cold rolling, the sample 3# with lots of dislocation cell substructure was annealed at 675°C. The sizes of grain get smaller than that of sample 2# result of the recrystallization, so the fine grain strengthening effect is obviously stronger than that of sample 2#. Compared to sample 2#, the quantity of carbides in sample 4# decreased significantly, which resulted that solid solution strengthening effect was stronger. As a result, the tensile strength of sample 4# is 1.27 times of that of sample 2#, the elongation of sample 2# is 1.61 times of that of sample 4#, which results that the product of tensile strength and elongation of sample 4# is smaller than that of sample 2#.

At last, we also put forward the improvement measures. By tempering at lower temperature, a large number of disperse small carbides can precipitation from the martensite, which may increase significantly the elongation and decrease slightly the tensile intensity. This may be an effective method of improving the strong plastic product of continuous annealing plate.

4. Conclusion

The microstructures of the third generation automotive steel in four different working procedure conditions were observed by means of OM and TEM. It was observed that performance of the cover annealing sample is better that of the continuous annealing sample, after their microstructure were analyzed. The higher strength plastic product of the cover annealing sample mainly due to the higher elongation. So, in order to enhance the strength plastic product of the continuous annealing sample, we could increase elongation significantly and at the same time, decrease the tensile intensity slightly by tempering at lower temperature, with a large number of disperse small carbides into the martensite. This should be an effective method to improve the properties of the third generation automobile steel.

References

1. F.C. Shi, Development of third generation automotive steel, *ShanXi Metallurgy* **38**, 1 (2015).
2. B. Han, S. Kuang and J.L. Cao, Effect of continuous annealing time on the microstructure and properties of medium manganese steel in warm-rolling condition, *AT&M Field of Vision* 10 (2014).
3. J.N. Deng, T.S. Xu, L.N. Qiu and Y.F. Shen, Investigation on mechanism of improved toughness for 7.9Mn-1.4Si-0.07C steel, *Journal of Materials and Metallurgy* **13**, 4 (2014).

A Forming Limit Curve Measurement of Sheet Metal Based on Digital Image Correlation

J. Li[1,2], J. Liang[3,†], Y. Z. Jiang[1,2], H. L. Yu[1,2], T. S. Cui[1,2], H. Zhao[1,2] and H. Hu[3]

[1]*Chongqing Changan Automobile Co. Ltd., Chongqing, 400023, China*
[2]*State Key Laboratory of Vehicle NVH and Safety Technology, Chongqing, 401120, China*
[3]*State Key Laboratory for Manufacturing Systems Engineering, School of Mechanical Engineering, Xi'an Jiaotong University, 710049*
[†]*Email: liangjin@mail.xjtu.edu.cn*

The Forming Limit Curve (FLC) is a key indicator of the sheet metal forming property. It is inconvenient to use general ways like tool microscope to obtain FLC after corroding and forming mesh on sheet metal's surface. Therefore, a fast and accurate FLC detection system is proposed in this paper. First, the image sequences are captured by the high-speed camera automatically while the sheet metal is forming during the cup drawing test. Second, by tracking the speckle images of the sheet metal's surface with Digital Image Correlation (DIC) and binocular stereo vision, 3 d (three dimensional) coordinates and the full field deformation and strain on sheet metal's surface are acquired accurately in the process of sheet metal forming. And then, combined with the space interpolation, the method based on the strain history makes data points extracted from cutting line more accurate, so that the calculated curve is closer to reality. The method provides experimental evidence for sheet metal's forming theory research and validation and material selection. Finally, this method is also applied in the measurement of magnesium alloy of auto parts.

Keywords: Forming limit curve; Digital image correlation; Time-strain; Binocular stereo vision; Sheet forming.

1. Introduction

FLC (Forming Limit Curve), an important indicator for sheet metal forming, which is also known as FLD (Forming Limit Diagram) and used to determine the deformation degree of material in the condition of stretch, bulging, or both stretch and bulging, can provide evaluation criteria for material selection in various forming process. As a result, it is very practical to use FLC to analyze the formability of sheet metal. In automobile stamping industries, it is widely used to refer to FLC for sheet metal's performance prediction and material selection.

Compared with the theoretical calculation and numerical simulation [1-2], the experimental method to measure ultimate strain and calculate the FLC of sheet metal can more truly reflect the reality of the sheet metal forming capability [3-4]. The experimental study of sheet metal forming mainly uses computer image processing and grid technology to obtain the strain distribution of sheet metal's surface [5-6]. Strain calculation results near the crack area are adopted instead of ultimate strain of the specimens in actual state of rupture. This method can only measure the final state of sheet metal forming. Therefore, the ultimate strain just approximately equals to the real limit state of rupture. In addition, complex grid preparation, low measuring efficiency, and the rebound data are also existing problems.

In this paper, a method by DIC (Digital Image Correlation) and binocular stereo vision technology to measure sheet metal's FLC is realized [7-10]. Binocular stereo vision measurement system is fixed on sheet metal forming machine to record the images in the whole process of sheet metal sample from initial state to rupture during the experiment, by correlation matching and 3 d reconstruction, the deformation field and strain field are calculated. Draw cutting lines and fit data in the last image before crack, interpolate in sheet deformation on the area of Necking to get the ultimate strain value under the conditions of different loading paths. Finally, draw FLC based on the value of ultimate strain. The main researches are as follows:

(1) A measurement system on sheet metal forming is designed, including:
   (a) Acquisition images of sheet metal sample in the whole deformation process;
   (b) Correlation matching in measurement area between images of different states and two perspectives;
   (c) Calculation of the deformation and strain.

(2) FLC calculation. According to National standard value of ultimate strain of sheet metal forming and the relevant theoretical research at home and abroad, confirm the necking area and draw FLC accurately.

(3) Do experiments on magnesium alloy By Nakajima Hemispherical punch, the magnesium alloy's FLC is calculated.

2. Measurement Scheme

The experimental method of FLC measurement needs to stamp sheet metal according to different loading paths and ways, and calculate the strain of sheet metal in each state. After a series of analysis and calculation, the final ultimate strain value can be obtained. In the method of measuring FLC by DIC, sheet metal cupping is the main way of forming test. By using the spherical punch to press sheet metal sample into concave die, make the sheet metal form to necking until cracking, and finally reach at the limit state of rupture. In the cup drawing experiments, two high-speed cameras record the whole process of sheet metal stamping.

The scheme and workflow of the FLC measurement is shown in Fig. 1. First, before the start of the experiment, according to the national forming limit standards of cup drawing experiment, prepare some different size of the specimens, which are proceed by wire-electrode cutting and sprayed speckle pattern by black and white matte painting on its surface. At the same time, calibrate binocular stereo accurately. After the installation of specimen and the lubrication of its forming surface, do experiments in the forming test machine with a diameter of 100 mm punch bulging. Meanwhile, collect a series of images of sheet metal forming by binocular stereo in the real-time deformation process. Then, by the method of DIC and the principle of computer stereo vision, calculate the correlation and match the couple images collected by the two cameras in one state, then carry out images recognition and matching among all state. Identified camera parameters are used to reconstruct corresponding image points, so the three-dimensional coordinates of all the nodes in the measurement area can be obtained. The displacement field and strain field of all nodes in different state are also calculated. Finally, confirm limit strain state of sheet metal forming by analyzing the collected images, fit the section line data under the limit state, and get the ultimate strain value of sheet metal forming under this load condition. Collect ultimate strain value under all the different path load condition, fit the limit strain values and draw the FLC.

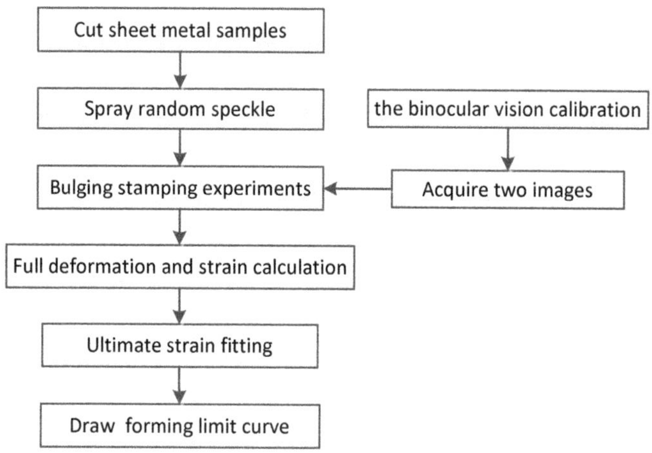

Fig. 1. Flow chart of system scheme.

According to the measurement scheme above, the sheet metal forming limit measure system is designed, whose structure is shown in Fig. 2. Experimental measurement device mainly consists of two industrial CCD cameras, the LED light source, a calibration board, a controller (control the camera and light source) and a computer and other hardware. Sheet metal forming machine stamping head die with open structure, is easy to acquire images in the process of sheet metal forming. Fix the binocular stereo measurement system on the concave die upper part of sheet metal forming machine, adjust the angle of two cameras and light intensity to make sure that a series of real-time images of the sheet metal samples can be acquired from the initial to the rupture. The sheet metal forming machine numerical control system can realize communication with speckle measurement system. At the start and end of a stamping testing it can automatically control the camera trigger.

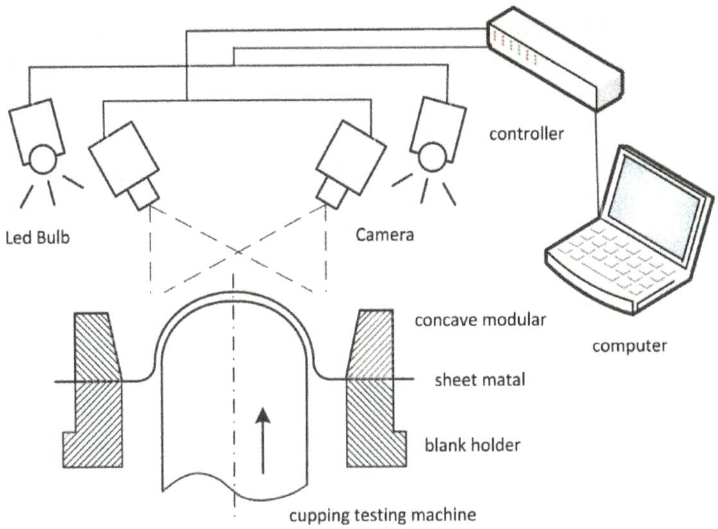

Fig. 2. Sheet metal forming limit measurement system schematic diagram.

3. FLC Calculation

Speckle match and the strain field calculation in the couple images deformation process is shown in Fig. 3, during the process of deformation, measurement area in red begin to match between left and right image in each state, and then match rest states with original one. The method using DIC and binocular stereo vision to calculate deformation and strain can look up reference [11-12].

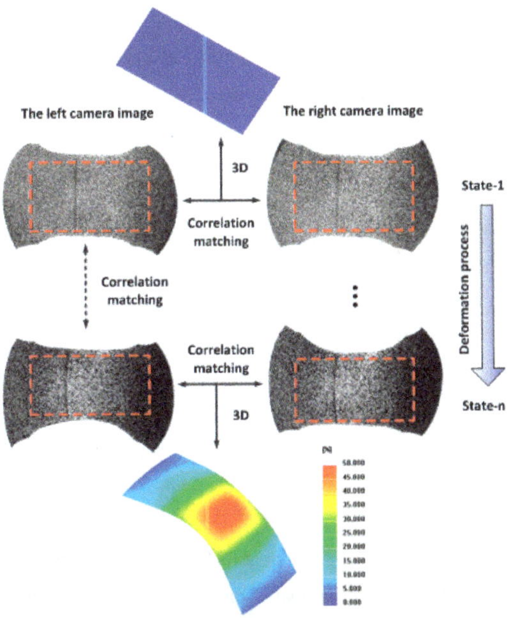

Fig. 3. Strain calculation by DIC and binocular stereo vision.

However, with the increasing of the deformation, the deformation state and original state would make a big difference so that the two states could not be matched. Based on the continuity of adjacent state, the deformation state and its later states match with last state, which has been matched, then add it up [13]. Thus, the whole displacement of all nodes in the stamping process can be obtained. Based on Cauchy - green tensor and triangle node strain, Lagrange strain value of each node can be calculated. Then the strain of each location of measurement area also can be interpolated.

With the strain field calculation results, the FLC can be calculated as follows:

(1) Find out the critical state of rupture, draw cutting line perpendicular to the direction of crack.

In the case of minimum principal strain less than zero, ensure that the section line is drawn along the length direction of the specimens (principal strain direction). Otherwise, make sure that section line is perpendicular to the direction of crack (deviation in - 25 ° ~ 25 °).

Draw the first section line by the maximum strain points, the second and third section lines on both sides of the first line that is away from 2 ~ 2.5 mm, whose direction is same as the direction of the maximum principal strain in last image before the fracture.

(2) Based on spatial interpolation and the strain history [14-15], screen the cutting line data:

Select a number of data points spacing (0.5 mm) on the section lines, draw corresponding strain-time curves of these data points.

As is shown in these strain-time curves, the data points near the crack area continue to deform after the specimen rupture, while others away from the crack area (outside necking) stop to deform. Take these data points which no longer deform as the value of ultimate strain of FLD candidate.

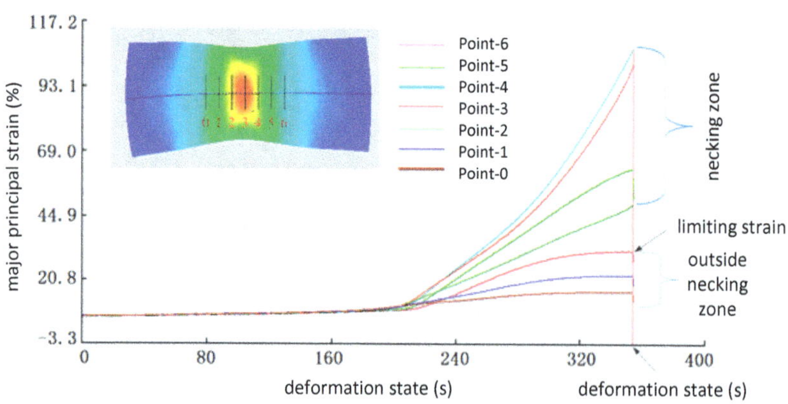

Fig. 4. Cutting line data extraction based on the strain history.

(3) Fit candidate data with parabolic and inverse parabolic described as the following two equations:

$$f(x) = 1/\left(ax^2 + bx + c\right) \quad (1)$$

$$f(x) = ax^2 + bx + c \quad (2)$$

In the equation, x stands for the length of section line, $f(x)$ stands for the node's strain value of the specimen's surface. When strain value is maximum principal strain, use inverse parabolic fitting as shown in equation (1) while minimum use parabolic in equation (1).a, b and c are unknown coefficients.

Search the extreme data in fitting curve, the maximum value of maximum principal strain is expressed as ε_1, and the minimum value of minimum principal strain is expressed as ε_2. ($\varepsilon_1, \varepsilon_2$) constitute a ultimate strain point. Three requirements should be noticed in the fitting process mentioned above:

a. Each cutting line can get a coordinate of ultimate strain at most.
b. Eliminate the points that do not meet the standard requirement.
c. Distance between the two nodes corresponding to the maximum principal strain and minimum principal strain extreme value should be less than three times the specimen's thickness.

(4) According to the ultimate strain points in the previous step, fit appropriate curve or form a stripe shaped area, and then sheet metal's FLC is obtained. And weed out ultimate strain points that do not conform to the requirements to correct FLC.

4. Magnesium Aluminum Alloy Measurement Experiment

Based on the Theoretical research and measurement scheme above, as shown in Fig. 5, a FLC measurement system is developed, which Mainly including the binocular stereo vision system, cupping testing machine and speckle match and FLC fit system. With this system, magnesium alloy FLC measurement has been done.

First, according to the requirement of sheet metal forming cup drawing experiment, prepare magnesium aluminum alloy sheet samples. Three sets of sheet metal forming tests were carried out respectively. There are 9 different samples in each group. For same shape and size of the specimen, change lubrication and friction condition between specimen and punch. There are three lubrication conditions: dry friction, grease lubrication, grease lubrication and polyethylene film lubrication. Collect whole images of each sample from the start of stamping to the end, and calculate the images of each state by DIC. For each sample, the full strain information from stamping deformation to burst can be obtained. The sheet metal samples after stamping are shown in Fig. 6.

By analysis of the images that are continuously shot during the sheet metal forming process, find out the image that sheet is looming rupture. Judge crack position and

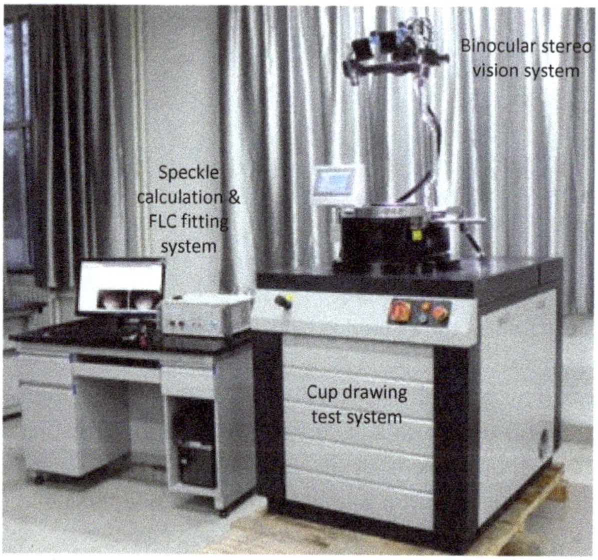

Fig. 5. FLC Measuring equipment.

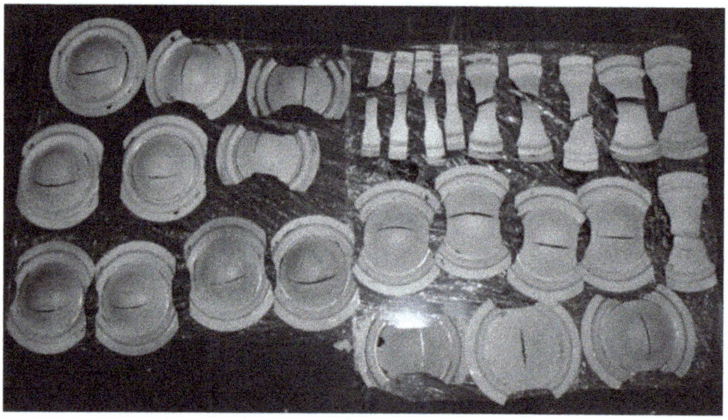

Fig. 6. Cup drawing of specimen after stamping.

direction of sheet metal sample on this image. Then return to the image of the start state, in order to ensure that designated section of the line can keep parallel with boundary. Along the direction perpendicular to the crack, draw 3 ~ 5 section lines. The results are shown in Fig. 7.

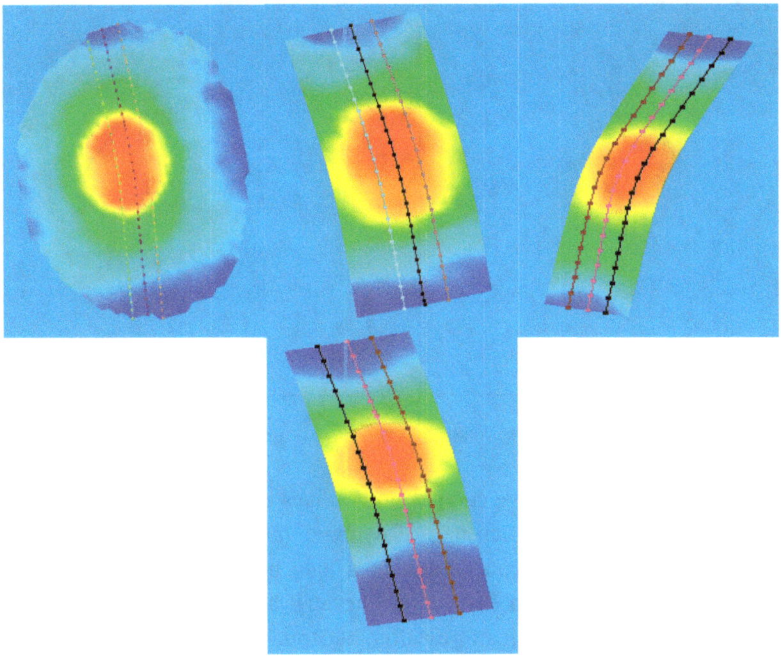

Fig. 7. Do cutting line on cup drawing specimen calculation results.

Screen out the section of line data under time-strain condition, and then fit these data with inverse parabolic and parabolic, as shown in Fig. 8. The vertical axis coordinates of two vertexes is ultimate strain point in this kind of sheet metal forming condition.

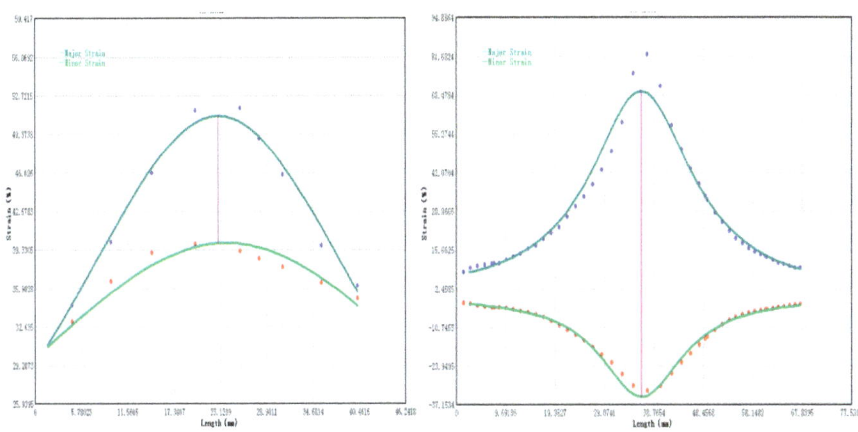

Fig. 8. Inverse parabolic and parabolic fitting.

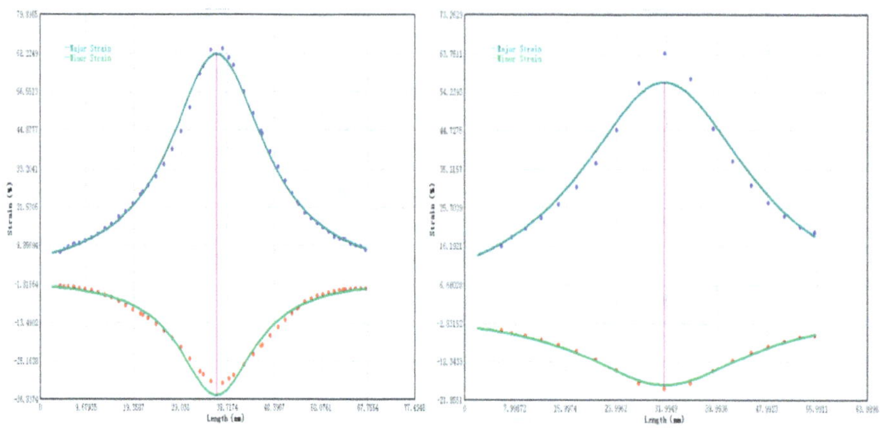

Fig. 8. (*Continued*)

Each section line fitted above can get limit values of maximum principal strain and minimum principal strain, and each pair of extreme value of maximum principal strain and minimum principal strain can determine a data point in FLC curve. Draw three section lines on each sample, so 81 section lines can be obtained after the experiment. Accordingly, there are 81 coordinates of ultimate strain points. Draw the ultimate strain data in Coordinate system whose minimum principal strain is set as horizontal axis and maximum principal strain as the vertical axis. By fitting data of ultimate strain point, the final FLC of the sheet metal is calculated. The magnesium aluminum alloy of FLC is shown in Fig. 9.

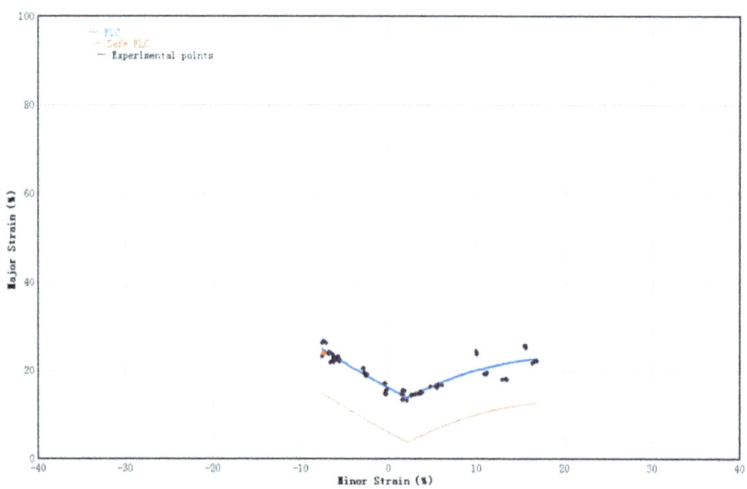

Fig. 9. Magnesium aluminum alloy forming limit curve.

5. Conclusion

Since grid method for measuring FLC has a disadvantage of low efficiency, a sheet metal FLC measurement system is designed by using DIC and binocular stereo vision in this paper. In the whole process of sheet metal cupping test, speckle images are acquired by high-speed CCD camera during the whole period. Confirm measurement area, then match by DIC and reconstruct by binocular stereo vision to get strain field. Solve out the sheet metal's ultimate strain values in the neighboring state of rupture. Eventually use the system to measure FLC of auto parts materials magnesium aluminum alloys.

The method proposed has the following advantages:

(1) In an experiment, each sample can get more data, including fracture limit curve, strain distribution of states from start to the rupture; ultimate strain value of calculation is more reliable;

(2) With simple speckle pattern preparation, fast measurement process, high efficiency and measurement precision, the result is more close to the real formability of sheet metal materials.

Acknowledgments

This work was supported by the Research Project of the State Key Laboratory of Vehicle NVH and Safety Technology under Grant NVHSKL-201506.

References

1. T. Pepelnjak and B. Barisic, Computer-assisted engineering determination of the formability limit for thin sheet metals by a modified Marciniak method, *J. Strain Anal. Eng. Des.* **44**, 459 (2009).
2. M. Firat, A numerical analysis of sheet metal formability for automotive stamping applications, *Comp. Mater. Sci.* **43**, 802 (2008).
3. S. Ahmadi, A. R. Eivani and A. Akbarzadeh, Experimental and analytical studies on the prediction of forming limit diagrams, *Comp. Mater. Sci.* **44**, 1252 (2009).
4. S. Panich, F. Barlat, V. Uthaisangsuk, S. Suranuntchaia and S. Jirathearanatd, Experimental and theoretical formability analysis using strain and stress based forming limit diagram for advanced high strength steels, *Mater. Des.* **51**, 756 (2013).
5. M. Geiger and M. Merklein, Determination of forming limit diagrams - a new analysis method for characterization of materials' formability, *CIRP Ann. manuf. Technol.* **52**, 213 (2003).
6. B. Q. Shi and J. Liang, Circular grid pattern based surface strain measurement system for sheet metal forming, *Opt. Laser. Eng.* **50**, 1186 (2012).
7. P. Vacher, A. Haddad and R. Arrieux, Determination of the forming limit diagrams using image analysis by the corelation method, *CIRP Ann. manuf. Technol.* **48**, 227 (1999).
8. H. Hu, J. Liang, Z. Tang, X. Guo and L. G. Li, Digital speckle based strain measurement system for forming limit diagram prediction, *Opt. Laser. Eng.* **55**, 12 (2014).

9. K. F. Wang, J. E. Carsley, B. Y. He, J. J. Li and L. H. Zhang, Measuring forming limit strains with digital image correlation analysis, *J. Mater. Process Tech.* **214**, 1120 (2014).
10. C. D. Schwindt, M. Stout, L. Iurman and J. W. Signorelli, Forming Limit Curve Determination of a DP-780 Steel Sheet, *Procedia Mater. Sci.* **8**, 978 (2015).
11. H. Hu, J. Liang, Z. Z. Tang and G. Lu, Image correlation method for full-field deformation measurements during metal sheet welding processes, *Optik Int. J. Light Electron Opt.* **124**, 5193 (2013).
12. Z. Z. Tang, J. Liang and Z. Z. Xiao, Three-dimensional digital image correlation system for deformation measurement in experimental mechanics, *Opt. Eng.* **49**, 103601 (2010).
13. Z. Z. Tang, J. Liang, Z. Z. Xiao and C. Guo, Large deformation measurement scheme for 3D digital image correlation method, *Opt. Laser. Eng.* **50**, 122 (2012).
14. M. Merklein, A. Kuppert and M. Geiger, Time dependent determination of forming limit diagrams, *CIRP Ann. manuf. Technol.* **59**, 295 (2010).
15. H. Z. Li, J. S. Cisneros, X. Wu, X. Chen, X. Xie, N. Xu and L. X. Yang, Benchmark 1 – Nonlinear Strain Path Forming Limit of a Reverse Draw: Part B: Physical tryout report, *AIP Conf. Proc.* **1567**, 27 (2013).

Forming Limit Diagram at High Temperature for Hot Stamping Steel

M. T. Ma[1,†], J. P. Zhang[1], L. F. Song[1] and H. Li[2]
[1]*China Automotive Engineering Research Institute Co. Ltd, Chongqing, 401122, China*
[2]*Beijing University of Technology, Beijing, 100124, China*
[†]*E-mail: mingtuma@126.com*

This paper measured the flow characteristics, friction coefficient and forming limit diagram (FLD) at high temperature for hot stamping steel. The flow behaviour of hot stamping steel at high temperature was tested by using of Gleeble-3500 machine. Double power function and Hollomon equation were applied to describe its flow behaviour. The result indicated that strain rate and temperature has notable effects on work hardening of hot stamping steel. Self-regulating setups were used to test the friction coefficient and FLD at high temperature. Then the FLD results and affecting factors were analyzed and the suggestion for FLD calculation at high temperature was advanced.

Keywords: Hot stamping steel; Flow characteristics; High temperature friction coefficient; High temperature FLD.

1. Introduction

Formability is the deformation ability of materials when undertaking external loadings or forces before necking or fracture. FLD is a significant way of evaluating the formability of materials. In a forming limit diagram FLD, the major strain is taken as the y-axis and the minor strain is taken as the x-axis. The FLD represents the limit strain of materials under different stress states before plastic instable damage happens in the coordinate system. In the reference [1], the physical meaning and testing methods of FLD have been discussed in detail. Since different stress states are involved in FLD, samples with different width have been adopted which makes the testing work very heavy and complex. The left side of FLD corresponds to the deep drawing stress state when the metallic sheets are stamped, and its failure mode is local necking. In forming limit diagram, the lowest position corresponds to the plane strain state of sheets, under which the deformation is difficult and the limit strain is the lowest. On the right side of FLD, the sheet is under biaxial tensile stress state and the failure mode is the diffuse necking and ductile fracture. According to these characteristics, FLD simple drawing method is proposed in the literature [2]. Literature has put forward the calculation equation of FLD considering the plastic anisotropic ratio [3]. After the calculation of FLD_0, the curve on the left side of FLD is obtained by drawing a line of 45 ° to the horizontal axis from FLD_0. The curve on the right side of FLD is gotten through drawing a line of 20 degree angle to the horizontal axis from FLD_0. The whole FLD is calculated in such a way and applied roughly to the estimate formability of the sheet. Although this result is comparable to the tested result to some extent, it is only available at room temperature. Hot stamping of the sheet is occurred in the autensite state. High temperature FLD is a very important basis for the simulation and evaluation of formability of hot stamping sheets. However, there are many difficulties to determine the FLD at high temperatures due to the existence of too many influence factors. The purpose of this article is to measure the high temperature FLD of the sheet and at the same time to discuss the influence of friction coefficient on forming limit diagram.

2. Materials and Experiments

The testing material was hot stamping steel sheet with the thickness of 1.8 mm. Its chemical composition was shown in table 1.

Table 1. Chemical compositions of the hot stamping steel (wt %).

element	C	Si	Mn	Cr	Mo	Al	B	Ti	P	S
content	0.23	0.17	0.95	0.24	0.21	0.03	0.0019	0.023	0.01	0.002

High temperature tensile test was conducted on Gleebe-3500. The temperature range is 600 °C to 900°C and the strain rate ranged from 0.01 s^{-1} to 1 s^{-1}. When the sample was heated to the set temperature, 5 min heat preservation was carried out to ensure complete austenitizing. Then the sample was tensioned and the strain-stress curves were recorded. According to the flow stress curves, the double power function constitutive equation ($\sigma = K\varepsilon^n\dot{\varepsilon}^m$. K-the hardening coefficient, n-work hardening exponent, m-the strain rate sensitivity index) was selected to describe the flow behaviour of hot stamping steel under different strain rates and temperatures. High temperature friction coefficient of hot stamping steel was tested on self-regulating friction coefficient tester as shown in Fig. 1. The material of friction head was H13 die steel. The samples were heated to 600 °C, 650 °C, 700 °C, 750 °C and 800 °C respectively in the furnace and pulled out with the speed of 100mm/s.

FLD experiment at high temperature was also carried out on self-regulating forming device (see Fig. 2). To prevent oxidation, the sample was heated in the furnace full of protection gas. When the temperature of the sample reached the set value, it was quickly transferred to the device, which was also added with heating accessory to guarantee the experiment temperature. The velocity of the forming was 10 mm / s.

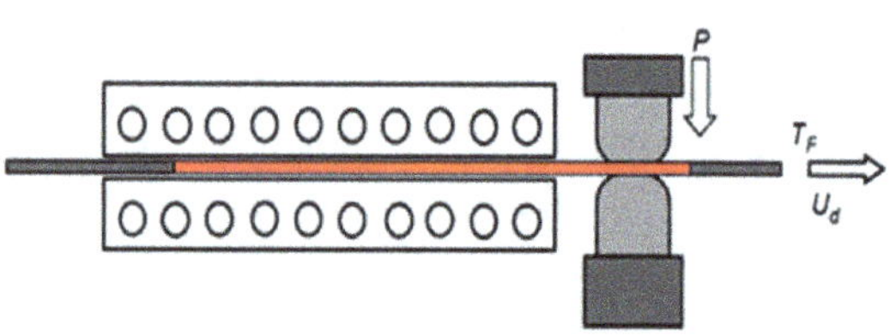

Fig.1 Sketch map of friction coefficient testing device

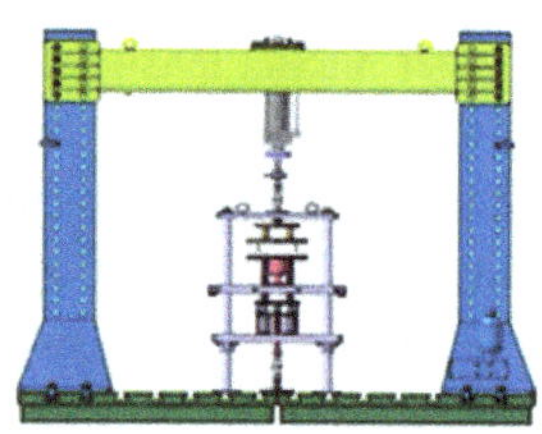

Fig.2 Sketch map of FLD testing device

3. Results and Discussion

3.1. *Tensile experiment at high temperature*

According to the experiment results, the constitutive equation of double power function was used. During the data processing, the assumption that K value and strain rate sensitivity index at the same temperature didn't change as the change of strain rate was set. The fitting results were listed in table 2. The comparison of fitted curves and the tested results (as shown in Fig. 3) indicated that they had good consistency. Under the same deformation temperature, strain hardening exponent increased with the increasing of strain rate. But with the increasing of the deformation temperature, it had a reduced tendency under the same strain rate. At higher temperatures, the strain hardening ability and the deformation resistance of the material decreased. Strain rate sensitivity index usually decreases as temperature rises, and to a certain extent its reduction tends to be mild. However, the regression of the experiment data had differences with this knowledge, which may be related to the method or the assumption of regression.

Table 2. The fitting values of the double power function

Tension Tem.	Hardening para.	Strain rate			Tension Tem.	Hardening para.	Strain rate		
		0.015/s	0.1/s	0.4/s			0.015/s	0.1/s	0.4/s
600°C	n	0.4171	0.40	0.4326	800°C	n	0.2387	0.193	0.2797
	m		0.151			m		0.1271	
	K		692.148			K		331.0601	
700°C	n	0.2475	0.305	0.3719	900°C	n	0.1662	0.2458	0.2105
	m		0.2067			m		0.1276	
	K		664.012			K		240.615	

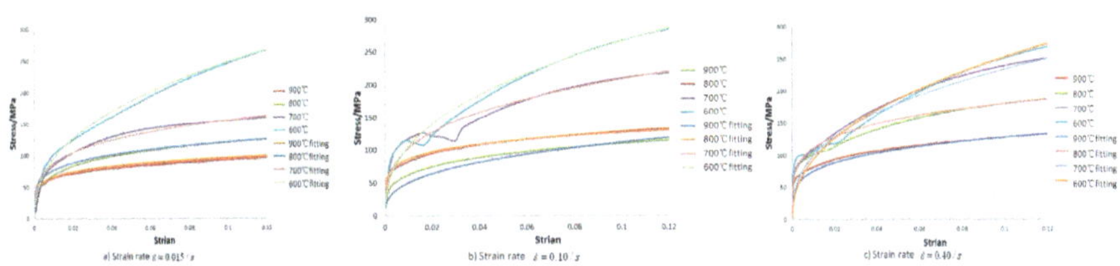

Fig. 3. The flow curves at different temperatures and different strain rates.

At the strain rate of $0.015s^{-1}$, the regression equations were listed as equation (1)-(4):

$$600°C:\ \sigma = 562\varepsilon^{0.382} \quad (1)$$

$$700°C:\ \sigma = 264\varepsilon^{0.232} \quad (2)$$

$$800°C:\ \sigma = 198\varepsilon^{0.218} \quad (3)$$

$$900°C:\ \sigma = 134\varepsilon^{0.160} \quad (4)$$

From these equations it can be seen that the material softening was prominent at high temperatures. The deformation ability get improved but the hardening ability lowered down, namely the work hardening exponent n and strengthening coefficient K in Hollomon equation decreased with the increase of temperature. At 600°C, n value was definitely too high, which may be associated with the error of the experiments. The flow curve at 600°C distinctly deviated from others as shown in Fig. 3a. Therefore, it was only for reference.

3.2. *Friction coefficient*

The friction coefficient results were listed in table 3. It was indicated that with the increase of temperature the friction coefficient increased. At high temperatures, the friction coefficient was 2 times larger than that in the room temperature. This probably related with the oxidation layer on the metal surface and the larger hardness difference between the sheet and the indenter at high temperature. The friction coefficient had a significant impact on the formability and the result of FLD.

Table 3. The friction coefficient of the steel at different temperatures.

room temperature	600°C	650°C	700°C	750°C	800°C
0.14	0.38	0.41	0.45	0.50	0.53
0.12	0.36	0.44	0.47	0.49	0.54
0.11	0.33	0.43	0.47	0.48	0.57
0.123	0.357	0.427	0.463	0.490	0.547

3.3. *FLD testing results*

When the temperature rises, the deformation resistance decreases and the formability was improved, thus FLD exhibits larger safety zone. However, high temperature leads to high coefficient on the metal surface, which causes the decrease of safety zone and the change of cracking location of samples during experiments (as shown in Fig. 4) [4]. In the cold forming process, the crack position is at the peak of the dome, where the sheet is undertaken biaxial tensile stress. But for hot forming, the deformation at the peak is restricted by the friction force. The fracture position moves to one side of the dome, which endures multiple stress states of deep-drawing and biaxial stress due to the influence of the friction at the peak of the dome. In the present work, the similar results were obtained as illuminated in Fig. 5. The high temperature FLD results of experimental materials, 22MnB5 and Arcelormittal were shown in Fig. 6 [4]. The results behaved some different, but the tendency kept consistent. Levy [2] and Ma[3] had proposed the calculation equation of FLD_0:

$$FLD_0 = (23.3 + 14.23t)(n / 0.21) \tag{5}$$

Where n is work hardening exponent and t is the thickness of the sheet in mm.

This equation mainly uses to calculate the FLD_0 at room temperature. At high temperature, it is deficient only considering the effect of n value. On the one side, as temperature rises, the work hardening exponent of materials decreases, which is harmful to the formability. From the other side, temperature increasing makes the decreasing of deformation resistance of material which is benefit to the formability. But as increasing of temperature, the surface stick flow will be increased due to the severe oxidation of surface, which is bad for the formability. How the three factors influence the FLD_0 of materials at high temperature needs to be discussed in the advanced work. It has got that the relationship of the high temperature FLD_0 and temperature by linear fitting based on the experimental data obtained at different high temperatures (see equation 6) [4]:

$$FLD_0 = 0.0997 + 0.003T \tag{6}$$

Where T is for temperature and its unit is °C.

In table 4 the n value, experimental FLD_0 and the calculated FLD_0 of 22MnB5 at different temperatures were listed. It can be seen that the n value decreases but FLD_0 rises with the increase of temperature. The climbing of FLD_0 is the comprehensive effects of work hardening, materials softening and the friction.

Table 4. The comparison of experimental FLD_0 and calculated FLD_0 of 22MnB5.

Temperature/°C	Thickness/mm	n value	calculated FLD_0	experimental FLD_0
600	1.8	0.382	0.2797	0.305
700	1.8	0.232	0.3097	0.345
800	1.8	0.218	0.3397	0.415
900	1.8	0.160	0.3697	0.465

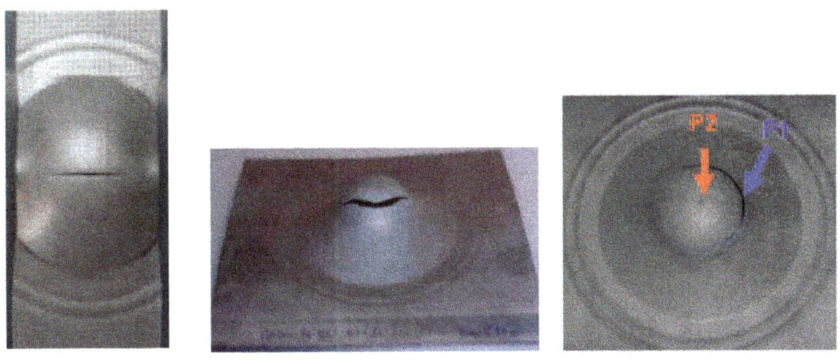

(a) cold forming. (b) hot forming. (c) hot forming.

Fig. 4. The cracking position of hot forming and cold forming during FLD experiments.

Fig. 5. The cracking position in the high temperature FLD experiment.

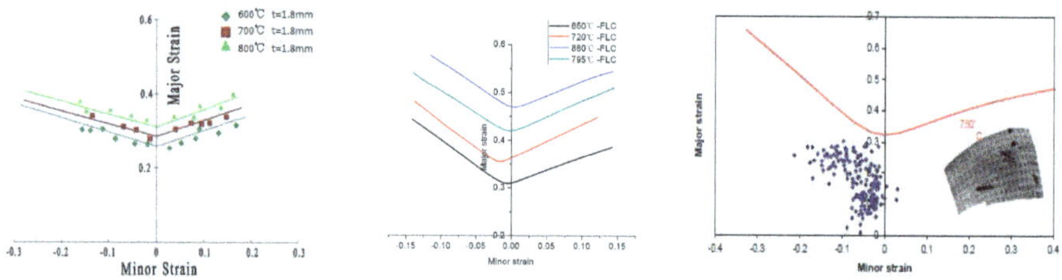

a) experimental steel. b) 22MnB5. c) Arcelormittal.

Fig. 6. The testing results of high temperature FLD.

4. Conclusion

This paper has measured flow curves and high temperature FLD of hot stamping steel at different temperatures as well as analyzed the factors that influenced the formability of the material. The result indicated that both strain rate and temperature has obvious effect on the flow characteristics of hot stamping steel. The factors affecting the formability of the material involves the work hardening character and ability, softening due to temperature rising and the state and value of friction coefficient at certain temperature. Their quantitative contribution to the formability of hot stamping steel at high temperature needs more experiments and study.

References

1. M.T. Ma and B.R. Wu, *Dual-phase steel Physical and Mechanical Metallurgy*, (Metallurgical Industry Press, 2009).
2. S.B. Levy, A comparison of empirical forming limit curves for low carbon steel with theoretical forming limit curves, in *IDDRG WG3*, (Eger, Hungary, 1996).
3. M.T. Ma, *Advanced automotive steel*, 1st ed., (Chemical Industry Press. 2008).
4. L. Durrenberger, Forming Limit Curves at high temperatures, Arcelormittal exchange data. 2012.

Study on Laser Welding Procedures of Hot-Rolled Strip for 1500MPa Grade Hot Stamped Steel

J. P. Zheng[†], R. T. Zhong, Z. J. Huang, Y. H. Hu and W. Y. Liu

Research and Development Center of Wuhan Iron & Steel Co., Ltd.
Wuhan, Hubei, P. R. China
[†]E-mail:cm_lb1987@126.com

Laser welding is widely used for the on-line joining of hot-rolled steel strips for hot stamped steel by steel plants recently. However, the high breaking rate caused by the emergence of hard martensite in weld was a fatal disadvantage. In this paper, the welded joints of hot-rolled strips for 1500 MPa grade hot stamped steel were made through laser filler wire welding and laser welding, and a series of PWHT experiments were further conducted on laser welding joints. The influence of filler wire and PWHT on the microstructure and mechanical performance of welded joints was analyzed. The results showed that weld appearance was refined by wire filling process, while the microstructure of weld was martensite and cupping test results could not meet the requirements still. As for welded joints with PWHT, the microstructure and mechanical performance were not refined when the PWHT temperature was 500 °C. But the cupping test results were much better at 600 °C and 800 °C compared with the as welded states, accordingly, the weld metal microstructure was tempered martensite and ferrite.

Keywords: Laser welding; Hot-rolled strip; Hot stamped steel; Post Weld Heat Treatment.

1. Introduction

As China has become the world's largest automobile production and consumption country, energy consumption and environmental pollution problems have become increasingly prominent. Lightweight of Automobile is one of the effective ways of improving the dynamic performances, reducing fuel consumptions and exhaust emissions of vehicles. It mainly includes the automotive lightweight materials, structural design optimization and advanced manufacturing technologies [1-2]. Research showed that the corresponding fuel efficiency could be increased by 6 % ~ 8 % with the weight of the vehicle decreased by 10% [3]. Therefore, the use of high strength steel plates instead of low strength ones will be the inevitable trend of the development of automotive industry [4-5].

Many steel plants have been developing different classes of advanced high strength steel (AHSS) which are mainly used for automobile bearing pieces, safety and strengthen components [6]. The 1500 MPa grade hot stamped steel is a new developed high performance steel, which was based on traditional C-Mn steel and a certain concentration of boron was added in to delay the nucleation of ferrite and bainite. Though the carbon content is low in this steel, the hardenability is greatly increased with the addition of Mn, Si, Cr and B elements [7]. The microstructure is ferrite and pearlite with the tensile strength about 500 ~ 600 MPa for this material before hot stamping, while the microstructure transforms into martensite and the tensile strength is improved to above 1500 MPa after the hot stamping process, so the light weight and ultra high strength components can be obtained [8].

The on-line joining for hot-rolled steel strips is one of the key technologies for hot stamped steel producing, which joins the head of the current strip to the end of the previous strip to benefit for continuous manufacturing. Laser welding is widely used as a joining method by steel plants recently, unfortunately, the high breaking rates caused by the emergence of hard martensite in weld metal and heat affected zone (HAZ) is a fatal disadvantage which greatly restrains the rolling capacity of steel plants. So study on the laser welding performance of the hot rolling strips to reduce the breaking rates is with significant importance. In the present study, a series of laser welding experiments and post weld heat treatments (PWHT) were performed on hot-rolled strips for 1500 MPa grade hot stamped steel, and the influence of filler wire and PWHT on the welded joint's microstructure and performance was analyzed to provide guidelines for the on-line joining processes.

2. Material and Experimental Procedures

The tested material in this study was 4.5 mm thick hot-rolled strips for 1500 MPa grade hot stamped steel. The chemical compositions of the tested steel are listed in Table 1 Fig. 1 shows the initial microstructure of tested steel, which consists of ferrite and pearlite.

Table 1. The chemical compositions of tested material (wt%).

C	Si	Mn	P	S	Al	Cr	N
0.23	0.31	1.28	0.011	0.0022	0.041	0.28	0.002

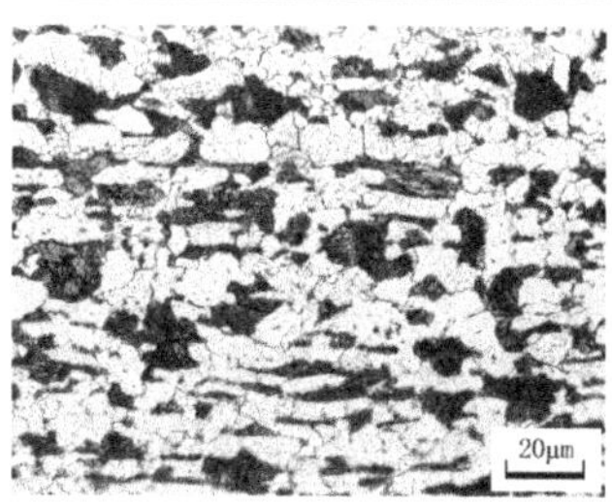

Fig. 1. The microstructure of tested material.

An HSL21 type laser welding machine produced by MIEBACH was used to conduct the on-line welding process, and A SC-G13133 type chamber electric furnace was used to carry out the heat treatments in this study. The welding procedure and heat treatment parameters were shown in Table 2. The first group was welded with laser filler wire welding method and the filler wire was 50S-6 solid welding wire with 1.0 mm in diameter. The second group was welded by laser welding method without filler wire. In order to investigate the effect of PWHT temperature on joints performance, the second group welded joint specimens were also heat treated for 8 min at 500 °C, 600 °C, 800 °C individually and then cooled to room temperature in air.

Table 2. The laser welding and heat treatments parameters.

NO.	Welding power/kW	Welding speed/ m·min^{-1}	Preheating power/kW	Post heating power/kW	Wire feeding speed/ m·min^{-1}	Heat treatment
1	11.5	5	14	5	9	As welded
2	11.5	5	14	5	/	As welded
3	11.5	5	14	5	/	500 °C×8min
4	11.5	5	14	5	/	600 °C×8min
5	11.5	5	14	5	/	800 °C×8min

Cupping test is a universal test method to evaluate the stamping formability of steel sheets, also known as Erichsen test. The Erichsen test value (IE value) can be used to indicate the formability of sheets. The greater the IE value is, the better the formability will be [9]. The test instrument used in this study was a GBW-60B automatic cupping test machine. The specimen size was 90mm×90mm, which were cleaned with alcohol and evenly coated with type 97 deep drawing lubricants on the surface before test. The cupping speed was about 10mm/min and the specimen clamping force was 10 kN. At last, the 4 % nitric acid and alcohol solution was used to reveal the microstructure, and an OLYMPUS GX71 optical microscope was used to observe the microstructure. A MC010-HRS-150 auto hardness indictor was used to test the micro-hardness, and the load was 0.2 kgf.

3. Results and Analysis

3.1. *Analysis of as welded joints made through laser filler wire welding and laser welding methods*

(a) The cupping test samples.

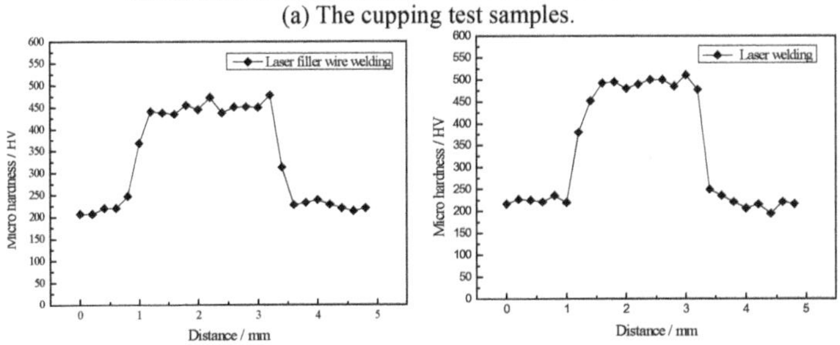

(b) The micro-hardness test results.

Fig. 2. The cupping test samples and micro-hardness distributions of welded joints made through laser filler wire welding and laser welding methods.

Fig. 2 shows the cupping test samples and micro-hardness distributions of as welded joints. It can be seen that the cracking positions were perpendicular to the weld seam, and the measured Erichsen values were 7 mm and 7.5 mm respectively. According to the micro-hardness test results, the base metal hardness was about 200 HV, the average hardness for the weld made with and without filling wire was about 460 HV and 500 HV respectively. The hardness distributions suggested that the HAZ and weld zone were seriously hardened over the laser welding process. No obviously difference in Erichsen value and hardness distribution was found between these two kinds of weld seam, it demonstrated that the weld mechanical performance was not improved by the laser filler wire welding method.

Fig. 3 and Fig. 4 exhibit the transverse section morphology and microstructure of as welded joints. It can be seen that both the welding seam appearances were good, and no obvious crack, shrinkage or other defect was found in the weld, but the weld seam made through laser filler wire welding seemed to be more filled than the laser welding one. There was little difference in microstructure between the two types of welded joints and almost all martensite structure was obtained in weld zone and HAZ. The reason may be that the weld formed by melting filler wire took a small proportion in the total weld zone because of the narrow welding gap (about 0.25 mm), as a result, the weld chemical compositions were not obviously changed and the microstructure could not be refined. Nevertheless, the filler wire did play a role in reducing depressions and refining weld appearance especially when the plates for welding were with low cutting precision or large welding gap.

(a) Weld morphology. (b) Weld zone. (c) HAZ.

Fig. 3. The weld morphology and microstructure of welded joints made with laser filler wire welding.

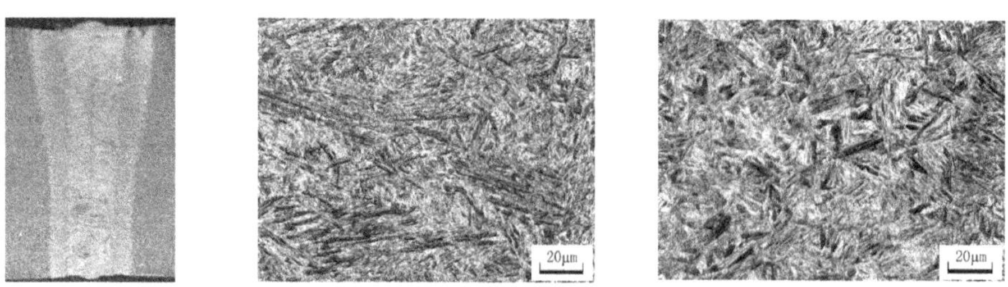

(a) Weld morphology. (b) Weld zone. (c) HAZ.

Fig. 4. The weld morphology and microstructure of welded joints made with laser welding.

3.2. *Analysis of laser welding joints under different PWHTs*

Fig. 5 shows the cupping test samples and micro-hardness distributions of welded joints under PWHTs. When PWHT temperature was 500 °C, the measured Erichsen value was 8mm and the average hardness of the weld metal was around 430HV, which seemed not obviously influenced by PWHT compared with the as welded one. The Erichsen value increased to 9.5mm and weld hardness slightly declined to about 330HV as PWHT temperature improved to 600 °C. When temperature reached 800 °C, the Erichsen value further improved to 11mm, correspondingly, the hardness distribution tended to be uniform and the weld hardness dropped to about 250HV. In summary, with the heat treatment temperature increased, the weld and HAZ hardness decreased while the Erichsen value increased gradually. The mechanical performance of welded joints was obviously improved by PWHT especially when the temperature was above 500 °C.

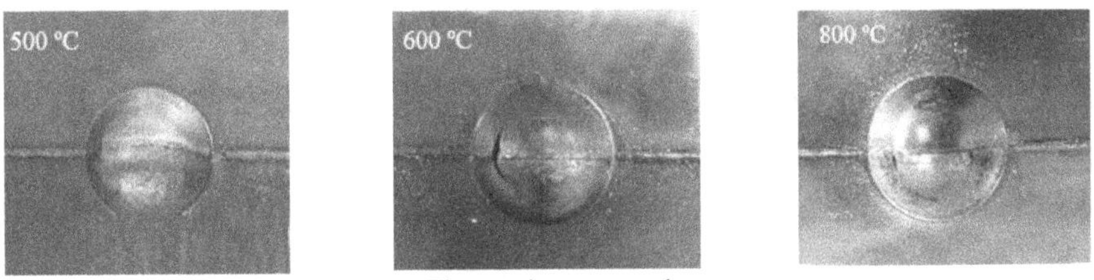

(a) The cupping test samples.

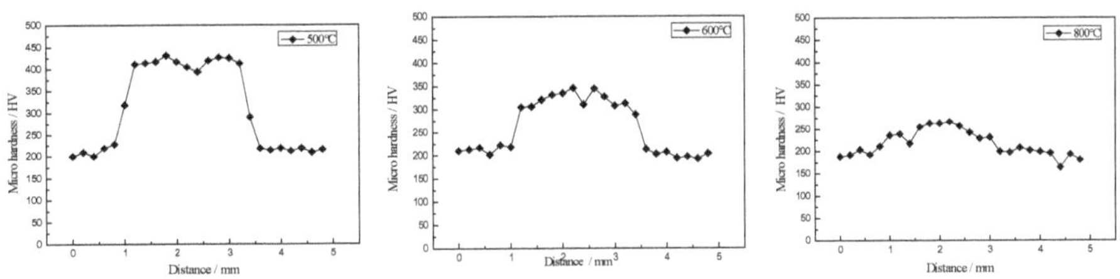

(b) The micro-hardness test results.

Fig. 5. The cupping test samples and micro-hardness distributions of welded joints with different PWHTs.

Fig. 6 shows the microstructure of the welded joints with different PWHTs. It can be seen that the microstructure of weld zone and HAZ was still martensite when the heat treatment temperature was 500 °C, there was no change compared with the as welded joints for almost no transformation was occurred under the relatively low temperature and short holding time. The metallographic structure was refined with the PWHT temperature increased, it mostly transformed to tempered martensite at 600 °C and to ferrite at 800 °C, and the evolutions of microstructure fitted in well with the mechanical test results as shown previously.

4. Conclusion

In this paper, a series of laser welding experiments were carried out on hot-rolled strips for 1500 MPa grade hot stamped steel, and the effects of filler wire and PWHT on the microstructure and performance of welded joints were analyzed. The main conclusions can be drawn as follows:

(1) For the as welded joints, the Erichsen value and hardness distribution was not significantly influenced while the weld appearance was refined with the use of laser filler wire welding procedure. The microstructure of weld zone and HAZ was all martensite for both laser filler wire welding and laser welding joints.

(2) With the increase of PWHT temperature, the weld and HAZ hardness decreased while Erichsen value increased gradually, and the mechanical performance of welded joints was obviously improved by PWHT especially when the temperature was above 500 °C. The metallographic structure for weld zone and HAZ of laser welding joints was still martensite when heat treated at 500 °C, and it transformed into tempered martensite at 600 °C and ferrite at 800 °C, respectively.

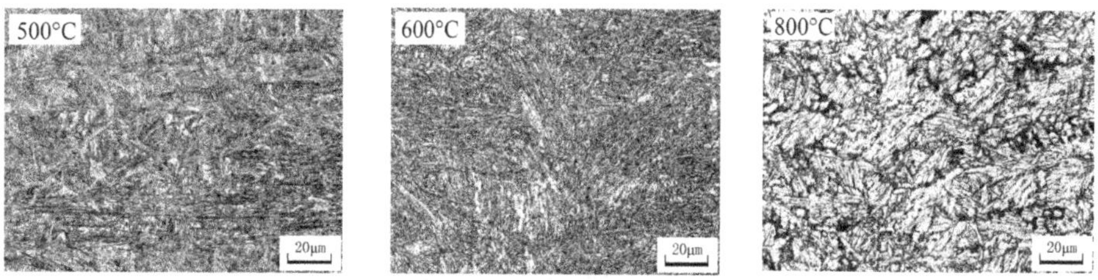

(a) The microstructure of weld zone.

(b) The microstructure of HAZ.

Fig. 6. The microstructure of weld zone and HAZ in welded joints with different PWHTs.

References

1. S. W. Lim, Advanced Technology for Lightweight Automobile, *J. Korean Soc. Heat Treat.* **21**, 164 (2008).
2. M. Ma, H. Yi and H. Lu, On the Light weighting of Automobile, *Eng. Sci.* **11**, 20 (2009).
3. W. S. Heggie and R. Sandri, Energy saving hydro-pneumatic power plant for the automobile, *Fluidics Q.* **12**, 1 (1980).

4. T. Altan, Hot-stamping boron-alloyed steels for automotive parts, *Stamping Journal* **19**, 10 (2007).
5. A. Bandivadekar, Reducing the fuel use and greenhouse gas emissions of the US vehicle fleet, *Energ. Policy* **36**, 2754 (2008).
6. N. Ma, G. Z. Shen, S. H. Zhang and P. Hu, Material Performance of Hot-forming High Strength Steel and Its Application in Vehicle Body, *J. Mech. Eng.* **47**, 60 (2011).
7. C. H. Kim, J. K. Choi, M. J. Kang and Y. D. Park, A study on the CO_2 laser welding characteristics of high strength steel up to 1500 MPa for automotive application, *J. Achiev. Mater. Manuf. Eng.* **39**, 79 (2010).
8. H. Karbasian and A. E. Tekkaya, A review on hot stamping, *J. Mater. Process. Tech.* **210**, 2103 (2010).
9. M. Naderi, M. Ketabchi, M. Abbasi and W. Bleck, Analysis of microstructure and mechanical properties of different high strength carbon steels after hot stamping, *J. Mater. Process Tech.* **211**, 1117 (2011).

Performance of Laser Weld Joints for QP980 and DP590 Steel Blanks with Different Thickness

X. G. Qin†, B. Yang and J. C. Duan

Dongfeng Motor Corporation Technical Center, Wuhan, 430058

†*E-mail: qinxg@dfmc.com.cn*

The hardness and microstructure of tailor welded blanks QP980 and DP590 with different thicknesses were tested and analyzed, and also the strength and the deformation of the weld zone were studied by tensile test and erichsen test. The results show that the microstructure of the welded joint is martensite, and the microstructure of the heat affected zone is similar to that of the base metal, but the grain size is coarser. The hardness distribution in the two sides of the weld is different. The hardness of the weld is higher than that of the base metal, while the hardness of the QP980 heat affected zone decreases by 17.6%, and the hardness of the heat affected zone of DP590 decreases by 9.56%, relatively compared to the base steel blanks. The erichsen value of the welded plate is lower than that of the base metal of any side, and the greater the friction force, the smaller the value of the cup. In addition, erichsen value is reduced by 9% when the test is carried out on the step surface compared to the reverse flat surface.

Keywords: QP980; Tailor Welded Blanks; Microstructure; Erichsen test; Micro hardness.

1. Introduction

With the development of science and technology, more and more requirements are put forward for sheet metal forming parts of automobile, aircraft and other manufacturing industries. Among them, the requirements of reducing weight, since reducing energy consumption is the focus of attention. The tailor welded plate is a new processing technology to meet the requirement. Tailor welded plate is a flat plate which is formed by welding the flat plate material with different thickness, different material or different surface coating before forming [1-2]. After the stamping forming of tailor welded blanks, the weight and energy saving can be reduced at the same time to meet the different requirements of the parts [3-4].

Q&P steel as a typical third-generation steel high strength steels, tailored blanks has been used in vehicle structures, such as the B-pillar and so on. In this paper, the QP980 and DP590 with laser welded together, study its microstructure after welding, at the same time detect the hardness, tensile strength and Erichsen value. The organization and performance of tailor welded blank forming relationship, which has practical significance for the actual production of high-strength steel tailored blanks, was analyzed.

2. Experiment Methods

The experiment was based on materials of QP980 t1.8mm and high strength steel DP590 t1.4mm which were supplied by Shanghai Baosteel. Welding parameters: laser power

3.2KW, welding speed 5m/s, laser focal point deviates thick plate direction of 0.4mm. The chemical compositions of the parent materials QP980 and DP590 are shown in table 1.

Table 1. Chemical composition of materials (mass%).

	C	Si	Mn	P	Cr	Alt
QP980	0.21	1.36	1.89	0.013	0.024	0.04
DP590	0.12	0.46	1.80	0.010	0.019	0.05

2.1. *Laboratory equipment*

Micro hardness testing of the weld zone of the tailor welded blanks was operated by FM800+ARS900. Microstructure was observed by Axio Observer. Tensile test is launched by Z250. Erichsen test was carried out on the digital display machine.

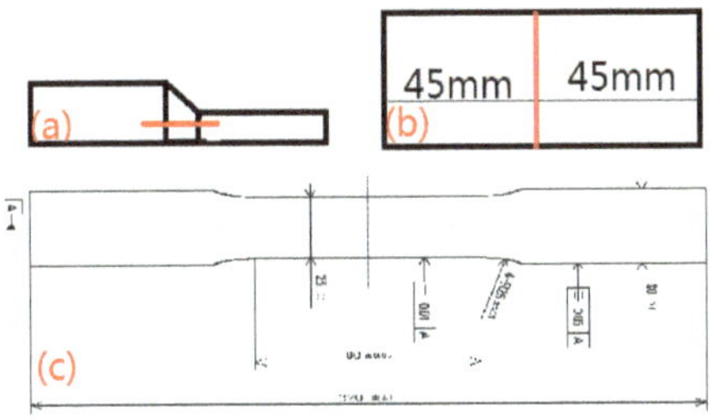

Fig. 1. Schematic drawing of sample.

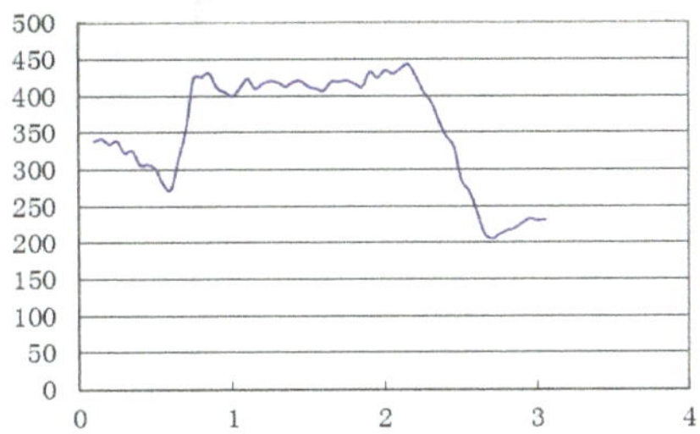

Fig. 2. The microhardness curve.

2.2. *Test sample*

Figure 1 (a) is shown as a schematic diagram of the hardness curve; (b) for the erichsen test specimen, the welding seam around the 45mm; (c) for tensile test specimens. Weld seam is located in the middle of the sample.

3. Experimental Results Analysis

3.1. *Micro hardness results and analysis*

Tailor welded blanks hardness testing starts from the QP980 base material, and points are made every 0.05mm, according to GB/T4330.The load of each point for the 500g, to maintain the load 15S, the test results shown in Figure 2.

It can be known from Figure 2 that the width of weld zone is about 3mm. The left side of the material is QP980, the hardness is about 340HV, and when it comes to the heat affected zone of QP980, hardness began to decline, and the minimum fell to 280HV, with the hardness decreased by 17.6%. From HAZ to weld center, hardness rises rapidly, and in the weld it reached the maximum value 447HV, then the hardness decreased gradually and at the distance away from the weld center about 1mm, the hardness reached the lowest 208HV, and finally gradually reached 230HV at parent material. Here the hardness decreasing is mainly because of the presence of HAZ, where the hardness decreased by 9.56%.

3.2. *Tensile results and analysis*

It can be seen from the table 2 that the strength of the weld is higher than that of the base metal, for tailor welded blanks, the fracture position is at DP590. In order to test the strength of the weld, QP980 and QP980 are welded by the same parameters, and then do the tensile test. As shown in Figure 3, the fracture position is at the base metal, so the weld is not the weak zone of the welded plate.

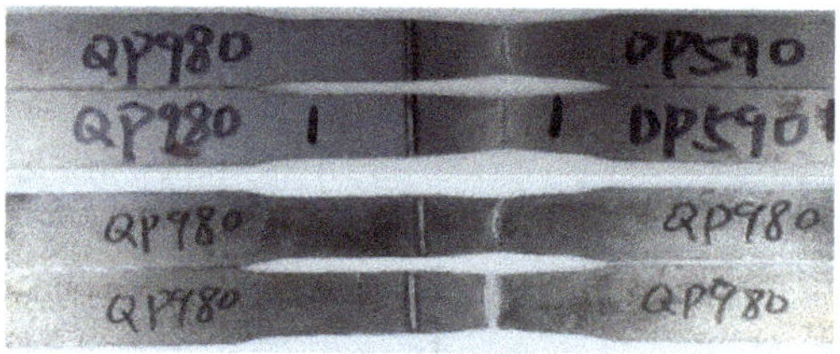

Fig. 3. Fracture specimen.

Table 2. The results of tensile tests.

Material	Tensile strength Rm(MPa)
QP980-DP590	655,657
QP980-QP980	975,975

3.3. *The microstructure analysis*

Figure 4 (a) is the microstructure of welded joint. In the process of laser welding, the line energy input is big, and the welding temperature gradient is big. The main way diffusing heat outward is conduction, especially for carbon steel welding. Because the welding cooling speed is fast, and the boundary of each area is obvious. From Figure 4 (a), it can be seen that the weld is obviously divided into DP590 base metal-DP590 heat affected zone-the weld-QP980 heat affected zone-QP980 five regions of the parent material, the columnar crystals in the middle of the welded joint are obviously cast.

Figure 4 (b) is metallographic photos of DP590, and the microstructure is ferrite and martensite; (c) is the microstructure of DP590 hot affected zone, it was obvious that the grains are coarse, which this is the main reason of the lower hardness of heat affected zone; (d) is weld metallographic photos, and the microstructure is martensite basically; (e) and (f) corresponding to the heat affected zone and base metal metallographic photos. Because the laser focus to QP980 from 0.4mm, and due to large heat input and slow heat diffusion, it can be equivalent to tempering, which can be proved from microstructure, (e) have been a part of the microstructure of tempered martensite, (e) the grain size was significantly larger than (f). The two common reasons together caused hardness of heat affected zone significantly attenuating.

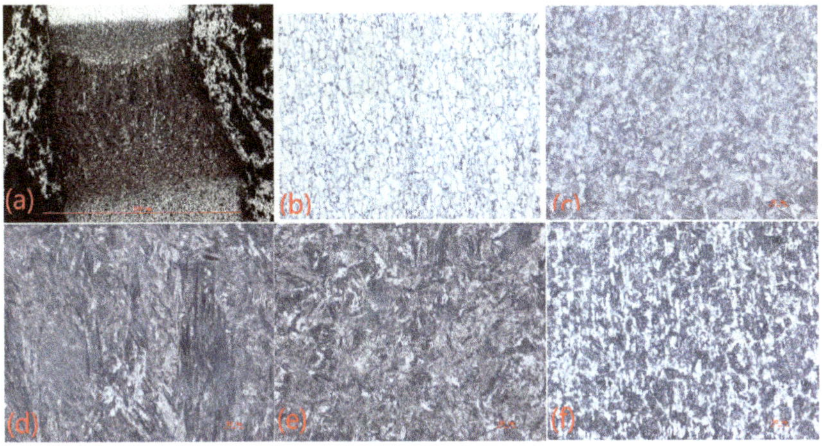

Fig. 4. Microstructure of laser tailor welded blanks.

3.4. *Erichesen test results and analysis*

From table 3, we can see that the cup of welded plate is lower than that of the base metal, and the welding seam is formed from a straight line into a parabolic shape. And the weld offsets to the thick plate. In the process of cupping test, because of the existence of the welding seam, and the high hardness of the weld, the deformation is poor, and in the process of cup test of the drawing, the weld is easy to crack. With the increasing of force, because of the smaller thickness and the lower strength, and thus smaller deformation stress, DP590 first enter the plastic zone, resulting in plastic deformation and work hardening, and the strength increasing. With the increase of drawing height and bending degree, the thick plate also begins to enter the plastic deformation stage, and the strength increases, and the sheet is subjected to a larger tensile stress. In this kind of deep drawing and hardening, the tensile strain of the thin plate is larger than that of the thick plate. Therefore, the larger the proportion of thin plate, the greater the value of the cup, the better the bulging, such as (a) and (b). Contrast (b) and (c) values of the cups we can know that, under the same conditions, the friction force increases, the value of the cup will decrease. Contrast (c) and (d) the value of the cup we can know that, because the position of the head of the ball is different, the difference between the value of the cup is 0.7, which shows that the level of the step has great impact on the value of the cup. In actual stamping process, attention should be paid to the orientation of the weld in the die.

Table 3. Erichsen test results.

Material	Serial	IE	The ball head contact surface
QP980-DP590	a	8.12	Plane
QP980-DP590	b	8.29	Plane
QP980-DP590	c	7.81	Plane +Grid
QP980-DP590	d	7.60	Steps
QP980		9	

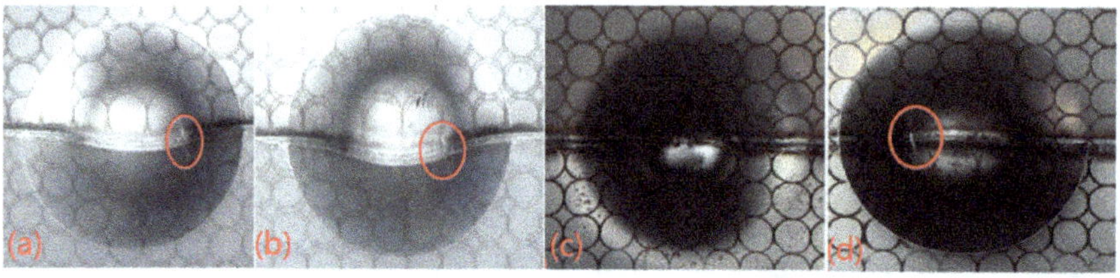

Fig. 5. The results of Erichsen Test.

4. Conclusion

(1) The weld microstructure of QP980 t1.8 and DP590 t1.4 laser tailor welded blanks are martensite and bainite. The microstructure of QP980 heat affected zone is martensite and austenite, but its grain size is larger than that of base metal. The microstructure of DP590 heat affected zone is martensite and ferrite, and the grain size is coarse.

(2) The hardness distribution in both sides of the weld is different. The highest hardness occurs in the weld seam, which reaches 450HV, while hardness of HAZ of QP980 declines by 17.6%, and hardness of HAZ of DP590 declines by 9.56%.

(3) The erichesen test value of the welded plate is less than that of the base material, it means the bulging ability of the welded plate is lower than that of the base material. And the larger the deformation ratio of the thinner blank, the higher the erichsen value, the better the bulging ability of the plate.

(4) The surface state of the ball contact part in the blank affects the erichsen value of the welded plate. The greater the friction, the smaller the value of the cup, Erichsen value on the step surface is smaller than the flat surface by 9%, so in the actual stamping process, it is best to pay attention to the direction of the weld step.

References

1. H.X. Cui, K. Chen and T.C. Zuo, CO_2 laser weldability of galvanized steel sheets, *Trans. China Weld. Inst.* **25**, 115 (2004).
2. J. Tusek, Z. Kampus and M. Suban, Welding of tailored blanks of differernt Materials, *J. Mater. Process. Technol.* **119**, 180 (2001).
3. T. Meinders, A.V.D. Berg and J. Huetink, Deep drawing simulations of tailored blanks and experimental verification, *J. Mater. Process. Technol.* **103**, 65 (2000).
4. Y. Heo, Y. Choi, H.Y. Kim and D. Seo, Characteristics of weld line movements for the deep drawing with draw beads of tailor welded blanks, *J. Mater. Process Technol.* **111**, 65 (2001).

Research on the Microstructures and Mechanical Properties of Cold Rolled Hot-Dip Galvanizing CP980 Steel

Y. Han[†], S. Kuang, X. M. Qi, C. Q. Xie, H. S. Liu and G. H. Liu

Shougang research institute of technology, Beijing, 100043, China

[†]*E-mail: hanyun@shougang.com.cn*

In this study, cold rolled hot-dip CP980 steel was designed and effects of galvanizing parameters on microstructures and mechanical properties of the steel were also investigated. The results show that microstructures of the experimental steel consist of ferrite, bainite and martensite. Mechanical properties of the steel mainly depend on the phase fraction which may be affected by the galvanizing process. Heating and slow cooling parameters play important roles in the phase fraction evolution. Moreover, there are a lot of fine micro-alloy precipitates distributed in the matrix which is useful to increase the yield strength and also refine the grain size. Excellent mechanical properties of the CP980 steel can be gained with 997MPa (TS), 751MPa (YS) and 11.0% (A50 EL). In addition, the hole expansion test results show that CP980 steel has a better hole expansion ratio compared to DP980 steel.

Keywords: Cold rolled complex Phase 980 (CP980); Microstructures; Mechanical properties; Precipitates; Hole expansion ratio.

1. Introduction

Energy conservation and emissions reduction and increasing the safety of the car body have become the development trends of the automotive industry. As a result, more and more high strength steels (HSS) and advanced high strength steels (AHSS) have been widely used in the white car body for safety components [1-5]. Because of the high yield strength, high yield ratio, homogeneous microstructure, cold rolled complex phase (CP) steel shows great superiority and good prospects for the production of components with complicated shape [6].

Furthermore, the chemical composition of CP steel is also important. It can be known that some quantities of Nb, V or Ti usually be added to cause the precipitation strengthening effect in DP steels [7, 8]. However, researches on the microstructures and mechanical properties of CP steels are seldom reported.

In this work, a new cold rolled hot-dip galvanizing CP980 steel was designed and the effects of the galvanizing processes on the microstructure and mechanical properties were also investigated.

2. Experimental Scheme

The chemical composition of the experimental steel is designed and given in Table 1. It can be seen that the carbon content is low and some micro-alloy elements are added. This experimental steel was fabricated by 50 kg vacuum induction furnace and casted in to square ingots.

Table 1. The chemical compositions of experimental steel (wt%).

Steel	C	Si	Mn	Cr+Mo+Al	Nb+Ti
Experimental steel	0.08-0.12	0.2-0.5	1.9-2.5	<1	<0.08

Finally, the ingots were hot rolled and cold rolled to 1.5mm thickness. Galvanizing cycles were simulated by a multi-purpose simulator in the laboratory and the galvanizing parameters are given in table 2.

Table 2. The parameters of galvanizing cycle.

Process No.	Preheating temperature	Heating temperature	Soaking temperature	Slow cooling temperature	Galvanizing temperature
1	220°C	780°C	780°C	680°C	460°C
2	220°C	800°C	800°C	700°C	460°C
3	220°C	820°C	820°C	720°C	460°C
4	220°C	780°C	780°C	--	460°C

The tensile samples were cut from the experimental plates and the tensile tests were conducted on universal materials tester at room temperature. The specimens were polished and etched in Lepera to observe their microstructures and MA islands. In order to study the precipitation behavior of micro-alloy element, the carbon extraction replicas were analyzed by JEM-2000F TEM and energy dispersive spectroscopy.

3. Experimental Results and Discussions

3.1. *Mechanical properties of experimental steels*

Mechanical properties of samples are shown in Table 3. It can be seen that, for the samples experienced slow cooling section (Nos. 1-3), the tensile strength decreases while the temperature increases. Meanwhile, the yield strength and yield ratio is accordingly enhanced. However, the mechanical properties, especially the yield strength, cannot match the standard of CP980. As for the No. 4 sample without slow cooling, the mechanical properties change obviously. Yield strength and yield ratio is increased and well matched the CP980 standard. In addition, after the online skin pass process, the yield strength can be further enhanced which may exceed 800MPa. It implies that heating temperature and the slow cooling section play important roles in the mechanical properties of the experimental steel.

Table 3. Mechanical properties of the experimental steel.

Sample No.	Rp0.2/MPa	Rm/MPa	A50mm/%	Yield ratio
1	623	1033	14.5	0.60
2	635	1021	14.0	0.62
3	663	1018	13.0	0.65
4	751	997	11.0	0.75

3.2. *Microstructure of experimental steels*

In order to clarify the effects of galvanizing parameters on mechanical properties of experimental steels, the microstructures of samples Nos. 1 and 4 are shown in Fig. 1. Fig. 1a shows the sample No. 1 which experienced the slow cooling section and Fig. 1b shows the sample No. 4 without slow cooling. It can be seen that two samples both have fine microstructure but also some obvious distinction. The microstructure consists of grey ferrite, black bainite and bright white MA islands. However, sample No. 4 has much more black bainites and microstructure of No. 1 is more similar to the DP steel character. In addition, the volume fraction of the MA islands in experimental steel is measured and listed in Table 4. It can be seen that the mean MA volume fraction of samples No. 1 and 4 is 17.52% and 12.46%, respectively. As for sample No. 4, after the soaking section, austenite was rapidly cooled to the bainite transformation area so that little ferrite can be formed. Meanwhile, without the diffusion of solution atoms from ferrite to austenite during the slow cooling section, the austenite hardenability is partly reduced. As a result, the MA volume fraction is decreased accordingly.

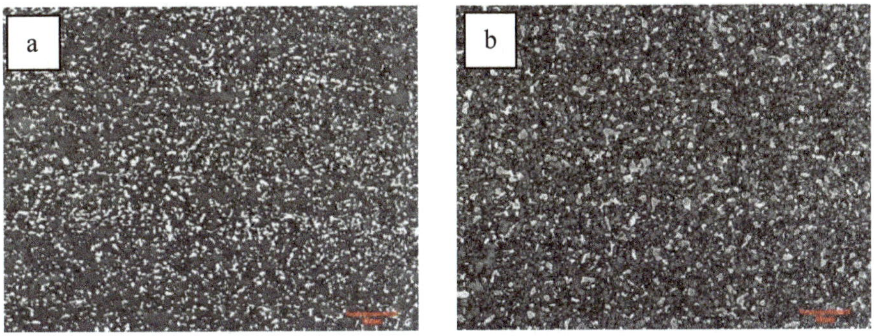

Fig. 1. Microstructure and MA islands distribution of experimental steels. (a) sample No. 1; (b) sample No. 4.

Combined with the data shown in Table 3, it can be concluded that mass fraction of bainite is helpful to increase the yield strength and yield ratio and MA volume fraction is more important to the tensile strength.

Table 4. MA ratios of the experimental steel.

Sample No.	MA ratio/%	Mean MA ratio/%
1	17.71 17.55 17.30	17.52
4	12.56 12.06 12.75	12.46

3.3. *Precipitates analysis of experimental steel*

It has been known that Nb and Ti elements were added in the experimental steel. In order to study the precipitation behavior, the carbon extraction replicas were analyzed and the results are given in Fig. 2. Figs. 2a and b show the morphology and distribution of precipitates in samples Nos. 1 and 4. It is obvious that the precipitates distribute evenly in the ferrite grain and mainly have the globular shape. Moreover, the size of them is about 10nm or much smaller. It is well known that the large amount of small size precipitates will produce strong strengthening to the ferrite and finally strengthen the steel. In addition, they also play important roles in the grain refinement of the steel.

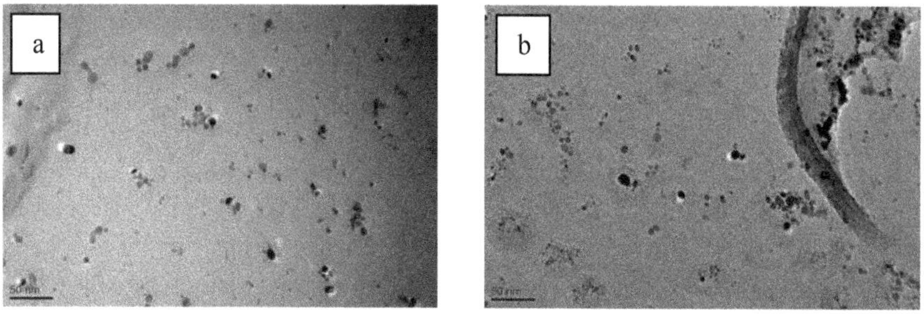

Fig. 2. Carbon extraction replicas analysis of experimental steel (a) sample No. 1; (b) sample No. 4.

However, there is nearly no difference between the two samples. It seems that precipitation of the steel mainly depends on the heating temperatures which are same for the samples. In addition, Fig. 3 shows the typical dispersive spectroscopy analysis of the precipitates. It is obvious that the typical peaks are Ti, Nb and C element which represents the TiC, NbC or [Ti, Nb] C phase.

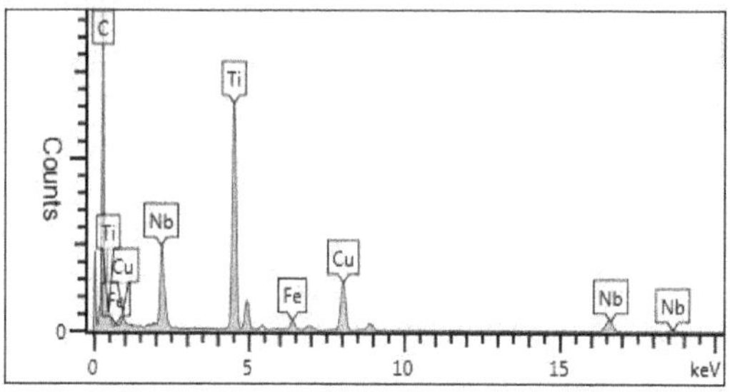

Fig. 3. Energy dispersive spectroscopy analysis of the precipitates.

3.4. *Hole expansion test*

Hole expansion test is regarded as an effective method to evaluate the flanging properties of steel. Hence, hole expansion tests were performed for the experimental steel. The results are shown in Table 5 and compared with DP steel.

Table 5. Hole expansion results of experimental steels.

steel	Initial hole diameter/mm	Hole diameter after caracking/mm			Hole expanding ratio/%	Mean hole expanding ratio /%
		D1	D2	mean		
	10.00	12.74	12.72	12.79	27.50	
DP980	10.00	12.41	12.38	12.47	24.20	26.77
	10.00	12.84	12.87	12.78	28.30	
	10.00	14.17	14.25	14.15	41.90	
Experimental CP980	10.00	13.86	14.02	13.79	38.90	41.10
	10.00	14.22	14.17	14.36	42.50	

It can be seen that experimental steel (sample No. 4) has better hole expanding ratio. The mean hole expanding ratio of DP980 is 26.77% while experimental CP980 is 41.10%. Thus, for the same tensile strength grade, CP steel with high yield strength and high yield ratio may have better flanging properties than DP steel.

4. Conclusion

(1) Microstructure of experimental steel mainly consists of ferrite, bainite and martensite. Galvanizing process without slow cooling section tends to form mass fraction of bainite which is helpful to increase the yield strength and finally match the CP steel mechanical properties standard. Excellent mechanical properties of the CP980 steel can be gained with 997MPa (TS), 751MPa (YS) and 11.0% (A50 EL).
(2) There are a lot of fine micro-alloy precipitates distributed in the matrix which is useful to increase the yield strength and also refine the grain size.
(3) The mean hole expanding ratio of experimental CP980 is 41.10% which is higher than that of DP980. It implies that CP steel have better flanging properties.

References

1. Y.L. Kang, Theory and technology of process and forming for advanced automobile steel sheets, Metallurgical industry press, Beijing, 2009.
2. K.S. Park, K.T. Park, D.L. Lee and C.S. Lee, Effect of heat treatment path on the cold formability of drawn dual-phase steels, *Mater. Sci. Eng. A* **499-451**, 1135 (2007).
3. G. Buzzichelli and E. Anelli, Present status and perspectives of European research in the field of advanced structural steels, *ISIJ Int.* **42**, 1354 (2002).
4. Q.H. Han, Y.L. Kang, X.M. Zhao, C. Lu and L.F. Gao, Microstructure and properties of Mo microalloyed cold rolled DP1000 steels, *J. Iron Steel Res. Int.* **18**, 52 (2011).
5. A. Nouri, H. Saghafian and S. Kheirandish, Effects of silicon content and intercritical annealing on manganese partitioning in dual phase steels, *J. Iron Steel Res. Int.* **17**, 44 (2010).
6. R. Kuziak, R. Kawalla and S. Waengler, Advanced high strength steels for automotive industry, *Arch. Civ. Mech. Eng.* **8**, 103 (2008).
7. Y. Han, J. Shi, L. Xu, W.Q. Cao and H. Dong, Effect of hot rolling temperature on grain size and precipitation hardening in a Ti-microalloyed low-carbon martensitic steel, *Mater. Sci. Eng. A* **553**, 192 (2012).
8. G. Xu, X.L. Gan, G.J. Ma, F. Luo and H. Zou, The development of Ti-alloyed high strength micro-alloy steel, *Mater. Des.* **31**, 2891 (2010).

Experimental Study About the Effect of Heating Power of Spot Induction Hardening on Material Properties of 42CrMo

M. Cheng[1], Z. Wang[1,2,3,†], X. P. Qin[1,2,3], K. Gao[1] and Y. Wang[1]

[1]*School of automotive Engineering, Wuhan University of Technology, Wuhan, Hubei, China*

[2]*Hubei Key Laboratory of Advanced Technology for Automotive Components, Wuhan, Hubei, China*

[3]*Hubei Collaborative Innovation Center for Automotive Components Technology, Wuhan, China*

[†]*E-mail: wangzhou@whut.edu.cn*

In order to investigate the effect of heating power on the materials properties of 42CrMo steel during spot induction hardening (SIH) process, a series of SIH experiments were carried out under different heating powers. Optical microscope, micro-hardness tester, and XRD methods were used to analyze the microstructure, micro-hardness and domain size after SIH process. The results showed that the heating power had a significant effect on the materials properties of 42CrMo.

Keywords: Spot induction hardening; Heating power; 42CrMo steel; Phase transformation.

1. Introduction

Induction hardening is one of most popular heat treatments in which a metal part is heated by induction heating and then quenched [1-4]. The aim of induction hardening steels and cast irons is to provide a martensitic layer on specific areas of the specimen to increase the hardness, strength, fatigue, and wear resistance, while allowing the remainder of the part to be unaffected by the process [5]. The typical procedure for induction hardening consists of heating the component above the austenizing temperature, holding the temperature long enough for essentially complete austenitic microstructures to form, and rapidly cooling the alloy until it is below the onset temperature of martensite transformation (Ms) [6-8]. As induction hardening is a complicate surface hardening process involving electromagnetic, heat transfer, and metallurgical phenomena, many investigations have been carried out on this process during last four decades with the aim of understanding and improving this useful surface hardening treatment [9-11]. Compared with other surface strengthen methods, induction hardening shows favorable features of controlled heating depth, energy saving and low cost [12]. Spot induction hardening (SIH), which belongs to selective hardening methods, was first proposed by Chen in 2013 [13] and the following work was carried out by Gao and Zhu [14-17]. The main characteristic of SIH is adopting a single-turn copper coil with a small size, which can be assembled to a five-axis cooperating CNC (computer numerical control) machine tools to realized a small region heat treatment of curved surface or some other complicated shapes. The effect of induction hardening is affected by various aspects, such as the air gap between specimen and inductor, heating power (HP), the flow rate of cooling water and so on. 42CrMo (American grade: AISI 4140) is one of the representative medium carbon and low alloy steel with high strength, toughness and good hardenability [18]. Furthermore, there are many excellent performances of this steel such as small deformation when hardening, high temperature with high strength and high creep rupture strength [19]. Due to its good balance of strength, toughness and wear resistance, 42CrMo high-strength steel is widely used for many general purpose parts including automotive crankshaft, rams, spindles, etc [20]. In this paper, a series of SIH experiments were carried out under different HP on 42CrMo, the influence of HP on temperature field, phase transformation and hardness distribution of 42CrMo specimens were analyzed systematically.

2. Materials

The specimens for SIH experiments was austenitized at 850 °C for 20 mins, quenched with oil and tempered at 650°C for 2 hours with yield strength of 700 MPa, tensile strength of 940 MPa and the hardness of 300 HV 0.3. The chemical composition is listed in Table 1. The dimensions of all specimens were 100 mm in length, 100 mm in width and 10 mm in thickness.

Table 1. The chemical composition of the samples for hardening treatment (wt.%).

C	Cr	Mo	Si	Mn	Ni	P	Al	S	Fe
0.44	1.21	0.22	0.28	0.81	0.07	0.02	0.03	0.02	Balance

3. Experimental Setup

The whole SIH system mainly consists of induction hardening subsystem and data acquisition subsystem. Fig. 1 shows the SIH facility setup. The induction hardening subsystem includes high frequency induction heating power supply (1), CNC system (2), cooling sprayer (3), magnetizer (4), inductor (5), transformer (6) and specimen stage (8). The data acquisition subsystem includes infrared thermometer (7) and data processing computer (9). The induction hardening treatment was performed by SIH system.

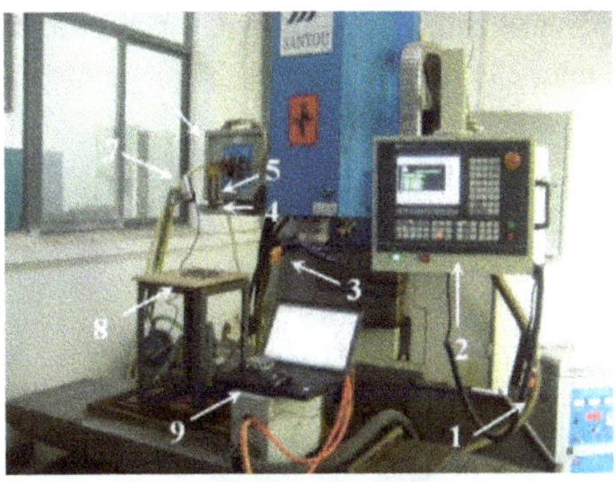

Fig. 1. SIH facility setup. 1. High-frequency induction heating power supply, 2. CNC system, 3. Cooling sprayer, 4. Magnetizer, 5. Inductor, 6. Transformer, 7. Infrared thermometer, 8. Specimen stage, 9. Data processing computer.

A series of SIH experiments were carried out on AISI 4140 specimen via SSF-50 high-frequency induction heating equipment whose maximum output power is 50 kW. The specimen was cooled by spraying water and the spraying pressure is in the range from 0.1 MPa to 0.3 MPa. In order to investigate the effect of heating power on the material properties of 42CrMo steel during SIH process, three groups of experiments were carried out on 42CrMo specimen under powers of 30 kW, 40 kW and 50 kW.

Table 2. Experimental scheme of SIH.

Specimen	A_1	A_2	A_3
Heating power (kW)	30	40	50
Air gap between specimen and inductor		1 mm	
Heating time		6 s	
Spraying pressure of cooling water		0.2 MPa	

Equipment inspection was carried out on each part of SIH system before induction hardening experiment. All the specimens were put at the center of specimen stage during SIH process. The distance between inductor and specimen was 1mm. The HP of specimen A1, A2 and A3 are 30 kW, 40 kW and 50 kW respectively. The heating time was 6 s and the specimens were cooled by spraying water immediately after 6 s in all specimens.

4. Results and Discussion

Fig. 2 shows the surface appearance of specimen after SIH treatment for 6 s. The dumbbell shaped areas and the rectangular areas in Fig. 2a, Fig. 2b and Fig. 2c represent the heat affected region and induction hardened region after SIH process. It can be seen that the heat affected region increases with HP increasing. In the condition of 30 kW HP, the length and width of heat affected region are about 40 mm and 16 mm. In the condition of 40 kW HP, the length and width of heat affected region are about 44 mm and 18 mm. In the condition of 50 kW, the length and width of heat affected region are about 46 mm and 20 mm. In terms of induction hardened region, where 100% martensite appears on surface, the length and the width in the conditions of 30 kW, 40 kW and 50 kW are 20 mm and 6 mm, 23 mm and 8.5 mm, 26 mm and 11 mm, respectively.

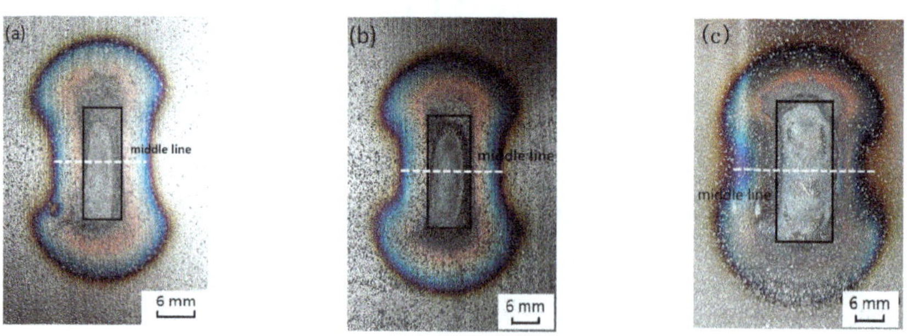

Fig. 2. Surface appearance of SIH after SIH with different power (a) P = 30 kW (b) P = 40 kW (c) P = 50 kW.

All specimens treated by SIH were cut with dimension of 10 mm × 8 mm × 8 mm for metallographic test. The metallographic test specimens were ground, polished and etched via 4% nitric acid and 96% ethanol reagent mixed in proportion. Fig. 3 shows the metallographic test specimen in 30 kW condition.

Fig. 3. The metallographic test specimen in the condition of P = 30 kW.

4.1. *Temperature field results*

In order to analyze the temperature variation during SIH process, the real time temperature measurement process was carried out on the center point of specimen via infrared thermometer and data processing computer. Fig. 4 shows the experimental time-temperature curves of center point of specimen. It can be seen that there are three stages including heating stage, stable stage and cooling stage. The heating rate decrease with temperature increasing in heating stage in all HP conditions. The times of reaching stable temperature in 40 kW and 50 kW HP conditions are nearly the same with value of 1.2 s. However, the time of reaching stable temperature in 30 kW condition is 3 s. The stable temperature in the conditions of P = 30 kW, 40 kW, 50 kW are about 891°C, 953°C and 1038°C respectively. By comparing the time-temperature curve in these three HP conditions, it can be seen that the heating stage can be divided into two temperature rising processes. In the first heating process the temperature increased quickly and in the second heating process the temperature increased slowly in all three HP conditions. The heating rates in the first heating process were nearly the same when HP condition changes and the heating rate in the second heating process increased with HP increasing. The times of stable stage in the conditions of P = 30 kW, 40 kW, 50 kW were about 3 s, 4.8 s and 4.8 s. After heating for 6 s, the cooling water was carried out on specimen and the temperatures decreased drastically. The cooling rates were very similar because the flow rate of cooling water were the same in these three HP conditions.

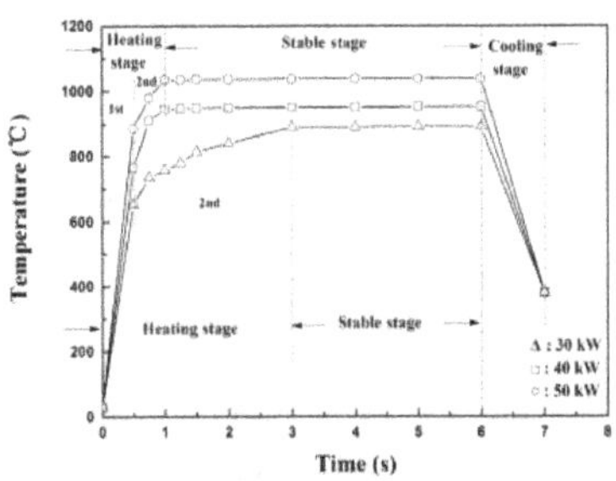

Fig. 4. The temperature-time curves of the heating center in the conditions of P = 30 kW, 40 kW and 50 kW.

4.2. *Microstructure*

In order to analyze the microstructure of specimen after SIH treatment, the specimen was cut along the middle line of specimen as shown in Fig. 2. Fig. 5 shows an overview of phase distribution in the cross section. The center region of specimen receives the maximum heat so the martensite structure region is a crescent shape. According to the result of optical microscope shown in Fig. 5, the maximum width and depth of hardened region increase with HP increasing. The maximum width of hardened region in the conditions of P = 30 kW, 40 kW, 50 kW were about 6 mm, 9 mm and 11 mm. The maximum depth of hardened region in the conditions of P = 30 kW, 40 kW, 50 kW were about 1.6 mm, 2.6 mm and 3.2 mm. Compared with the maximum depth of hardened region in the condition of P = 30 kW, an increase of 62.5% was reached in the condition of P = 40 kW. However, there was only an increase about 23% was reached when the HP increased from 40 kW to 50 kW.

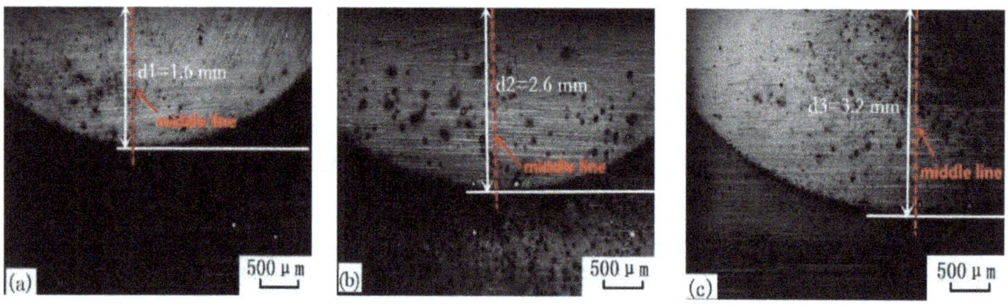

Fig. 5. Microphotograph of hardened layer depth after SIH on different power (a) P = 30 kW (b) P = 40 kW (c) P = 50 kW.

Fig. 6 illustrates the microstructure of SIH treated specimen in different region. Fig. 6a shows the matrix region microstructure of the specimen, which is typical pearlite phase. Fig. 6b shows the microstructure of transitional region between hardened region and matrix region in the condition of P = 30 kW. The upper part is martensite and the lower part is the mixture of martensite and pearlite. Fig. 6c shows the microstructure of hardened region in the condition of P = 30 kW, which is typical martensite phase. Fig. 6d and Fig. 6e show the microstructure of transitional and hardened region in the condition of P = 40 kW. Fig. 6f and Fig. 6g show the microstructure of transitional and hardened region in the condition of P = 50 kW. By comparing with these six diagrams, it can be seen that the microstructure in transitional and hardened regions are nearly the same in different HP conditions.

In order to investigate the statistical domain size of hardened region after SIH, the Scherrer equation, which is one classic XRD method, was used as following:

$$D = \frac{K\gamma}{B\cos\theta}$$

Where D is the mean size of the ordered (crystalline) domains, B is the line broadening at half the maximum intensity (FWHM) in radiants, K is a dimensionless shape factor, with a value close to unity, γ is the wave length of X-ray, and θ is the Bragg angle [21].

Fig. 6. Metallographic images of matrix region of all specimens (a), transitional region (b) and hardened region (c) of P = 30 kW, transitional region (d) and hardened region (e) of P = 40 kW, transitional region (f) and hardened region (g) of P = 50 kW.

Fig. 7 shows the XRD patterns of quenched and tempered 42CrMo in matrix region and hardened region via SIH with different HP. Three characteristic peaks of α-Fe in {110}, {200} and {211} diffraction directions can be observed which corresponds to the majority phase in material. By comparing XRD patterns of hardened area and matrix area in Fig. 7, it can be seen that all the characteristic peaks of α-Fe become less sharp after SIH process were performed on the surface. As the characteristic peaks of α-Fe in {200} and {211} diffraction directions have a low peak intensity, which may result in a significant error in domain size calculation, the characteristic peak of α-Fe in {110} diffraction direction were selected in the following analysis. The full width at half maximum (FWHM) of matrix region is about 0.214°. The FWHM of hardened surface region are 0.559°, 0.436° and 0.353° under the HP of 30 kW, 40 kW and 50 kW. According to the Scherrer equation, the domain size of matrix region is 549 nm and the domain sizes of hardened surface region are 147 nm, 174 nm and 185 nm under the HP of 30 kW, 40 kW and 50 kW. It can be seen that the domain size in matrix area is largest, which is due to the heat treatment before SIH. According to Section 2, the martensitic hardening in oil was carried out after austenitization at 850°C for 20 mins and tempering was conducted at 650°C for 2 hours. The austenitizing temperature and time as well as tempered temperature and time are high and long enough, so the domain size of matrix region is much larger than that of hardened region. In terms of hardened region, the domain size increases with HP increasing, which is due to the heating time and temperature during SIH. It is can be seen from Fig. 4 that the stable stage time of temperature in the conditions of P = 30 kw, 40 kW, 50 kW were about 3 s, 4.8 s and 4.8 s, and the stable temperature in the conditions of P = 30 kW, 40 kW, 50 kW are about 891°C, 953°C and 1038°C respectively. The more heating time and the higher heating temperature are, the larger domain size is.

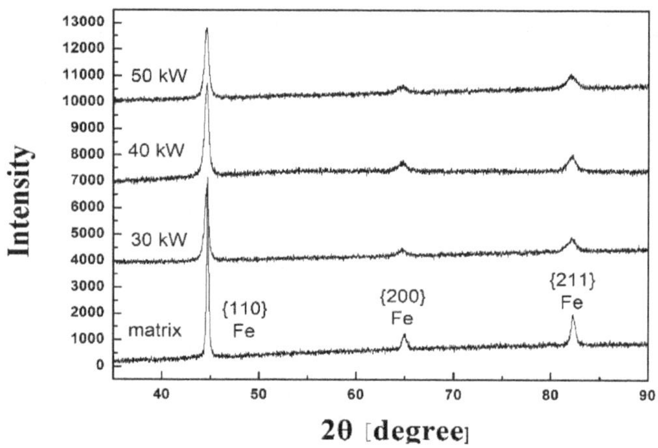

Fig. 7. XRD patterns of quenched and tempered 42CrMo in matrix region and hardened region via SIH with different HP.

4.3. *Micro-hardness*

In order to obtain the micro-hardness distribution in depth, the micro-hardness measurements were carried out on the polished cross section of treated specimen using HUAYIN HV-1000 with 500 g and a diamond indenter. The measured points are from the top surface to the core region along the middle line shown in Fig. 5 and the first point is at the depth of 0.15 mm. The interval distance between adjacent chosen points is about 0.2 mm. Fig. 8 shows the micro-hardness depth profiles in different HP conditions. It can be seen that the larger HP makes the larger hardened depth. The region where the micro-hardness is larger than 610 HV is considered as 100% martensitic region. The depths of 100% martensitic region in HP = 30 kW, 40 kW and 50 kW are about 1.6 mm, 2.6 mm and 3.2 mm. It is very interesting to see that increasing HP from 30 kW to 40 kW, the depth of 100% martensitic region increases 1.0 mm. However, increasing HP from 40 kW to 50 kW, the depth of 100% martensitic region increases only 0.6 mm. This phenomenon is because same increment of HP does not make the same increment of 100% martensitic region depth in the deeper and wider region. It is reasonable to estimate that the increment of 100% martensitic region depth becomes less than 0.6 mm if HP increases from 50 kW to 60 kW. In terms of the transitional region, the depths of this region in HP = 30 kW, 40 kW and 50 kW conditions are nearly the same with value of 0.8 mm. This phenomenon is due to the depth of transitional region only relating to the material own thermal property. From the experimental data it can be concluded that choosing larger HP can obtain larger depth of 100% martensitic region but the depth of transitional region is nearly the same in all HP conditions.

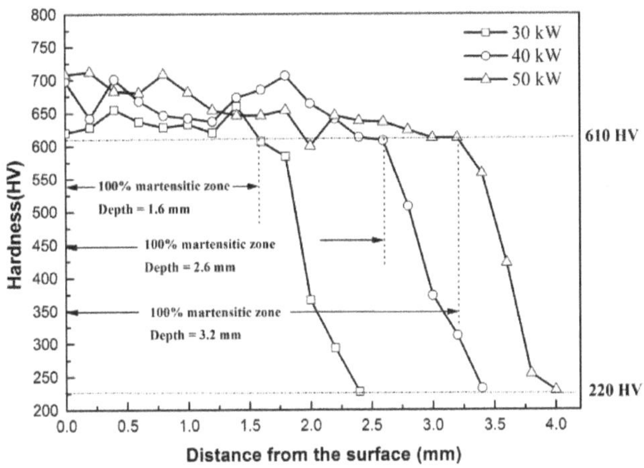

Fig. 8. The depth profiles of micro-hardness of different HP.

5. Conclusion

The investigation of the effect of heating power (HP) on spot induction hardening process (SIH) of 42CrMo steel was carried out in this work via temperature, microstructure and micro-hardness analysis. The following conclusions have been reached:

(1) There are three stages including heating stage, stable stage and cooling stage during SIH process. And the heating stage can be divided into quick heating and slow heating processes in all HP conditions. The stable heating temperature increases with HP increasing.

(2) SIH is an effective method to strengthen the surface region of 42CrMo. According to experiment date, the length and the width of 100% martensitic region on surface in the conditions of 30 kW, 40 kW and 50 kW are 20 mm and 6 mm, 23 mm and 8.5 mm, 26 mm and 11 mm. The area of the 100% martensitic region increased with HP increasing obviously.

(3) The statistic domain size of matrix region is about 549 nm and the statistic domain size of hardened surface region are 147 nm, 174 nm and 185 nm under

the HP of 30 kW, 40 kW and 50 kW. The dramatic difference of domain size between matrix and hardened surface region is due to the heat treatment before SIH. The domain size of hardened surface region increases with HP increasing, which is due to the heating time and temperature during SIH.

(4) The HP influences the depth of 100% martensitic region but nearly does not influence the depth of transitional region. The depths of 100% martensitic region in HP = 30 kW, 40 kW and 50 kW are about 1.6 mm, 2.6 mm and 3.2 mm. In terms of the transitional region, the depths of this region in HP = 30 kW, 40 kW and 50 kW conditions are nearly the same with value of 0.8 mm.

Acknowledgments

The financial support of the National Natural Science Foundation of China (NSFC) (No. 51405356 and No. 51175392) and the Research Fund for the Doctoral Program of Higher Education of China (No. 20130143120015) are gratefully acknowledged.

References

1. R. Valery, L. Don, Cook, Raymond, Black and Micah, *Handbook of Induction.* (Heating, CRC Press, 2002).
2. V. Deshmukh and S. Sahay, An Integrated Modeling Approach for Induction Hardening Process, *SAE Technical Paper* **28**, 81 (2010).
3. S. Sacki, H. Leiner, M. Zuber and H. Clemens, Induction Hardening vs Conventional Hardening of a Heat Treatable Steel, *The Minerals, Metals & Materials Society and ASM International* **45**, 5657 (2014).
4. J. Montalvo-Urquizo, Q. Liu and A. Schmidt, Simulation of quenching involved in induction hardening including mechanical effects, *Computational Materials Science* **79**, 639 (2013).
5. V. Rudnev and D. Loveless, Induction Hardening: Technology, Process Design and Computer Modeling, *Comprehensive Materials Processing* **12**, 489 (2014).
6. Z.Q. Cui and Y.C. Tan, *Metal materials and heat treatment*, China Machine Press (2007).
7. I. Magnabosco, P. Ferro, B. Tiziani and F. Bonollo, Induction heat treatment of a ISO C45 steel bar: Experimental and numerical analysis, *Computational Materials Science* **35**, 98 (2006).
8. Y. Totik1, R. Sadeler, H. Altun and M. Gavgali, The effects of induction hardening on wear properties of AISI 4140 steel in dry sliding conditions, *Mater. Des.* **24**, 25 (2003).
9. P. Karban and onátová, Continual induction hardening of steel bodies, *Mathematics and Computers in Simulation* **80**, 1771 (2010).
10. T. Palin-Luc, D. Coupard, C. Dumas and P. Bristiel, Simulation of multiaxial fatigue strength of steel component treated by surface induction hardening and comparison with experimental results, *Int. J. of Fatigue* **33**, 1040 (2011).

11. C. Li, and B. Lynn Ferguson, Stress Generation in an Axle Shaft during Induction Hardening, *Research Gate* **14**, 1196 (2014).
12. S. Zinn and S. L. Semiatin, *Elements of Induction Heating: Design Control and Applications*, (ASM International, 1988).
13. H. Chen, X.P. Qin, Z.Wang and K.Gao, Finite Element Analysis of Coupled Electromagnetic Thermal Field for 3D Spot Induction Hardening, *China Surface Engineering* **26**, 79 (2013).
14. K. Gao, X.P. Qin, Z. Wang, H. Chen, S.X. Zhu, Y.L. Liu and Y.L. Song, Numerical and experimental analysis of 3D spot induction hardening of AISI 1045 steel, *J. Mater. Process Tech* **214**, 2425 (2014).
15. K. Gao, X.P. Qin, Z. Wang and S.X. Zhu, Effect of spot continual induction hardening on the microstructure of steels: Comparison between AISI 1045 and 5140 steels, *Mater. Sci. Eng. A* **651**, 535 (2016).
16. S.X. Zhu, Z. Wang, X.P Qin, H.J. Mao and K.Gao, Prediction of Phase Transformation and Hardness Distribution of AISI 1045 Steel After Spot Continual Induction Hardening, *J. Mater. Eng. Perform.* **24**, 3719 (2015).
17. S.X. Zhu, Z. Wang, X.P. Qin, H.J. Mao and K. Gao, Theoretical and experimental analysis of two-pass spot continual induction hardening of AISI 1045 steel, *J. Mater. Process Tech* **229**, 814 (2015).
18. Y.Y. Li, S.D. Zhao, S.Q. Fan and G.H. Yan, Study on the material characteristic and process parameters of the open-die warm extrusion process of spline shaft with 42CrMo steel, *J. Alloy. Compd.* **571**, 12 (2013).
19. J.J. Gu, Y.Q. Qin, Z.L. Chen and Q.H. Lu, Effects of High Frequency Induction Heating Quenching and Tempering on Microstructure and Hardness of 42CrMo Steel, *Material & Heat Treatment* **39**, 160 (2010).
20. Y.C. Lin, M.S. Chen and J. Zhong, Effects of deformation temperatures on stress/strain distribution and microstructural evolution of deformed 42CrMo steel, *Mater. Des.* **30**, 908 (2009).
21. A.W. Burton, K. Ong, T. Rea and I.Y. Chan, On the estimation of average crystallite size of zeolites from the Scherrer equation: A critical evaluation of its application to zeolites with one-dimensional pore systems, *Microporous Mesoporous Mate*r. **117**, 75 (2009).

Comparison Research for Spring Backs of Dual-Phase Steels with Three Different Strength Grades

M. T. Ma[1,†], D. G. Kim[2], L. F. Song[1], Y. Zhao[1] and G. Fang[1]

[1]*China Automotive Engineering Research Institute CO., LTD, Chongqing 401122, China*
[2]*Posco global research and develop center, Incheon-si, 406-840, Republic of Korea*
†*E-mail:mingtuma@126.com*

The spring backs of dual-phase steels with three different strength grades were investigated. The mechanical properties were measured, microstructures were observed, spring backs of L type and V type bending samples were measured. The influence of yield strength, flow stress, tensile strength on the spring back of dual-phase steels is analyzed. The results show that the spring back angles of L type samples increase with increasing strength of dual-phase steel. The spring back angles of V type samples increase with increasing bending angles and strength. The relation test results are of reference for prediction and compensation of spring back. The causes of complex spring back for V-type bending samples of three are analyzed.

Keywords: Dual-phase steel; Spring back; Mechanical property; Binder force.

1. Introduction

The development of automobile lightweight and security technology [1] lead to the application of high-strength steel, ultra high-strength steel and advanced forming technique [2]. The dual-phase steel is high strength and high formability steel, which has a stable production process and consistent properties: it is an important stamping steel for the automobile industry [3]. The spring back of stamping parts usually increases with increasing strength of the plate during stamping. The study of spring back for dual-phase steel contribute to application of high strength dual-phase steel, promoting stamping process and property of dual-phase steel, improving stamping simulation precision. So the study of spring back for dual-phase steel is always a project which people are interested in. The spring back in forming process at different positions of the sidewall of hat-type structure for dual-phase steel is analyzed in literature [4]. The various stress state of U-type structure for dual-phase steel when it forming is analyzed in literature [5]. The empirical model of spring back for controlling V-type bending forming of plate is proposed in literature [6]. The comparison measurement and study for spring back with L and v-type of dual phase steel with different strength levels has not been reported. The purpose of this article is to measure the mechanical properties of dual phase steel and the spring back of L type and V type specimens, discuss factors that influence the spring back, provide reference model for computer simulation under similar forming mode.

2. Steels and Methods Used in the Testing

The testing steel is the dual phase steel plate (DP590, DP780, DP980) from a company industrial production. The chemical constituents, and the tensile samples are cut from three direction which is parallel, perpendicular and 45-degree to the rolling direction. The gauge length is 20×90mm. The thicknesses of samples are respectively 1.4mm, 1.8mm, 1.6mm. The three samples are cut in each direction. We get average value from three samples. Strain hardening index n values for uniform deformation zone was calculated by Holloman equation regression analysis, microstructure of test materials with three surface (vertical rolling direction, parallel to the rolling direction, perpendicular to the rolling surface) was observed by microscope Leica-DMI 3000 type, and the phase constitution and martensitic morphology analysis were conducted by SEM (FE quanta 20 type). The sample size of spring-back test was 140mm×70mm×plate thickness for the L-type 90° bending test, and 160mm×70mm×plate thickness for the V-type bending test, the spring-back angle was measured as representation parameter for spring-back test, and the test apparatus and results of L-type bending test were shown in Fig. 1, the apparatus and results of V-type bending test were shown in Fig. 2, also the test fixture was shown in Fig. 3.

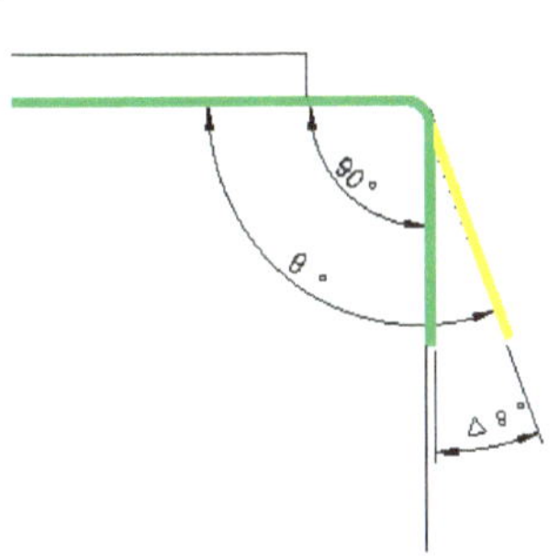

Fig. 1. Bending test device of L-type sample and the measurement diagram of spring back angle.

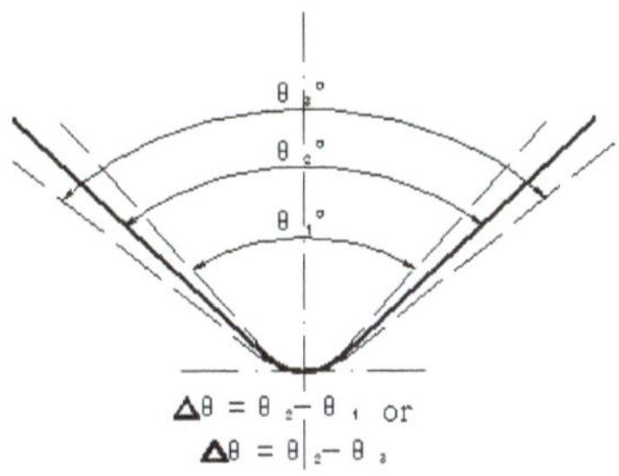

Fig. 2. Bending test device of V-type sample and the measurement diagram of spring back angle.

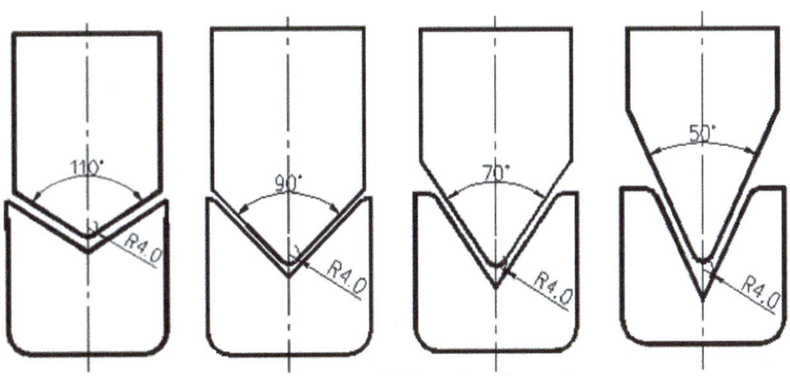

Fig. 3. The fixture of different bending angle.

3. Test Result and Analysis

3.1. *Chemical composition*

The chemical composition of test three steels (DP590, DP780, DP980) were listed in Table 1.

Table 1. The chemical composition of test steels (wt.%).

Steel	Chemical composition (w_t%)							
	C	Si	Mn	Cr	Cu	Al	P	S
DP590	0.084	0.28	1.79	0.011	0.019	0.038	0.012	0.008
DP780	0.110	0.30	1.81	0.202	0.021	0.034	0.017	0.009
DP980	0.170	0.40	2.32	0.390	0.015	0.039	0.013	0.009

3.2. *Mechanical properties*

The mechanical property of DP590 was listed in Table 2, DP780 and DP980 were listed in Table 3. From the listed results, the mechanical properties of these three types of dual phase steels showed isotropy with a slightly different of r value.

Table 2. Mechanical property of DP590.

Steel grade	Thickness, mm	Serial number	El.,%	0.2% YS., MPa	TS., MPa	Young's modulus (E) MPa	Strain hardening exponts (n2-20)	Plastic strain ratio (r2-20)	Poison's ratio (μ)
DP590	1.4	1#-0	26.44	385	654	197656	0.20	0.91	0.29
		1#-45	24.76	395	659	206002	0.19	0.79	0.26
		1#-90	26.81	387	656	201550	0.19	1.09	0.30
	1.8	2#-0	28.67	389	635	195328	0.19	0.89	0.29
		2#-45	25.33	408	634	207762	0.18	0.89	0.29
		2#-90	26.67	406	648	203501	0.19	1.07	0.28

Table 3. Mechanical property of DP780 and DP980.

Steel grade	Thickness, mm	Serial number	El.,%	0.2% YS., MPa	TS., MPa	Young's modulus (E) MPa	Strain hardening exponts (n2-6)	Plastic strain ratio (r2-6)	Poison's ratio (μ)
DP780	1.4	3#-0	16.00	584	864	200502	0.13	0.83	0.27
		3#-45	14.83	584	860	198784	0.13	0.94	0.29
		3#-90	15.17	571	859	203606	0.13	0.89	0.30
DP980	1.6	4#-0	8.83	823	1026	201073	0.11	0.78	0.28
		4#-45	9.17	827	1023	200970	0.10	0.87	0.30
		4#-90	8.33	829	1032	205353	0.10	0.77	0.28

The constitutive equations of three type steels were obtained by regression within the scope of 0.5%-3% strain along the rolling direction.

DP590 (1.4mm):

$$\sigma = 1050\varepsilon^{0.187} \tag{1}$$

DP590 (1.8mm):

$$\sigma = 948\varepsilon^{0.164} \tag{2}$$

DP780 (1.4mm):

$$\sigma = 1201\varepsilon^{0.135} \tag{3}$$

DP980 (1.6mm):

$$\sigma = 1606\varepsilon^{0.131} \tag{4}$$

With the constitutive equations, it is observed that the strengthening coefficient increased along with increasing strength, but the strain hardening index decreased. The thickness of DP590 showed a significant influence on strain hardening index.

3.3. *Microstructure of dual phase steel*

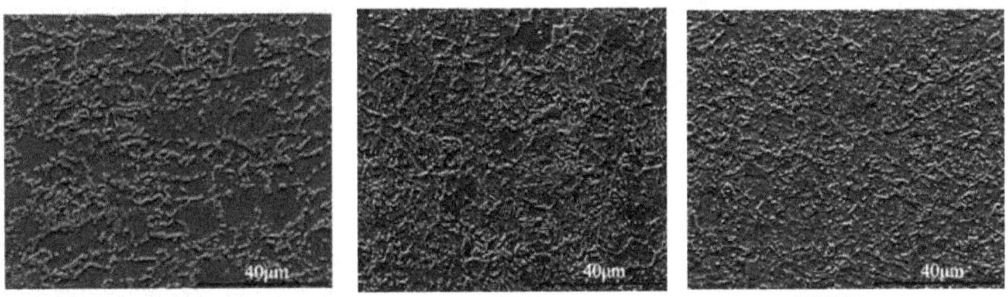

DP590 DP780 DP980

Fig. 4. Microstructure of dual phase steel.

The microstructure of DP590, DP780, DP980 from SEM were shown in Fig. 4, the martensite within the microstructure was fine and evenly distributed, and the volume fraction of martensite increased with increasing strength.

3.4. *L type and V-type spring-back test*

The relationship between spring-back angle of L-type sample and yield strength, flow stress with 3% strain was shown in Fig. 5. It can be seen from the Fig. 5 that the spring-back angle of L-type sample and the yield strength presented a linear relations, also a linear relation with the flow stress of 3% strain, the two curves showed a close slope. The loading force was simple during the spring-back testing for L type sample, the spring-back angle and strength maintained a simple linear relationship.

The regression of flow stress with 3% strain and spring-back angle was carried out considering the effective strain was close to 3% during stamping process for many automotive components. The relationship of the spring-back angle of L-type with tensile strength, strengthening parameter K in Holloman equation were displayed in Fig. 6, also a linear relationship between these parameters, but the curves slope was different.

The linear regression result of Fig. 5 and Fig. 6 were displayed in eq.(5)~eq.(8), these results also exhibited that regression curves slope decreased and intercept increased along with chosen increasing flow stress. The influence of K in Holloman equation on L-type bending test was in agreement with material strength, also consistent with the research result of reference [7]. The stress mode was simple of spring-back during the L type sample bending test, therefore, the relationship between spring-back and plate properties was discussed, it is difficult to find out related factors to control spring-back under this deformation mode.

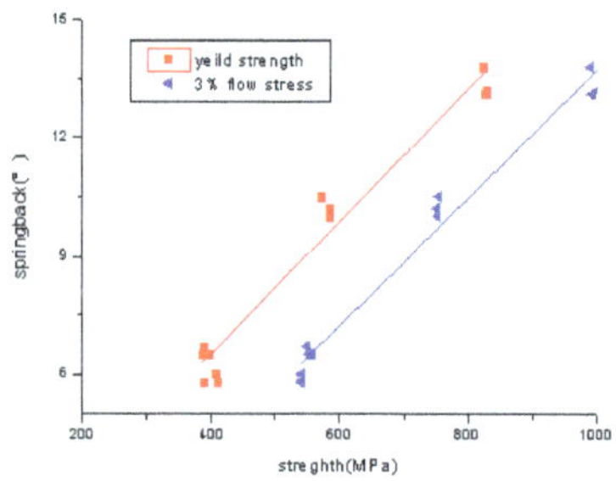

Fig. 5. Relationship of the spring-back angle with yield strength and 3% flow stress.

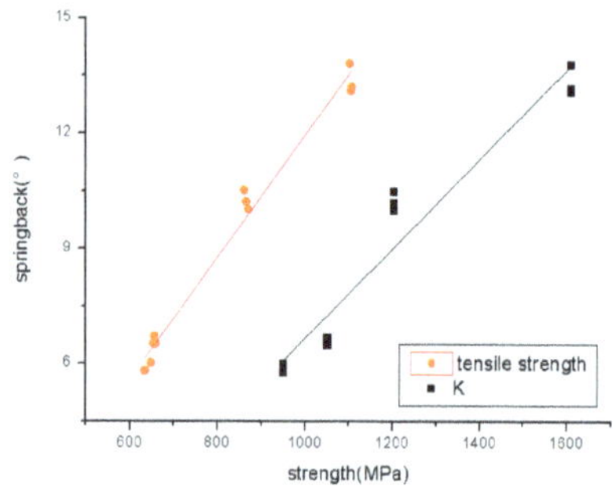

Fig. 6. Relationship of the spring-back angle with tensile strength and K.

$$\theta = 0.0167\sigma_s - 0.1798 \tag{5}$$

$$\theta = 0.0162\sigma_3 - 2.4876 \tag{6}$$

$$\theta = 0.0159\sigma_b - 3.9686 \tag{7}$$

$$\theta = 0.0117K - 5.0170 \tag{8}$$

Result of V-type spring-back test was shown in Fig. 7, the spring-back angle of DP590, DP780 and DP980 were enlarged with the increase of bending angle, DP590 and DP780 showed negative spring-back, DP980 displayed all positive spring-back, the relationship between spring-back angle and bending angle of these three tested steels were linear.

From a macro point of view, there was a unstressed layer within plate during the bending process, which was called middle layer, the outer side of the middle layer was subjected to tensile stress during bending, and the layer inner side was subjected to press stress, elastic recovery occurred in the side suffering tensile stress, which lead to generate positive spring-back and the bending angle decreased.

The inner of middle layer was subjected to press stress during bending, and the inner side suffered tensile stress after bending due to the elastic recovery, the torques generated from elastic recovery force in the both side of middle layer all generate positive spring-back.

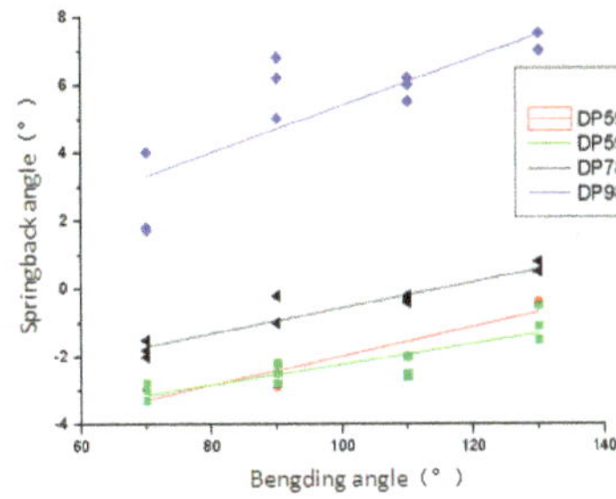

Fig. 7. Curve of spring-back angle and bending angle.

From a micro point of view, the dual-phase steel microstructure consisted of hard martensite and soft ferrit, the martensitic island would be subjected to tensile stress generated by ferrit deformation in outer side of middle layer during bending, which lead to elastic deformation along the tensile direction. Deformed martensitic island would generate recovery force after removing loading, the recovery force would compress ferrit phase, force status of these microstructures, also make the plate to generate positive elastic recovery after bending.

The inner side of the middle layer, martensite island would suffer compress during bending, removing loading would generate elastic recovery, which make ferrit suffer tensile stress, the combination of these two micro torques would generate positive spring-back.

Due to the different martensitic island deformation constraint during compress and tensile process, the martensitic island would easily deformed during tensile process, so the elastic recovery force generated in the both side of the middle layer could be different, positive spring-back would occur based on general research. Actually, the plate bending and spring-back were a complex process, reference [4] analyzed spring-back of hat type structure bending process, which was shown in Fig. 8, negative spring-back generated in some position. Reference [5] originally analyzed the DP600 forming process of U types structure, the forming was a complex process, so the spring-back could not be easily measured and analyzed, shown in Fig. 9.

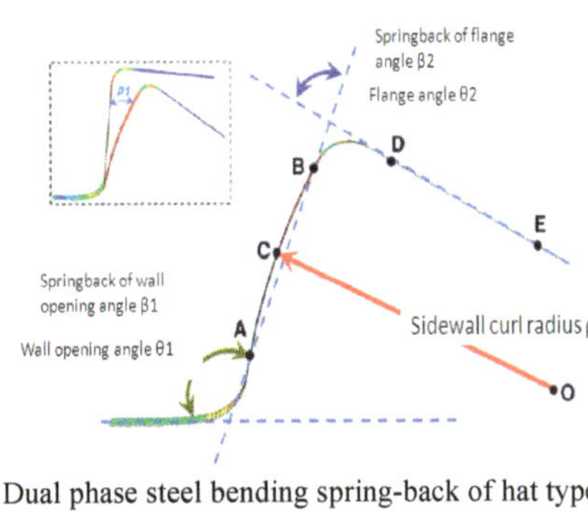

Fig. 8. Dual phase steel bending spring-back of hat type structure.

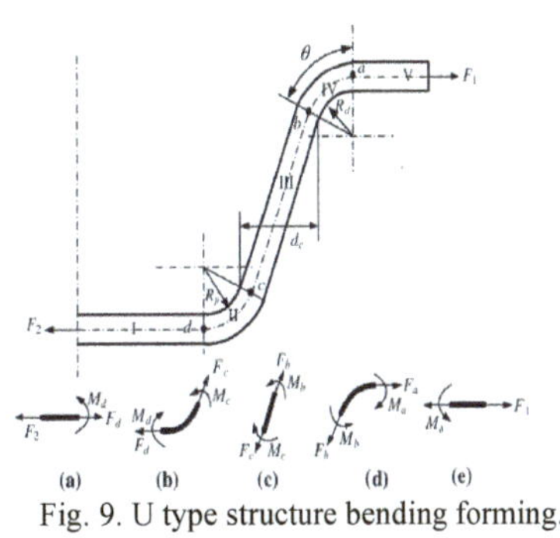

Fig. 9. U type structure bending forming.

The deformation process of plate free bending was shown in Fig. 10 [8]. The stress status in some area would generate negative spring-back after final bending forming, and stress status in some other area would generate positive spring-back, which was related to balance result of stress status.

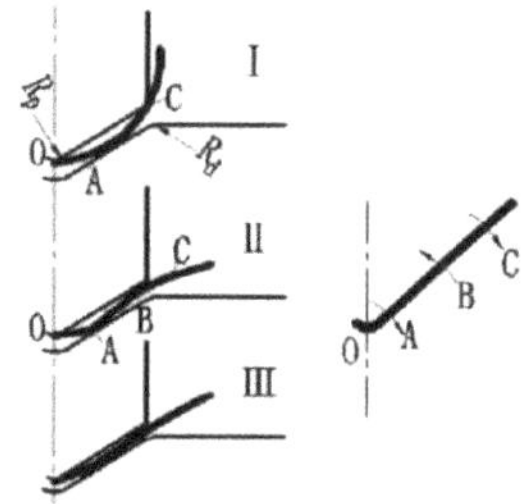

Fig. 10. Deformation process of plate bending.

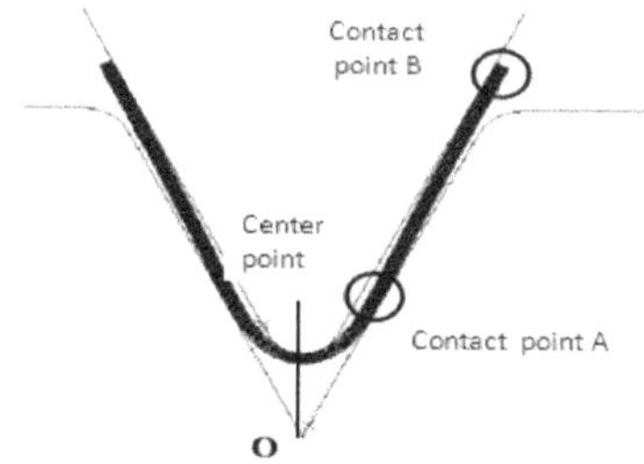

Fig. 11. Plate partly thickness reduction during bending.

In this paper, the bending punch radius was 4mm, the bottom of dies was sharp corner, during bending, in order to control the spring-back after forming process, the chosen stamping force could be high, the specimen in the top of the punch would generate partly plastic deformation, also thickness reduction would occur, and the deformation would generate special stress status, and the partly stress status and local plastic deformation distribution will be changed after bending process, negative spring-back maybe occur in this area. As shown in Fig. 11, the material between the touch point A and B would suffer tensile stress in the inner side and compress in the outer side, means bending forming which lead to circumferential stress change of some nodes from center point O to contact point A.

In the compared three types dual phase steels, the hardness of DP590, DP780 and DP980 were respectively 200HB, 234HB and 311HB. During the bending process, DP590 and DP780 would generate partial deformation and thickness reduction, which may lead to negative spring-back. Due to the high strength of DP980, it is diffcult to generate partial deformation, negative spring-back would not occur.

From the microstructure, the difference of the martensite and ferrit hardness for DP980 far less than that for DP590, and the martensite volume of DP980 was high, due to the plastic strain between the martensite and ferrit was incompatible for DP980, the interphase stress generated was lower than that of DP590 and DP780, especially when the bending angle was large, it is not easy to generate negative spring-back for DP980.

After an overall consideration, the spring-back test result in Fig. 6 would be obtained, there need further study work to quantitatively analyze above result. Based on the microstructure parameter measurement, form the simulation of microstructure deformation to stress status simulation in macro deformation process, it is possible to simulate the spring-back during bending process.

4. Conclusion

The mechanical properties and microstructures of these three dual phase steels were tested in this paper, also the L-type and V-type spring-back test were conducted, the result showed that:

(1) The regression hardening index decreased and strengthening coefficient increased by Hollomon equation along with the increasing strength of dual phase steel.

(2) The measured result of L-type spring-back test showed that the spring-back angle increased with higher yield stress, tensile strength, flow stress in the strain of 3% and K value in the Hollomon equation, and displayed a linear relationship.
(3) The measured result of V-type bending test showed that these three type steels displayed a quite complex trend; in terms of DP590, a small bending angle generated negative spring-back, a large bending angle displayed less negative spring-back. A small bending angle also generated negative spring-back, and increased bending angle generate less negative spring-back, and close to zero for DP780. According to the test result of DP980, a small bending angle generate positive spring-back, a large bending angle showed an increased spring-back angle.

These spring-back characteristics of dual phase steel were related to phase constitution, and strength level. From the macro analysis, the spring-back characteristics were related to the complex loading model, plastic deformation and partially thickness reduction resistance during bending process, as well as complicated stress status in different position of bending deformation sample introduced by former mentioned factors.

References

1. M. T. Ma, H. L. Yi and H. Z. Lu, Review of automotive lightweight, *Chinese Engineering Science* **11**, 9 (2009).
2. M. T. Ma, H. Z. Lu and Z. G. Li, Automotive lightweight and high strength advanced forming technology, in *Proceeding of Chinese mechanical engineering society annual meeting*, 2008.
3. M. T. Ma and B. R. Wu, *Dual phase steel-physical and mechanical metallurgy*, 2th edn, (Metallurgical industry press, 2009).
4. P. Chen and M. Koc, Simulation of springback variation in forming of advanced high strength steels, *J. Mater. Process Tech.* **190**, 189 (2007).
5. D. Bekar, E. Acar, F. Ozer and M. A. Guler, Robust Springback Optimisation of DP600 Steels for U-Channel Forming, in *Proceedings of the World Congress on Engineering 2011*, (London, U.K, 2011).
6. Z. Tan, B. Pesson and C. Magnusson, An empiric model for controlling springback in V-die bending of sheet metals, *J. Mater. Process Tech.* **34**, 449 (1992).
7. D. J. Zhang, Z. S. Cui and Y. Q. Li, The influence of materials reinforcement model for plate springback, *J. Shanghai jiaotong university* **40**, 1671 (2006).
8. X. W. Ji, Research on high strength steel bending spring-back and control, master thesis, Chongqing University (2012).

Prediction of Dynamic Flow Behavior of a Dual-Phase Steel

J. P. Zhang[a], J. Zhou[b] and M. T. Ma[c]

China Automotive Engineering Research Institute CO., LTD.
Automotive Lightweight Engineering Research Center
Chongqing, 401122, China
Chongqing Automotive Lightweight Engineering Research Center
Chongqing 401122, China
E-mail: [a]zwwjp@163.com, [b]zhoujia@caeri.com.cn, [c]mingtuma@126.com

To study the flow behavior of the dual-phase steel under dynamic loading, high speed tensile experiments are employed with DP590 and the strain sensitivity is discussed. On this basis, BP neural network and constitutive models are respectively established to simulate the dynamic flow behaviors of the DP steel under the strain rates of $0.1s^{-1}$, $50s^{-1}$, $100s^{-1}$, $200s^{-1}$ and $500s^{-1}$. Then the flow curves of the material loaded at $100s^{-1}$ and $200s^{-1}$ are predicted by BP neural network model and constitutive models respectively. The predicted results by the aforementioned models and the tensile results under the strain rate of $100s^{-1}$ and $200s^{-1}$ are compared. The prediction accuracy is validated according to the error.

Keywords: Dynamic flow behavior; Dual-phase steel; BP neural network; Constitutive model.

1. Introduction

DP steels exhibit low yield strength ratio, high initial hardening rate and good match of strength and ductility, combining the advantages of high strength of Martensite and good ductility of Ferrite, which are widely applied in the automobiles as bumpers, doors, wheels and other safety components. To be largely used as safety components of automobiles, the dynamic mechanical properties of DP steels are very important since its dynamic mechanical property is different from that in quasi-static state [1-3]. In the development process, the crashworthiness of safety components is considered strictly, which is originally evaluated by numerical simulation for the purpose of saving cost and shortening development period. In order to ensure the precision and veracity of simulation, it is essential to precisely describe the materials deformation behavior under quasi-static and dynamic loadings.

Considering the test period and cost, three to five dynamic strain rates are adopted to represent the dynamic characteristics of materials. However, the actual deformation refers to a large range of strain rates, generally from quasi-static to 500s-1. Usually, the flow behavior of materials under dynamic loadings is described by constitutive models, which can describe the deformation behavior of materials under any strain rate on the basis of the experimental data. The most commonly used models involve Johnson-Cook, Cowper-Symonds and modified Johnson-Cook equations [4-8]. Besides, BP network is another way of predicting the deformation models of materials [9].

---

[a] Corresponding author.

In the present work, the mechanical properties of a dual-phase steel DP590 under five different strain rates are studied. The dynamic flow behaviour is predicted based on the two aforementioned approaches. The results are compared with the experimental data and the error is calculated.

2. Experiments

In this work, DP590 steel sheet with thickness of 1.4 mm is taken as the test material. The high speed tensile tests are carried out on ZWICK HTM 5020 high speed testing machine under room temperature, and the strain rates involve $0.1s^{-1}$, $50s^{-1}$, $100s^{-1}$, $200s^{-1}$ and $500s^{-1}$. During the experiments, the strain and force are obtained via two types of strain gauges which are respectively glued to the clamping portion and parallel portion of the specimen.

3. Constitutive Equations and BP Neural Network Models

Ignoring the temperature effect, the Johnson-Cook model can be expressed as equation (1):

$$\sigma = \left(A + B\varepsilon^{n}\right)\left(1 + C\ln\dot{\varepsilon}\right) \tag{1}$$

Where A, B, C and n are material constants.

Cowper-Symonds (CS) model can be described as equation (2):

$$\sigma = \left(A + B\varepsilon^{n}\right)\left(1 + \left(\frac{\dot{\varepsilon}}{D}\right)^{\frac{1}{P}}\right) \tag{2}$$

Where A, B, D, n, and P are material constants.

The modified Johnson-Cook model can be written as equation (3):

$$\sigma = B\varepsilon^{n}\left(1 + C\varepsilon^{n'}\ln\left(\frac{\dot{\varepsilon}}{\dot{\varepsilon}_0}\right)\right) \tag{3}$$

Where B, C, n, and n′ are material constants.

Artificial neural network is a large scale distributed parallel processing system, which can simulate the memory and learn functions of the human brain, unnecessary to understand the variation of the input and output parameters. In this work, three layers BP artificial neural network is established as shown in Fig. 1. It includes input layer, hidden layer and output layer. The establishment of BP neural network model involves four steps:

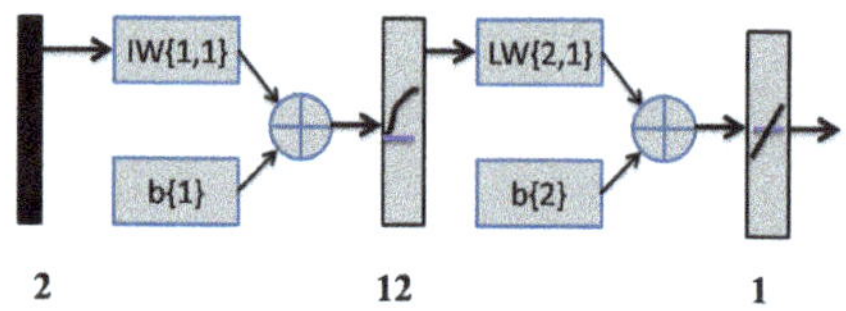

Fig. 1. Structure of the BP artificial neural network for DP590.

(1) Data sampling and unitary treatment

Usually, three kinds of data involving strain, stress and their corresponding strain rates are sampled as original data. Then unitary treatment is employed according to the following equation (4) in order to avoid overlarge network system error and prevent most nerve cells from supersaturated state.

$$p_n = 2\times(p - p_{\min})/(p_{\max} - p_{\min}) - 1 \tag{4}$$

Where p_n, p, p_{min} and p_{max} are normalized sample data, the original sample data, the minimum and maximum of the original sample data, respectively.

(2) Prediction network establishment

As shown in Fig. 1, the cells of the three layers are 2, 12 and 1 respectively. On the basis of precision and efficiency of the functions, transfer, training and learn functions are selected.

(3) Prediction network training

After the establishment and initiation of BP artificial neural network, training process is conducted based on the sample data and repeated adjustment is progressed to ensure the error smaller than the demand of precision.

(4) Prediction of dynamic mechanical properties of materials

In this paper, the mechanical properties of DP590 sheet under five different strain rates are measured through HTM5020. Set strain rate and strain as input vectors, and treat stress as the output vector. Then the data at four strain rates are taken as learning samples for modeling of artificial neural network, while the data of the rest strain rate are used to validate the accuracy of prediction.

4. Results and Discussion

The true strain-true stress curves of DP590 under different strain rates are shown in Fig. 2. For DP590, its yield strength is about 370MPa deformed at $0.1s^{-1}$. When the strain rate reaches $50s^{-1}$, the yield strength remarkably increases. With the strain rates rise, the strength of materials continue increasing because of the strain rate strengthening effects, which indicates that DP590 possesses positive strain rate sensitivity.

As shown in Fig. 3, two strain rates of $100s^{-1}$ and $200s^{-1}$ are chosen to verify the accuracy of the constitutive models and the prediction results of BP neural network. It is quite easy to figure out that the curves based on J-C model and C-S model seriously depart from the experimental ones. Oppositely, the curves calculated by modified J-C model and predicted by BP neural network fit well with the experimental ones.

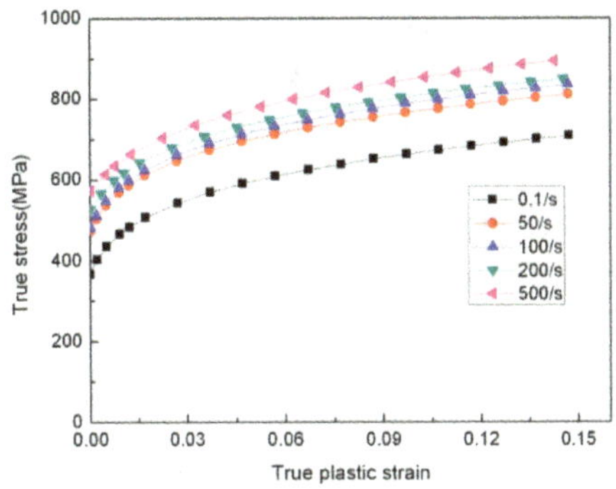

Fig. 2. True strain-true stress curves of DP590 under different strain rates.

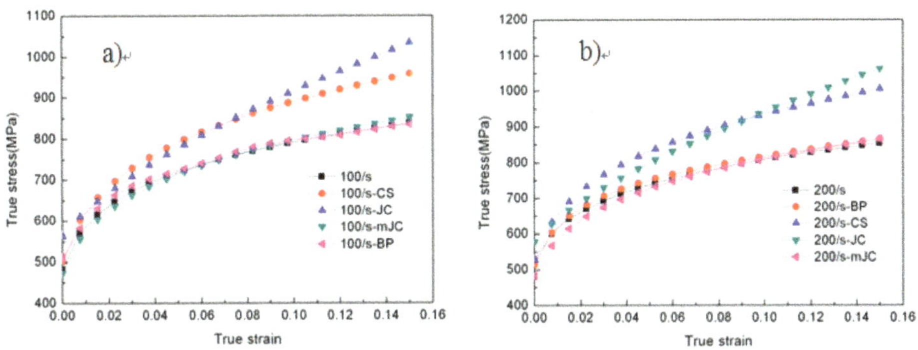

Fig. 3. Comparison of true strain-true stress curves of DP590 under different strain rates: a) 100s-1; b) 200s-1.

Table 1. Comparison and error analysis of true stress among the results of the constitutive model calculation and BP neural network prediction.

Strain rate/s^{-1}	True strain	True stress/MPa	J-C model calculation error/%	C-S model calculation error/%	Modified J-C model calculation error/%	BP Prediction error/%
100	0.03	481.70	17.01	3.90	-1.74	6.57
	0.06	671.70	5.71	8.17	-1.20	2.24
	0.09	737.43	9.66	9.82	-0.25	0.46
	0.12	779.95	14.29	10.67	0.41	1.18
	0.15	811.96	18.85	11.16	0.90	-0.26
200	0.03	526.20	10.05	0.45	-8.19	-2.36
	0.06	692.89	5.28	10.03	-2.32	1.97
	0.09	756.07	9.89	12.22	-0.90	1.48
	0.12	797.25	14.87	13.26	-0.03	1.07
	0.15	828.33	19.69	13.83	0.60	1.11

The stress relative error under the two strain rates are listed in table 1. Under the strain rate of 100s^{-1}, the J-C model exhibits the worst accuracy compared with the experimental data. Its average relative error is about 15%. The average relative error of the calculation result from C-S model is large as well, which is about 9%. For the modified J-C model and BP neural network model, the results match well with the experimental one. Their average relative errors are about 1% and 1.8% respectively. Under the strain rate of $200s^{-1}$, the modified J-C model and BP neural network model are still the better ones for describing the flow behaviour of DP590. The average relative errors of J-C model, C-S model, modified J-C model and BP neural network model are 14%, 11%, 2% and 1.6% respectively. Therefore, the modified J-C model and BP neural network model are more suitable for describing and predicting the dynamic flow behaviour of DP590.

5. Conclusion

DP590 exhibits obvious strain rate sensitivity when tensioned under dynamic loadings. To describe the dynamic flow behaviour of DP590 under different strain rates, the constitutive models including J-C model, C-S model and the modified J-C model and the three-layer BP neural network model are established. By comparing the results calculated by constitutive models and BP neural network model with the experimental result obtained at $100s^{-1}$ and $200s^{-1}$, it is validated that modified J-C model and BP neural network model can precisely predict the dynamic flow behaviour of DP590. The predicted curves of J-C model and C-S model depart seriously from the experimental ones.

References

1. H. Yoshida, A. Uenishi, Y. Kuriyama and M. Takahashi, Material characteristic at high strain rates for optimizing car body structures for crash events, in *Nippon Steel Technical Report* **88**, 22 (2003).
2. H. Huha, S. B. Kim, J. H. Song and J. H. Lim, Dynamic tensile characteristics of TRIP-type and DP-type steel sheets for an auto-body, *Int. J. Mech. Sci.* **50**, 918 (2008).
3. J. G. Qin, R. Chen, X. J. Wen, Y. L. Lin, M. Z. Liang and F. Y. Lu, Mechanical behaviour of dual-phase high-strength steel under high strain rate tensile loading. *Mater. Sci. Eng. A* **586**, 62 (2013).
4. G. R. Johnson and W. H. Cook, A constitutive model and data for metals subjected large strains, high strain rate and high temperatures. in *Proceedings of the seventh international symposium on ballistics*, (Hague, Netherlands, 1983).
5. K. Xu, C. Wong, B. Yan and H. Zhu, A High Strain Rate Constitutive Model for High Strength Steels, in *SAE World Congress* (Detroit, Michigan, 2003).
6. N. Singh, E. Cadoni, M. Singha and N. Gupta, Dynamic tensile behavior of multi phase high yield strength steel, *Mater. Des.* **32**, 5091 (2011).
7. R. Smerd, S. Winkler, C. Salisbury, M. Worswick, D. Lloyd and M. Finn, High strain rate tensile testing of automotive aluminum alloy sheet, *Int. J. Impact Eng.* **32**, 541 (2005).

8. J. V. Slycken, P. Verleysen, J. Degrieck, L. Samek and B. DeCooman, High-strain-rate behavior of low-alloy multiphase aluminum- and silicon-based transformation-induced plasticity steels, *Metall. Mater. Trans. A* **37**, 1527 (2006).
9. Y. Sun, W D. Zeng, Y Q. Zhao, Y. L. Qi, X. Ma and Y. F. Han, Development of constitutive relationship model of Ti600 alloy using artificial neural network, *Comp. Mater. Sci.* **48**, 686 (2010).

The Microstructures and Mechanical Properties of DP Steels

J. P. Zhang[a], G. Fang[b] and M. T. Ma[c]

China Automotive Engineering Research Institute CO., LTD.

Automotive Lightweight Engineering Research Center, Chongqing, 401122, China

Chongqing Automotive Lightweight Engineering Research Center, Chongqing, 401122, China

E-mail: [a]zwwjp@163.com, [b]fg_caeri@163.com, [c]mingtuma@126.com

In the present work, three DP steels, DP590, DP780 and DP980 are analyzed in terms of microstructures and mechanical properties. The microstructures of the steels are investigated from three different directions and compared to each other. The tensile properties, work hardening, bake hardening of DP steels are studied by means of tensile experiments and heat treatment. The results indicate that the volume ratio of martensite to ferrite of DP steels varies as the increase of the strength. DP steel exhibits little anisotropy and the increase of strength leads to the decrease of ductility and work hardening exponent. DP steel exhibits high initial work hardening rate when the strain is lower than 3% and decreases sharply as the strain increases. Pre-strain has better contribution to the strength increase compared with baking. The most effective strengthening process for DP steels is pre-strain + baking.

Keywords: DP steels; Microstructure; Mechanical properties; Bake hardening; Work hardening.

1. Introduction

Dual phase (DP) steel in possession of the advantages of low yield strength ratio, high initial hardening rate and good match of strength and ductility, is one of the earliest developed advanced high strength steels used for automotive body [1-4]. It is usually used as the automotive components with the need of high strength and crashworthiness and good formability, such as wheel, bumper, suspending system and other strengthening parts [5, 6]. DP steel is composed of two phase, martensite and ferrite. The shape and amount of the two phases determines the mechanical properties of the DP steels and vary with its chemical compositions and the heat treatment process [7, 8].

In the present work, three DP steels with different strength levels are analyzed in terms of microstructure and mechanical properties. By metallographic analysis, the microstructure evolution in three directions for each material is investigated. Then, a series of tensile experiments and heat treatment experiments are carried out to study the basic mechanical properties, work hardening effects and bake hardening abilities of the DP steels. Finally, the strengthening contributions from different processes are also discussed.

---

[a] Corresponding author.

2. Experiments

Three DP steels, DP590, DP780 and DP980 are selected as experimental materials. Their thickness and chemical compositions are shown in table 1. To investigate the microstructures of DP steels, the pieces from three perpendicular directions are cut, which are rolling direction, transverse direction and nominal direction respectively. After milling and polishing, the samples are eroded by the solution of 4ml HNO_3 and 96ml C_2H_5OH for few seconds and observed via Leica DMI 3000M microscope.

Table 1. Thickness and chemical compositions of DP steels.

Material	Thickness(mm)	Chemical composition (wt %)							
		C	Si	Mn	Cr	Cu	Al	P	S
DP590	1.4	0.084	0.28	1.79	0.011	0.019	0.038	0.012	0.008
DP780	1.4	0.110	0.30	1.81	0.202	0.021	0.034	0.017	0.009
DP980	1.6	0.170	0.40	2.32	0.390	0.015	0.039	0.013	0.009

The tensile experiments are carried out with a speed of 5mm/min on SANS CMT5305 tensile tester. The sample size refers to A80 as recommended in GB/T 228. In order to study the baking and work hardening effects on the mechanical properties of DP steels, some samples are baked at 170℃ for 30 minutes or deformed with a 2% pre-strain.

3. Results and Discussion

3.1. *Microstructures*

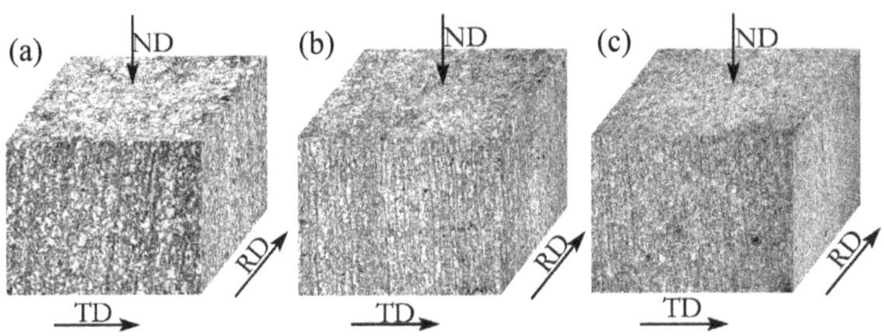

Fig. 1. The microstructure of DP steels: (a) DP590; (b) DP780; (c) DP980.

As shown in Fig. 1, the microstructure of DP steels is mainly composed of white ferrite and gray martensite. Ferrite presents polygonal form and martensite appears with a typical form of island. Tiny banded structures along the rolling and transverse directions are observed while the microstructure elongation in the nominal direction is undiscovered. As increase of the strength level of DP steels, the volume ratio of martensite in the microstructure increases and the sizes of martensite get smaller and smaller.

3.2. *Mechanical properties*

The engineering strain-stress curves of DP steels are shown in Fig. 2. DP590 exhibits the best ductility while its strength is the lowest among the three DP steels. Moderate strength and ductility are observed for DP780. As for DP980, its yield strength and tensile strength are about 820MPa and 1020MPa and the fracture elongation is about 8%.

To study the anisotropy of the mechanical properties, tensile experiments from rolling direction (0°), transverse direction (90°) and 45° along the rolling direction are carried out. The tensile properties of DP steels along three different directions are listed in table 2. The results indicate that for each DP steels the yield strength $R_{p0.2}$, tensile strength R_m and work hardening exponent n are nearly the same in the three directions. The plastic strain ratio r is almost less than 1. The degree of planar anisotropy for three steels are 0.21, -0.08 and -0.095 respectively which illuminates that DP steel has little anisotropy. Besides, when the strength level of DP steels increases, the work hardening exponent decreases and the degree of planar anisotropy approaches 0.

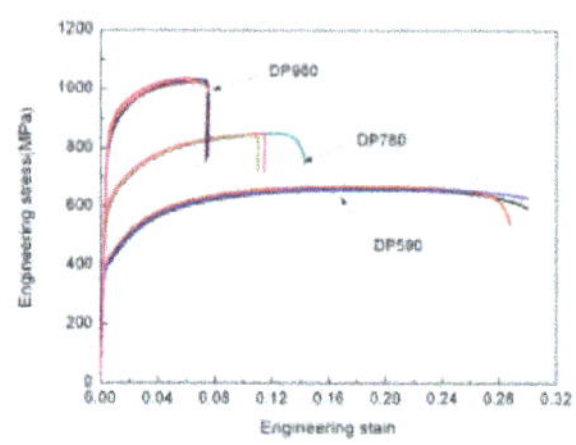

Fig. 2. The engineering strain-stress curves of DP steels.

Table 2. Tensile properties of DP steels along three different directions.

Materials	direction	$R_{p0.2}$	R_m	n	r
	0°	385	654	0.2	0.91
DP590	45°	395	659	0.19	0.79
	90°	387	656	0.19	1.09
	0°	584	865	0.13	0.83
DP780	45°	584	860	0.13	0.94
	90°	571	859	0.13	0.89
	0°	823	1026	0.11	0.78
DP980	45°	827	1023	0.1	0.87
	90°	829	1032	0.1	0.77

Work hardening and aging strengthening effects are investigated by several processes as shown in table 3. The yield strength of materials after baking, 2% pre-strain, baking+2% pre-strain and 2% pre-strain+baking is considered and compared with that of raw material. After aforementioned processes the yield strength of DP steels all get improved which illustrates that baking and pre-strain are effective ways of strengthening for the materials.

Table 3. The yield strength change of DP steels after heat treatment and work hardening.

Materials	$R_{p0.2}$/Mpa				
	raw material	baking	2% pre-strain	baking+2% pre-strain	2% pre-strain +baking
DP590	385	398	515	523	545
DP780	584	613	723	728	763
DP980	823	865	1000	1005	1019

3.3. *Discussion*

The plot of derivation of the stress on the strain versus strain is illustrated in Fig. 3. It is apparent that for three DP steels the initial work hardening rate is very large which is about 3 to 6 times than that in large strain state. As the deformation goes, the work hardening rate decreases sharply.

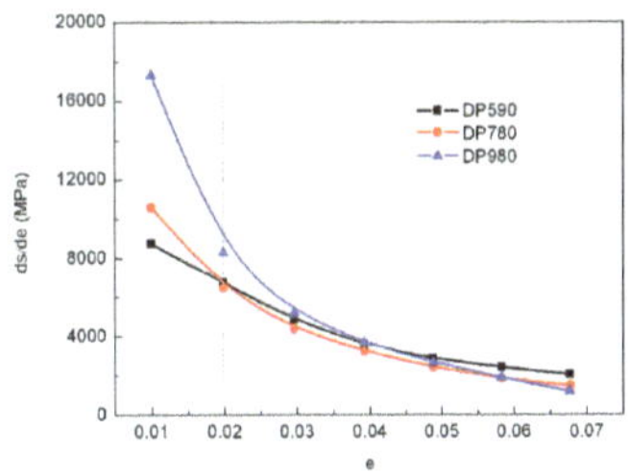

Fig. 3. The variety of work hardening rates of DP steels.

Among three DP steels, DP 980 performs the highest initial work hardening rate, while DP780 and DP590 have the similar work hardening rate. When the plastic strain is over 3%, the work hardening rates of the materials tend to be close to each other.

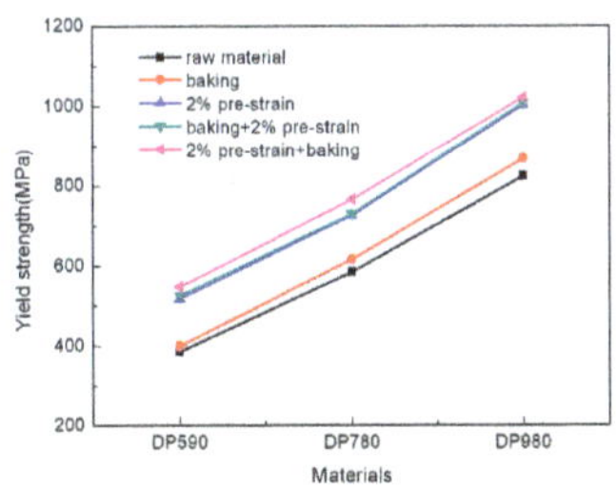

Fig. 4. Comparison of yield strength of DP steels after different processes.

It is noted in table 3 that all of the four processes have the strengthening effect on DP steels. The strengthening contribution of the processes is illustrated in Fig. 4. The baking process has the least strengthening effect while 2% pre-strain+baking process possesses the most powerful strengthening effect. Between them, 2% pre-strain and baking+2% pre-strain processes have nearly the same strengthening contribution, indicating that baking process without work hardening has little contribution to strengthening and 2% pre-strain before baking is beneficial to exerting the strengthening effects of both processes.

4. Conclusion

(1) DP steels are mainly composed of ferrite and martensite. With the increase of the strength level, the volume ratio of martensite in the microstructure increases and its size become smaller.

(2) The strength increasing of DP steels leads to the decrease of ductility and work hardening exponent. The degree of planar anisotropy of DP steels approaches 0 which indicates its mechanical property has little anisotropy.

(3) The work hardening rate of DP steels when the plastic strain is less than 3% is much larger than that in the large strain state. As the deformation goes, the work hardening rate decreases. DP980 exhibits the highest initial work hardening rate while DP780 and DP590 have the similar rate.

(4) Both of baking and pre-strain are effective approaches of improving the strength of DP steels, but pre-strain has much better contribution on strengthening. Among four strengthening processes, pre-strain combining the later baking process is the most effective way of strengthening for DP steels.

References

1. Z.P. Xiong, A.G. Kostryzhev and N.E. Stanford, Effect of deformation on microstructure and mechanical properties of dual phase steel produced via strip casting simulation, *Mater. Sci. Eng. A* **651**, 291 (2016).
2. W.J. Dan, Z.Q. Lin, S.H. Li and W.G. Zhang, Study on the Mixture Strain-hardening of Multi-phase Steels, *Mater. Sci. Eng. A* **552**, 1 (2012).
3. R.H. Wagoner, J.H. Kim and J.H. Sung, Formability of advanced high strength steels, *Int. J. Mater. Forming* **2**, 359 (2009).
4. W.R. Wang, C.W. He and Z.H. Zhao, The limit drawing ratio and formability prediction of advanced high strength dual-phase steels, *Mater. Des.* **32**, 3320 (2011).
5. R. Kuziak, R. Kawalla and S. Waengler, Advanced high strength steels for automotive industry, *Archives of civil and mechanical engineering* **8**, 103 (2008).
6. M.T. Ma and B.R. Wu, *Dual-phase steel: the physical and mechanical metallurgy*, (Beijing: Metallurgical Industry Press, 1988).

7. T.Baudin, C. Quesnea, J. Jura and R. Penelle, Microstructural characterization in a hot-rolled two-phase steel, *Mater. Charact.* **47**, 365 (2001).
8. Z.Z. Zhao, G.C. Jin, N. Feng, T. Di and A.M. Zhao, Microstructure evolution and mechanical properties of 1000 MPa cold rolled dual-phase steel, *Transactions of Nonferrous Metals Society of China* **19**, 563 (2009).

Research on Microstructure and Property of Hot-Rolled 700MPa Grade Automobile Sheet

H. G. Gu[a], Y. G. Wang[b], J. C. Jin[c], Y. G. Liu[d] and Q. Q. Lu[e]

Technology center of Ma'anshan Iron and Steel Company, Ma'anshan, P.R. China
E-mail: [a]gohero1234@163.com, [b]272507498@qq.com, [c]jinjiachun1988@163.com, [d]lyghotmail@hotmail.com, [e]172451373@qq.com.

In response to the need of lightweight of new energy automobile, the ultra-high strength automobile sheet with thickness less than 3.0 mm and yield strength of 700 MPa grade has been successfully developed by Masteel. This steel was designed with low C-Mn-Mo-Ti-Nb alloy system and appropriate TMCP process, with excellent cold bending property, welding performance, balance of strength and ductility and good adaptability to industrial production. Component, process, microstructure and property were analyzed systematically. While the main strengthening mechanism of this material are precipitation and fine-grain strengthening by a large number of nanoscale (Ti, Nb) (C, N) distributed dispersively, the welding performance and cold bending property are effectively guaranteed by the low carbon and Nb, Ti microalloying design.

Keywords: 700MPa Grade; Lightweight; Nb, Ti microalloying Template.

1. Introduction

As the automotive demand of energy, environmental protection, safety is becoming more and more strict, lightweight has become an inevitable choice for the development of automotive industry. The main ways to lightweight include structural optimization, advanced processing technology and application of advanced lightweight materials, as shown in figure 1.

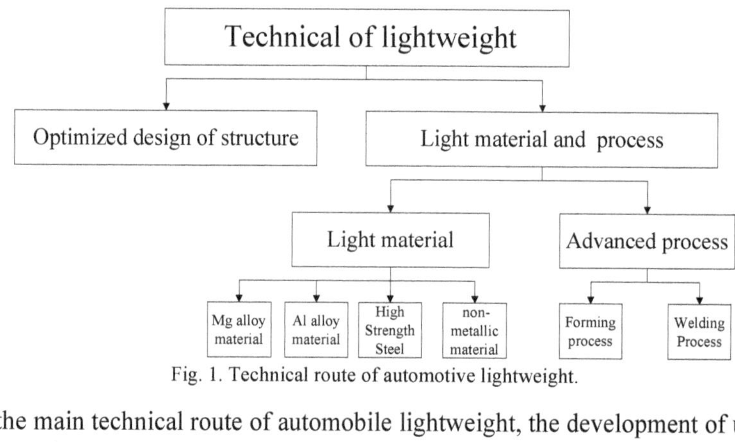

Fig. 1. Technical route of automotive lightweight.

As the main technical route of automobile lightweight, the development of ultra-high strength steel to material thinning has proved to be an effective way and has been increasingly applied extensively.

[a] Corresponding author.

The effect of material thinning is shown in the formulae below [1]:

$$\frac{\text{t1}}{t2}=\sqrt{\frac{\sigma\text{s2}}{\sigma\text{s1}}} \tag{1}$$

In the formulae: t1, t2, σs1, σs2 represent respectively the yield strength and thickness of the two kinds of materials.

However, the thin gauge rolling of ultra-high strength steel face the contradiction of steady production and microstructure property control, which restrict each other and can not be avoided. This ultra-high strength steel with thickness less than 3.0 mm and yield strength of 700 MPa grade was mainly developed for the new energy automobile chassis. The alloy design, TMCP and microstructure property control were researched in this paper.

2. Experimental Materials and Methods

2.1. *Alloy design*

The chemical composition of the experimental steel is shown in table 1.

Table 1. The chemical composition of the experimental steel, wt%.

C	Si	Mn	Nb	Ti	S	P
0.04~0.12	0.10~0.60	1.00~1.80	0.02~0.06	~0.12	≤0.020	≤0.020

In order to ensure the experimental steel's high strength and high toughness, machinability and welding performance at the same time, carbon was controlled at a lower level. Even if can enlarge the α area, improve the activity of carbon, accelerate the formation of proeutectoid ferrite, solution strengthen to a certain degree, Si was designed at a low level as the strict requirements of the automobile surface quality. Nb, Ti design is to form carbonitride which can dissolve and precipitate under different conditions by the mechanism of fine-grain strengthening and precipitation strengthening [2]. Micro alloying elements added in HSLA steel had changed the traditional concept that the carbon equivalent is the main factor of the ferrite-pearlite steel reinforcement. In comparison to the traditional design, micro alloying can reduce carbon content by 50% or more, which made the steel's toughness and welding performance better.

The function of the microalloy carbonitride deposition during TMCP included many[3-5]. Microalloy carbonitride undissolved during soaking can stop austenite grain coarsening based on the mechanism of particle pinning grain boundary. As a result, a tiny soaking austenitic grain can be formed at last. In the rolling process, because of the particles pinning grain boundary and subboundary, the precipitated phase induced by strain can prevent the recrystallization of the deformed austenite. After the rolling, the residual micro alloying elements in the austenite can further precipitate in ferrite, coming into being a significant precipitation strengthening effect.

But the effect of micro alloying elements and the added quantity are not linear relationships. Generally speaking, Nb less than 0.06% can make an obvious fine gain strengthening, over 0.06% it has no obvious effect. Literature [6] points out that in order to lower the precipitation temperature of TiN, Ti is better to be controlled at less than 0.12%. Therefore, this experimental steel had a upper limit of Nb and Ti.

2.2. *Smelting and TMCP process*

Frist molten steel was converter smelted, than be treated in the vacuum. At last slab was heated at about 1250°C. For the experimental steel, heating process was actually a solution process of Nb, Ti. Thus, it was better to ensure a higher temperature during reheating.

For the experimental steel, it was a large challenge to roll the ultra-high strength steel to such a thin gauge. When rolling thin gauge, the temperature of the strip end would decrease quickly. Thus, rolling force and axial force would increase , which induced strip can not be rolled steadily [7]. Therefore, A high temperature rolling method was adopted to improve the ability of dislocation motion, inhibited the rapid increase of deformation resistance, thereby increasing the stability of rolling process. Then water cool and coil at about 600°C [1] which was the best precipitation temperature for Nb, Ti.

The experimental steel's mechanical property, welding performance and microstructure were analyzed as below.

3. Results of Experiment and Analysis

3.1. *Mechanical properties*

The mechanical properties of the experimental steel are shown in table 2 and figures 2, 3.

Table 2. Mechanical properties of the experimental steel.

NO.	σ_s/MPa	σ_b/MPa	A/%	180° cold bending t=a
1	738	801	19.5	qualified
2	729	790	18.5	
3	717	772	19	
4	715	765	20.5	
Technical requirements	≥700	≥750-950	≥11	qualified

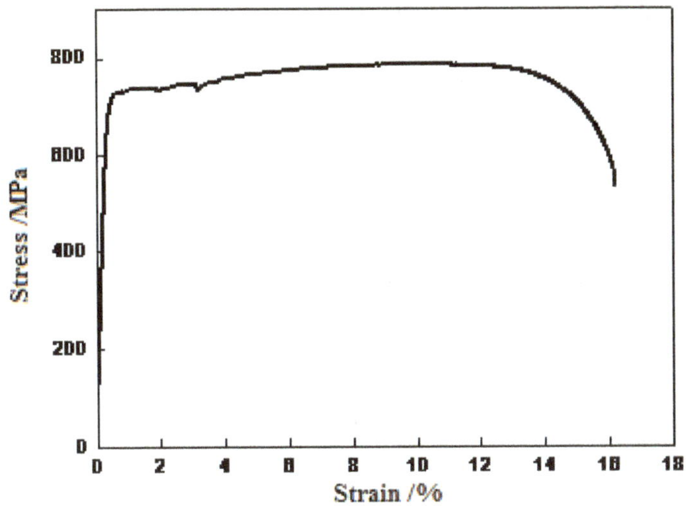

Fig. 2. Typical tensile curve.

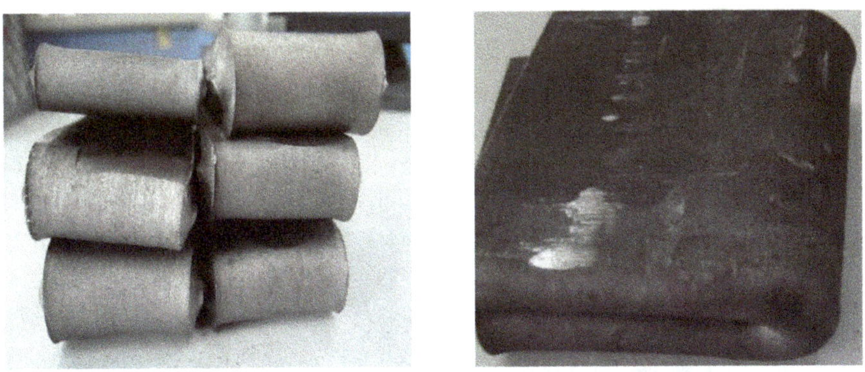

180° cold bending once. 180° cold bending twice.

Fig. 3. Feature of 180° cold bending.

From table 2, the mechanical performance of experimental steel meets the technical requirement. And it has a good balance of strength and plastic, especially has high elongation, which can effectively meet the requirement of cold forming.

3.2. *Welding performance*

The welding performance of material was evaluated by two kinds of welding methods, Y type groove welding and CTS lap welding.

According to the chemical composition in table 1 and the formula of cold crack sensitivity P_{cm}:

$$P_{cm} = C + \frac{Si}{30} + \frac{Mn}{20} + \frac{Cu}{20} + \frac{Ni}{60} + \frac{Cr}{20} + \frac{Mo}{15} + \frac{V}{10} + 5B(\%) \quad (2)$$

After calculation, the experimental steel's cold crack sensitivity index was 0.17, which represent good welding ability. As shown in figures 4, 5, the two different ways of welding experiment proved the experimental steel's welding performance is good.

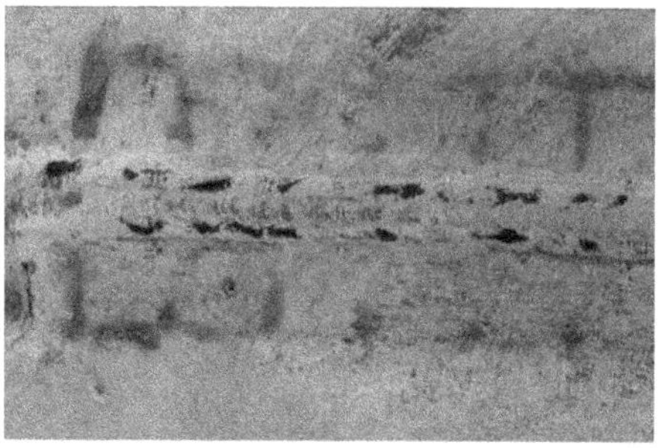

Fig. 4. Surface morphology of Y type groove welding after magnetic powder coloring. (Crack rate of weld surface, cross-section and root, Cf=0).

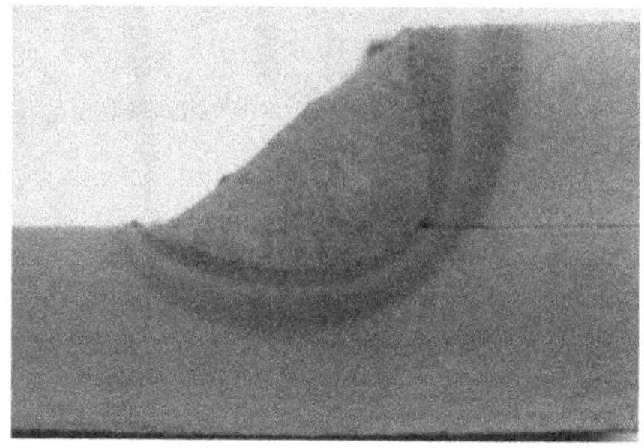

Fig. 5. Cross section morphology of CTS lap welding (Crack rate of upper plate (C1) and lower plate (C2) are both 0).

3.3. *Analysis of structure*

3.3.1. *Analysis of microstructure and inclusion*

The microstructures of the experimental steel are shown in figure 6. The rolling microstructures are main ferrite and a small amount of granular bainite. And the bainite distributes dispersively between the ferrite. Ferrite grain was fine and uniform, and the size was measured by cutting line method, resulting in about 5 μm.

OM SEM

Fig. 6. Typical microstructure of the steel.

The strengthening of fine grain follows the formula of Hall – Petch, namely:

$$\sigma_g=k_y d^{-1/2} \tag{3}$$

In the formula: σg –strengthening of fine grain, ky—coefficient, 15.1～18.1Nmm-3/2 for large angle grain boundary; d—the average diameter of ferrite grain.

With the measured grain size, the calculation result of strengthening contribution from fine-grain is about 235 MPa, proving a good effect of fine-grain strengthening.

The rating results of non-metallic inclusions indicate that the steel is pure, as shown in table 3. The inclusions in steel are mainly oxides of B and D type, which are little and small.

Table 3. The rating results of inclusions.

A fine	A coarse	B fine	B coarse	C fine	C coarse	D fine	D coarse	DS
0	0	0.5	0	0	0	0.5	0	0

3.3.2. *Research of the precipitated particles*

For this experimental steel, the alloy and controlled rolling process were designed mostly based on the mechanism and theory of precipitation strengthening.

The second phase particle were analyzed by transmission electron microscopy (TEM), as shown in figure 7.

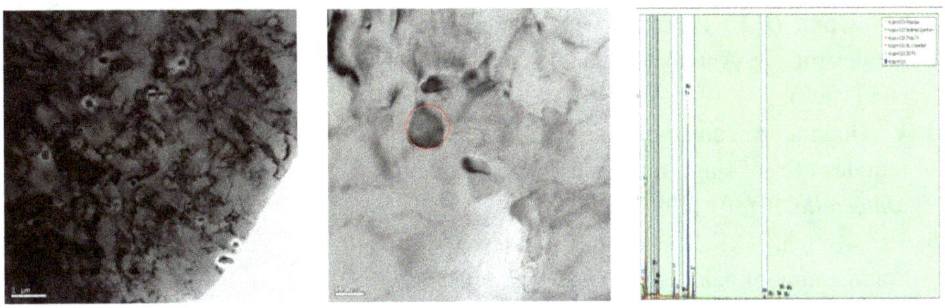

Distribution of second-phase. Form of second-phase. EDS of second-phase.

Fig. 7. The distribution, form, EDS of second-phase.

The types of the second-phase particles are round and distributed dispersively. According to the EDS picture, the particles mainly contain Ti and Nb elements. With the analyze of the particles' form and EDS, the second-phase can be concluded as carbonitride of Nb, Ti, which are hard and difficult to be deformed. It is suitable for the theory of Gladman, etc. Based on calculation results of literature [8-10], the precipitation strengthening contribution in this experimental steel is about 150 MPa which means precipitation strengthening is one of the main strengthening mechanism.

4. Conclusion

(1) The ultra-high strength steels with thickness less than 3.0 mm that yielding strength of 700 MPa grade has been successfully developed by Masteel. It includes a low C-Mn-Mo-Ti-Nb alloy system design, reheating and hot rolling at high temperature, quick cooling and coiling at appropriate temperature, which fully consider the stability of production and material performance.

(2) The product developed by Masteel is pure. And the rolling structure is ferrite and a small amount of granular bainite. The yield strength is about 710 MPa and the

elongation is more than 18%, which prove a good balance of strength and plastic. By different welding procedure qualification, the developed steel proves a good welding performance which can fully meet the needs of the downstream customers.

(3) This experimental steel integrates a variety of reinforcement method. Through the theoretical calculation, fine-grain strengthening and precipitation strengthening are the main reinforcement mechanism.

References

1. J.X. Lu, The production technology of 700 MPa grade high strength micro alloy steel, Master thesis, Northeastern University (2004).
2. F.H.Samuel, S.Yue, J.J. Jonas and K.R. Barnes, Effect of Dynamic Recrystallization on Microstructural Evolution during Strip Rolling, *ISIJ international* **3**, 216 (1990).
3. F.M. Haggag, T.S. Byun, J.H. Miraglia, P.Q. and K.L. Murty, Indentation-engergy-to-fracture (IEF) Parameter for characterization of DBBT in carbon steels using nondestructive Automated ball indentation (ABI) technique, *Scripta Materialia* **4**, 645 (1998).
4. Y. Huarna, A.Yamamoto and H. Tsubakino, Effect of sulfur on rolling contact fatigue life of high-manganese precipitation-hardening austenitic steel, *Scripta Materialia,* **9**,1301 (1998).
5. S.B. Singh, H.K.D and H. Bhadeshia, Qunatitative evidence of mechanical Stabilization of bainite, *Mater. Sci. Technol* **7**, 610 (1996).
6. X.D. Huo, X.P. Mao and S.C. Xia, The nano deposition in high strength and Timicro alloying steel by the CSP process, *Journal of Beijing university of science and technology* **33**, 941 (2011).
7. Z.C. Zhang, T.T. Sun and X.M. He, The key hot striptandem rolling process control of thin pickling plate, *Steel rolling* **30**, 18 (2013).
8. H.R. Gu. J. Zhang and X.L. Yang, Research on Microstructure and Property of High hole expansion and ferrite/bainite dual phase steel, *Journal of Wuhan university of science and technology* **34**, 419 (2011).
9. D.Y. Zhu, J.X. Gao, Q.L. Chen, Z.L. Tian, Y. Peng and J.H. Chen, The development and application of high strength steel at the yield strength of 700 MPa, *Metallurgy Supplement* **3**,8 (2011).
10. C.J. Wang, Q.L. Yong, X.P. Mao and Z.D. Li, The influence of Ti and Mn contents on Ti separation characteristics and strengthening mechanism in the CSP process, *Acta Metallurgica Sinica* **47**,1541 (2011).

Modeling & Simulation

Accurate Evaluation of Automotive Body Safety with Increasing Use of Advanced High Strength Steels

K. Takada*

Automobile R&D Center, Honda R&D
Haga-gun, Tochigi-ken, 321-3393, Japan
E-mail: kenji_takada@n.t.rd.honda.co.jp

N. Ma

Engineering Technology Division, JSOL Corporation
Nishiku, Osaka, 550-0001, Japan
E-mail: ma.ninshu@jsol.co.jp;ma.ninshu@jwri.osaka-u.ac.jp

P. Farahat

Dassault Systems K.K.
ThinkPark Tower, 2-1-1 Osaki, Shinagawa-ku, 141-6020 Tokyo, Japan
E-mail: Pierre.FARAHAT@3ds.com

With increasing use of advanced high strength steels to automotive bodies, the safety evaluation with the computer aided engineering (CAE) has some difficulties in the accuracy. In this report, some obvious phenomena existing in advanced high strength steels such as Bauschinger effect, strain dependent anisotropy and fracture criteria are discussed using the results obtained from both experimental measurements and computer simulations. Then, computational material models with consideration of Bauschinger effect, strain dependent anisotropy and ductile damage criterion were implemented in to commercial software such as LS-DYNA and ABAQUS. Using the newly developed computational material models, collapse characteristics and fracture strength of high strength steel frames in automotive bodies due to the impact loading were accurately simulated and the detail results are represented.

Keywords: Advanced high strength steels; Bauschinger effect; Strain dependent anisotropy; Ductile damage criterion; Safety evaluation; Computer simulations; Experimental measurements.

1. Introduction

The Cockcroft-Latham fracture criterion was applied to predict the fracture of high-strength steels. In order to improve the simulation accuracy, local anisotropic parameters based on the plastic strain (strain dependent model of anisotropy) were measured using the digital image grid method and they were incorporated into Hill's anisotropic yield condition by authors. In order to confirm the validity, uniaxial tensile tests and three-point bending test were performed. It was found that the accuracy of the predicted rupture strain was improved if the local anisotropic r-value was used. While the Bauschinger effect is strong in high strength steels. Conventionally, material properties used in crash

* Corresponding author.

analysis are based on the isotropic hardening role and the Bauschinger effect is ignored in crash analysis fields. However, many recent studies on static crash analysis of hat-section problems have shown that the amount of energy absorb is higher compared with the experimental results. In this study, Chaboche model and modified ANK model are found to be more effective for the accurate safety evaluation of advanced high strength steel parts in crash analysis.

2. Cockcroft-Latham Fracture Criterion

The Cockcroft-Latham fracture criterion [1] shown in Eq.(1) is the integral damage value using the maximum principal stress and the equivalent strain. It needs only one parameter to express the amount of ductile damage.

$$\int_0^{\overline{\varepsilon_f}} \sigma_{\max} d\bar{\varepsilon} = C_1 \tag{1}$$

where σ_{max} is the maximum principal stress, $\bar{\varepsilon}$ is the equivalent strain, $\bar{\varepsilon}_f$ is the equivalent strain at which the fracture occurs, and C_1 is the material constant to express the limit of ductile damage [2]. Marciniak-type biaxial stretching tests [3] for high-strength steels were performed to measure the material constant of the Cockcroft-Latham fracture criterion.

2.1. *Digital image grid method*

The digital image grid method [4], which is used in uniaxial tensile tests, is schematically shown in Figure 1. Table 1 shows the material properties of 980MPa and 1180MPa grade.

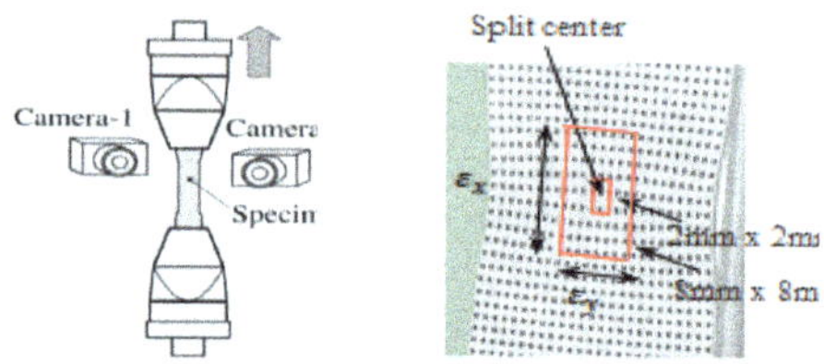

Fig. 1. Digital Image Grid system.

Table 1. Tensile properties of high strength steels.

Material grade	980MPa	1180MPa
Gauge length [mm]	50	50
r-value	0.92	0.86
Damage limit C_1	612MPa	589MPa

After testing, the digital images were analyzed using software to calculate the major and minor strains. The strain in the thickness direction was calculated using the constant volume role. The local anisotropic r-value R_{local} is given by Eq.(2).

$$R_{local} = \varepsilon_y/\varepsilon_z = -\varepsilon_y/(\varepsilon_x + \varepsilon_y) \tag{2}$$

Figure 2 shows the local anisotropic r-value-strain curves of high-strength steels measured using the digital image grid method. Each diagram has four curves which are 2mm x 2mm, 4mm x 4mm, 8mm x 8mm and 16mm x 16mm grid pitch. It is assumed that the local anisotropic r-values are approximately constant values up to the hardening coefficient n-value and decrease gradually after the n-value.

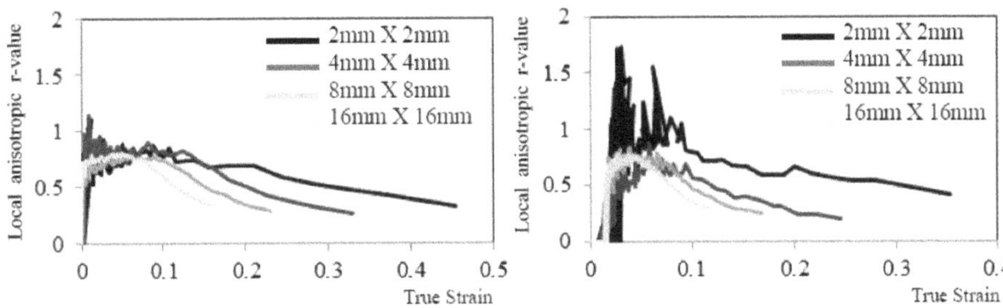

(a) 1180MPa grade (b) 1500MPa grade

Fig. 2. Local anisotropic r-value – true strain curves.

2.2. *Improvement of existing yield condition with strain-dependent anisotropy model*

There are several yield functions for anisotropic materials. In this study, Hill's anisotropic yield condition, which is often used for steel materials, was selected, as shown in Eq.(3). The strain dependent model of the local anisotropic r-value was incorporated into the Hill's anisotropic yield condition by the authors [4, 5]. In other words, in order to deal with the strain dependency of the local anisotropic r-value, H, G, F and N were modified as shown in Eq. (4).

$$\sqrt{F(\sigma_y-\sigma_z)^2+G(\sigma_z-\sigma_x)^2+H(\sigma_x-\sigma_y)^2+2N\sigma_{xy}^2+2L\sigma_{yz}^2+2M\sigma_{zx}^2}=\sigma_Y(\overline{\varepsilon}^p) \tag{3}$$

$$\left.\begin{aligned}
&H=\frac{R_{00}(\overline{\varepsilon}^p)}{1+R_{00}(\overline{\varepsilon}^p)},\ G=\frac{H}{R_{00}(\overline{\varepsilon}^p)}=\frac{1}{1+R_{00}(\overline{\varepsilon}^p)},\ F=\frac{H}{R_{90}(\overline{\varepsilon}^p)}=\frac{R_{00}(\overline{\varepsilon}^p)}{(1+R_{00}(\overline{\varepsilon}^p))R_{90}(\overline{\varepsilon}^p)}\\
&N=(R_{45}(\overline{\varepsilon}^p)+1/2)(F+G)=\frac{(1+2R_{45}(\overline{\varepsilon}^p))(R_{00}(\overline{\varepsilon}^p)+R_{90}(\overline{\varepsilon}^p))}{2(1+R_{00}(\overline{\varepsilon}^p))R_{90}(\overline{\varepsilon}^p)}\\
&L=M=1.5
\end{aligned}\right] \tag{4}$$

2.3. *Uniaxial tensile specimen simulation*

The uniaxial tensile tests and simulations based on a JIS No. 5 tensile specimen, shown in Figure 3, were performed. To investigate the validity of the Cockcroft-Latham fracture criterion, a user subroutine including the Cockcroft-Latham fracture criterion was implemented into LS-DYNA. Figure 4 shows the nominal stress–nominal strain curves of the experiments, the CAE results with and without local anisotropic r-value, respectively. When considering the local anisotropic r-value, it was found that the accuracy of the rupture strain was greatly improved.

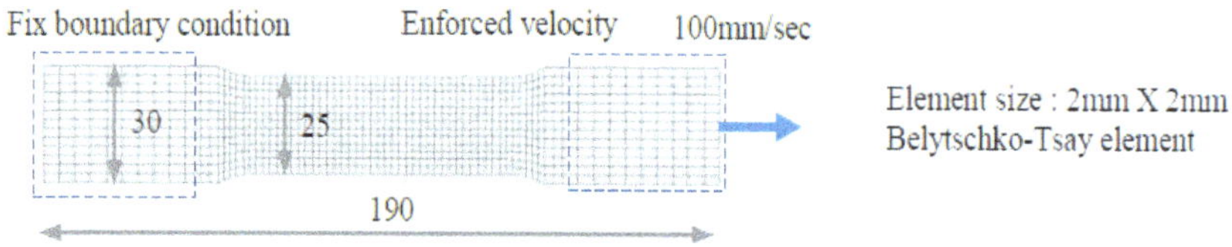

Fig. 3. Dimensions of the uniaxial tensile specimen model.

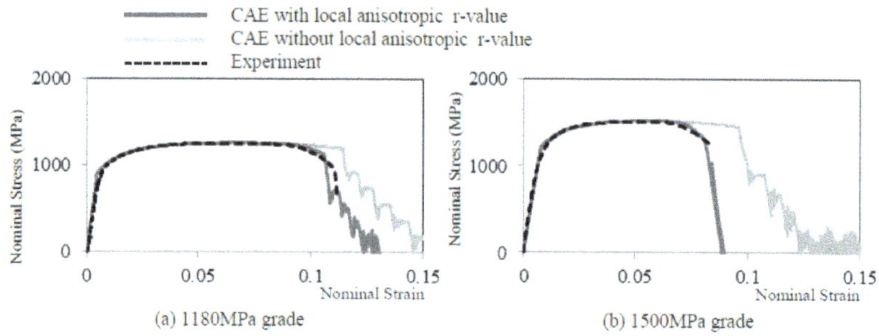

Fig. 4. Comparison of nominal stress-nominal strain curves between the experiments and CAE results.

2.4. *Three-point bending frame simulation*

Three-point bending tests were performed as shown in Figures 5 and 6. To accurately simulate the strength of the frame and mechanical deformation during the three-point bending, the strain-dependent anisotropic material model was only employed.

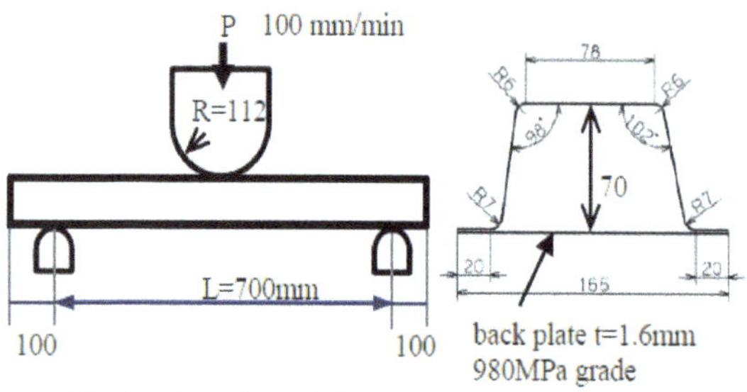

Fig. 5. Dimensions of three point bending frame.

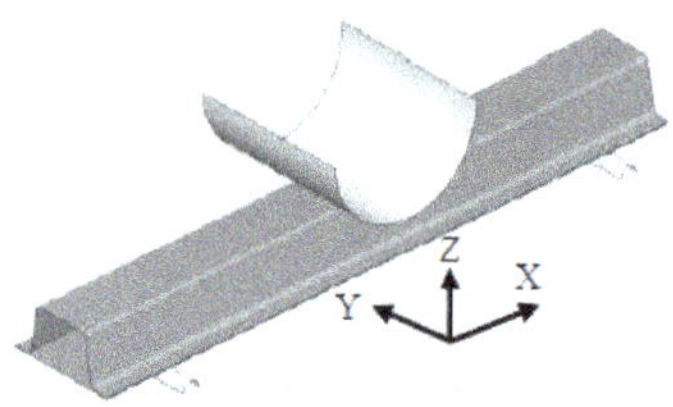

Fig. 6. CAE model of three-point bending.

The high strength steels used in this study was 1180MPa. Figure 7 shows the force-stroke curves obtained by the experiments and the CAE, respectively. It was found that the CAE results correlated well with those of the experiments.

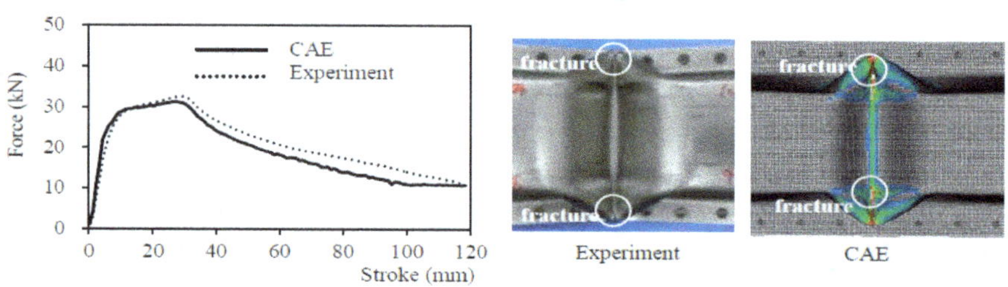

Fig. 7. Force-stroke curves and deformation modes of 1180MPa grade.

3. Bauschinger Effect

Many reports on the static crash analysis of hat section frame show that the energy absorption of high-strength steels of CAE is larger than test results. Conventionally, material properties used in crash analysis are based on the isotropic hardening role. To study the Bauschinger effect in crash analysis, authors have performed many tension-compression tests since 1999 and investigated many Bauschinger models.

3.1. *Axial crush tests and calculations*

The axial crush tests and simulations with hat section frame were performed to investigate the energy absorptions of the frames. The steel grades used in this test were 590MPa and 780MPa grade. Figure 8 shows the specification of the hat section frames and Table 2 shows the material properties of 590MPa and 780MPa grades.

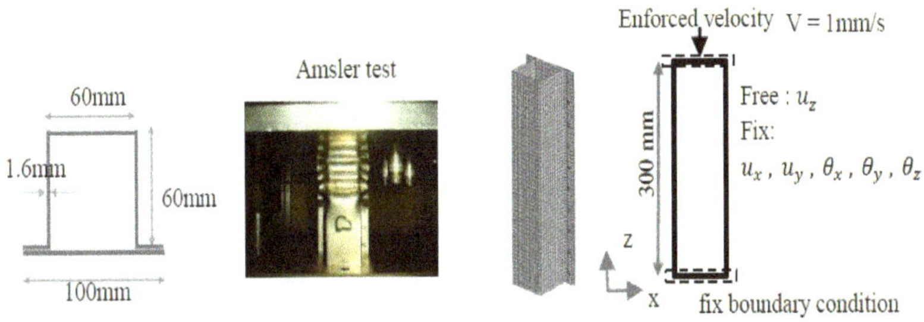

Fig. 8. Dimensions of the hat section frame and boundary condition of CAE.

Table 2. Material properties for two classes steels.

	Thickness	YS	TS	n-value
590MPa	1.6mm	370MPa	568MPa	0.151
780MPa	1.6mm	463MPa	752MPa	0.114

3.2. *Modeling of Bauschinger effect*

Figure 9 shows the deformation mechanism of thin-walled prismatic frame [6]. The deformation of thin-walled prismatic frame is progressed by the folding of the wall. Initial hinge line (1-C-2) moves to point B (1-B-2) by the folding of wall, which is named as moving hinge line. In the initial condition, the point B is on the middle of the wall. However, when the initial hinge line (1-C-2) moves to point B, the tensile stress occurs at the surface of the point B (reverse face of point B becomes compression). In the next step, when the moving hinge line (1-B-2) moves to point A (1-A-2), the compression stress occurs at the surface of point B (revers face point B becomes tensile). In other words, in the case of axial crush problem of thin-walled frame, the moving hinge line progresses reversed the stress from the tensile state to compressive stage and the Bauschinger effect needs to be taken into account.

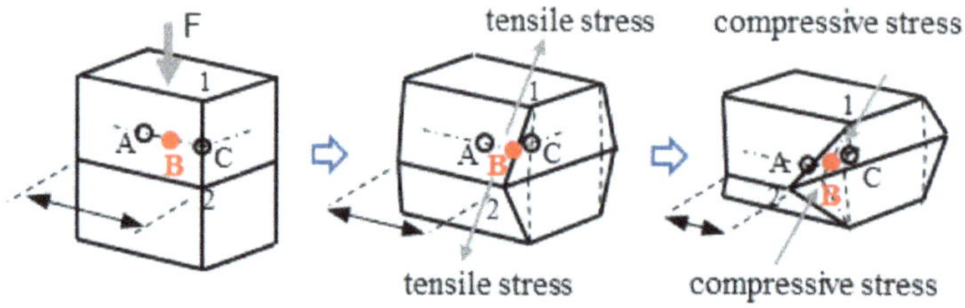

Fig. 9. Moving hinge line in deformation mechanism.

Figure 10 shows the energy absorption of point B. If the material property is isotropic hardening, the energy absorption of point B is expressed in Figure 10(a). While if the material property of point B is complex hardening (Bauschinger effect), the energy absorption is expressed in Figure 10(b). In other words, the energy absorption of isotropic hardening is larger than complex hardening.

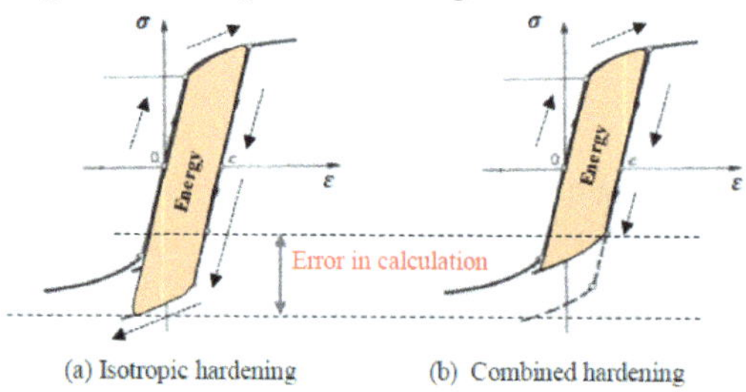

Fig. 10. Energy absorption of isotropic and combined hardening.

3.3. *Tensile and compression tests*

Figure 11 shows the measuring system for Bauschinger effect. Basically, in the tensile and compression test, the buckling of the test specimen occurs in compression state. In this system, the buckling prevention back plate was used to prevent the buckling. Figure 11(a), (b) shows the tensile and compression curves of 590MPa and 780MPa grades obtained by tests.

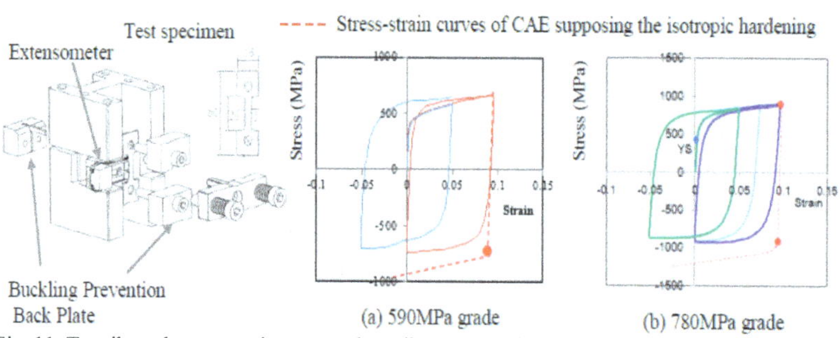

Fig. 11. Tensile and compression test and tensile-compression curves of 590MPa and 780MPa grade.

3.4. *Bauschinger effect models*

The Bauschinger effect on spring back prediction has been studied in plastic forming field. There are many models such as Armstrong and Frederick model, Chaboche mode [7],

ANK model, ZANG model and etc., proposed for reproducing the Bauschinger effect. Chun et al (2002) [8] proposed the ANK model which captures the permanent softening in the cyclic loading by evolving the second back stress. In this study, Chaboche and ANK model were investigated and modified ANK [9] was developed to predict the stress-strain behaviors of high strength steels as shown in Eqs. (5), (6) and (7).

$$\sigma_t = \sigma_0 + K\left(1 - e^{-N\bar{\varepsilon}^{pl}}\right) - \frac{C_1}{\gamma_1}\left(1 - e^{-\gamma_1\bar{\varepsilon}^{pl}}\right) - C_2\bar{\varepsilon}^{pl} + (\alpha_1 + \alpha_2) \tag{5}$$

$$\dot{\alpha}_1 = \frac{C_1}{\sigma^0}(\sigma - \alpha_1)\dot{\bar{\varepsilon}}^{pl} - \gamma_1\alpha_1\dot{\bar{\varepsilon}}^{pl} \tag{6}$$

$$\dot{\alpha}_2 = \begin{bmatrix} \frac{C_2}{\sigma^0}(\sigma - \alpha_2)\dot{\bar{\varepsilon}}^{pl} & for\ initial\ loading \\ OFF & for\ reversal\ loading \end{bmatrix} \tag{7}$$

where σ_t is stress tensor, σ_0 is yield stress, C_i and γ_i are material parameters that should be defined through cyclic loading tests, α_1 is first back stress, α_2 is second back stress.

Figure 12 shows the force-stroke and energy absorption curves of hat section frame with 590MPa and 780MPa grades with Chaboche model. When considering the Bauschinger effect, it was found that the accuracy of the energy absorption was greatly improved.

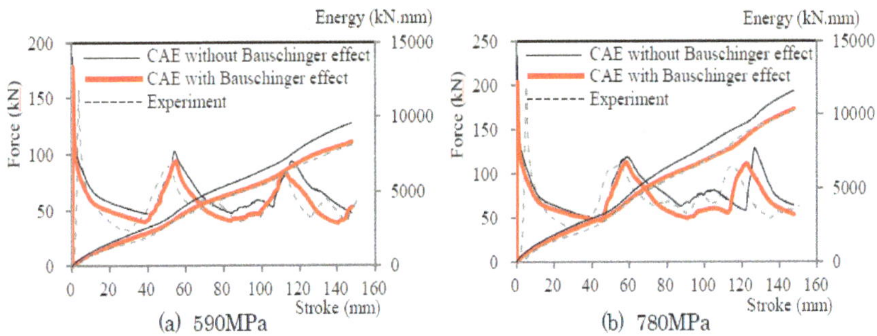

Fig. 12. Force-stroke and energy-stroke curves with the Chaboche model.

Figure 13 shows the identified results with modified ANK model. It was found that modified ANK coincided well with the tensile and compression tests. It will be supposed that this model is also useful for stamping simulation.

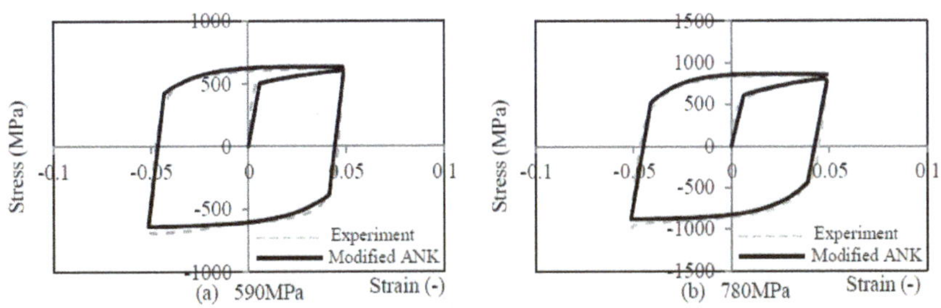

Fig. 13. The identified results for 590MPa and 780MPa grade with modified ANK model.

4. Conclusion

Measured results indicated that the local anisotropic r-value gradually decreases after the strain is larger than the hardening coefficient n-value due to the effects of diffused necking and local necking. A strain-dependent model of the local anisotropic r-value was incorporated into Hill's anisotropic yield condition. When taking into consideration the local anisotropic r-value, it was found that the fracture phenomenon can be more accurately predicted. To evaluate the energy absorption of automotive parts of advanced high strength steels due to axial crash, considering Bauschinger effect can greatly improve the simulation accuracy.

References

1. M. G. Cockcroft and D. J. Latham, Ductility and the workability of metals, *J. I. Met.* **96**, 33 (1968).
2. H. Takuda, T. Hama, K. Nishida, T. Yoshida and J. Nitta, Prediction of forming limit in stretch flanging by finite element simulation combined with ductile fracture criterion, *J. Lumin.* **9**, 137 (2009).
3. Z. Marciniak and K. Kuczynski, Limit strains in the processes of stretch-forming sheet metal, *Int. J. Mech. Sci.* **9**, 609 (1967).
4. K. Takada, K. Sato and N. Ma, Fracture Prediction for Automotive Bodies using a Ductile Fracture Criterion and a Strain-Dependent Anisotropy Model, Fracture Prediction for Automotive Bodies using a Ductile Fracture Criterion and a Strain-Dependent Anisotropy Model, *SAE Int.* **8**, 3 (2015).
5. N. Ma, Accurate simulation on failure and springback of sheet metal forming, *Sokeizai* **54**, 21 (2013).
6. T. Wierzbicki and W. Abramowicz, in *The Mechanics of Deep Plastic Collapse of Thin Walled Structures*, (Wiley, New York, 1989).
7. J. L. Chaboche, Time independent constitutive theories for cyclic plasticity, *Int. J. Plasticity* **2**, 149 (1986).
8. B. K. Chun, J. T. Jinn and J. K. Lee, Modeling the Bauschinger effect for sheet metals, *Int. J. Plasticity* **18**, 571 (2002).
9. K. Takada, K. Sato, P. Farahat, C. Yangwook and A. Yashin, Prediction of compressive characteristic of hat section high-strength steel frame taking Bauschinger effect into account, in *2015 SIMULIA Community Conference*, (Berlin, Germany, 2015).

Crystal Plasticity Finite Element Model Based on Dislocation Density Evolution Coupling with Thermal Activation and Damage Mechanism

Y. Liu, Y. G. Zhu, L. Ying and P. Hu[†]

State Key Laboratory of Structural Analysis for Industrial Equipment, Faculty of Vehicle Engineering and Mechanics, Dalian University of Technology, Dalian, China

[†]*E-mail: pinghu@dlut.edu.cn*

Here in this paper, a multiscale framework based on crystal plasticity is proposed coupling with thermal activation mechanism, as well as the continuum damage mechanism. The microscopic hardening phenomenon is revealed by a dislocation density evolution model, which is constructed according to the temperature. The physical-based exponential function of shear strain rate is posed to describe the thermal material behavior replacing the general phenomenological power-law equation. A 3D spatial distribution of stress, strain and damage is presented in the finite element method, parameters of which are previously determined by a RVE calculation and fitting test compared to the experimental data. Finally, some discussions of stress heterogeneity and texture evolution are proposed and conclusions are made.

Keywords: Crystal plasticity; Hot forming; 22MnB5; Dislocation density; Damage; Thermal activation.

1. Introduction

22MnB5 is increasingly widely used in vehicle industry for constructing different car components, such as engine hood, panel structure, bumper and trunk with good material property. In order to produce the car components using 22MnB5, the forming limit should be considered and at both elevated and low temperature, the total strain could be enhanced.

Here in hot stamping or warm forming process, the high temperature range is obtained and thermal effects or thermodynamics should be inserted in the material behavior. The traditional models, including phenomenological model and isotropic model lack the description of both the inner metal microstructure and anisotropic behavior in local behavior. Thus, the prediction of the material property is of great interest to consider the inner mechanism and crystallography in microscopic phenomenon.

Microstructural components of metal determine the orientation-dependent properties of polycrystalline aggregates. To better obtain the realistic material behavior of 22MnB5 at various outer conditions, here in this paper, the crystal plasticity extend finite element method (CPFEM) is used to fully describe the actual physical-based phenomenon. As for more complex outer and inner effects, such as the temperature, strain rate, recrystallization, and dislocation mobility, the mechanical response could be influenced and deviated enormously from the ordinary condition [1-3]. Complicated model are needed to describe these effects, and constitutive laws are revised to satisfy more complicated cases. These early attempts are used to simulate the texture and lattice reorientation, however, complex grain interaction and complicated out boundary cannot be revealed explicitly.

The representative volume element (RVE) is simulated to obtain the thermal-associated parameters by the fitting test Along with the jump test, all the parameters could be confirmed in the RVE simulation and later used in the CPFEM simulation of actual thermal shear test. The practical specimen of 22MnB5 is simulated and localized deformation and void evolution are shown in the 3D large plastic strain simulation [3]. This article is constructed as follows: In chapter 2, the CPFEM framework is introduced; in chapter 3, the thermal constitutive equations and the damage functions are briefed; then, the parameters are obtained in the RVE simulation; in chapter 4, the simulation of actual thermal shear test is shown; conclusion and some discussions are brought in chapter 5.

2. Multi-level Modeling of CPFEM

The multi-scale model of CPFEM is briefed as follows to obtain the anisotropic and heterogeneous local mechanical behaviors. The continuum mechanics of large displacement and strain theories are considered in this paper to determine the realistic plastic deformation of 22MnB5 in shear experiment. Both the stress equilibrium and strain compatibility are satisfied in the finite element method in a weak numerical sense [4-5]. Different mechanisms are inserted, such as temperature field, thermodynamics, damage evolution and dislocation density mobility. The crystal plasticity based theory is coupled in the User-defined Material Mechanical Behavior (UMAT) of one commercial software ABAQUS to be secondly developed in the FORTRAN language.

2.1. *Macro-homogenization*

Taylor model is considered first to test the material behavior prediction of crystal plasticity theory, which lack exhibition in mechanical fields distribution. Then the finite element method is coupled with crystal plasticity to further determine the realistic mechanical behavior of 22MnB5.

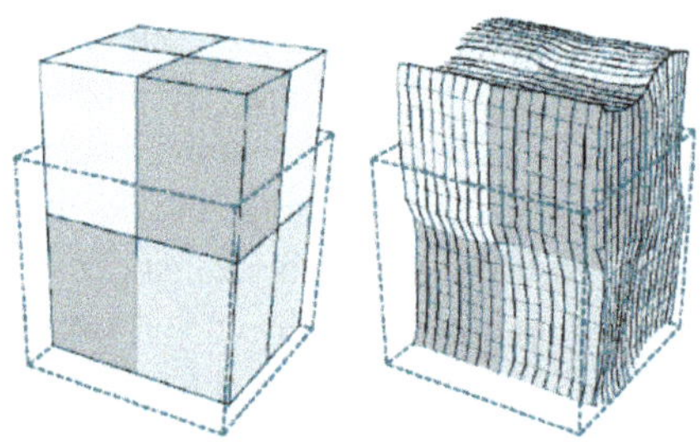

Fig. 1. The full constraint Taylor model and full field explanation of finite element model.

2.2. *Meso-continuum mechanics*

This section describes the meso-scale continuum mechanics for large displacement and large strain deformation. We introduce a thermal configuration in Eq. (1) to couple elasto-plastic deformation with thermal deformation and deformation gradient of each integration point is decomposed into several terms, including elastic and plastic deformation gradient and another isothermal gradient.

$$F = F^e F^p F^\theta \tag{1}$$

where, $\mathbf{F}^e$ represents for lattice rotation and distortion.

$$F^e = RU = VR \tag{2}$$

And $\mathbf{F}^p$ accounts for the irreversible incompressible plastic deformation gradient related to pure slip. $\mathbf{F}^\theta$ is the isothermal deformation gradient correlated to the temperature dependence, $\mathbf{F}^\theta = [1 + \mu(\theta - \theta_0)]\mathbf{I}$. The current and initial temperature states are θ and θ_0, and $\mathbf{I}$ is the unit tensor. The constitutive equations in continuum mechanics and the large strain deformation process is used. The plastic deformation evolution in time discretized form is deduced by time derivative in Eq. (3).

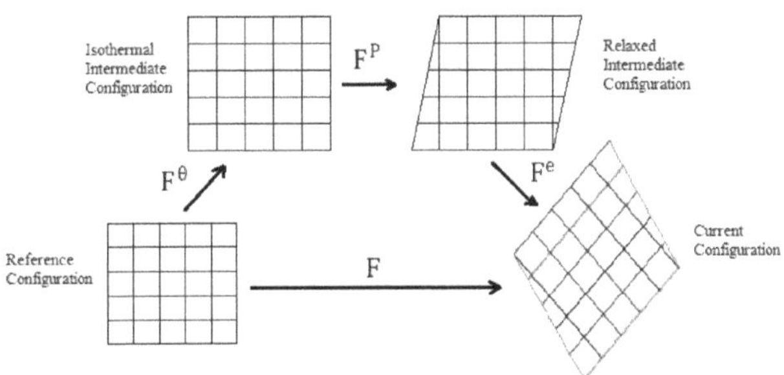

Fig. 2. The four configurations included in the large strain continuum mechanics.

$$\dot{F}^p = L^p F^p \tag{3}$$

where, $\mathbf{L}^p$ could be expressed as the microscopic form by shear strain.

$$L^p = \sum_\alpha \dot{\gamma}^\alpha \overline{P}_0^\alpha, \ \overline{P}_0^\alpha = S_0^\alpha \otimes m_0^\alpha \tag{4}$$

Here, the Schmidt tensor is constructed by the initial slip direction and slip normal, $\mathbf{s}_0^\alpha$ and $\mathbf{m}_0^\alpha$. The geometrical deformation of crystallography is described by the velocity gradient and the reorientation and stretching rate is used in a time integration form.

$$L = \dot{F} \cdot F^{-1} = D + \Omega \tag{5}$$

2.3. *Micro-dislocation density based model*

The thermal activation mechanism is proposed as the physical-based model in the exponential Eq. (6).

$$\dot{\gamma}^{\alpha} = \frac{\rho_m v_{id} b^2}{2} \exp\left(\frac{-\Delta F}{k\,\theta}\right) \exp\left(\frac{\left(\tau^{\alpha} - s^{\alpha}\right)\Delta V}{k\,\theta}\right) \tag{6}$$

where ΔF is the Helmholtz free energy, ρ_m the mobile dislocation density, v_{id} the dislocation vibration frequency, b the Burgers tensor, k the Boltzmann constant. This shear strain rate function is related to the resolved shear stress (RSS) τ^{α} and critical resolved shear stress (CRSS) s^{α}. As the RSS is higher than CRSS, the shear strain rate is activated, otherwise $\dot{\gamma}^{\alpha} = 0$. Since geometrical necessary and statistical stored dislocation density are related to the CRSS, it could be decomposed into several components, including friction, Hall-Patch, forest and debris dislocation density [6] in Eq. (7).

$$s(\dot{\varepsilon},\theta) = s_{0,f}(\theta) + s_{0,HP}(\theta) + s_{for}(\dot{\varepsilon},\theta) + s_{deb}(\dot{\varepsilon},\theta) \tag{7}$$

Several thermal parameters are considered in this equation and they can be obtained in the RVE simulation process.

3. RVE Determination

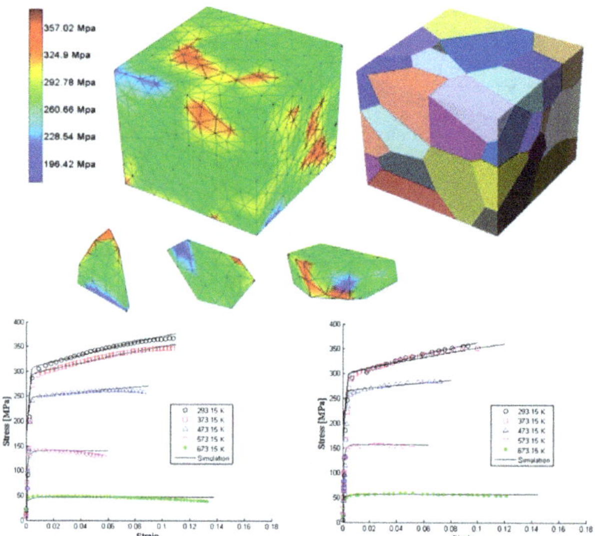

Fig. 3. The RVE simulation and schematic curve fitting process.

In the CPFEM simulation, the representative volume element (RVE) is used to determine the thermal parameters, damage parameters and dislocation-density associated parameters. The schematic determination process is illustrated in Fig. 3.

4. Shear Experiment Simulation

The shear experiment is simulated in CPFEM and the mechanical fields are shown in Fig. 4. As presented, the local strain distribution in the small circle is similar to the realistic deformation, which is revealed as the element rotation. And the large strain multitude distribution, described as the white space, is presented as the same direction as the shear experiment.

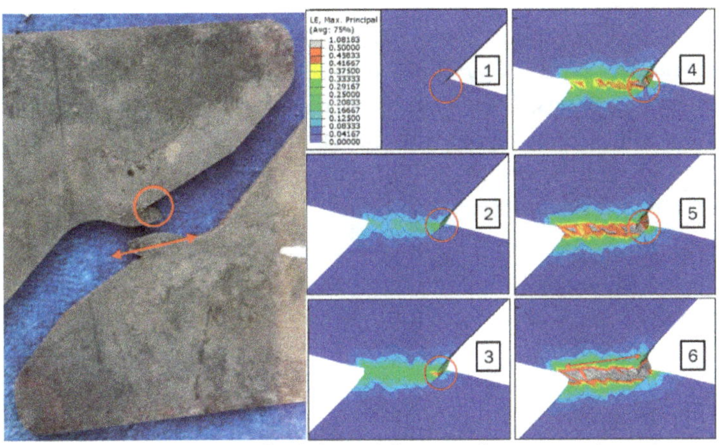

Fig. 4. The CPFEM simulation and shear experiment comparison.

5. Conclusion

The CPFEM is verified in this paper that it could be used to fully simulate the local mechanical fields and the multiscale simulation is able to present the orientation-dependent deformation of polycrystalline aggregates. The crystallographic slip system is expressed as the orientation mismatch among adjacent grains. The realistic hot stamping process is our simulation goal in further study.

Acknowledgments

This research is financially supported by the Key Project of the National Natural Science Foundation of China (No. 11272075), China Postdoctoral Science Foundation (2014M561223) and the Fundamental Research Funds for the Central Universities (DUT16RC (4) 28).

References

1. B. Peeters, S. R. Kalidindi, P. V. Houtte and E. Aernoudt, A crystal plasticity based work-hardening/softening model for b.c.c. metals under changing strain paths, *Acta Mater.* **48**, 2123 (2000).
2. C. C. Tasan, J. P. M. Hoefnagels, M. Diehl, D. Yan, F. Roters and D. Raabe, Strain localization and damage in dual phase steels investigated by coupled in-situ deformation experiments and crystal plasticity simulations, *Int. J. Plasticity* **63**, 198 (2014).
3. J. Kim and J. W. Yoon, Necking behavior of AA 6022-T4 based on the crystal plasticity and damage models, *Int. J. Plasticity* **73**, 3 (2015).
4. L. Chen, J. Chen, R. A. Lebensohn, Y. Z. Ji, T. W. Heo, S. Bhattacharyya, K. Chang, S. Mathaudhu, Z. K. Liu and L. Q. Chen, An integrated fast Fourier transform-based phase-field and crystal plasticity approach to model recrystallization of three dimensional polycrystals, *Comput. Method Appl. M.* **285**, 829 (2015).
5. D. Raabe and F. Roters, Using texture components in crystal plasticity finite element simulations, *Int. J. Plasticity* **20**, 339 (2004).
6. J. R. Mayeur, I. J. Beyerlein, C. A. Bronkhorst and H. M. Mourad, Incorporating interface affected zones into crystal plasticity, *Int. J. Plasticity* **65**, 206 (2015).

Effect of Material Damage on Thermal Forming Limit Prediction for Hot Stamping

W. Q. Liu[1], P. Hu[2], L. Ying[2,†] and D. T. Wang[2]

[1]Department of Engineering Mechanics, State Key Laboratory of Structural Analysis for Industrial Equipment, Dalian University of Technology

[2]School of Automotive Engineering, State Key Laboratory of Structural Analysis for Industrial Equipment, Dalian University of Technology

Dalian, P.R. China

†E-mail: yingliang@dlut.edu.cn

In the hot stamping process, the formability of high strength steel is limited by the evolution of void damage inside materials. Thus, to evaluate the formability of hot metal blank, the damage evolution must be fully considered. In the present work, the extended Gurson-Tvergaard-Needleman (GTN) damage model for normal anisotropic materials is used to characterize material damage behavior. The prediction method of forming limits considering void damage is developed with the framework of Marciniak and Kuczynski (M-K) theory. To confirm the validity of the proposed prediction method, the forming limit strains contained by Nakazima experiments are compared with the prediction results and good accordance is acquired. Then, the thermal forming limit diagrams (TFLDs) at different temperatures are established. Simultaneously, the effect of void damage on forming limits and the void growth process under typical deformation modes are investigated.

Keywords: Hot stamping; GTN yield function; M-K model; Numerical prediction; TFLD.

1. Introduction

Hot stamping of high strength steel is an attractive technology to achieve lightweight and guarantee security of car body. [1] But, similar to traditional cold stamping, the formability of hot sheets is still limited by the damage evolution inside materials. Thus, to evaluate metal sheet formability, the material damage evolution during deformation must be fully considered.

To couple the influence of material damage, some works have been done within the framework of the M-K model. Needleman et al. derived the constitutive relation coupling porous damage based on Gurson model and gave a prediction algorithm in conjunction with the basic M-K model. [2] Using another set of algorithms, the effects of yield criteria, void shape and hardening law on forming limits were investigated by Kim and his collaborators, and the predicted results were compared with the experimental results. [3-4] However, there are few reports on forming limits investigation of hot stamping using the M-K model coupling with damage theory.

In the present work, the TFLD prediction method combined the extended void damage model is developed within the framework of M-K theory. To confirm the validity of the predicted results and calibrate initial imperfection parameters, the high temperature

Nakazima test for boron steel 22MnB5 is conducted. Finally, the TFLDs at different temperatures are established by numerical prediction. The effects of void on forming limits and the void growth process under typical deformation modes are also studied.

2. Theoretical Model and Numerical Algorithm

2.1. *Damage model for normal anisotropic material*

The damage model proposed by Gurson and modified by Tvergaard and Needleman which is well known as GTN model is adopted in this paper and given as: [5-7]

$$\varphi = \left(\frac{\sigma_{eq}}{\Sigma_{eq}}\right)^2 + 2C_v q_1 \cosh\left(\frac{q_2\sigma_{kk}}{2\Sigma_{eq}}\right) - \left(1 + q_3 C_v^2\right) = 0 \tag{1}$$

where C_v is the void volume fraction, σ_{eq} is the macroscopic von Mises equivalent stress, σ_{kk} is the macro hydrostatic stress, Σ_{eq} is the flow stress of matrix material, the parameters q_1, q_2, q_3 are introduced by Tvergaard to improve the prediction accuracy.

The hot boron steel 22MnB5 in austenite phase exhibits normal anisotropic behavior while planar anisotropy can be neglected. Thus, to accurately describe the damage evolution of austenitic 22MnB5, the GTN model must be further extended. Thus, the Hosford anisotropic yield criterion has been adopted to replace the von Mises equivalent stress in Eq. (1): [8]

$$\sigma_{eq}^M = \frac{r}{1+r}(\sigma_1 - \sigma_2)^M + \frac{1}{1+r}(\sigma_1 - \sigma_3)^M + \frac{1}{1+r}(\sigma_2 - \sigma_3)^M \tag{2}$$

where r is the mean anisotropy parameter, M is the yield exponent.

2.2. *Numerical algorithm for TFLD*

According to related flow rules, the expression of major strain increment $d\varepsilon_1$ is given by:

$$d\varepsilon_1 = d\lambda' \left\{ 2\left(\frac{1}{1+r}\right)^{\frac{2}{M}} \left[r(\sigma_1-\sigma_2)^M + \sigma_1^M + \sigma_2^M\right]^{\frac{2}{M}-1} \left[r(\sigma_1-\sigma_2)^{M-1} + \sigma_1^{M-1}\right] + \frac{1}{2}C_v q_1 q_2^2 (\sigma_1+\sigma_2) \right\} \tag{3}$$

Base on the hypothesis that matrix material is non-compressible, the relationship between $d\varepsilon_1$ and dC_v can be derived by some mathematical derivation:

$$d\varepsilon_1 = \frac{3dC_v}{2C_v q_1 q_2^2 (\sigma_1+\sigma_2)(1-C_v)} \left\{ 2\left(\frac{1}{1+r}\right)^{\frac{2}{M}} \left[r(\sigma_1-\sigma_2)^M + \sigma_1^M + \sigma_2^M\right]^{\frac{2}{M}-1} \left[r(\sigma_1-\sigma_2)^{M-1} + \sigma_1^{M-1}\right] + \frac{1}{2}C_v q_1 q_2^2 (\sigma_1+\sigma_2) \right\} \tag{4}$$

To couple the damage evolution during metal sheet deformation, the M-K theory is further modified by the extended GTN model. A schematic drawing is shown in Fig. 1.

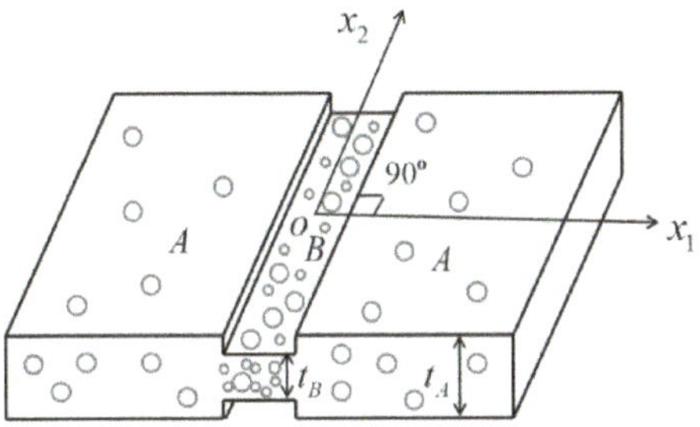

Fig. 1. Illustration of metal sheet having void damage and initial thickness imperfection.

The flow behavior of matrix material 22MnB5 is described by the modified Norton-Hoff hardening law:

$$\sigma(\varepsilon,\dot{\varepsilon},T) = K(\varepsilon_0+\bar{\varepsilon}_M)^{n(T)}\dot{\bar{\varepsilon}}_M^{m(T)}\exp(\beta/T) \tag{5}$$

$$n(T) = n_0\exp(-c_n(T-T_0)) \tag{6}$$

$$m(T) = m_0\exp(c_m(T-T_0)) \tag{7}$$

where $K,\beta,\varepsilon_0,m_0,n_0,c_m,c_n$ are material constants, T_0 is the reference temperature.

According to the key hypotheses of the M-K theory, the force equilibrium relation in direction x_1 must be satisfied in the whole deformation process. And noting that the matrix material supports the load, the equilibrium equation can be written as:

$$\phi_B(\varepsilon_0+\bar{\varepsilon}_{MA})^{n(T)}\dot{\bar{\varepsilon}}_{MA}^{m(T)}(1-C_v^A) = \phi_A(\varepsilon_0+\bar{\varepsilon}_{MB})^{n(T)}\dot{\bar{\varepsilon}}_{MB}^{m(T)}(1-C_v^B)f_0\exp(\varepsilon_3^B-\varepsilon_3^A) \tag{8}$$

3. High Temperature Nakazima Test

Based on Nakazima principle, a high temperature forming limits testing apparatus has been developed as shown in Fig. 2a and an optimized test procedure is given in Fig. 2b. The Nakazima test is conducted at 800°C with an average strain rate of 0.1 s^{-1}.

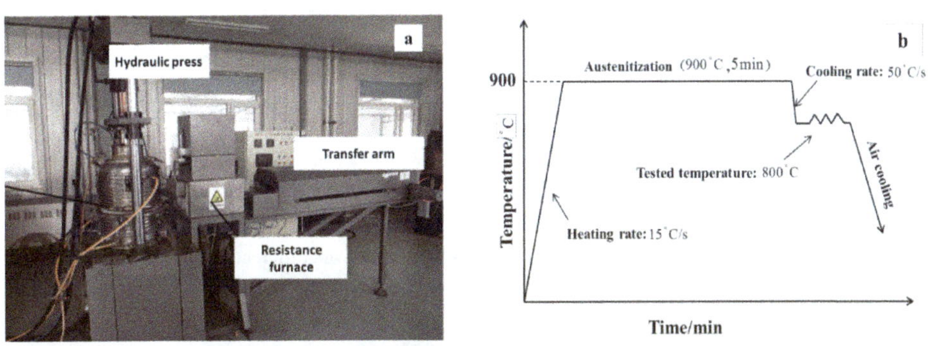

Fig. 2. High temperature Nakazima apparatus and test procedure.

4. Prediction Results and Analysis

The most optimal initial parameters can be defined by comparing the experiment and numerical prediction results. After repeated calculation, the values, $C_{v0}^{A} = 0.002$, $C_{v0}^{B} = 0.0095$ and $f_0 = 0.99$ are adopted, where C_{v0}^{A} is a same order of magnitude for steels. The predicted TFLD and experimental data at 800 °C are plotted in Fig. 3a and good consistency has been obtained. The void evolution under typical deformation modes is shown in Fig. 3b. It shows that the void growth rate of zone B is obvious larger than zone A and it has a dramatic acceleration around the occurrence of necking.

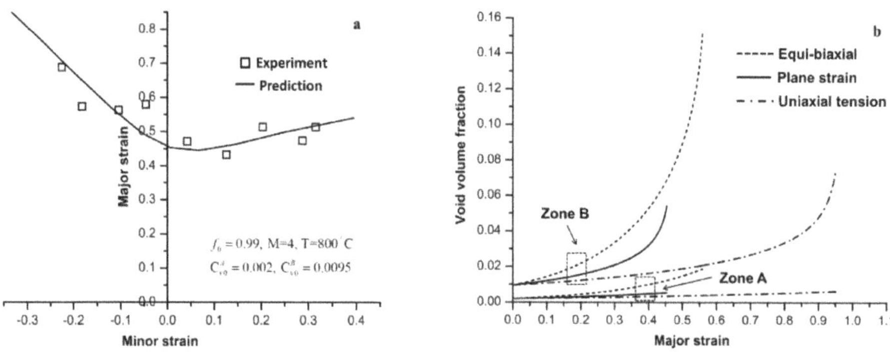

Fig. 3. TFLD at 800°C and void evolution under typical deformation modes.

The predicted TFLDs under different temperatures are plotted in Fig. 4a. It shows that temperature has a significant impact on the formability of 22MnB5 and positive correlation is observed. The predicted FLDs with variable void parameters are shown in Fig. 4b. It shows that the forming limits obviously decrease with the increasing of initial void volume fraction. On the other hand, the process of localized necking will accelerate with the increasing of difference between C_{v0}^{B} and C_{v0}^{A} by comparing curves b and c whose value of C_{v0}^{A} is equal.

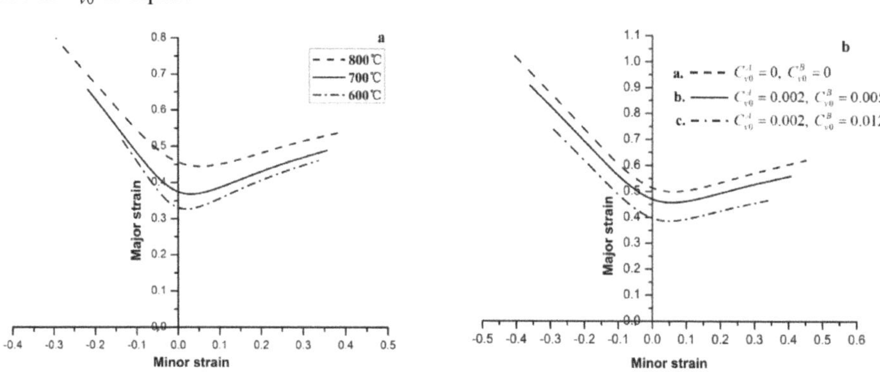

Fig. 4. TFLD prediction results at different forming temperatures and with different initial void volume fraction.

5. Conclusion

Considering the void growth inside material, the M-K theory is further extended by a series form of GTN model. The high temperature Nakazima test at 800°C is performed to confirm the validity of the proposed algorithm. Furthermore, the influence of temperature, void volume fraction on TFLD is analyzed. The positive correlation between temperature and forming limits is observed while the opposite conclusion is observed between void volume fraction and forming limits.

Acknowledgments

This work is funded by the Key Project of the National Natural Science Foundation of China (No. 11272075), China Postdoctoral Science Foundation (2014M561223) and the Fundamental Research Funds for the Central Universities (DUT16RC(4)28).

References

1. H. S. Liu, Z. W. Xing, J. Bao and B. Y. Song, Investigation of the Hot-Stamping Process for Advanced High-Strength Steel Sheet by Numerical Simulation, *J. Materi. Eng. Perform.* **19**, 325 (2010).
2. A. Needleman and N. Triantafyllidis, Void growth and local necking in biaxially stretched sheets, *J. Eng. Mater. T.* **100**, 164 (1978).
3. Y. S. Kim, H. S. Son, S. H. Yang and S. R. Lee, Prediction of forming limits for voided anisotropic sheets using strain gradient dependent yield criterion, *Key Engineering Materials* (Trans Tech Publ, 2003).
4. S. G. Choi, H. S. Son and Y. S. Kim, Prediction of Forming Limits for Anisotropic Sheets with Ellipsoidal Voids, *Key Engineering Materials* (Trans Tech Publ, 2004).
5. A. L. Gurson, Continuum theory of ductile rupture by void nucleation and growth: Part I—Yield criteria and flow rules for porous ductile media, *J. Eng. Mater. T.* **99**, 2 (1977).
6. V. Tvergaard and A. Needleman, Analysis of the cup-cone fracture in a round tensile bar, *Acta. Metall.* **32**, 157 (1984).
7. A. Needleman and V. Tvergaard, An analysis of ductile rupture in notched bars, *J. Mech. Phys. Solids.* **32**, 461 (1984).
8. W. F. Hosford, in *yield loci of anisotropic cubic metals, Proceedings of the seventh North American metalworking research conference* (1979).

A New Method in Cracking Prediction in Hot Stamping of High Strength Steel

H.S. Liu†, G.T. Zhou and K.Y. Jiang

Fujian Provincial Key Laboratory of Special Energy Manufacturing, Huaqiao University, Xiamen, 36102, P.R. China

†E-mail: liuhs@hqu.edu.cn

Hot stamping of high strength steel (HSS) can significantly improve overall mechanical properties of part and thus meet the increasing demands for weight reduction and safety standards in vehicles. However, cracking prediction using traditional forming limit curve (FLC) in hot stamping is challenging. In this paper, to predict HSS cracking in hot stamping a temperature-dependent forming limit surface (FLS) is developed by simulations combined with experiments. Different from the FLC the newly developed FLS, where temperature and phase transformation are included, suits the hot stamping of HSS. A finite element (FE) thermo-mechanical coupled numerical model of the hot stamping process is developed and implemented under ABAQUS/Explicit platform. Finally the developed FLS is used to predict crack initiation in a hot stamping. Results show that effectiveness of the developed FLS is verified in cracking prediction for hot stamping of HSS.

Keywords: High strength steel; Hot stamping; Cracking; Finite element.

1. Introduction

For the traditional cold forming of sheet metal, the FLC serves as an effective tool to describe how much the material can be deformed without cracking [1]. To predict cracking, the Gurson-Tvergaard damage model that takes into account the influence of triaxial stress state is modified [2]. Depending on the grade, failures in HSS are not always predictable by the traditional FLCs. Such FLCs represent the forming limit based on localized deformation or necking, which has proven successful in predicting failures in cold forming of sheet metal. However FLC is ineffective in hot stamping since temperature is not considered in the FLC [3].

In this work, in order to predict the cracking of HSS in hot stamping of HSS, M-K model is used to include the dependency of all procedures of the cracking on temperature. FE analysis of the biaxial tension which resembles M-K analysis is performed to establish FLCs at elevated temperatures. A FLS is established using these FLCs. Therefore, the proposed novel FLS includes temperature and martensitic transformation effects. A coupled thermo-mechanical FE model of the hot stamping of HSS is developed under ABAQUS/Explicit platform to obtain major and minor strains on hot stamped part for different process parameters. At last, a FLS is developed for hot stamping of HSS.

2. Determination of Temperature-dependent FLS

In order to establish a FLS, many hot stamping tests at different temperatures need to be carried out. Considering the stress versus strain curves at different temperatures[4], based on the M-K analysis [5], a FE model of isothermal biaxial tension is developed as shown in Figure 1. In this model 186526 Tet elements are used. In order to improve computation accuracy, more and smaller elements are concentrated at the groove. The isothermal hot biaxial tension of a plate with a groove is simulated at different temperatures using the developed FE model. In these simulations, when the strain increment i.e. $d\varepsilon_x$ in an element in the center of region B is larger than 5 times the strain increment in an element[5], it means the material point damages and crack occurs, then the analysis is ceased. Upon failure, the average major and minor strain pair (ε_{major},ε_{minor}) in the element is used to determine a point on the FLS in a space of major and minor strains. Three isothermal hot biaxial tension tests at 700°C, 800°C and 900°C are also done, respectively, to validate the simulation results and the developed FE model of the isothermal tension. Through simulations major strain ε_{major} and minor strain ε_{minor} at the material point where crack initiates for different temperatures are obtained and listed in Table 1.

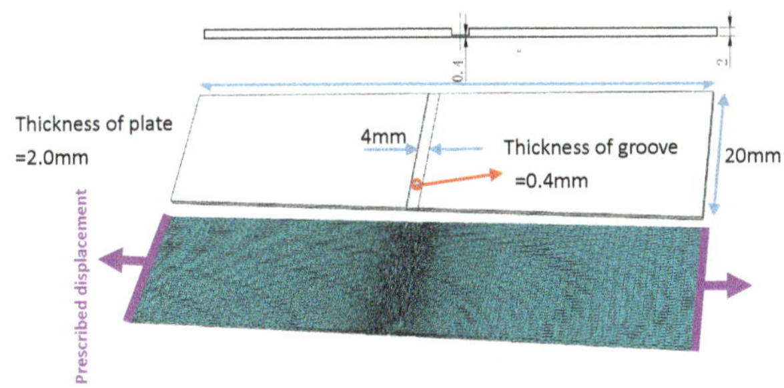

Fig. 1. Geometry and FE model of the plate with a groove.

A FLS for the hot forming of HSS is obtained by fitting all the coordinates (ε_{major}, ε_{minor}) in the space of major and minor strains as shown in Figure 2. The 3D surface manifests that temperature has significant effect on FLC of HSS: higher temperature higher maximum major and minor strains due to that the plasticity of HSS is drastically improved as temperature increases.

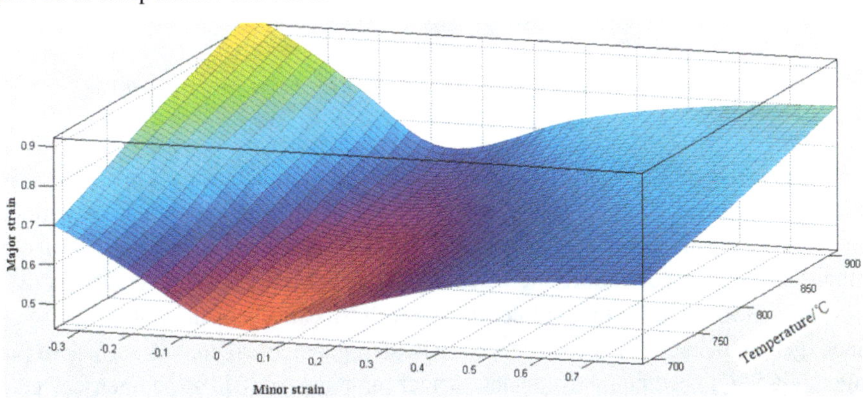

Fig. 2. Temperature-dependent FLS for hot forming of HSS.

Table 1. Major and minor strains ε_1 and ε_2 for different temperatures at material points of initial crack

Temperature/°C	700		800		900	
	ε_1	ε_2	ε_1	ε_2	ε_1	ε_2
	-0.21	0.61	-0.25	0.76	-0.3	0.9
	-0.18	0.585	-0.15	0.67	-0.18	0.8
	-0.1	0.518	-0.075	0.57	-0.08	0.7
	-0.05	0.48	-0.045	0.56	-0.045	0.66
Major and	0	0.45	0	0.515	0	0.625
minor strains	0.045	0.46	0.05	0.537	0.115	0.635
	0.075	0.485	0.1	0.543	0.24	0.67
	0.178	0.525	0.2	0.575	0.4	0.7
	0.31	0.575	0.35	0.63	0.565	0.775
	0.45	0.618	0.5	0.675	0.74	0.792
	0.65	0.64	0.65	0.701	0.78	0.806

3. Application of the Developed FLS

In this section, experiment and modeling of HSS hot stamping for forming a box-shaped part are performed to validate the developed FLS. A quarter of model as shown in Figure 3 for the hot stamping for a box-shaped part is used due to symmetry. The blank and tools are modeled with 8-node brick elements and tools are considered to be rigid. FE meshes of blank, punch and die used in the hot stamping model is presented.

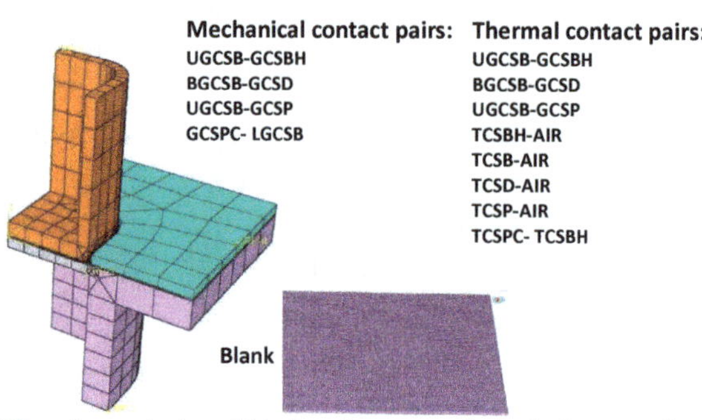

Fig. 3. FE meshes on tools and blank and contact pairs in the hot stamping modeling.

Mechanical contact between blank and tools is defined using kinematic contact method and Coulomb friction law. Surfaces of the blank and tools must be defined in order to treat the mechanical and thermal contacts involved in the modeling of hot stamping process. In addition to the mechanical contact, the contact involved in the hot stamping of HSS includes thermal contact as well. Six mechanical contact surfaces are defined: geometrical contact surface of die (GCSD), geometrical contact surface of punch counter (GCSPC), geometrical contact surface of punch (GCSP), geometrical contact surface of blank holder (GCSBH), upper geometrical contact surface of blank (UGCSB) and lower geometrical contact surface of blank (LGCSB) illustrated in Figure 3. In order to treat the thermal contact, other five surfaces in contact with air need to also be defined: thermal contact surface of die (TCSD), thermal contact surface of punch (TCSP), thermal contact surface of punch counter (TCSPC), thermal contact surface of blank holder (TCSBH) and thermal contact surface of blank (TCSB) illustrated in Figure 3. Contact pairs are defined as presented in Figure 3. For different contact surfaces, different ways [6] are used to treat heat transference. Simulations are performed under different processing parameters: blank holder force and forming temperature. Strains ε_{major} and ε_{minor} are also average values which are calculated from integration points within a single element. For punch stroke of 34mm and forming temperature of 900°C and blank holder force of 88 kN modeling and experiment of hot stamping process are performed. Through simulations equivalent strain distribution on hot formed part at the end of hot forming is shown in Figure 4.

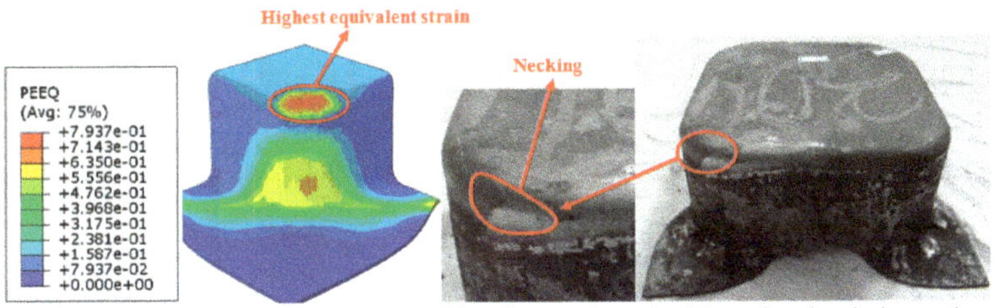

Fig. 4. Positions of high equivalent strain and necking for punch stroke 35mm by simulation and experiment.

Highest equivalent strain occurs at the round corner of the box-shaped part due to the material at the round corner undergoes largest deformation. Correspondingly serious necking marked with a loop is also found at the round corner as shown in Figure 4. The results demonstrate the simulation and experimental results have good agreement.

4. Conclusion

In this work, A temperature-dependent FLS for the hot stamping of HSS are determined by performing experiments and simulations of isothermal biaxial tension of a plate with a groove. The FLS is applied in the hot stamping of HSS for a box-shaped part. In this case, from the comparison between the cracks obtained by experiments and simulations, it is concluded that the FLS is reliable in cracking prediction in the hot stamping of HSS.

Acknowledgments

This work is funded by the Fujian province science and technology plan, (No. 2008H0085), the National science and technology plan, (No. 2012BAF12B15), the Education department of Fujian province, (No. JA13018) and the Promotion Program for

Young and Middle-aged Teacher in Science and Technology Research of Huaqiao University (Z15J0002).

References

1. K. Bandyopadhyay, S. Basak, S.K. Panda and P. Saha, Use of stress based forming limit diagram to predict formability in two-stage forming of tailor welded blanks, *Mater. Des.* **67**, 558(2015).
2. Z.Y. Chen and X.H. Dong, The GTN damage model based on Hill'48 anisotropic yield criterion and its application in sheet metal forming, *Comp. Mater. Sci.* **44**, 1013(2009).
3. N. Selvakumar, M.J.S. Mohamed, R. Narayanasamy and K. Venkateswarlu, Forming limit diagram and void coalescence analysis of AA5052 coated with molybdenum-based ceramic nanocomposites, *Mater. Des.* **52**, 393(2013).
4. H.S. Liu, W. Liu and J. Bao, Numerical and experimental investigation into hot forming of ultra-high strength steel sheet, *J. Mater. Eng. Perform* **20**, 1(2011).
5. Z. Marciniak and J.L. Duncan, Mechanics of sheet metal forming, Edward Arnold, London, 1992.
6. H.S. Liu, B. Jun and Z.W. Xing, Modeling and FE simulation of quenchable high strength steels sheet metal hot forming process, *J. Mater. Eng. Perform* **20**, 894(2011).

Development of Press-Hardening Simulation System for Ultra-High Strength Steel Parts

R. F. Liu† and N. S. Ma

JSOL Corporation

2-2-4 Tosabori, Nishiku, Osaka, 550-0001, Japan

†E-mail:rongfeng.liu@jsol.co.jp

The hot stamping process is strongly coupled with multi-physical phenomena of thermal transfer, temperature dependent material properties, material phase transformation in thermal cycles and large mechanical deformation. An integrated sheet metal forming simulation system JSTAMP/NV was designed and developed considering multi-physical characteristics occurred in the hot stamping process. Using a typical center pillar, formability was evaluated and hardness was predicted.

Keywords: Hot-stamping; Simulation system; Hardness prediction; Thermal cycles.

1. Introduction

In order to reduce the white bodies' weight and improve the crashworthiness as well, hot stamping parts with ultra-high strength up to about 1500MPa have been increasingly employed in automotive bodies. The strong requirements from industries pushed development of hot stamping simulation system. Comparing with the research on a specified academic issue, designing and developing a simulation system are much more difficult works because both academic issues and engineering issues must be solved and integrated [1]. On the other hand, the development of the simulation system needs many innovations so as to give a robust solution to users based on a stable design in handling various sub-systems and a lot of testing works. In this paper, authors presented the design and developing concept of a simulation system. Then formability was evaluated and hardness of a hot stamping part was predicted.

2. Design and Development of Hot Stamping Simulation System

A hot stamping system JSTAMP and its concept as shown in Fig. 1 were designed and developed. In this system, a commercial FEM solver LS-DYNA and a dedicated pre-post processor are incorporated together. The dedicated pre-post processor can create the simulation model and evaluate the simulation results for hot stamping. The thermal mechanical coupling feature in LS-DYNA is employed to simulate the hot stamping process. The stress and strain obtained by thermal mechanical simulation can be used to evaluate the formability. The thermal cycles and cooling rate due to thermal conduction simulation can be employed to predict hardness and then the strength of a hot stamped part.

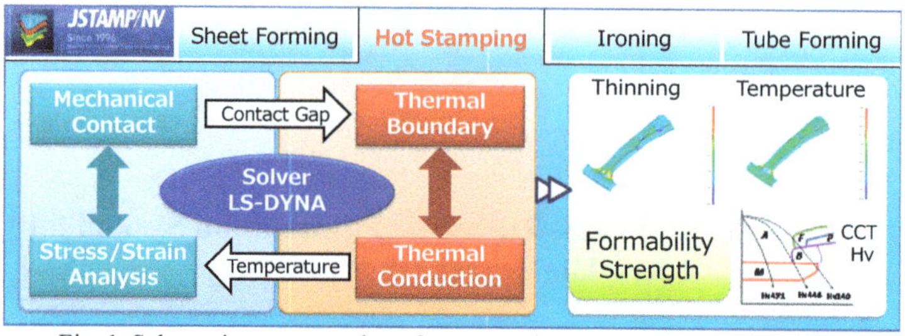

Fig. 1. Schematic representation of concept of stamping simulation system.

A full integration shell element with assumed strain for mechanical analysis, an advanced thermal shell element with three nodes through the thickness proposed by Bergman and Eldenburg [2], are employed.

Generally, the loading time for explicit method is from about 0.001sec to 1.0sec. The practical hot stamping may take several 10 seconds. The velocity scaling technique is used to accelerate forming velocity so as to improve the simulation efficiency without losing the accuracy [3]. It must be noted that the velocity scaling technique scaled the strain rate in explicit mechanical solution. If strain rate is considered in the simulation, its effect must be carefully processed. The material properties such as thermal conductivity, heat transfer coefficient of convection and radiation relating to time, must be also scaled if the velocity scaling techniques is employed.

For hot stamping steel 22MnB5, the effect of cooling curves on and the generation of martensitic microstructure and the vicker's hardness *Hv* were measured by Naderi et al [4]. The relation between the Vicker's hardness and average cooling rate from austenite state temperature (approximately 880°C) to the martensitic transformation finish temperature Mf (233°C) can be quantitatively represented. When the cooling rate is more than a critical cooling rate 25 (K/sec), a fully martensitic microstructure can be obtained and hardness keeps high value. When the cooling rate changes from 5 (K/sec) to 25 (K/sec), only a partial martensitic microstructure can be obtained and hardness varies greatly. If the vicker's hardness *Hv* is known, the tensile strength *TS* can be predicted using following Eq. (1).

$$TS = a \times Hv + b \tag{1}$$

Where, *a* and *b* are coefficients which are about 3.127 and 21.2MPa by fitting data in ASTM-E140 ISO18265.

3. Simulation for Formability Evaluation

3.1. *Simulation model*

A center pillar model provided by Oberpriller et al [5] is selected as an example of hot stamping simulation. In the real hot stamping processes, a blank sheet was heated in a furnace from room temperature 20°C (293K) to austenization temperature 940°C (1173K) and then transported to tools for setting up, forming and quenching. The FE model and simulation stages are schematically represented in Fig. 2 and Fig. 3, respectively.

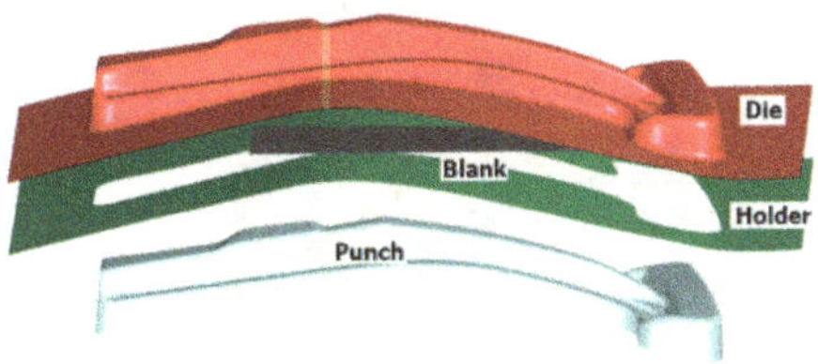

Fig. 2. FE model for hot stamping simulation.

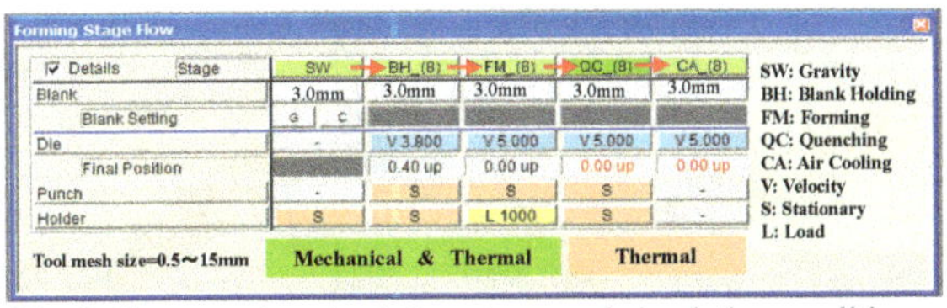

Fig. 3. Hot stamping simulation stages and simulation conditions.

The material properties used in hot stamping simulation are listed in Table 1. The stress-strain relation and its temperature dependence are shown in [5].

Table 1. Material properties and simulation parameters.

Density	ρ=7.80E+3[kg/m^3]
Young's modulus	E=1.00E+5[MPa]
Poisson's ratio	υ=0.3
Stress-strain-temperature relation	See stress-strain curves
Thermal expansion coefficient	α= 1.3E-5[1/K]
Thermal conductivity	λ=32[J/(mKs))]
Heat capacity	C_p=650[J/(kgK)]
Convection heat transfer coef.	h_{air}=160[J/(m^{2}Ks)]
Radiation heat transfer coef.	C_0=5.67E+8[J/(m^{2}Ks)]
Contact pressure p[MPa] vs. Heat transfer coef. h_{tool}	P=0.00[MPa], h_{tool}=1300[J/(m^{2}Ks)] P=20.00[MPa],h_{tool}=4000[J/(m^{2}Ks] P=35.00[MPa],h_{tool}=4500[J/(m^{2}Ks)]
Friction coefficient	μ=0.4
Full thermal contact gap size	Lmax=0.001mm
Non thermal contact gap size	Lmax=0.4mm

3.2. *Simulation result*

The temperature distributions taken at the time of 0.5sec, 1.08sec and 20sec, just when blank holding stage, stamping stage and die quenching stage ended, are represented in Figs. 4 and 5, respectively. In Fig.4, the temperature is lower at the positions where blank and tools contacted. The maximum temperature decrease is about 60 degrees and the lowest temperature at the edges of section A-A of blank is about 750°C (1023K). When the 1.08sec forming stage finished, the blank is stamped to a designed shape and the hot blank was cooled down to about 750°C (1023K) - 500°C (773K). The highest temperature 750°C (1023K) is still above A1 temperature (723°C) and the lowest temperature 500°C (773K) is higher than the martensite start temperature Ms (410°C).

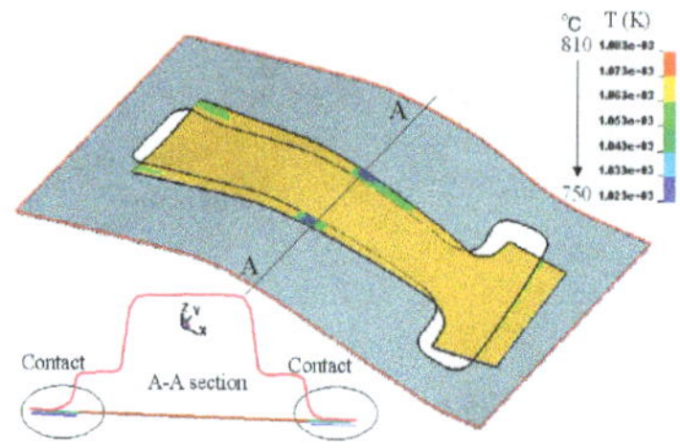

Fig. 4. Temperature after 0.5sec blank holding stage.

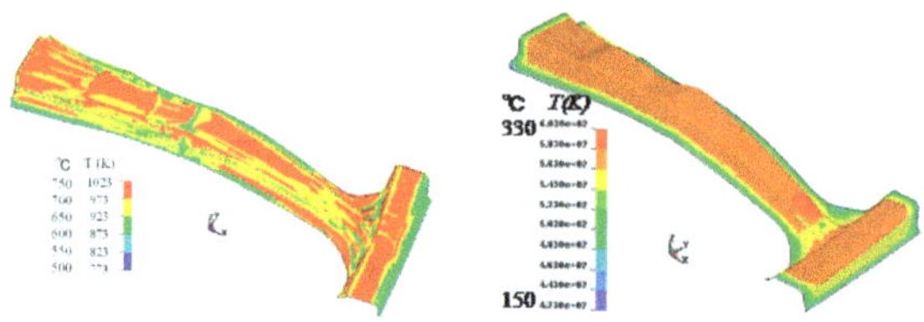

Fig. 5. Temperature after 1.08sec forming stage and 20sec die quenching stage.

The blank temperature after 20sec die quenching was between 330°C (603K) and 150°C (423K). The highest temperature 330°C (603K) existed at a limited region and is a little bit higher than the martensite finish point Mf (233°C). It can be estimated that the martensitic transformation was not finished at this region and a little bit die quenching time may be necessary.

The thickness distribution along a section P1a of the hot stamped center pillar is shown in Fig. 6. The blue line (JSTAMP) denotes the thickness by hot stamping simulation. The vertical red lines are the variation of measured thickness and cycle marks (•:Experiment) indicates the averaged measuring thickness [5]. The thickness obtained by simulation agreed very well with the measured values.

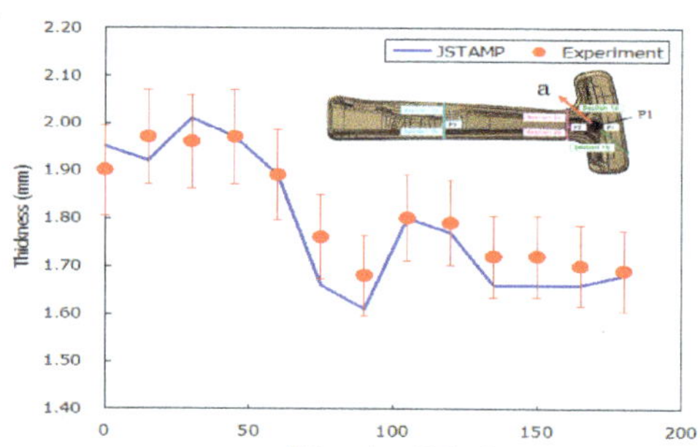

Fig. 6. Thickness distribution computed and measured along section P1a (initial thickness=1.95mm).

Based on the CCT diagram and the relation between cooling rate and vicker's hardness *Hv*, the vicker's hardness *Hv* at two typical points A and B was predicted using the cooling curves obtained by simulation as show in Fig. 7. The predicted hardness at two typical points A and B is about 428 and 464, respectively. The tensile strength at two typical points A and B is predicted to be about 1360MPa and 1472MPa using Eq. (1), respectively.

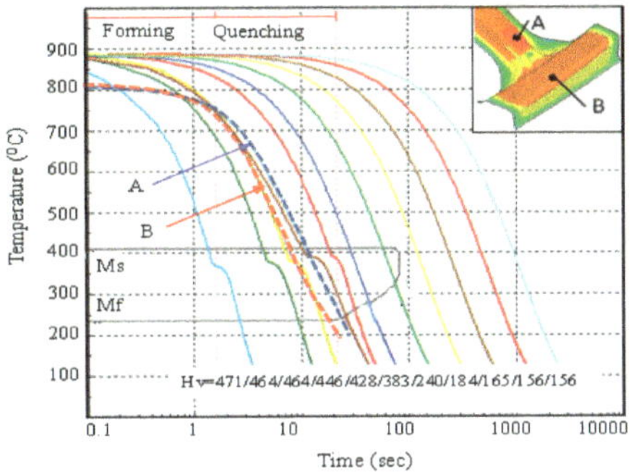

Fig. 7. Hardness prediction based on CCT diagram and cooling curves.

4. Conclusion

A hot stamping simulation system was designed and developed for industrial applications based on the analysis of characteristics of main physical phenomena at multi hot stamping processes. Multi stage hot stamping simulation for a center pillar was efficiently performed using the developed system. The thickness and hardness were accurately predicted compared with experimental results using the developed simulation system.

References

1. G. Sergia, A. Demurov, G. Petrosyan and R. Jobava, On one approach to scientific CAD/CAE software developing process, *Int. J. Comput. Sci. Issues* 7, 9 (2010).
2. G. Bergman, and M. Oldenburg, A finite element model for thermal-mechanical analysis of sheet metal forming, *Int. J. Numer. Mech. Eng.* **59**, 1167 (2004).
3. N.S. Ma and Y. Umezu, Application of Explicit FEM to Welding Deformation, *J. Light Met. Weld. Constr.* **46**, 142 (2008).
4. Naderi and Malek, Hot stamping of ultra-high strength steels, Doctoral Theses, RWTH, Aachen (2007).
5. B. Oberpriller, L. Burkhardt and B. Griesbach, Benchmark-3, *continuous press hardening part B: Benchmark Analysis* (2008).

Design and Development of Hot Forming Front Bumper Beam

X. G. Qin[†], T. Zhang and B. Yang
Dongfeng Motor Corporation Technical Center, Wuhan, China
[†]*E-mail: qinxg@dfmc.com.cn*

In vehicle crashworthiness simulation analysis, anti-crash performance of the front bumper beam assembly of auto body made of 22MnB5 by hot forming and high strength steel by cold forming is compared. Software PAM-CRASH is employed to analyze the energy absorption and the acceleration value of beams with different profiles and thicknesses during the frontal impact process. And the structure of the hot forming bumper beam is optimized to get more light-weight effect. The results show that the use of hot forming beam can not only improve the energy absorption, but also can effectively reduce the weight of the body in white.

Keywords: Hot forming; Beam; CAE; Optimization; Light weight.

1. Introduction

1)The front structure of the vehicle is required to be able to rely on its own deformation to slow down the impact force and absorb the impact energy in order to protect the occupants. Front anti-collision beam assembly is the first important part in energy absorption, which usually consists of outer plate, inner plate and energy absorption box together. To improve the performance of the front bumper beam assembly, it is key to improve the energy absorption capacity [1-2]. At present, the manufacture generally used increases the thickness of the material or uses high performance materials to improve the energy absorption of the assembly. The European Materials Association's survey showed that weight reduction of the vehicle per 100Kg, can save fuel by 6%-8% [3]. Therefore, high strength steel body is an effective way to improve the safety performance of collision experiment and lightweight. However, with the increase of the strength of the steel plate, some problems begin to appear, for example the forming difficulty of the parts, forming spring-back and so on. The emergence of the hot forming technology solves these problems, it can not only satisfy the forming performance of the parts, but also can meet the light weight requirement and improve the safety during collision [4].

In this paper, the impact of the anti-collision performance of a vehicle with common steel and a hot forming front bumper beam is analyzed. Firstly, with reduction of thickness of material and material change of the energy absorbing box, the performance of hot forming beam is not worse than that of the general steel. And then more optimization is done to achieve lighter weight. The optimized model of the crash beam is verified in the vehicle crash simulation analysis.

2. Simulation Analysis and Results

2.1. *Establishment of collision analysis model*

According to the requirement of C-NCAP experiment, the finite element model of the whole vehicle is established. Vehicle impacts the rigid wall at the speed of 64KM/H. The vehicle finite element model is shown in Fig. 1.

(a) (b)

Fig. 1. (a) Top view of the vehicle; (b) Side view of the vehicle.

2.2. *Outer plate thickness selection*

In the original design, materials of the energy absorbing box and the outer plate are both D340DP. In the following optimized designs, material of outer plate is changed to 22MnB5. The materials used for the outer plate in these cases are shown in Table 1.

Table 1. Materials and thicknesses of the outer panel.

	Materials	Thickness(mm)
General steel	D340DP	2.0
CASE1	22MnB5	2.0
CASE2	22MnB5	1.8
CASE3	22MnB5	1.5

After material information is inputted into the vehicle model, the calculation results are shown in table 2. The measurement position of each parameter is shown in Fig. 2.

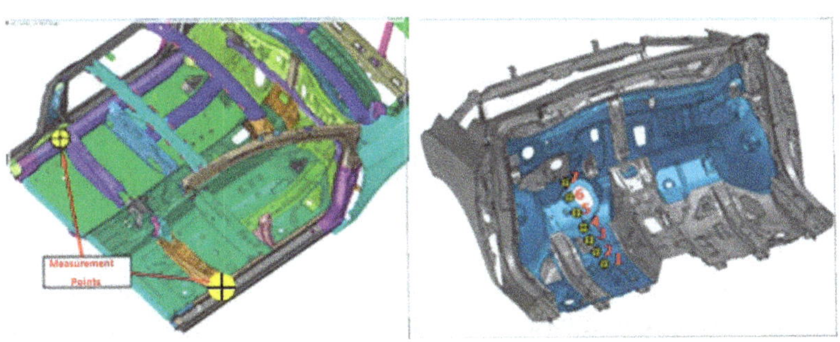

Fig. 2. Diagram of measurement.

From table 2 we can know that, in the frontal impact process, compared with the general steel plate design, the percentage of energy absorption by bumper beam, crash box and front longitudinal rail together in the hot forming design is slightly increased, but the disadvantage is that the peak value of acceleration at the B pillar is increased obviously. In consideration of lightweight effect, the 22MnB5-t1.5 case of the outer panel of the beam is selected.

In order to reduce the peak value of the acceleration at the B pillar, the strength of the energy absorbing box material need to be changed.

Table 2. Calculation results from different cases.

	Energy absorption by Beam+Crash box+FLR (%)	B pillar acceleration (g)		Firewall intrusion (mm)		
		left	right	5	6	7
General steel	44.65	43.2	44.5	64.3	58.84	40.13
CASE1	48.45	46.5	52.6	63.32	63.38	47.85
CASE2	48.73	47.6	48.6	61.65	61.50	45.83
CASE3	47.75	46.2	50.3	69.46	64.58	43.35

2.3. *Energy absorbing box material*

In order to achieve the goal of reducing the peak acceleration value of B pillar, we designed CASE4, CASE4 material scheme: the material of the outer plate of the beam is the same, as 22MnB5, and the energy absorbing box material is changed from D340DP to D260DP, and the material thickness is unchanged.

The calculation results are shown in table 3. As can be seen from the results, in tcase4, with little change in the energy absorption ratio, the peak value of acceleration and the intrusion of the firewall decreased significantly. It shows that this scheme is feasible.

Table 3. Calculation results after material change of energy absorbing box.

	Energy absorption by Beam+Crash box+FLR (%)	B pillar acceleration (g)		Firewall intrusion (mm)		
		left	right	5	6	7
General steel	44.65	43.2	44.5	64.3	58.84	40.13
CASE3	47.75	46.2	50.3	69.46	64.58	43.35
CASE4	46.37	41.8	46.8	65.92	59.26	37.80

2.4. *Lightweight optimization*

On the premise of not reducing the energy absorption of the parts, the structure of the outer plate of the front bumper beam is optimized to reduce more weight.

There are three kinds of solutions, and the change of the three schemes is shown in Fig. 3.

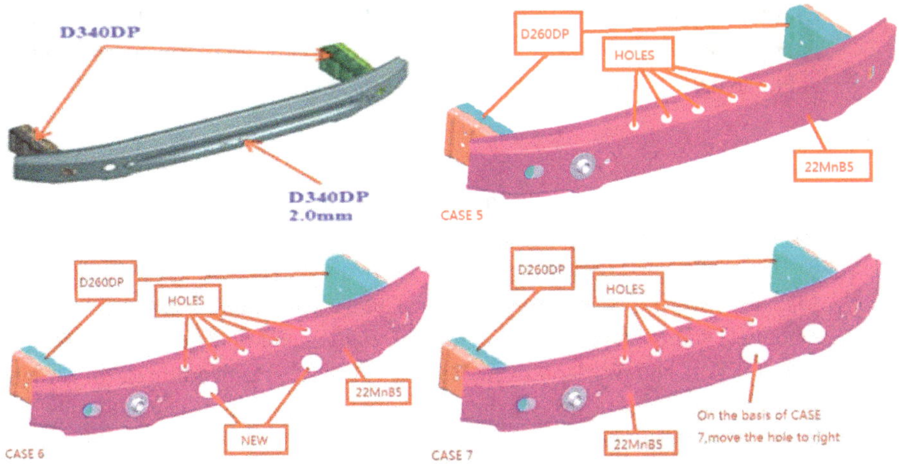

Fig. 3. Schematic diagram of common steel plate design and three kinds of hot forming schemes.

The results of the vehicle CAE analysis of the three schemes are shown in table 4. From the results of the simulation analysis, when CASE6 gets a weight loss by 1.31kg, its the energy absorption ratio, B pillar acceleration and firewall intrusion are still slightly better than the cold forming solutions.

By changing the position of the hole in case 7 compared to case6, the transfer path of the force is changed, and the absorption energy proportion increased obviously, but the B pillar acceleration basically has not changed, with a substantial decline in the amount of invasion, so CASE7 solution is the best for application.

Table 4. Calculation results of light weight solutions.

	Weight (Kg)	Energy absorption by Beam+Crash box+FLR (%)	B pillar acceleration (g)		Firewall intrusion (mm)		
			left	right	5	6	7
Generalsteel	4.3	44.65	43.2	44.5	64.3	58.84	40.13
CASE5	3.16	47.46	45.5	44.1	64.67	58.66	40.41
CASE6	2.99	45.27	42.8	46.0	61.79	59.00	39.23
CASE7	2.99	46.61	43.5	43.9	62.68	57.61	37.90

3. Conclusion

The following conclusions can be obtained through the design and analysis of the hot forming anti-collision beam:

1) The anti-collision performance of the parts can be improved by using hot forming steel and technology, with improvement of the energy absorption ratio from the subsystem in the whole vehicle collision process, and reduction of B pillar acceleration and intrusion of the firewall.

2) Hot forming bumper beam is one of the effective methods of lightweight. In this model, the weight of the outer plate of the anti-collision beam is reduced from 4.3kg to 2.99kg, and the weight loss ratio reaches 30.5%.

References

1. H.T. Sun, P. Hu, N. Ma, G.Z. Shen, B. Liu and D.L. Zhou, Study on application of hot forming high strength steel of body parts based on crash, *Chin. J. Mech. Eng.* **23**, 8 (2010).
2. N. Ma, P. Hu, G.Z. Shen, Model and numerical simulation of hot-forming. International symposium on automobile steel, *Beijing Metall. Ind. Press* 2009.
3. C. Park, C. Kan and W. Hollowell, Investigation of crashworthiness of structural composite components in frontal and side NCAP tests, SAE Technical Paper 2013-01-0650, 2013, doi:10.4271/2013-01-0650.
4. S. Kumar and P. Deb, Improving Side Crash Performance of a Compact Car via CAE, SAETechnical Paper 2014-01-0546, 2014, doi: 10.4271/2014-01-0546.

Statistic Modeling and Optimization of the Sheet Performance in Resistance Heating Process of Hot Stamping

X. L. Li†, C. Zhang, L. Chen and M. Wang

Department of Material Science and Engineering,
Hubei University of Automotive Technology, Shiyan, Hubei, CN
†Email: nanliu@outlook.com

A superior performance is the primary target to which the hot-stamping process engages. Resistance heating is an effective method to higher quality that has been proved by researches. Generally, finding the available domain for the match of factors (ADMF) is the key step to obtain a superior performance. However, the ADMF is not clear about a special superior set-up system. The ADMF of principle operational factors was primarily detected by the full factorial design and the response surface methodology (RSM). These statistic models were confirmed by experiments. A heating specification was prepared based on the statistic models.

Keywords: Hot-stamping; Resistance heating; Response surface.

1. Introduction

Lightweight of automobile has already been validated in all over the world for serval years for which environmental protection and energy saving require. Hot stamping technique, as an effective approach to obtain lightweight of automobile with the rapid development, has been stepping into people's sight. It is the basic for hot stamping to make the austenitizing process complete and evenness by heating equipment. The most common heating technologies implemented are the hearth heating, resistance heating and induction heating. Among these, the most important differences lie in the heating properties. As a proven technique, which is widely used, hearth heating is being limited cause of its low heating rate and relatively high cost [1]. The realization of economic advantage of heating blank by materials resistance using joule effect has spurred the growth of resistance heating over the recent years. The composition and properties of heating workpiece are strongly influenced by the heating rate. Control of heating rate is important in heating process where typically high rate is desirable.

During the heating process, a high heating rate contributes to fine grain and a limited time of heating. Thus, the distributed uniformity of structure which influences the strength and ductility heavily cannot be obtained without fitness holding time. Meanwhile, the endpoint temperature also has a huge impact on the grain size. To obtain the strength-ductility balance, it's necessary to control the relevant process parameters to maximize the strength and ductility simultaneously on which the quality of a hot-stamping is based. It has been proved by serval researchers that efficient use of the statistical design of experimental techniques is a scientific way to combine the scientific methodology with the productive procedure [2~4]. Hence, in this investigation, the design of experiments was used to conduct the process parameters. The study was carried out in two steps. In the first step, second order quadratic model for the prediction of strength and ductility was developed from the date obtained by conducting the experiments. In the second step, the process parameter were optimized using response surface methodology (RSM) optimization technique to obtain the maximize strength and ductility.

2. Scheme of Investigation

2.1. *Develop the design matrix*

The central composite design of second order was found to be the most efficient tool in RSM to establish a mathematical relation of the response surface using the least possible number of experiments without losing it accuracy [5]. Considering the ranges of factors, a three-factor, three-level and central composite design matrix was used to optimize the experimental conditions. With a view to achieve the above-mentioned aim, statistically designed experiments based on the factorial technique were used to reduce the cost and time, as well as to obtain the required information about the main and interaction effects on the response parameters. Table 1 presents the ranges of factors considered and Table 2 shows twenty sets of coded conditions used to form the design matrix. Thus, the twenty experiments runs allowed the estimation of linear, quadratic and two-way interactive effects of the variables. The method of designing such matrix is dealt elsewhere [6,7]. For the convenience of recording and processing experimental date, the upper and lower levels of factors have been coded values of any intermediate values can be calculated.

Table 1. Important parameters and the working range.

Parameter	Notation	Level		
		Low (-1)	Mid (0)	High (+1)
Heating rate/(°C • s^{-1})	HR	20	50	70
Endpoint temperature/(°C)	ET	850	900	950
Holding time /(s)	HT	20	35	50

2.2. *Conducting the experiments*

The test material used in the investigation was Usibor 1500P steel, widely used for the hot stamping process. The 30mm×100mm×1.6mm heating sample was designed considering the power consumption. Meanwhile, the tensile specimens were designed according the size of heating sample. The shorter gauge length causes a huge error to the outcome of ductility during the tensile test. We choice a method of Mathematical Morphology to obtain a high accuracy, with the help of Mathematical Morphology, ductility can be quantified by pixels in the image that captured by Industrial View Camera. The experiments were conducted at random order to avoid systematic errors infiltrating the system.

Table 2. Experimental design matrix.

No.	Coded value			Original value		
	HR	ET	HT	HR	ET	HT
1	-1	-1	-1	20	850	20
2	1	-1	-1	70	850	20
3	-1	1	1	20	950	20
4	1	1	-1	70	950	20
5	-1	-1	1	20	850	50
6	1	-1	1	70	850	50
7	-1	1	1	20	950	50
8	1	1	1	70	950	50
9	-1	0	0	20	900	35
10	1	0	0	70	900	35
11	0	-1	0	45	850	35
12	0	1	0	45	950	35
13	0	0	-1	45	900	20
14	0	0	1	45	900	50
15	0	0	0	45	900	35
16	0	0	0	45	900	35
17	0	0	0	45	900	35
18	0	0	0	45	900	35
19	0	0	0	45	900	35
20	0	0	0	45	900	35

3. Prediction Model for Product of Strength and Plasticity

In this study, the response surface model-building technique was utilized to predict high product of strength and plasticity in terms of the process parameters for hot-stamping. The details of the model-building technique are discussed below.

In the practical applications of RSM, it is necessary to develop a fitting model for the response surface, and it is typically driven by some unknown physical mechanism. For prediction, the RSM is practical, economical, and relatively easy for use [8]. RSM consists of the experimental strategy for exploring the space of process independent variables, empirical statistical modeling for developing an appropriate relationship between the outcome and the process input, and optimization methods for finding the levels or values of the process variables that produce desirable values of the responses [9]. In this investigation, to correlate the process parameters and the product of strength and plasticity, a second order quadratic model is developed to predict the product. Representing the product of strength and plasticity "P", the response function P can by expressed as P=f(R,C,T), where, R is the heating rate, C is the endpoint temperature, T is the holding time. The model chosen includes the main effects and interaction effect of all factors. For the three factors, the selected polynomial can by expressed as

$$Y = b_0 + b_1R + b_2C + b_3T + b_{11}R^2 + b_{22}C^2 + b_{33}T^2 + b_{12}RC + b_{13}RT + b_{23}CT \quad (1)$$

To estimate the regression coefficients, a series of experiments were designed based on experimental design techniques. With the help of Minitab 15.0, the significant coefficients (at 95% confidence level) were confirmed, the final model was developed using only these coefficients and the final mathematical model to estimate product of strength and plasticity is given below:

$$P = 3.74R + 9.57C + 14.76T - 0.025R^2 - 0.005C^2 - 0.037RT - 0.0 - 4470.18 \quad (2)$$

The method of analysis of variance technique (ANOVA) was used to test the adequacy of the model and the results are presented in Table 3. The determination coefficient (R^2) indicates the goodness of fit for the model. In this case, the value of the determination coefficient ($R^2 = 0.93$) indicates that more than 90% of the total variations are explained by the model. The value of adjusted determination coefficient (adjusted $R^2 = 0.89$) is also high, which indicates a high significance of the model. Meanwhile, the value of probability of K, R, R^2, C^2, RT, RT in Table 3 for the model is less than 0.05, which indicates that the model is significant. Also, there are main factors such as C and T with the probability more than 0.05, but the main factors is the base of model, it is not necessary to remove the main factors from model unless there is a clear significance of its interaction effect. So the terms including R, C, T, R^2, C^2, RT and CT were found to be significant.

Lack of fit is not significant as desired [10]. The normal probability plot of the residual product is shown in Figure 1 (a), which reveals that the residuals are falling on the straight line, indicating that the errors are distributed normally [11]. All the above considerations indicate an excellent adequacy of the regression model. Each observed value compared with the experimental values is shown in Figure 1 (b).

Table 3. ANOVA results for product.

Source	Coefficients	Standard error	T value	P value
K	147.266	2.067	71.251	0.000
R	4.468	1.949	2.293	0.041
C	-3.674	1.949	-1.886	0.084
T	-1.818	1.949	-0.933	0.369
R^2	-15.792	3.445	-4.584	0.001
C^2	-12.678	3.445	-3.680	0.003
RT	-13.820	2.179	-6.343	0.000
CT	-11.022	2.179	-5.059	0.000
Standard deviation	6.16215	R^2		93.07%
Lack of fit	0.083	Adjusted R^2		89.03%

4. Result and Discussion

The developed mathematical model can be effectively used to predict the range of parameters used in the investigation by substituting their respective values in coded form. Based on these models, the main and interaction effects of the process parameters on the product of strength and plasticity were computed and plotted in graphical form as shown in Figure 2.

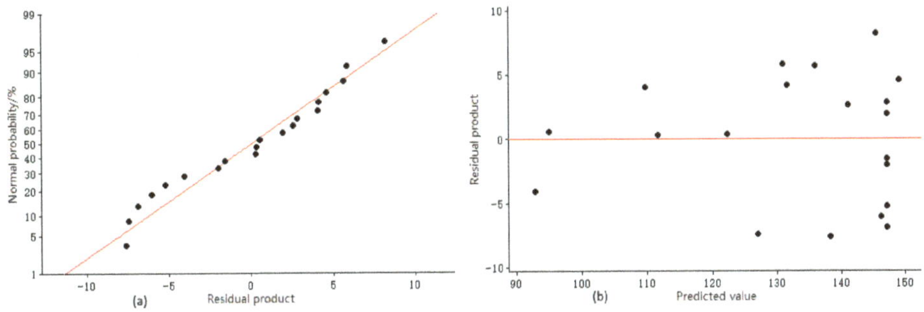

Fig. 1. Normal probability plot of product and residual product plot of predicted value.

4.1. *Effect of heating rate and holding time on product*

Figure 2 (a) shows the effect of heating rate and holding time on product of strength and plasticity. From the figure, it can be seen that the product decreases from the bottom right corner to the left three corners. It indicated that there are vertexes on response surface. Meanwhile, the heating rate has directly relationship with the product of strength and plasticity, i.e., when there is a short holding time, if the heating rate increase, the product increases, and vice-versa. However, the holding time has inversely relationship with the product, i.e., when there is a high heating rate, if the holding time decreases, the product increases, and vice-versa.

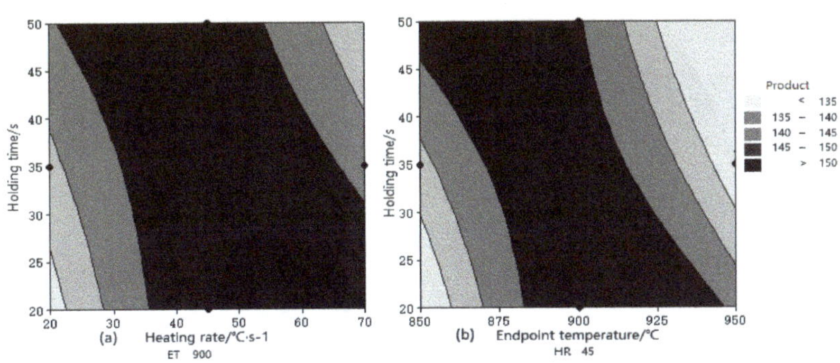

Fig. 2. Effect of parameters on product.

The holding time has inversely relationship with the product, i.e., when there is a high heating rate, if the holding time decreases, the product increases. At higher heating rate such as 50°C /s~70°C /s, the nucleation rate can be accelerated by high degree of over-heat, it ensures finer initial grain size. Furthermore, there is limited time for the grain to grow, with the condition of short heating time, finer grain can be obtained. But, with the increasing of holding time, the original fine grain continues growing, leading to the consequence of coarse grain. So the product of strength and plasticity decreases clearly. At lower heating rate such as 20°C /s~40°C /s, low degree of over-heat can't proves grains enough impetus to grow faster, with the holding time increasing, the normal grains becomes more and more uniform, that is the key reason why product increases with the extending holding time. Moreover, if the heating rate below a certain level between 40°C /s~50°C /s, there will not be finer grain even if the longest holding time. That shows the significant effect heating rate causes during the heating process.

4.2. *Effect of endpoint temperature and holding time on product*

Figure 2 (b) shows the effect of endpoint temperature and holding time on product of strength and plasticity. According to Table 3, the interaction of endpoint temperature and holding time has a significant effect on the product (Mean Coefficients=-11.022, P value=0). As to shorter holding time, the product increases with the increasing of endpoint temperature, and there should be lower endpoint temperature to cooperate with long holding time. That is to say, temperature has a positive effect on the growth rate of grain, if the growth rate is low, there should be longer holding time to ensure the high product, and vice-versa. Meanwhile, a phenomenon of insensitivity between product and holding time has been observed at 900°C. That may indicate the thermal equilibrium when the temperature up to 900°C.

5. Optimizing the Process Parameters

To obtain the influencing tendency of the process parameters on product of strength and plasticity further, the three-dimensional diagrams are plotted under certain processing conditions. Eqn. (3.2) is plotted in Figure 3 as surface plots for each process parameters.

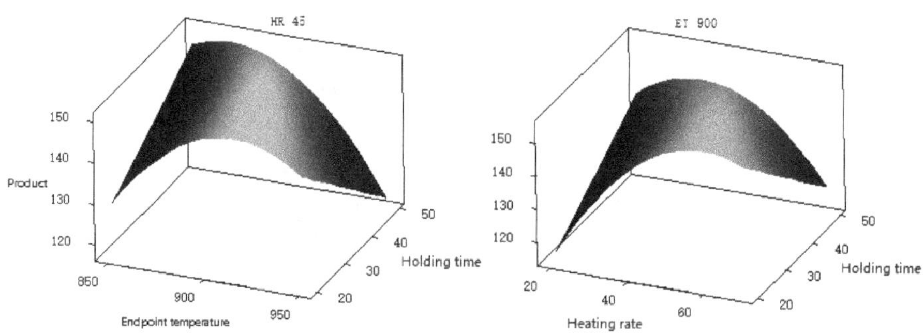

Fig. 3. Response graphs.

It is clear from the Figure 3 that the product of strength and plasticity falls and increases with the process parameters such as heating rate, endpoint temperature and holding time. The valley of the response plots gives the maximum product. These response valleys can help in the prediction of the response at any zone of the experimental domain [12]. According to the second order response surfaces, the plot will be more complex than it occurs with first order models. Once the stationary point is confirmed, it is usually necessary to determine the response surface nearby the point. Contour plot (Figure 2) always be the most effect way to classify this. [13] Contour plots play a very important role in the study of response surface. By solving Eqn. (3.2) and analyzing the profile of the response surface and the corresponding contour plot, the value of product obtained is 155.5 MPa%, which is found to be maximum and is located at the heating rate of 59°C/s, endpoint temperature of 915°C and the holding time of 20s. Three confirmation experiments were conducted to compare the experimental results with the prediction. The mean experimental product of strength and plasticity was obtained as 149 MPa%. The error percentage of 4.8% showed an excellent prediction of the model.

6. Conclusion

It is experimentally seen that the use of design of experiments coupled with the statistical fitting model strategy is simple, effective and efficient way to develop high quality hot-stamping process. Optimization of the processing variables and responses in the resistance heating process is achieved through appropriate construction of the process model in RSM. Based on the experimental results, the following conclusions have been drawn.

1) A mathematical model is developed using response surface methodology to estimate the product of strength and plasticity. The model can by effectively used to predict and optimize the performance of hot stamping sheet with resistance heating method.
2) The interaction of three factors is proved to have significant effect on the product of strength and plasticity, even the single effect is not that remarkable. The interaction of heating rate and holding time has the highest effect (Coded Coefficients=-13.82， P value=0), and the single effect of holding time has the lowest effect (P value=0.369).
3) The maximum product of strength and plasticity of 155.5 MPa% can be obtained using the process parameters such as heating rate of 59°C /s, endpoint temperature of 915°C and holding time of 20s.

Acknowledgments

This work is funded by Hubei Key Laboratory of Advanced Technology for Automotive Components (2013-7), the National Nature Science Foundation of China (51205116), the Natural Science Foundation of Hubei Province (2014CFB628), the Outstanding Young Scientific & Technological Innovation Team Plan of Colleges and Universities in Hubei Province (T201518).

References

1. J. Lechler, M. Merklein and M. Geiger, Determination of thermal and mechanical material properties of ultra-high strength steels for hot stamping, *Steel Res. Int.* **79**, 98 (2008).
2. P.K. Palani and N. Murugan, Development of mathematical models for prediction of weld bead geometry in cladding by flux cored arc welding, *Int. J. Advanced Manufacturing Technol.* **30**, 669 (2006).

3. N. Srimath and N. Murugan, Development of Mathematical Models for Prediction of Weld Bead Geometry in Cladding Mild Steel Valve Seat Rings by PTAW, *Procedia Engineering* **38**, 15 (2012).
4. A. K L N and V. Balasubramanian, Comparison of RSM with ANN in predicting tensile strength of friction stir welded AA7039 aluminium alloy joints, *Transactions of Nonferrous Metals Society of China* **19**, 9 (2009).
5. S. Suryakumar, K.P. Karunakaran and A. Bernard, Weld bead modeling and process optimization in Hybrid Layered Manufacturing, *Computer-Aided Design* **43**, 331 (2011).
6. D.C. Montgomery, *Design and analysis of experiments.* (John Wiley and Sons, 1991).
7. I. Miller and J.E. Freund, *Probability and statistics for engineers.* (Prentice Hall, 1990).
8. J.S. Kwak, Application of Taguchi and response surface methodologies for geometric error in surface grinding process, *Int. J. Machine Tools & Manufacture* **45**, 327 (2005).
9. U. Çaydaş and A. Hasçalık, A study on surface roughness in abrasive waterjet machining process using artificial neural networks and regression analysis method, *J. Mater. Process Tech.* **202**, 574 (2008).
10. I. Miller and J.E Freund, *Probability and statistics for engineers 3th.* (Prentice Hall, 2001).
11. M. Sharpe, Optimization of friction welding of tube to tube plate using an external tool, *Structural & Multidisciplinary Optimization* **42**, 449 (2010).
12. Atikah, H. Awang and R. Nani, Application of response surface methodology in the optimization of cutting conditions, University Malaysia Pahang (2009).
13. G.E.P. Box and K.B. Wilson, On the Experimental Attainment of Optimum Conditions, *Breakthroughs in Statistics* 270 (2012).

Study on Variable Gradient Characteristics Hot Stamping Under Non-Uniform Temperature Field

L. Kong†, C. Y. Liu and Y. Peng

National Engineering Research Center for Equipment and Technology of Cold Strip Rolling, Yanshan University Qinhuangdao, Hebei 066004, P.R. China
†E-mail: kongling@ysu.edu.cn

The traditional hot forming product has high strength, but the plasticity is usually low and comprehensive mechanical performance is not high. In order to improve the comprehensive mechanical properties, the variable gradient characteristics of forming technology has been drawing more and more attention and has become a research hot spot in hot forming control technology. The variable gradient characteristic is different from the traditional pressure processing and has its own characteristics in numerical simulation. It belongs to the problem of the intersection of pressure processing, material science and computational mechanics. In this paper, the main numerical simulation differences between the hot stamping process and traditional pressure processing are summarized. We take a non-uniform initial heating temperature field to form the variable gradient component by hot stamping as the research object, and a finite element model is established in order to analyze the key influence factors of the hot stamping to form the variable gradient characteristics parts. The initial sheet is divided into high temperature and low temperature regions. The simulation and the forming process satisfy certain stamping and quenching conditions. The simulation results shows that stamping speed, the die temperature and the holding time are the important influence parameters in the variable gradient characteristics hot forming process.

Keywords: Hot forming; Performance gradient distribution; Numerical simulation; Non-uniform temperature.

1. Introduction

The strong demand of automotive lightweight and safety performance has given rise to a large number of applications in high strength steel plates in the car, and the hot forming process of high strength steel became an important mean to achieve automotive lightweight and ensure safety performance. The red hot sheet metal forming and its quench cooling in the mould at the same time are the important characteristics in hot stamping which can get the high-strength sheet metal parts with 1500MPa–1800MPa. It's widely used in security parts, such as car door anti-collision, the front and rear bumpers and also used in the production of car body structure including column A, column B and center channel [1-2].

With the popularization and application of ultra-high strength steel and hot forming technology in engineering, people came to realize that auto parts aren't a uniform distribution in use state. Lots of collision experiments established that in order to ensure the overall stiffness of auto bodies and collision energy absorption effect, different parts should match different use requirements, and the performance of the final parts gradient that based on the load distribution form become the goal that automotive anti-collision core parts pursuing [3]. So hot forming technology researching with material gradient feature has broad prospects and significance.

2. Hot Stamping with Variable Gradient Features

Hot Stamping with variable gradient features base on hot forming technology, usually it establishes a non-uniform temperature field for the stable and multi-scale polyphase structure to implement the organization and performance of parts present gradient features. There are two kinds of methods to realize the variable gradient performance parts preparation: one kind is regulating the heating process to establish the non-uniform temperature field, to get the final rendering performance gradient distribution by steel plate initial microstructure differences. This method puts forward higher requirements for heating equipment. The German Benteler company has developed a new type of heat treatment furnace which can realize partial heating of the steel plate; another method is regulating the cooling process to establish the non-uniform temperature field. It achieves final microstructure difference of parts by the control of cooling speed in quenching stage for getting the gradient distribution performance [4-5]. This method requires the hot forming die needs to design special heating and cooling pipe, by reasonable control of steel plate parts in the process of quenching cooling rate to obtain the material microstructure presents gradient distribution. The cost of the mould is high and the service life span is short. In this paper the first kind of control method is adopted in the simulation.

3. The Simulation Characteristics of Variable Gradient Characteristics Hot Stamping

The deformation analysis of traditional sheet metal generally refers to triple coupling nonlinear geometry, physics and boundaries, mainly aiming at the numerical simulation of metal deformation at normal temperature. The plastic load path in hot stamping usually accompanied by a material of elastic and plastic instablity, partial introverted outside surface bifurcate deformation, until the break the form of shear zone. At present the numerical simulation of traditional automobile plate cold stamping process is mature, a series of commercial software are widely used in this field [6].

Firstly, the numerical simulation analysis of traditional cold forming is focused on the stress field and stain field, but the hot forming with high load, high temperature and high complex in workpiece temperature and microstructure transformation, which makes the temperature field and the phase transformation field become very important. Variable gradient features in hot forming process with non-uniform temperature field make it more complex. The numerical simulation of high strength steel plate in variable gradient features hot forming process is a several coupling process with thermal, mechanical and phase transformation, the analysis process is shown in Figure 1.

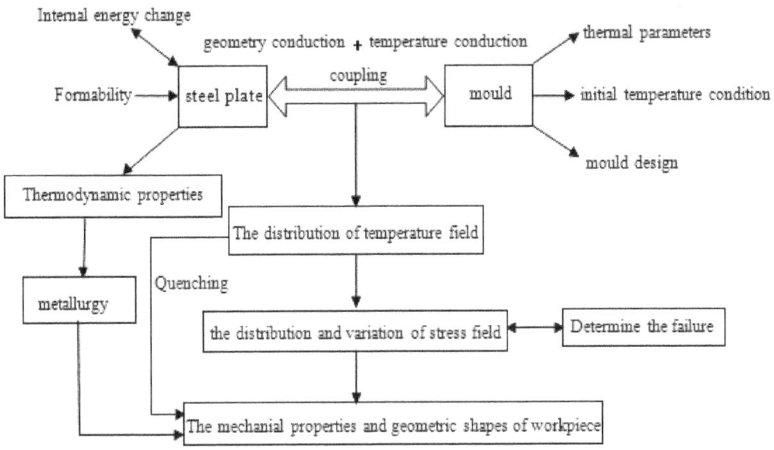

Fig. 1. Coupling analysis method of variable gradient features hot forming hot forming process.

Second, traditional stamping numerical simulation usually only involves material geometric large deformation, but gradient features hot forming simulation refers to the multiscale problems with space and time, involving more kind of simulation method, as shown in figure 2.

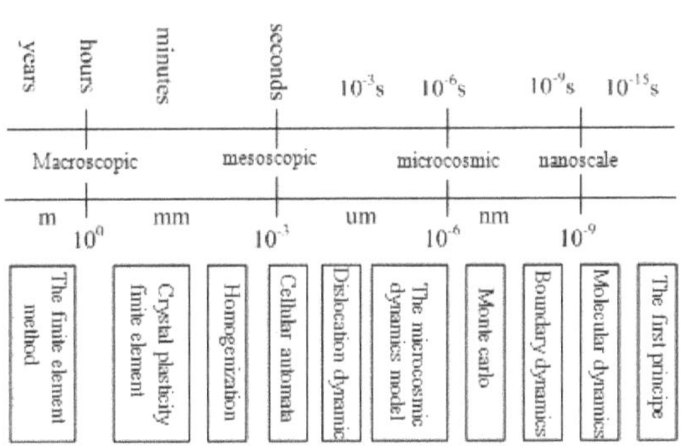

Fig. 2. Hot forming numerical simulation of the space and time scales.

4. The Variable Gradient Characteristics Hot Stamping Numerical Simulation

4.1. *The establishment of finite element model*

The finite element model was established based on U-shaped cross section of important safe parts B-pillar (Fig. 3). The bottom die was set to the rigid body and the upper die was stamped with fixed speed. The material of blank was BR1500HS with the thickness of 2mm. Initial temperature field of sheet metal can be divided into high and low temperature zone with initial temperature 850°C and 800°C. After heating, high temperature zone of blank have fully austenite and the microstructure of low temperature zone steel hold the ferrite and pearlite. The simulation used MSC. Marc software, adopt Thermal-Structural material model and generated mesh with solid shell element.

Fig. 3. The finite element model of variable gradient features.

In the simulation, the value of process parameters was selected under different temperature such as heat transfer coefficient, young modulus and specific heat in order to improve the accuracy of the finite element analysis. Part of the material parameters is shown in Table 1.

Table 1. The material parameters.

BR1500HS					
Density（kg/m3）			7800		
Poisson's ratio			0.3		
Friction coefficient			0.2		
Temperature (°C)	20	200	400	600	800
Elasticity modulus (GPa)	207	180	150	110	90
Specific heat（J/kg·K）	470	462	474	699	969
Heat transfer coefficient（W/m·K）	31	34	40	42	43

4.2. *The analysis of variable gradient features in hot forming numerical simulation*

The gradient organization of variable gradient hot forming process was based on the non-uniform temperature field. So in the paper the influence of punching speed, mold temperature and holding time on the non-uniform temperature field is analyzed. The conditions of simulation process are shown in Table 2.

Table 2. The conditions of simulation process.

Process parameter	Variable value		
Pressing speed mm/s	100, 160, 200	160	160
Mold temperature °C	25	25, 125, 225	25
Holding time s	6	6	0, 3, 6, 9

4.2.1. *Punching speed*

As shown in Figure 4, it's the relation between temperature extremum and pressing speed of high temperature and low temperature zone in simulation. It shows the forming temperature of parts increased with the increase of pressing speed, and the blank has been in a non-uniform temperature field in the whole pressing process. There is temperature difference at about 100°C between high temperature and low temperature zone to guarantee the cold speed difference. With the increase of pressing speed, temperature minimum of blank increase obviously, temperature difference decreased.

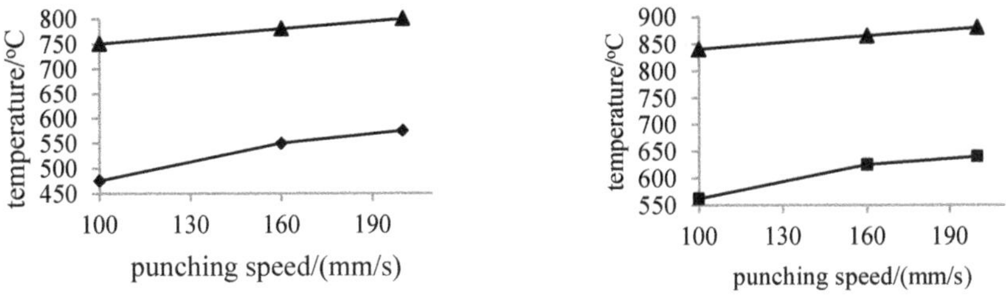

a) low temperature zone. b) high temperature zone.

Fig. 4. The relation between temperature extremum and pressing speed.

4.2.2. *Mold temperature*

The effect of hot forming mold is to ensure blank forming and metal sheet quenching take place at the same time. Mold cooling capacity had a significantly influence on temperature field. As shown in Figure 5, the higher the initial temperature of the mold, the earlier moment that temperature field of blank become steady state and the temperature difference of non-uniform temperature field were smaller. The martensitic transformation critical cooling rate is 30°C /s, the high temperature cooling speed must be greater than it in order to complete the transformation from austenitic to martensite fully. In low temperature zone, the temperature difference was small and cooling rate was low which results in pearlite and ferrite retained.

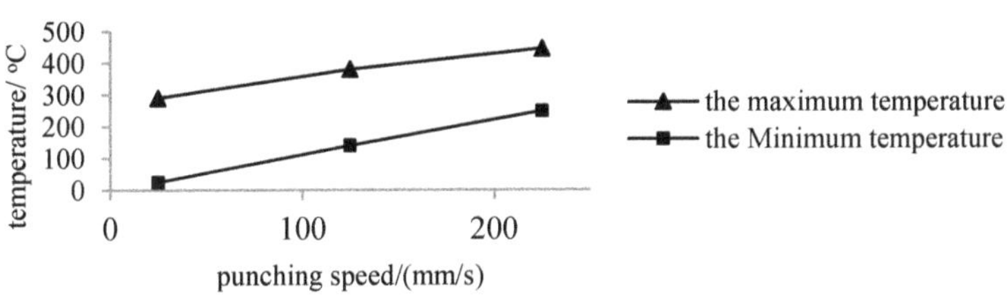

Fig. 5. The relation between temperature extremum and initial temperature of the mold.

4.2.3. *Holding time*

As shown in Figure 6, it is temperature distribution of blank when the pressing finished with different holding time (3 s, 6 s, 9 s). From the figure, when the pressing process ended, the high temperature zone of blank was 830°C, low temperature zone is about 780°C. After holding 3s time, with the effect of mold, temperature of blank decreased rapidly, the decline of high temperature and low temperature zone are more than 300°C, temperature difference between zones decreased, The lowest temperature appeared in the wing side and bottom of the u-shaped parts. This is mainly because the high thermal conductivity in the wing side and bottom of the u-shaped parts, caused by top die and bottom die clamping force. When the holding time from 6s increased to 9 s, The highest temperature of blank reduced from 325°C to 166°C and the temperature is lower than the end of the martensitic transformation temperature, the transformation of high temperature zone of the u-shaped parts ended, the temperature difference between high temperature and low temperature zone because of initial heating in the pressing quenching stage almost disappeared.

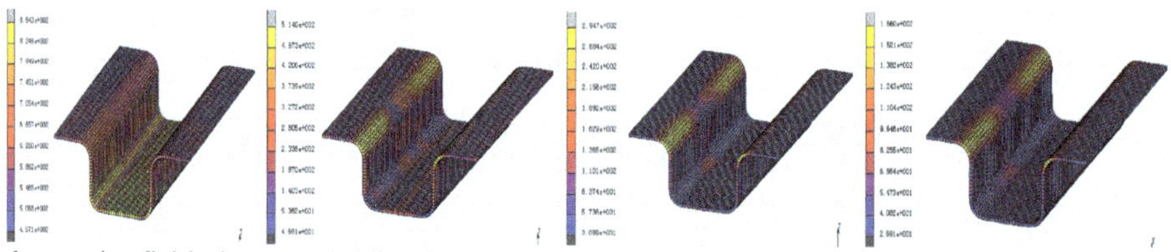

the pressing finished. holding time (3 s). holding time (6 s). holding time (9 s).

Fig. 6. Temperature distribution of blank.

5. Conclusion

In this paper, the gradient characteristics hot stamping method was studied, with the help of numerical simulation technology. It is found that the stamping speed, die temperature and holding time are the important parameters that impact the properties of the hot forming parts. Too large or too small stamping speed will have a great influence on the microstructure and the mechanical properties of the parts. To ensure the stability of the variable gradient characteristics hot stamping process, the die quenching temperature need to maintained at a lower temperature; too high temperature will impact on the formation of ideal martensite; because the lower cooling speed, will affect the quality of the variable gradient parts. Hence, in order to ensure the sufficient cooling speed, more cooling pipe need to be installed in the mold corresponding to the high temperature area, and the holding pressure quenching time should not be too long.

References

1. J.P. Lin, L.Y. Wang, H.B. Tian, G.H. Sun and Z.B. Wang, Research and progress of hot stamping of ultrahigh strength steel, *Hot Working Technology* **37**, 140 (2008).
2. J.G. Li, J.D. Zhang, H.L. Huang and S.Y. Zhang, Research status and development trend of high strength steel for automotive, *Mater. Rev.* **26**, 397 (2012).
3. M. Zhu, L. Ji and H. Ye, Optimization of yield stress distribution in B pillar reinforcement panel regarding side crashworthiness, *J. Jili Univ. Eng. Tech. E.* **41**, 1210 (2011).
4. K. Mori, Smart hot stamping of ultra-high strength steel parts, *T. Nonferr. Metal. Soc.* **22**, 496 (2012).

5. K. Mori, T. Maeno and K. Mongkolkaji, Tailored die quenching of steel parts having strength distribution using bypass resistance heating in hot stamping, *J. Mater. Process Tech.* **213**, 508 (2013).
6. W.K. Liang, L. Wang, Y. Liu, Y.L. Wang and Y.S. Zhang, Hot stamping parts with tailored properties by local resistance heating, *Procedia Eng.* **81**, 1731 (2014).

Analysis of Hot Forming of Ultra-High Strength Steel Hollow Part Using Variable Strength and Variable Cross Technology

W. Xiao, W. J. Tao, B. Zhu and Y. S. Zhang†

State Key Laboratory of Materials Processing and Die and Mould Technology, Huazhong University of Science & Technology, Wuhan 430074, China

†E-mail: zhangys@mail.hust.edu.cn

A new forming process of hollow torsion beam using high strength steel tube and resistance heating was developed to replace electrical heating furnace. In this process, a quenchable steel tube was rapidly resistance-heated to improve the formability. To obtain the effect of the forming processes on the performance of torsion beam, numerical simulation and experiments were carried out to analyze the deformed shapes of the CTBA part, damages, stress, and the temperature fields.

Keywords: Tube forming; Resistance heating; Numerical simulation; high strength steel tube; Hot stamping.

1. Introduction

For the reduction in weight of automobiles, high strength steel hollow products [1, 2, 3] are increasingly employed for automobile parts. On one hand, hollow structures are advantageous to rigidity. On the other hand, high strength hollow parts are useful for cabin ones to protect passengers for collision.

In the hot stamping process, the base material with strength around 500-600 MPa tube is heated up to around 930°C, stamped at around 900-800°C, and cooled down rapidly to obtain high strength over 1400 MPa through the martensitic phase transformation.

There are more process parameters to be optimized in the hot stamping process than the conventional cold or warm forming, because the hot stamping involves the stamping at the elevated temperature and the cooling as well. In addition, the unique nature of steel or phase transformation is involved.

In this paper, the hot stamping process and resistance heating was applied to the coupled torsion beam axle which is a component of the wheel suspension systems, and is formed from a tubular beam. The requirement for the coupled torsion beam axle part is both high strength and great dimensional accuracy in addition to the tubular formability. What's more, different strength should be distributed in different regions. As the part is quite complex, it is very challenging to attain required strength by the cold forming of AHSS. We use different quenching methods to solve the problem. The base material of the initial tubular beam is boron-added steel. The finite element analysis is used here to study the deformed shapes of the CTBA part, damages, and stress, under difference sets of temperature [4, 5, 6].

2. The Finite Element Simulations

2.1. *Materials & Parameters*

The tubular specimen is made of 22 MnB5 boron-added steel with the chemical composition of 0.188 % C, 0.37 % Si, 1.11 % Mn, and 21 ppm B. The boron steel is commercially available with good hardenability or subject to transform into martensite from austenite with moderate cooling speed. For the reliable finite element analysis of the hot stamping process, the mechanical and thermal data should be provided. Table 1 summarizes important thermal properties used for the simulation of the hot stamping process. Table 2 summarizes thermal-mechanical material properties for 22MnB5 steel.

Table 1. Thermal properties used for FE analysis for different steps.

Process	Parameter	Values
	Resistance heating temperature	930°C
Heating		
	Convection coefficient	0.02N/s/mm/°C
	Air temperature	20°C
Transfer		
	Convection coefficient	0.02N/s/mm/°C
	Conduction coefficient	7N/s/mm/°C
Forming	Convection coefficient	0.02N/s/mm/°C
	Die temperature	20°C
Quenching	Conduction coefficient	5N/s/mm/°C

Table 2. Thermal-mechanical material properties for 22MnB5 steel.

Temp [C]	20	100	200	300	400	500	600	700	800	900	1000
E [MPa]	212	207	199	193	166	158	150	142	134	126	118
v	0.284	0.286	0.289	0.293	0.298	0.303	0.310	0.317	0.325	0.334	0.343
K [W/Mc]	30.7	31.1	30.0	27.5	21.7		23.6		25.6		27.6
Cp [J/kg]	444	487	520	544	561	573	581	586	590	596	603

2.2. *Numerical analysis*

Finite element analysis was carried out to simulate the effect of process conditions, especially the cooling time during the quenching process on the deformation behavior. The FE software Deform3D-HT was used. The thermomechanical simulations could consider the heat transfer between tools and tubular specimens, the heat convection between tool and air, or specimen and air.

The finite-element model for our stamping analysis consisted of upper and lower dies, a boron steel tube, and inserts. The boron steel tube, 1200 mm×102 mm×2.8 mm (length × diameter × thickness), was modeled as two layered solid elements. Figure 1 shows the finite-element model and its boundary conditions.

When forming, the inserts at both sides intrude into the tube at a speed of 20 mm/s: the upper die moves in the z-axis direction at a speed of 20 mm/s, and the lower die is fixed. 70000 continuum elements with two element layers in the thickness direction were used for the blank, while rigid body element was assumed for other tools. A half of the specimen was modeled using the symmetry of the problem. Coulomb friction coefficient of 0.3 was assumed, other parameters are listed in table1 and table2.

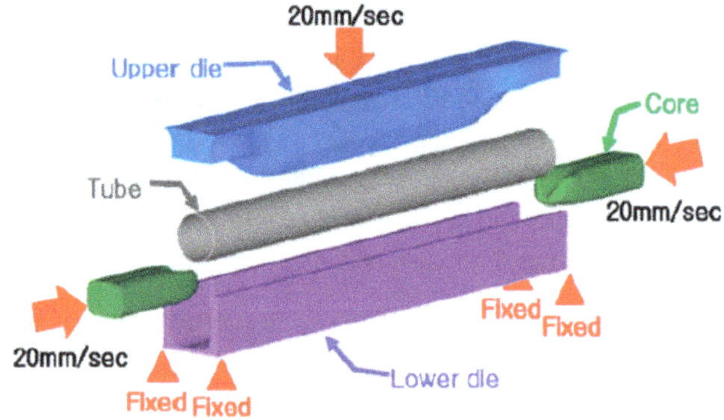

Fig. 1. Finite element modeling for the stamping of the CTBA part with boundary conditions.

3. Results

Hot stamping simulations of the tubular part showed that the highest and the lowest temperatures just after forming step were 905 and 629°C, respectively and the martensitic transformation started at 325°C. The direct hot stamping with die quenching showed good shape accuracy, only in some stress concentration areas show damage. Figure 2 shows damage after forming.

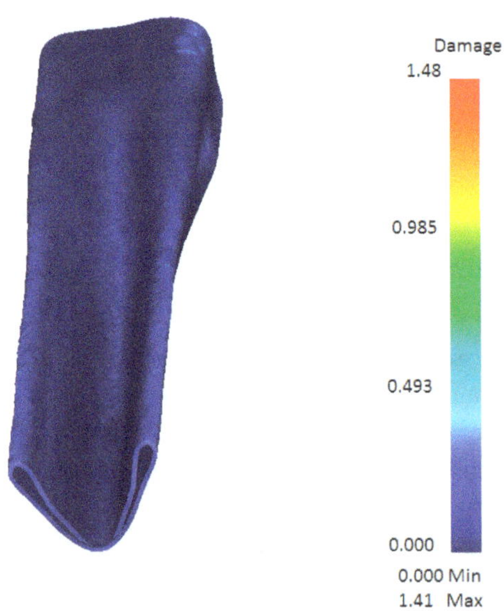

Fig. 2. Damage distribution in different regions.

The shape distortion after quenching and unloading is another problem frequently met in the real application. To evaluate the shape of formed part after hot stamping, the difference from the height of edge part and bottom of the center part (H) is measured.

Height difference between three-dimensional model and the simulated part is 5mm. The larger shape change could be explained by the incomplete martensitic phase transformation and phase transformation plastic strain.

After cooling 10s, distribution of temperature is shown in figure 3. The temperature changes always accompany the volumetric strain that also contributes to the evolution of the residual stress in the material. Therefore, when the part is unloaded it is distorted by the release of the residual stress.

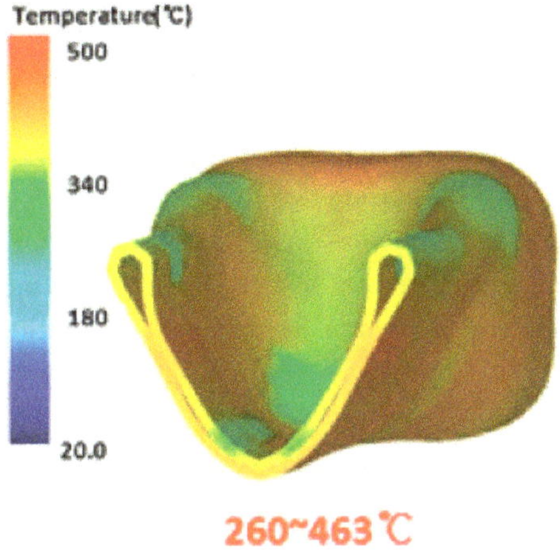

Fig. 3. Predicted distribution of temperature after cooling 10s.

Due to the different contact sequent between the specimen and the mold, the temperature field distribution in different parts of the specimen is relatively large; the inserts at both sides intrude into the tube make the maximum stress. Figure 4 shows stress after forming.

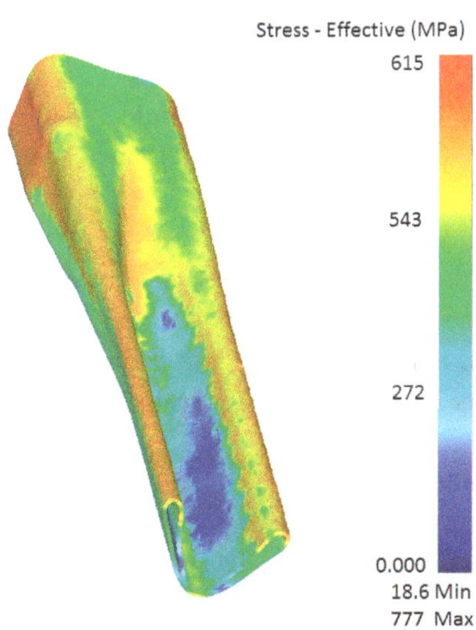

Fig. 4. Effective stress after cooling.

4. Conclusion

In this study, the deformation behavior of the tubular specimen for the CTBA part during the hot stamping process was studied. Using finite element simulations, we were able to study and predict the deformation change after hot stamping. Based on the simulations, the following conclusions are drawn.

1) Due to the different contact sequent between the specimen and the mold, the temperature field distribution in different parts of the specimen is relatively large, cause the maximum principal stress and the maximum principal strain distribute in different regions. The inserts at both sides intrude into the tube make the maximum stress, and the V-shape regions make the minimum stress.
2) The direct hot stamping with die quenching showed good shape accuracy, but the core at both sides intrude into the tube require high accuracy, it is important to position the tube.
3) The larger shape change could be explained by the incomplete martensitic phase transformation and phase transformation plastic strain.

Acknowledgments

This research work funded by the National Natural Science Foundation of China (Grant No. U1564203) and National Natural Science Foundation of China (Grant No. 51405171).

References

1. H.H Bok, J.W Choi and F. Barlat, Thermo-mechanical-metallurgical modeling for hot-press forming in consideration of the prior austenite deformation effect, *Int. J. Plasticity* **58**, 154 (2014).
2. J.K Park, Y.S Kim and O.S Seo, Improved hot-stamping analysis of tubular boron steel with direct measurement of heat convection coefficient, *International Journal of Automotive Technology* **14**, 717 (2013).
3. T. Maeno, K. Mori and K. Adachi, Gas forming of ultra-high strength steel hollow part using air filled into sealed tube and resistance heating, *J. Mater. Process Tech.* **214**, 97 (2014).
4. M.G Lee, S.J Kim, H.N Han and W.C Jeong, Application of hot press forming process to manufacture an automotive part and its finite element analysis considering phase transformation plasticity, *Int. J. Mech. Sci.* **51**, 888 (2009).
5. H.Y Kim, J.K Park and M.G Lee, Finite element simulation of hot stamping of tubular beam and its application to die channel design, *Materials Transactions* **53**, 838 (2012).
6. H.Y Kim, J.K Park and M.G Lee, Phase transformation-based finite element modeling to predict strength and deformation of press-hardened tubular automotive part, *The International Journal of Advanced Manufacturing Technology* **70**, 1787 (2014).

Process Design

Lightweight Technologies — Trends, Challenges, Solutions

M. M. Gharbi

Schuler Pressen GmbH
Bahnhofstrasse 41, 73033 Goeppingen
E-mail: mohammad.gharbi@schulergroup.com

The main motivation for the lightweight concepts in automotive sector is nation and international legislations to decrease the CO_2 emission. The use of lightweight technologies or lightweight materials or the combination of both is indispensable. In this paper we present lightweight technologies which could be used to produce parts with higher performances. Different technologies such as aluminium forming, Plastic forming, Hydroforming and Hot stamping is introduced, where the Trends, Challenges and Solutions are also discussed. Finally a new technology, Hot Tube Forming, based on the existing technologies (Hot stamping and Hydroforming) is presented.

Keywords: Lightweight; Hot stamping; Aluminium; Hot tube forming; Tool; Die technology.

1. Introduction

In the modern automotive industry, the need for lightweight design and construction is omnipresent. National and international regulations to reduce the CO_2 emission see a reduction of 25% - 30% in the coming 5 to 10 years. This goal could just be realized while using lightweight concepts.

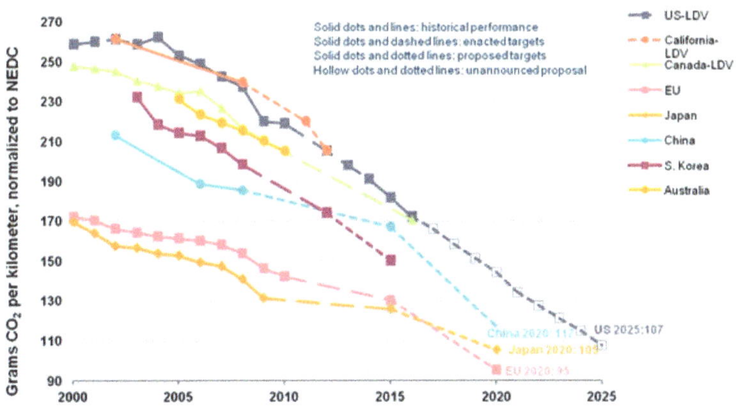

Fig. 1. National and international regulation to reduce CO_2 (Rebuffet, 2015).

A reduction of car body weight by 100 kg, results a further weight reduction of other car components by adjusting and adapting them to car body up to 16 kg. This results fuel reduction of 0.3 - 0.6 lit/100km, depending on car category. The mentioned strict emission guidelines stand against the demand for durability, crashworthy qualities and passenger safety. Lightweight concept has the potential to fulfill all of these demands by three avenues: technology, form and material lightweight design. The use of lightweight technology has the most impact following by form and material.

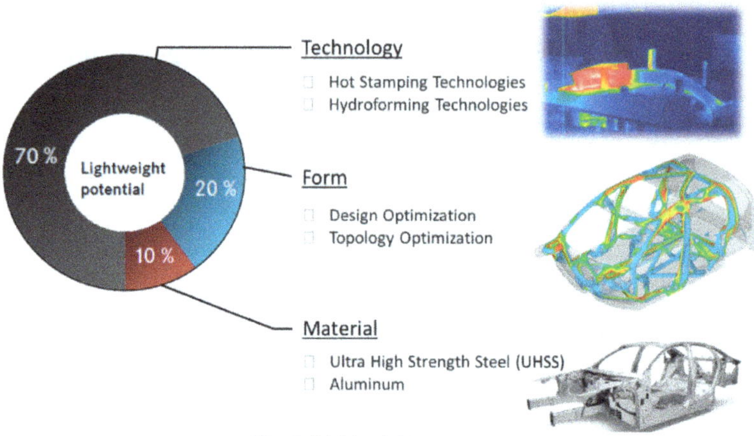

Fig. 2. Lightweight concepts.

In the most of the cases an intelligent combination of these factors is used and ensures the possibility to use the maximum potential of lightweight concepts. In the following, we consider on lightweight technologies, in some cases combined with lightweight materials, as a key solution to fulfill the functional requirements of the body in white (BIW). Generally, in BIW one could define three areas; exterior, safety cell, and front/rear area of deformation. The main challenge here is to use the right lightweight concepts to fulfill the desired characteristics, where the buckling resistance meets the strength.

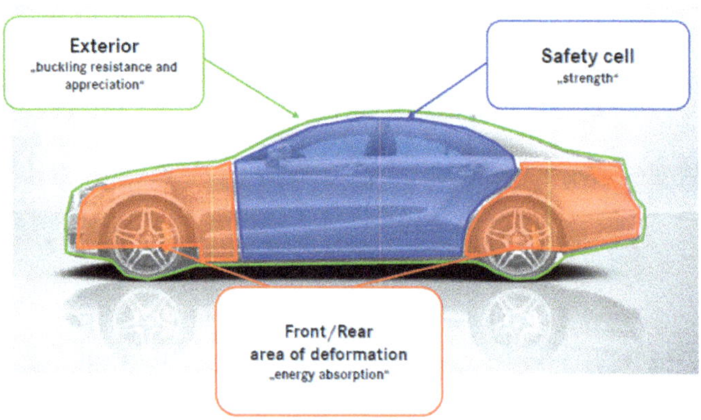

Fig. 3. Functional requirements in BIW.

2. Functional Requirements of the Body in White: Exterior Part Design

Considering exterior, where the buckling resistance is a key factor, the choice of the right material is driven by the functional requirements of BIW. For the exterior or "hang-on" parts, the buckling resistance and appreciation is important, therefore the thickness of the parts cannot be reduced. The specific weight of the material has to be reduced in order to guarantee the lightweight design. Usually for these parts, mild steel is a common material. For the new generation of car bodies, materials such as Aluminium (5xx and 6xxx and 7xxx series), plastics (Sheet Mold Compound: SMC, Carbon Fiber Reinforced Plastic: CFRP), steel sandwiches and Magnesium are also considered.

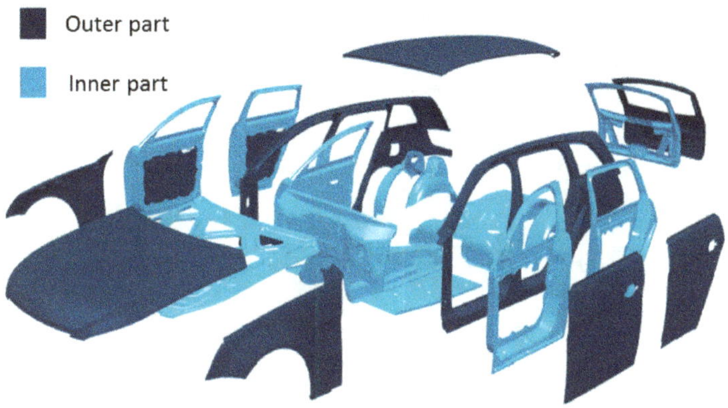

Fig. 4. Exterior parts in BIW.

The last two cases are under development. ThyssenKrupp Steel Europe introduced a new generation of sandwich material "LITECOR®" to the market. This material is a combination of two steel layers with a plastic material in between. The specific weight of this material is between Aluminium and steel, but exhibits higher bending stiffness as steel as reference. To this point, there is no industrial application of this material in car body for high volume production. On the other side, magnesium is not corrosion resistance and base on this fact, cannot be used easily for BIW. There are research activities in this area to develop an alloy which has a good corrosion resistance, but still these materials are not applicable for the serial production of exterior parts.

Aluminium and plastic-based materials such as CFRP and SMC are becoming in the recent years more and more application as exterior and hang-on parts. The following table shows a comparison of Aluminium with sheet molding compound (SMC). The material and process cost of the SMC is higher than Aluminium, where the tooling cost for producing part made of Aluminium is higher than SMC. Depending on the car category and the amount of parts produced in year, one can decide over one or other technology.

In the following, two above mentioned technologies, the Aluminium forming technology and the plastic forming technology are discussed in detail.

2.1. *Aluminium forming technology*

Aluminium, as a lightweight material, exhibits various characteristics, which makes this material more and more interesting for enhanced application. Superb damping properties, high corrosion resistance, high thermal conductivity, low specific weight, recyclability, variety surface function, and in some cases depending on technology used high strength and good formability are characteristics which make this material, as a lightweight material, usable.

Properties	Al	SMC*
Density [g/cm³]	2,7	2,0
E-Modulus [GPa]	70	10-50
Ductility [%]	15-25	flex.
Alpha [10^{-6}/K]	24	7-27
Material/process cost	low	high
Tooling cost	high	low
Styling	limited	high

* Sheet Molding Compound

Fig. 5. Comparison between Aluminium and SMC.

In the recent years, the use of Aluminium alloys (5xxx and 6xxx) in car body found its place. Almost all know car makers use Aluminium cold forming technology in platforms of their premium segments. The figure below shows a selection of the autos which used Aluminium parts in car body.

It is to mention, that based on using this material, a significant reduction of total weight up to 400 kg could achieve. The recent studies of material car body in north America - the second big market of automobile - shows, in the next 10 years the amount of Aluminium parts in car (including car body) will increase by factor 4, from 6% to 24%, where the amount of mild steel decrease from 71% to 33%. It means, more parts will be produced by Aluminium instead of mild steel. The need of weight reduction leads to an increasing trend of using different Aluminium alloys (6xxx, 7xxx) for producing car body components (Gharbi, 2016).

In order to be able to form complex shaped parts, to decrease spring back and to increase the material properties of the formed part, Aluminium forming at elevated temperature seems to be adequate solution. Forming processes supported by temperature are a very good example for lightweight enabling processes. The main difference to normal cold stamping is using temperature as a process parameter. Along with temperature, time becomes also a critical process parameter. In temperature assisted processes, time & temperature are linked together making process control complex and essential. Every single part has to be produced in exactly the same given path, which includes geometry, time and temperature in order to guarantee the reproducibility. Aluminium forming at elevated temperature could separate into two different processes: Aluminium warm forming and Aluminium hot forming.

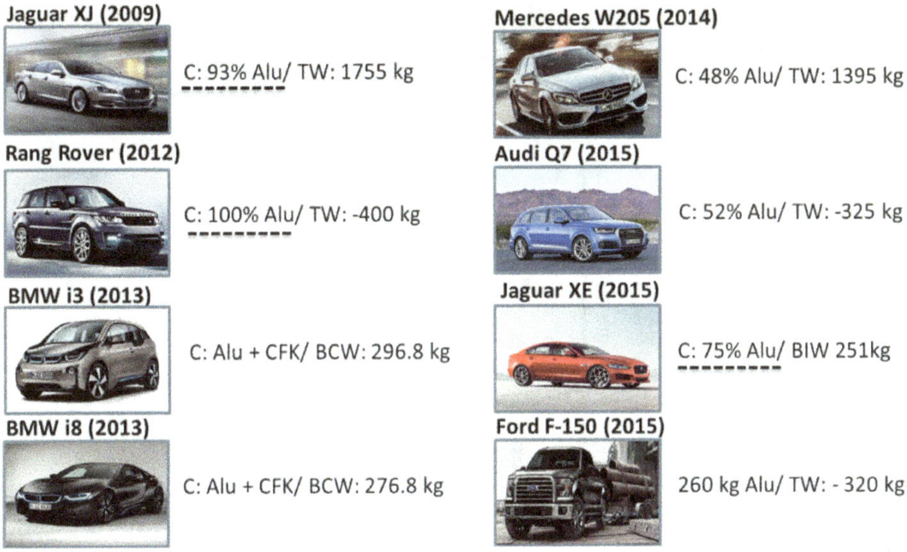

Fig. 6. Selection of autos with car body made from Aluminium, C: Car body, TW: total weight, BIW: Body in white, BCW: Body car weight (Ducker worldwide study, 2014).

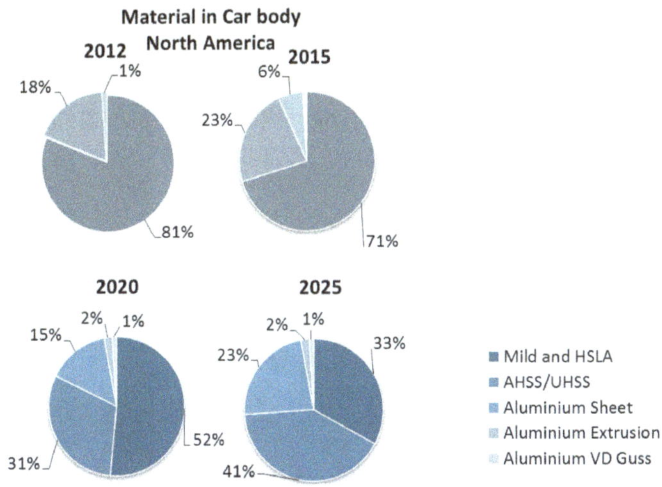

Fig. 7. Evolution of material in car body in north America (Ducker worldwide study, 2014).

2.1.1. *Aluminium warm forming technology*

In the Aluminium warm forming process, the blank is heated up to a temperature between 190°C to 250°C, depending on alloy, formed in a cold or heated die, and then cooled again in air, respectively. Heating the blank up to 250 degrees gives a moderate increase in formability. It is important that the total length of the heating cycle kept as short as possible, as the exposure of the blank to the elevated temperature change the microstructure of the Aluminium.

Main advantages of this process route are that the needed temperature is only moderate and there is no need for post aging. The achieved formability is enough for medium complex parts. On the other hand, the full potential of formability is not used (Gharbi, 2016).

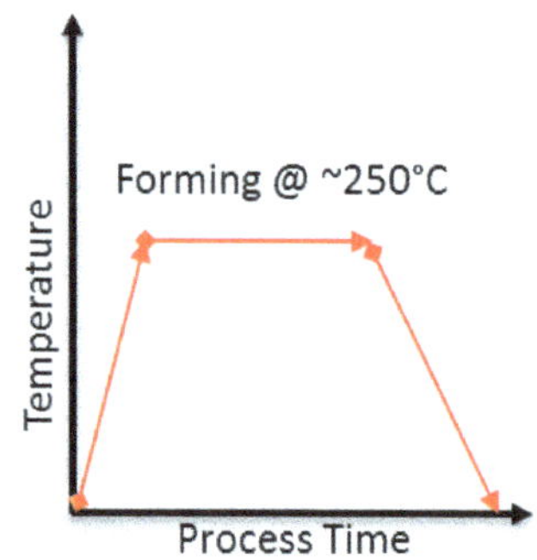

Fig. 8. Process route Aluminium warm forming.

2.1.2. *Aluminium hot forming technology*

In the Aluminium hot forming process the blank is heated to solution heat treatment, and kept there some seconds for soaking. It is then formed and quenched in one step in cooled dies. The followed artificial aging process gives the part its final strength. One main advantage is that the maximum formability available in the material is used. The quenching in the dies gives the part its final shape, where the spring back is reduced significantly.

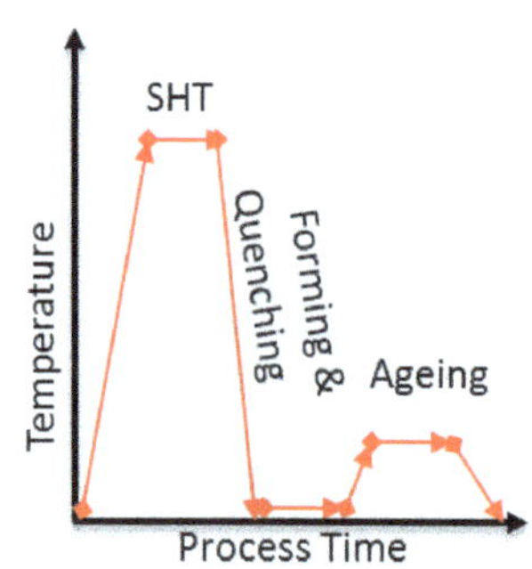

Fig. 9. Process route Aluminium hot forming.

An industrial application for high serial production faces different challenges while thinking on Aluminium forming at elevated temperature. On one side, the heating technology and heating up the blank to the desired temperature within a very short time is an important issue. Jet heating, conduction heating technology and induction heating technology seems to be promising techniques to heat up the blank. On the other hand, choosing of correct lubrication and surface treatment for dies is still an under developed issue. There are lots of researches with the focus on finding the right combination of the die surface treatment and lubrication, where others concentrating on use the forming technology without lubrication. This needs a high level die technology with an appropriate surface treatment. Furthermore, the question of additional ageing process in hot forming is not answered yet. One of the most important factors to provide the whole manufacturing line is to optimize the whole components of the production line to have the maximum output, while the quality of the product is acceptable. Schuler Pressen GmbH, the world leader in the metal forming solutions could offer the effective key solution, from die technology to the press and automation, to this kind of technologies (Gharbi, 2016).

2.2. *Plastic forming technology*

Below figure shows the development of the worldwide demand of CFRP parts (Carbon fibre reinforced plastic). From 2013 to 2014 there is a growth of 15 % with an expected growth in the future of 11 % per year, resulting in a total CFRP part demand of 175 kilo tons in 2021. The enablers for this growth are the EU restrictions considering the reduction of the CO_2 emissions and additionally effort of the USA to reduce the fuel consumption by reducing the weight. Key markets are especially the automotive as well as the aerospace sector.

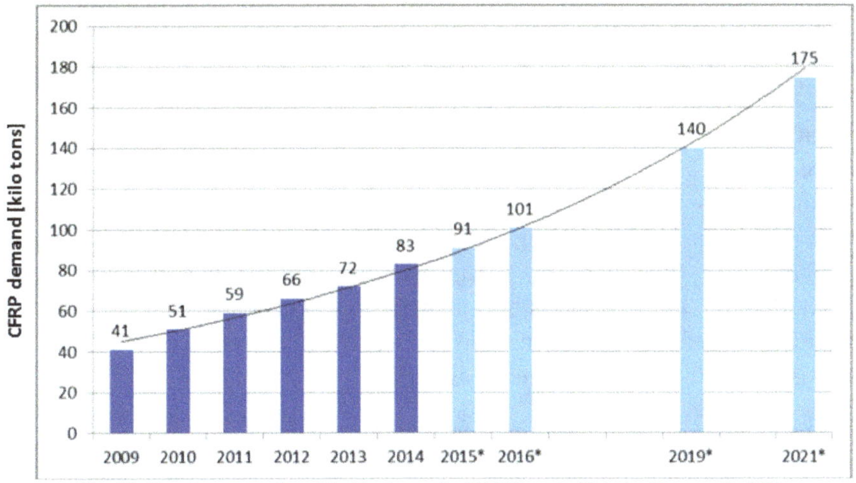

Fig. 10. Development of worldwide demand of CFRP parts (Schuler Pressen GmbH).

The illustration shows a prediction of the CC (carbon composites) demand separated for the different industry sectors. The demand for the automotive sector is expected to grow tremendously within the next year and achieving a higher demand by 2021 compared to aviation and defense.

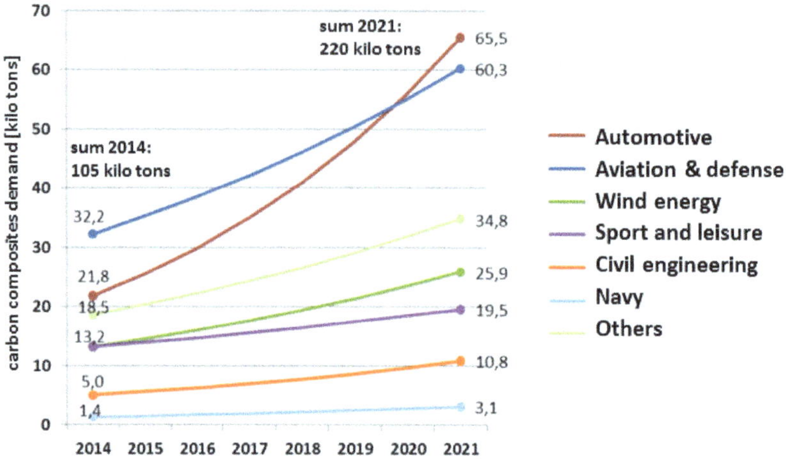

Fig. 11. Prediction of CC demands (Schuler Pressen GmbH).

For carbon fiber reinforced plastic (CFRP) parts, typically following processes are being used for series applications of the automotive industry: Long fiber and Endless fiber.

In the long fibers process only glass is used due to high price of carbon fibers and low material properties by max. fiber length of 50 mm and arbitrary fiber orientation. SMC is typically only with glass fibers (trunk lid, roof spoiler, etc.) but by way of exception also used with carbon fibers, so called C-SMC. An example of this technology is applied on trunk lid of BMW M4 Coupé.

In the endless fiber, one can utilize the maximum lightweight potential and to optimize material properties with high strength carbon fibers. RTM / HP-RTM often used with carbon fibers. This technology is implemented to produce roof of BMW i3 and i8, BMW M3/M4/M5. Organic sheets also used for this kind of components. Only small components like pedal using carbon fibers for series application. Very often this type of material is used with glass fibers due to cost reason.

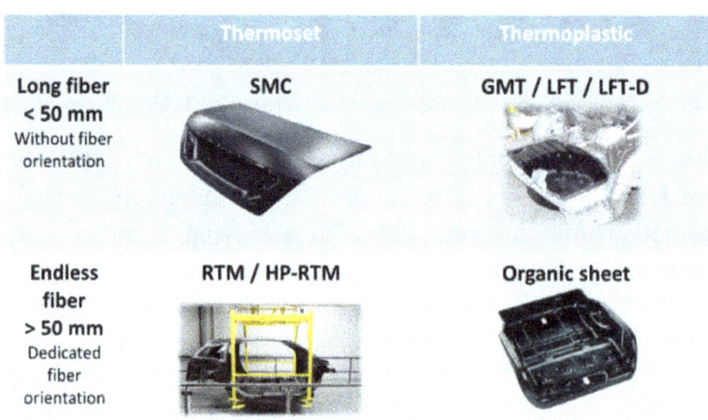

Fig. 12. Process route of CFRP production.

No matter which process route selected, using an efficient machine and system to produce parts with given quality is important. Since the production cost of such parts is relatively high, using an optimum concept to produce these parts is becoming more important. The short-stork press system, which is innovated by Schuler Pressen GmbH is one of the optimal solution to produce parts made of plastic.

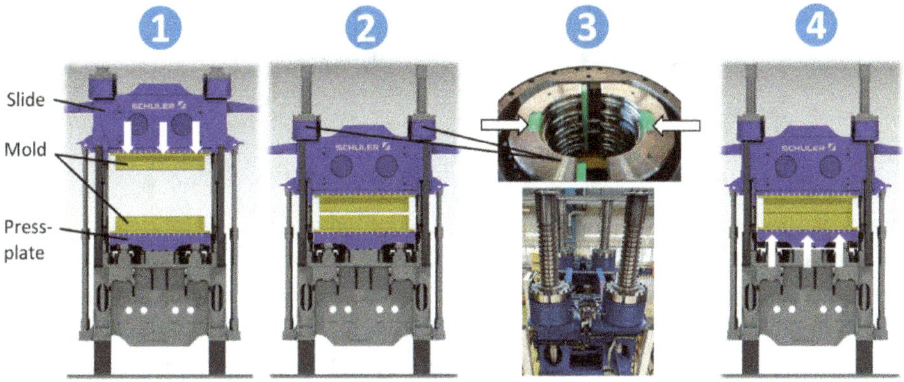

Fig. 13. Short Stork press of Schuler (Schuler Pressen GmbH).

The advantages of short-storke press can be summarized as:

- Significant reduction of machine height: Reduction of machine height by 40 % ,
- Accessibility from all four sides: Column machine design with lateral openings,
- Integrated parallelism control: No adaption for different mould heights required,
- Short pressure build-up time: With short stroke cylinders, and
- Reduced energy consumption: Significant less compression oil.

For the further optimization of this system and increase the efficiency of production it is important to have a paradigm change: Regulation of component properties, not the process parameters. To achieve this goal, a research project with contribution of Schuler Pressen GmbH is subjected (iComposite 4.0). The goals of this research project are:

- Increase of resource efficiency (processing time -30 %, piece costs – 50 %)
- Combination of specific process advantages at one complete system
- Customized mass production through data consistency and self-optimization
- Fast and easy operation of the system due to intuitive control panel with process visualization.

The main idea is to have a production function based manufacturing instead of geometry based manufacturing.

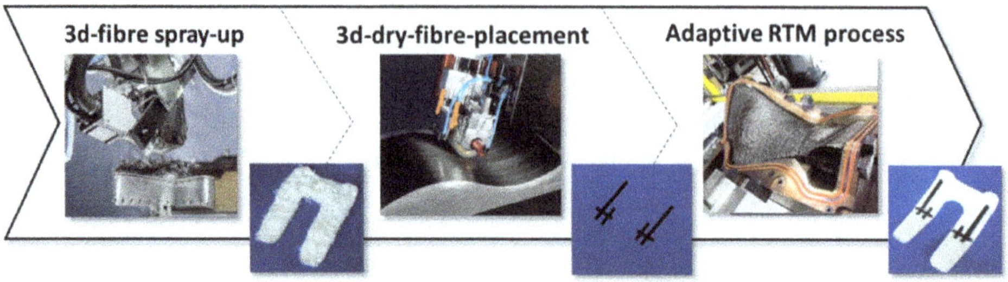

Fig. 14. Project iComposite 4.0 (Schuler Pressen GmbH).

3. Functional Requirements of the Body in White: Safety Cell

Considering safety cell, the strength and crash worthiness is the objective function. The need for an alternative material and new technology to form parts with higher strength is increased. On the other side the demand on part complexity and the design criteria are rises. In the following, two technologies with are mainly used to formed part which are mainly used for safety structure exhibiting high strength, the hot stamping and hydroforming is presented.

3.1. *Hot stamping technology*

Due to the demand for reduced vehicle weight, improved safety, and crashworthiness qualities, the need to manufacture automobile structural components from ultrahigh strength steels is apparent (Åkerström, 2006). Hot stamping was developed and patented (GB1490535, 1977) by a Swedish company (Plannja), which used the process for saw blades and lawn mower blades. In 1984 Saab Automobile AB was the first vehicle manufacturer who adopted a hardened boron steel component for the Saab 9000 (Berglund, 2008). The number of produced parts increased from 3 million parts/year in 1987 to 8 million parts/year in 1997. Since the year 2000 more hot stamped parts have been used in the cars and the number of produced parts/year has gone up to approximately 700 million parts/year in 2017 (Aspacher, 2008). The applied hot stamped parts in the automotive industry are chassis components, like A-pillar, B-pillar, bumper, roof rail, rocker rail and tunnel.

The hot stamping process currently exists in two different main variants: the direct and the indirect hot stamping method. In the direct hot stamping process, a blank is heated up in a furnace, transferred to the press and subsequently formed and quenched in the closed tool. The indirect hot stamping process is characterized by the use of a nearly complete cold pre-formed part which is subjected only to a quenching and calibration operation in the press after austenitization. (Merklein et al., 2008). Full martensite transformation in the material causes an increase of the tensile strength of up to 1500 MPa.

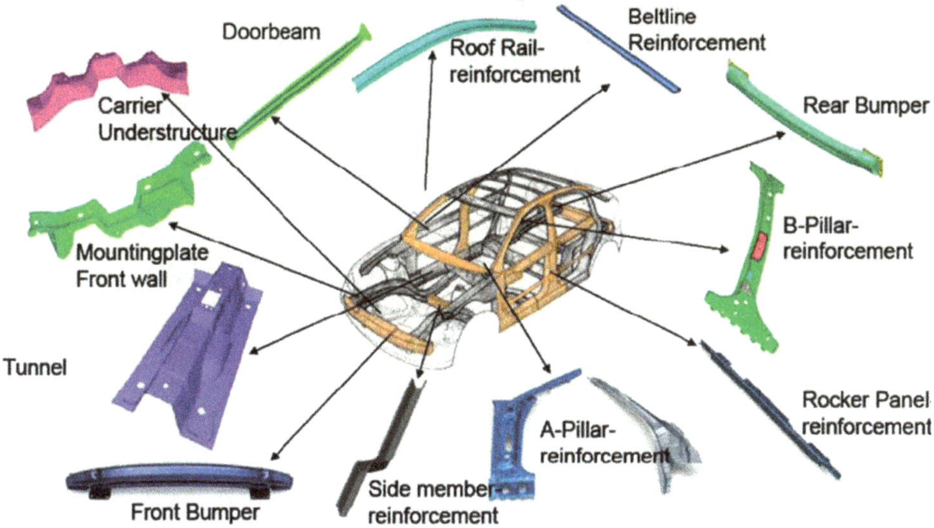

Fig. 15. Example of hot stamped parts.

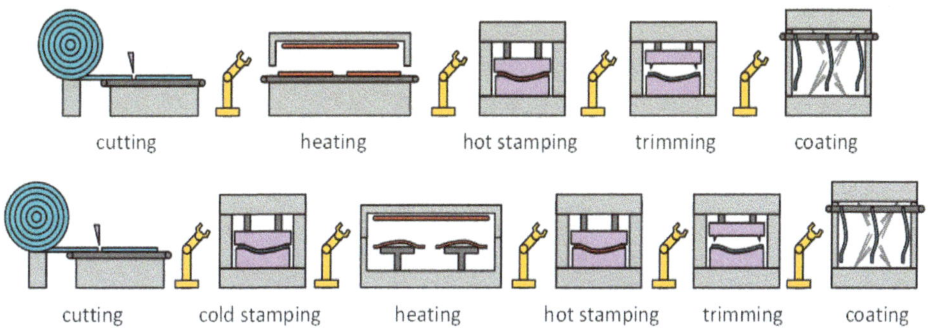

Fig. 16. Direct (top) and Indirect (bottom) hot stamping.

Investigation on alternative cold rolled medium Mn steel containing 9.76Mn-0.16C-l .37Al-0. 19Si (mass%) processed via a short time (3 min. soaking) continuous annealig-type treatment and then subjected to typical direct hot forming thermal cycles, with reheating at 650:800 °C could achieve an excellent mechanical properties (1330-1448 MPa tensile strength; 16.7-25.3% total elongation) of the hot formed-steel, were obtained with reheating at 700 °C. 22MnB5 steel grade is the most commonly used steel grade in hot stamping processes. The material exhibits a ferritic–pearlitic microstructure with a tensile strength of about 600 MPa. After the hot stamping process, the component finally has a martensitic microstructure with a total strength of about 1500 MPa. In order to achieve such a microstructure and hardness transformation, the blank has to be austenitized in a furnace for at least 5min at 950 °C. Afterwards, the blank is formed and quenched simultaneously by the water-cooled die for 5–10 s, depending on the part thickness. Due to the contact between the hot blank and the cold tool, the blank is quenched in the closed tool. If the cooling rate exceeds a minimum cooling rate of approximately 27 K/s, at a temperature of around 400 °C, a diffusionless martensitic transformation will be induced, which finally is responsible for the resulting high strength of the part. The martensite transformation begins at 425 °C (martensite start point Ms) and ends at 280 °C (martensite finish point Mf). The mechanical properties of steel after quenching change in dependence on its carbon content and consequently, the strength after quenching can be controlled by a proper adjustment of the carbon content (Somani et al., 2001).

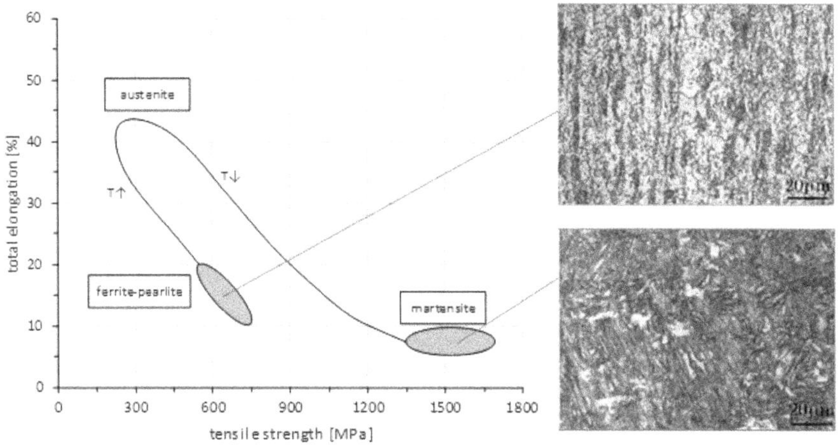

Fig. 17. Microstructure change during hot stamping process.

The desired phase transformation and hardenability is achieved by technically feasible cooling rates (Garcia Aranda et al., 2002). After the forming of the heated blank in the austenitic temperature range, the part is quenched in the closed tool until the entire martensite transformation of the part structure is completely. A cooling rate of more than 27 K/s is necessary for a full martensite microstructure of 22MnB5.

In hot forming process, the dimensional accuracy of produced components relies not only on an optimum friction level for the deformation of the workpiece, but it also gets affected by wear of the forming tools, which reduces the service life of the tool as well. It is desirable to enhance the durability of the tools by understanding the influence of contact conditions on tool wear.

3.1.1. *Cooling channels*

The quenching operation during the hot stamping process does not only influence the economy of the process but also the final properties of the component. The objective of the cooling ducts design is to quench the hot part effectively and to achieve a cooling rate of at least 27 K/s while martensite forms. The die cooling system is economical if a fluid coolant, such as water, is used, which flows through cooling ducts around the contours of the component. The heat flow in the formed component is dependent on the heat transfer from the component to the tool, the heat conductivity within the tool, and the heat transfer from the tool to the coolant (Karbasian and Tekkaya, 2010).

mechanism		equations
convection	free	$q_{conv} = h_{conv} \cdot A \cdot (T_1 - T_\infty)$
radiation		$q_{rad} = \varepsilon \cdot \sigma_{SB} \cdot A \cdot (T_1^4 - T_\infty^4)$
conduction	contact con-ductance	$q_{cont} = h_{cont} \cdot A \cdot (T_2 - T_1)$
conduction	solid	$q_{cond} = \frac{k}{L} \cdot A \cdot (T_3 - T_2)$
convection	forced (fluid cooling)	$q_{conv} = h_{conv} \cdot A \cdot (T_3 - T_{fluid})$

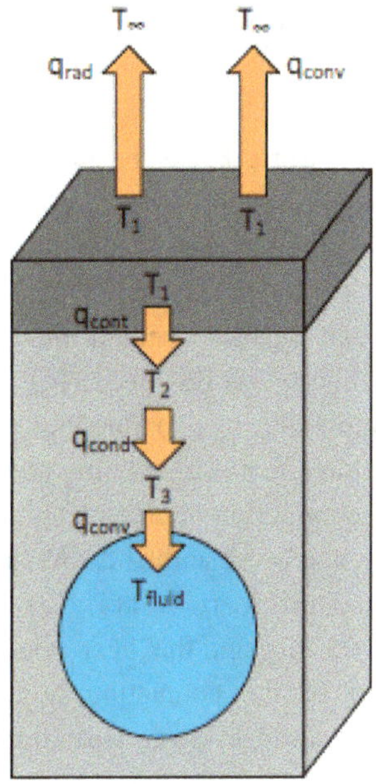

Fig. 18. Heat transfer phenomena during hot stamping process (Schuler Pressen GmbH).

For an optimum heat transfer between component and tool, the contact surface should not exhibit a scale or a gap. The heat conductivity within the tool can be considerably influenced by the choice of the tool material. Another important factor with respect to heat drain is the design of the cooling ducts, which is defined by the size, location, and distribution of the cooling ducts. The heat drain can be accelerated by using a coolant with a low temperature in order to increase the temperature difference between the coolant and the tool and therefore the resulting heat flux (Steinbeiss et al., 2007). The cooling holes can be drilled in the forming tool. For this method, the machining restrictions should be considered in the design of the hole position (Steinbeiss et al., 2007). Therefore, an optimal design of the cooling system as to heat transfer is not possible. Another alternative is to provide the cooling holes as pipes in the casting mold of the tool (Kuhn and Kolleck, 2006). The unrestricted design of the cooling system is the advantage of this method. Alternatively, the tools can be manufactured using the lasered blank segments, which are screwed and form the tool surface with integrated cooling holes (Freieck, 2007). This method is very cost-effective, but the lamellar design can have negative effects on the part surface and the heat transfer within the tool. Todays, additive manufacturing also assists design the cooling channels. Where the cooling channels cannot drill near to the active surface, one can substitute this segment with a 3D printed segment, which exhibits the same properties as conventional.

3.1.2. *Surface and contact pressure*

One of the most critical factors is the contact pressure between the blank or part and the die surface because it controls the heat transfer time very much and therefore the cycle time.

The experiments reveal that the interface heat transfer coefficient is proportional to contact pressure. It can be inferred that the contact condition at the interface between sheet metal and tool surfaces during the die quenching stage plays a very critical role in the cooling rate of the sheet metal (Hamasaiid et al, 2015).

Surface roughness and contact characteristics on the interface between blank and die, are critical. Surface irregularities promote the formation of gaps between the tool and blank, where air is retained. This lack of contact implies that heat transfer is not carried out completely by conduction phenomena, but through convection and radiation. Conduction is constrained only at some discrete points on the interface. The heat transfer through conduction and convection of the fluid in the cavities is negligible according to the reference because the heat transfer coefficient of the fluid is small. The radiant heat of the noncontact part is also very small (Abdulhay et al., 2001). Heat is primarily transferred through the heat conduction between for real contact spots between the asperities of the contacting bodies. Therefore, and as it is known, more surfaces roughness implies lower Heat transfer coefficient (HTC). In the case of hot stamping processes as a high pressure is applied, the contact spots should be increased by deforming or penetrating the asperities into the contacting surface. Consequently HTC increases when the forming pressure increases. A larger contact force will cause more deformation of the asperities, so the actual contact area increases, thus HTC also and the thermal contact resistance decreases. Higher pressure speeds up the blank cooling. At the beginning of contact, the temperature difference at the interface is large, so it is the thermal kinetic force, resulting in a rapid HTC increase. When blank temperature and die surface temperature approach, thermal kinetics becomes weaker and HTC decreases. Increasing the pressure entails higher values of HTC (Hamasaiid et al, 2015).

There are different possibilities to generate force to have high contact pressure during hot stamping process. Conventional springs, nitrogen springs, hydraulic cylinders in the die are some of the possibilities to have higher contact pressure. Schuler Pressen GmbH new cushion technology, PCH© Flex (Pressured Control Hardening), is special developed system for press hardening with interacting individual cylinder. This system enables to reduce cooling time while generating high contact pressure which can be verify individually to the part. PCH© Flex made it possible to have high controlled pressure to compensate the material tolerances and faster cooling rate, respectively (Aspacher et al., 2015).

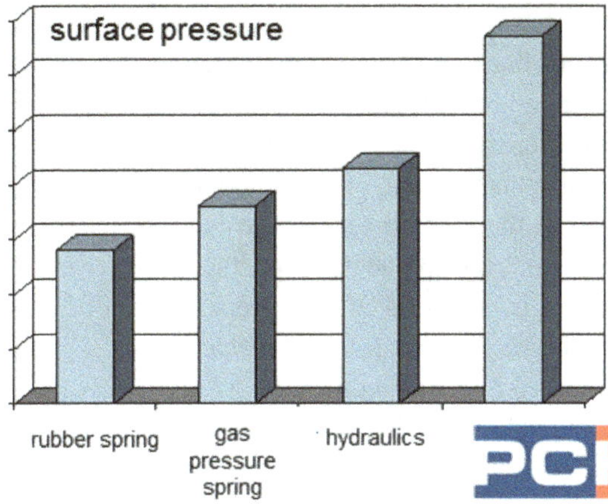

Fig. 19. Surface pressure affected by different strategy (Schuler Pressen GmbH).

3.1.3. *Wear*

The wear resistance of the tool has been measured by Dessain et al. (2008) using a strip drawing equipment which is adapted to high temperature tests. The resistance heating device allows heating up the Al–Si coated 22MnB5 strip. The heated strip slides over the die radius. For this test, the contact surface was made of abrasion and adhesion areas. The die abrasion was predominant and adhesion of Al–Si was observed at the end of the contact area with the blank. These built layers, formed very early at the beginning of the test, are known to form compacted layers on the tool surface during sliding of the strip (Hardell and Prakash, 2008) and exhibited low wear. Using the appropriate die material combined with an adjusted heat treatment to achieve the desired hardness and wear resistance is an important factor in hot stamping.

During hot stamping thermo-mechanical loads impact adhesive layer build-up on tools. Both, contact pressure and blank temperature influence the wear behavior of Aluminium-silicon coated 22MnB5 only marginal. A trend of an overall reduction of wear height at high contact pressures could be stated whereas a reduction of wear height as a result of decreasing blank temperature was only seen for low contact pressures. Also forming induced loads affect wear behavior by breakout of particles out of the Aluminium-silicon coating. This failure behavior occurs at a certain threshold temperature which shifts in dependency of compression strain rate. A reduction of real contact surface is the result, impacting adhesive layer build-up on hot stamping tools. A qualitative prediction of adhesion of the Aluminium-silicon coating on hot stamping tools is possible by taking this failure behavior into account (Karbasian and Tekkaya, 2010).

3.1.4. *Tailored properties*

The fully martensitic transformation during hot stamping of the parts leads to a tensile strength of up to 1500MPa and a low elongation of about 5%. To improve and optimize the crash and deformation characteristics, press hardened components with partial adjusted mechanical properties - the so-called Tailored Properties - are realized, which are based on a local enhancement of the ductility. The main purposes are the well-directed deformation in the loading case, the enhancement of the maximum absorbable deformation energy and the prevention of the crack growth and spreading. The manufacturing of a single part with tailored properties can be carried out using different process control strategies or using tailor welded blanks. Alternatively, the thermal process can be influenced by reducing the cooling rate below the specific 27 K/s to avoid a fully martensitic microstructure in defined areas of the component during the forming process, or by reducing the annealing temperature below the material-specific Ac_3 temperature, which leads to an incomplete austenitization. Both methods implicate a lower strength and therefore a higher ductility. The other areas of the component are quenched, following the commonly known press hardening time–temperature profile (Stoehr et al., 2009).

Next to the well-known manufacturing strategies, the partial tempering of press hardened parts offers the opportunity of generating tailored Properties without regarding the preceded process chain and provides the potential of adjusting all requested mechanical properties, independent of the required geometry. Partial tempering of press hardened steel by direct flame impingement is introduced by (Zimmermann et al., 2015). A highly flexible manufacturing approach enables the generation of constantly reproducible mechanical properties in dependence of the maximum annealing temperature induced by partial flame impingement. The use of laser softening of press hardened steel in high volume production lines is discussed by (Schaefer et al., 2015), where the laser light is not only dedicated to cutting or welding but can also be used for softening local areas of hot stamped structural steel parts working like a single sided surface heat source.

In order to generate different cooling conditions, usually the tools are heated to decrease the cooling rate of the material. In contrast, the other possibility is based on a change of the heating process. In this case several process strategies are conceivable. First of all two different areas within a blank can be achieved if one area is completely austenitized and the second area is not austenitized due to a temperature level below the austenitisation start temperature. As the result of this approach, after press hardening, a part is provided with two areas exhibiting different mechanical properties (e.g. soft, hard).Another approach uses the possibility of partial austenitisation of the blanks before press hardening. Depending on the transformation grade of the microstructure during the heating of the material, different final mechanical properties can be obtained after quenching. Based on this approach studies on a new press hardenable steel grade were carried out. (Marten et al., 2015) use this method, where the results obtained are compared with another steel grade with similar tensile strengths after a conventional press hardening process. For the partial austenitisation process, material simulations were performed to provide information on the theoretical process windows. By means of basic material tests the simulations were evaluated afterwards. Finally, demonstration parts were manufactured by using the two different process routes.

Behrens et al. (2015) uses subsequent cooling outside the press for adjusting the tailored properties. With the designed quenching device cooling rates of part sections are varied flexibly by applying a water-air spray cooling. The high potential of this process is demonstrated on behalf of the mechanical properties which are compared for conventionally produced parts and parts featuring tailored properties induced by the modified process sequence. Last but not least, (Liu et al., 2015) presents a new hot stamping process to make tailored properties based on air cooling. In high strength region, microstructure of blanks with full austenite are quenched from austenite to martensite by tools with cooling water channel, and after air cooling microstructure evolution involving ferrite, bainite and martenstie have been fulfilled in absorption region. And tailored cooling rates are formed during transition region because of tailored contact pressures.

No matter which method is used, the process to produced part with tailored properties should be stable and reproducible. It is important that, the overall process time stay almost the same in comparison with production normal parts.

3.2. *Hydroforming technology*

Automotive industry follows various strategies to reduce the weight of the vehicles. One of the options for lightweight design is application of high and ultra-high strength steel (HSS and UHSS) alloys. The combination of high strength and optimized part design allows both reduction of part thickness and improved crash behavior. One of the appropriate technologies to form high strength parts is hydroforming.

Fig. 20. Example of hydroformed parts used in automotive.

Hydroforming has been always a method which is not only used for smaller batch sizes, but especially for mass quantities. Even in Europe the automotive production is slightly increasing in the last few years. Typical products for hydroforming are beside exhaust, body parts and chassis (Haenelt and Freydank, 2016).

The big challenge in hydroforming high and ultra-high strength steel is springback. The factors causing springback and the effects of these factors on springback are summarized in following.

Factor	Effect on Springback
Length of part	The longer the part – the more springback
Thickness of the tube	A thicker tube is stiffer and has less springback
Diameter / Thickness	Tube diameter / wall thickness ratio plays a role in springback
Bending radius	The larger the bending radius - the smaller is the springback
Cross section / Geometry	The shape of the cross section
Forming process	Bending, preforming method, hydroform parameters etc.

Fig. 21. Springback in hydroforming.

Springback compensation of hydroformed parts is a rather recent challenge strongly related to the increasing application of HSS and UHSS materials. Compared to tool makers and press shops for sheet metal parts, most hydroforming companies cannot look on a longer history of experience and profound know-how of spring back compensation. The use of CAD models in FEA/CAE to predict the springback in hydroformed parts enable a reduction in number of iteration needed for tooling adjustment in press. That leads to notable time saving for tool making and tryout but there is an investment for simulation model development needed.

For an effective compensation of springback, each forming step needs to be carefully considered and compensated. This approach should implement in the early stage of the engineering phase, before the first tools have been milled and even before the final CAD tool design has been released. This methodology already implemented by Schuler Pressen GmbH. This company is well prepared to master future challenges of rapidly growing need for even higher strength steels and the related complication for springback compensation (Hertell et al, 2016).

Beside an appropriate methodology for design and engineering the parts, a stable process is a key factor to produced HSS and UHSS hydroformed parts. In conventional hydroforming, a preformed profile is loaded on a hydroforming press. In the following steps, the profile is filled with water or a non-compressive media, the dies are closed and the pressure is conducted to assure the forming process. Schuler Pressen GmbH is implement a modified version of this process, where in the first step the preformed profile is filled with water, while closing the dies, a nominal pressure is applied on the part following with the last step which the dies are completely closed and a high pressure to the calibration the form is applied. With this technique, a better geometrical accuracy can be achieved (Hertell et al, 2016).

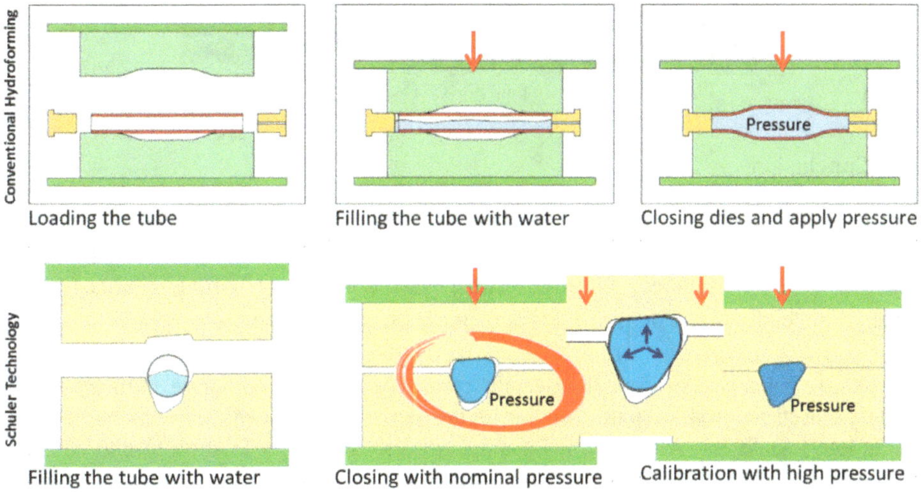

Fig. 22. Conventional hydroforming and Schuler hydroforming technology (Schuler Pressen GmbH), New Technology: Hot Tube Forming.

As mentioned before, the higher the strength of the material, the more is the springback acquired after forming. Despite of the fact that, the methodology and process route can fairly compensate the springback, but the strength of the applied material is still limited in hydroforming process. State of the art of used material is DP1000.

On the other side, the need of lightweight construction with advanced mechanical properties, strength and superb crashworthiness for today's auto is mandatory. The advantages of crash behavior of profiled shape parts formed by hydroforming technology make also the use of profiles in the new generation of car body more and more interesting.

One possibility is to combine the advantages of profile shape parts with high strength which can be achieved by hot stamping. The combination of these two technologies is practiced and patented by Schuler Pressen GmbH, a so called Tube hot stamping. For this process, a preformed boron tube is heated up to the austenitization temperature, after transformation to the press; a moderate inner pressure is applied by gas while dies are closing. Finally, the pressure goes up to the maximum and the final geometry is achieved. By quenching the part via cooled dies, it is possible to have the phase transformation from austenite to martensite and therefore a strength range up tp 1600 MPa could achieve.

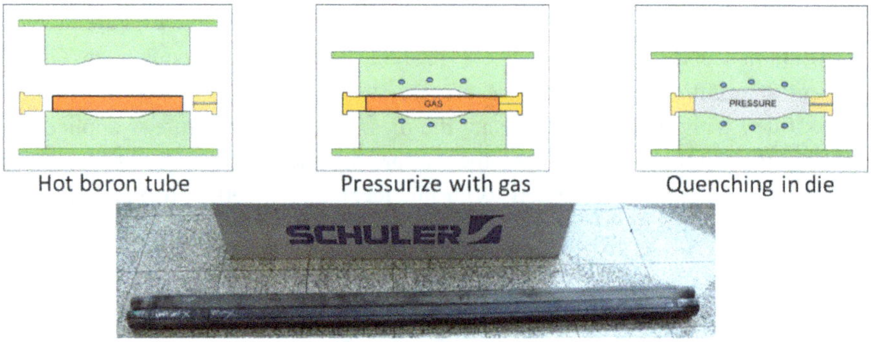

Fig. 23. Tube hot stamping process (Schuler Pressen GmbH).

4. Conclusion

The main motivation for the lightweight concepts in automotive sector is fundamental in decreasing the CO_2 emission. The use of lightweight technologies or lightweight materials or the combination of both is indispensable in today environmental conscience society, especially in the light of Paris agreement on climate change.

In this paper we present lightweight technologies for automotive industry, which could be used to produce parts for the higher performances vehicles with modern day technology; such as Aluminium forming, Plastic forming, Hydroforming and Hot stamping were discussed. We, therefore, foresee that:

The volume market for automotive industry will be in

- Hot stamping for the safety cell and body structure,
- High strength steel; formed by hydroforming technology,
- Aluminium cold formed part on exterior.

There will be a growing demand for

- Aluminium forming at elevated temperature for structure parts and hang-on parts,
- Plastic forming technology; SMC and CFRP,
- Tube hot forming,
- Magnesium forming technology; sheet and cast.

And the interesting technologies of the future are likely to be

- Roll forming technology,
- Steel sandwich for the hang-on parts.

In here, a new technology, Hot Tube Forming, based on the existing technologies (Hot stamping and Hydroforming) is presented.

Acknowledgments

The author thanks Mr. Palm, Mr. Nickler, and Mr. Winterhalter, from division hydraulic Schuler Pressen GmbH Germany and Mr. Hertell from Schuler Incorporated USA for the support.

References

1. B. Abdulhay, B. Bourouga and C. Dessain, Experimental and theoretical study of thermal aspects of the hot stamping process, *Appl. Ther. Eng.* **31**, 674 (2011).
2. P. Åkerström, Modeling and simulation of hot stamping, Phd theses, Luleå University of Technology, Sweden (2006).
3. J. Aspacher, Forming hardening concepts, in 1st International Conference on Hot Sheet Metal Forming of High-Performance Steel (CHS2), (Kassel, Germany, 2008).
4. J. Aspacher, D. Haller and P. Thom, From First Draft to Serial Production: Hot stamping Part design and Feasibility Study with Respect to Functionality and Optimization of Production Costs, in *Proceeding CHS2*, (Toronto, Canada, 2015).
5. B. Behrens, H. Maier, F. Nuenberger, J. Schroedler, L.Wolf and C. Gaebel, Hot Forming and Subsequent Cooling Outside the Press for Adjusted Tailored Properties of 22MnB5 Steel Sheets, in *Proceeding CHS2*, (Toronto, Canada, 2015).
6. G. Berglund, The history of hardening of boron steel in northern Sweden, in *1st International* Conference *on Hot Sheet Metal Forming of High-Performance Steel(CHS2)*, (Kassel, Germany, 2008).
7. C. Dessain, P. Hein, J. Wilsius, L. Penazzi, C. Boher, and J. Weikert, Experimental investigation of friction and wear in hot stamping of usibor 1500P, in *1st International* Conference *on Hot Sheet Metal Forming of High-Performance Steel(CHS2)*, (Kassel, Germany, 2008).
8. Ducker Worldwide study, North American Light Vehicle Aluminium Content Study - executive Summary, 2014.
9. U. Freieck, In zwei Wochen zum Serienwerkzeug. Blech **2**, 46 (2007).
10. L. G. Aranda, Y. Chastel, J. F. Pascual and T. D. Negro, Experiments and simulation of hot stamping of quenchable steels, *Adv. Tech. Plasticity* **2**, 1135 (2002).
11. M. Gharbi, 2016. Warmumformung von Aluminium blech aus hochfesten Legierungen. Blechnet 2016 (1), 26-27.
12. M. Gharbi, Hot formed Aluminium Parts, Todays Motor Vehicels, 48 (2016).
13. M. Gharbi, Automobil production, 75 (2016).
14. H. Haenelt and H. Freydank, Hydroforming, Advanced High Strength Steel, Springback, Reduced weight, Series production, in *Proceeding New Development in Hydroforming of Sheets, Tubes and Profiles*, (Fellbach, Germany, 2016).
15. A. Hamasaiid, C. Casas and I. Valls, Role of Tool Material in the Interfacial Problematic of Tool/Blank in Press Hardening Process, in *Proceeding CHS2*, (Toronto, Canada, 2015).

16. J. Hardell and B. Prakash, High Temperature Friction and Wear Behavior of Different Tool Steels During Sliding Against Al–Si-Coated High Strength Steel, *Tribol. Int.* **41**, 663 (2015).
17. K. Hertell, G. Moncelsi and W. Teufel, Springback Prediction and Reality in Hydroforming, in *Proceeding New Development in Hydroforming of Sheets, Tubes and Profiles*, (Fellbach, Germany, 2016).
18. H. Karbasian and E. Tekkaya, A Review on Hot Stamping, *J. Mater. Process Tech.* **210**, 2103 (2010).
19. D. Kuhn and R. Kolleck, Warmumformung – den höheren Festigkeiten folgend, *MM das* IndustrieMagazin **17**, 86 (2006).
20. P. Liu, K. Wang, Z. Wang, B. Zhu and Y. Zhang, A New Hot Stamping Process to make Tailored Properties Based on Air Cooling, in *Proceeding CHS2,* (Torento, Canada, 2015).
21. T. Marten, H. Block and T. Troester, Partial Hardening of New Press Hardenable Steels, in *Proceeding CHS2*, (Toronto, Canada, 2015).
22. M. Merklein, J. Lechler and T. Stöhr, Characterization of tribological and thermal properties of metallic coating for hot stamping boron-manganese steels, in *Proceedings of the 7th International Conference Coatings in Manufacturing Engineering*, (Chalkidiki, Greece, 2008).
23. O. Rebuffet, Materials in Car Body Engineering, in *Automotive Circle international (ACI)*, (Bad Nauheim, Germany, 2015).
24. M. Schaefer, D. Schuoecker, O. Spitzer and T. Harrer, Laser Softening of Press Hardened Steel in High Volume Production Lines, in Proceeding CHS2, (Torento, Canada, 2015).
25. M. C. Somani, L. P. Karjalainen, M. Eriksson and M. Oldenburg, Dimensional changes and microstructural evolution in a B-bearing steel in the simulated forming and quenching process, *ISIJ Int.* **4**, 361 (2001).
26. T. Stoehr, J. Lechler and M. Merklein, Investigations on different strategies for influencing the microstructural properties with respect to partial hot stamping, in *2nd International Conference on Hot Sheet Metal Forming of High-Performance Steel(CHS2)*, (Luleå, Sweden, 2009).
27. H. Steinbeiss, H. So, T. Michelitsch and H. Hoffmann, Method for optimizing the cooling design of hot stamping tools, *Production Engineering—Research and Development* **1**, 149 (2007).
28. F. Zimmermann, J. Spoerer and W. Volk, Partial Tempering of Press Hardened Steel by Direct Flame Impingement – the Review of an Alternative Approach for Tailored Properties, in *Proceeding CHS2*, (Toronto, Canada, 2015).

Innovation, Development and Implementation of the HFQ® Process

L. L. Wang[1], T. Dean[2] and J. G. Lin[1,†]

[1]*Mechanics of Materials Division, Department of Mechanical Engineering, Imperial College London, UK*

[2]*Department of Mechanical Engineering, University of Birmingham, UK*

[†]*E-mail: Jianguo.Lin@Imperial.ac.uk*

The material most immediately available for further light weighting body and chassis structures of mass produced automobiles is aluminium alloy and currently this is being used in relatively small quantities, mainly for premium cars. The main barrier to its wider use is its lower room temperature formability compared with that of deep drawing steel and, if to alleviate this, hot/warm forming is attempted, the low resulting mechanical properties. To overcome these drawbacks, the 'Hot Forming and cold die Quenching', HFQ® process was conceived and subsequent research and development has resulted in a technology that has been proven to enable considerable light weighting of chassis and body structures to be achieved, with productivity and cost acceptable for mass production. This paper is a description of the mechanics of the HFQ® process which provides a means for manufacturing components from sheet aluminium alloy with equivalent shape complexity and significantly lighter weight than of those from steel. The use of a suitable temperature and forming speed to enhance formability and means for obtaining high mechanical properties from heat treatable alloys are described. A number of components designed for both road and air production vehicles have been made using HFQ® and some of these are presented together with estimates of production costs.

Keywords: Light weight; Formability; High strength aluminum alloy.

1. Introduction

The need to lighten the weight of road vehicles is a result of the pollution they generate. Although manufacturing processes cause environmental pollution, by far the most pollution arises as exhaust emissions during the time a vehicle is driven on the road and as shown in Fig. 1, legislation, exemplified by that of the EU, has dictated that in ten years to date, the maximum allowable emission of carbon dioxide from an automobile is reduced from 161g/km to 95g/km, a decrease of 41%. Great strides have been made in increasing the efficiency of internal engines and emissions of carbon and nitrogen containing gases have been much reduced during the past twenty years or so. Another important factor affecting the amount of noxious exhaust gases generated is the weight of a vehicle. Power consumed and volume of exhaust gas emitted, in accelerating and holding high speed is increased as weight increases. Therefore major effort is being expended in minimising weight in current designs of vehicles. As body and chassis structures of a car, comprising largely of components made from sheet and tube, together account for the greatest proportion of total weight, potentially large rewards are to be gained in lightening them, through novel designs using less material or by using materials of low density. Fig. 2 lists the advantages and disadvantages of materials with potential for use in body and chassis structures. Early attempts at reducing weight were based on the use of high strength steels which allowed thinner gauges to be used for equivalent strength. Compared with conventional steel structures, a weight reduction of about 20% is obtainable. A disadvantage of these is that they are difficult to form so feasible shape complexity is reduced and spring back is extensive. Boron steel, which is formed at high temperature, overcomes these two disadvantages and as described in the comprehensive review paper by Karbasian and Tekkaya [1], chassis components such as; A and B-pillars, roof rails, rocker panels, bumpers and side impact beams, are being made in their millions. But part shapes, feasible by this method, are of limited complexity and because of surface degradation due to heating, A-class surfaces cannot be produced. Reinforced polymer composites are used for some premium range cars enabling weight reduction of typically 55%, but they are prohibitively expensive for use in mass produced vehicles. Magnesium is the ultimate low density structural metal and much research is being undertaken to enable it to be formed sufficiently to make useful components, but, as yet its use is very limited. In addition its great tendency to corrode hinders its widespread use. If it could be used fully in body/chassis structures, an estimated weight saving of about 50% could be achieved. Aluminium is currently the most useful material for reducing body/chassis weight, estimated to be about 40% if fully used for body/chassis structures. Currently it is used extensively in one or two premium grade cars and in a pick-up truck. But its utilisation is limited because it has low formability cold and if heated to improve this presents low mechanical properties, so that it is necessary to use a greater amount of metal to achieve a specified strength. In addition, due to its low formability, structures with many shape features that can be made as one part with steel, have to be made with several assembled aluminium parts.

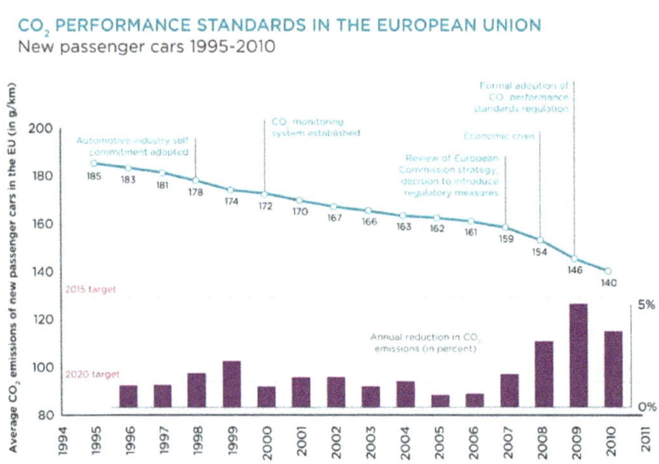

Fig. 1. Background to the Use of Aluminium in Automobiles.

•**Steel:** 97% body and chassis structural components currently are made of deep drawing steel.

• **Ultra high strength steel** – weight reduction up to 20%; fuel saving up to 10%. But very difficult to form.

•**Composites:** Weight saving: up to 55%; - Expensive

•**Aluminium:** Weigh saving: 40%; Fuel saving: 23%. Difficult to form cold; Expensive, Mainly due to high processing cost.

•**Magnesium:** Weight saving could be over 50%. Not used yet – Too expensive for sheets, difficult to form, Corrosion problems.

Fig. 2. Alternative Materials for Lightweight Body and Chassis Structures.

To overcome the drawbacks inherent in the use of aluminium alloys for body and chassis components, the novel process of: heating, forming and cold die quenching, given the acronym HFQ®, was conceived. A schematic of the mechanics of the process is given in Fig. 3. Essentially, the operations are: an aluminium alloy blank is heated to a temperature at which a metallurgical solid solution achieved. The blank is then transferred to unheated tools in which it is formed and held under pressure until cool. Rapid cooling within the tools inhibits precipitation within the alloy and a virtually complete solid solution is retained within the formed part, which for a heat treatable alloy can be aged to achieve ultimate mechanical properties.

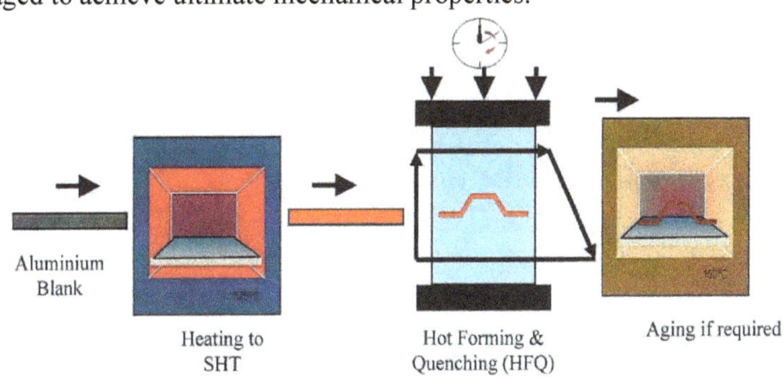

Fig. 3. Schematic of the HFQ® forming Process

The HFQ® process now has reached the stage at which volume production is possible and the research and development steps that have been undertaken leading to this are described in this paper, including results of laboratory tests to illustrate the various deformation characteristics that can be encountered and how these may be modelled for predicting process outcomes and the advantages to be gained in design of component shape and parts made.

2. Work-Piece Formability

It is well known that values of ductility obtained from uniaxial tensile tests do not necessarily translate to draw-ability in a sheet forming process and tests in which conditions are similar to those in shape forming operations should be used. To provide information relating more closely to that required for modelling and predicting outcomes in forming operations where biaxial stress and strain states predominate, novel in-plane biaxial tests have been undertaken to obtain formability data of metal sheets at various temperatures, strain rates and strain paths. The core parts of the novel biaxial formability testing system are a biaxial testing mechanism which was used to convert an input uniaxial force into an output biaxial force of different loading ratios, a Gleeble 3800 which controls heating and cooling processes, and the digit image correlation (DIC) system for strain measurement. AA6082 was used as the sample material, from which the designed cruciform specimens were machined. The material was heated to the solution heat treatment temperature of 535°C at a heating rate of 30 °C/s, soaked for 1 minute and then quenched to a designated temperature in the range of 370-510°C at a cooling rate of 100 °C/s. The tensile tests were conducted at constant strain rate in the range of 0.01-1 /s under different strain paths, i.e., uniaxial, plane strain and biaxial testing. Fig. 4 shows a typical specimen for biaxial tension and resulting forming limit curves [2]. The resulting forming limit curves show, in Figs. 4 (b) and (c) that, judging by the value of major strain at which failure occurred, formability increased with temperature and with increasing forming speed.

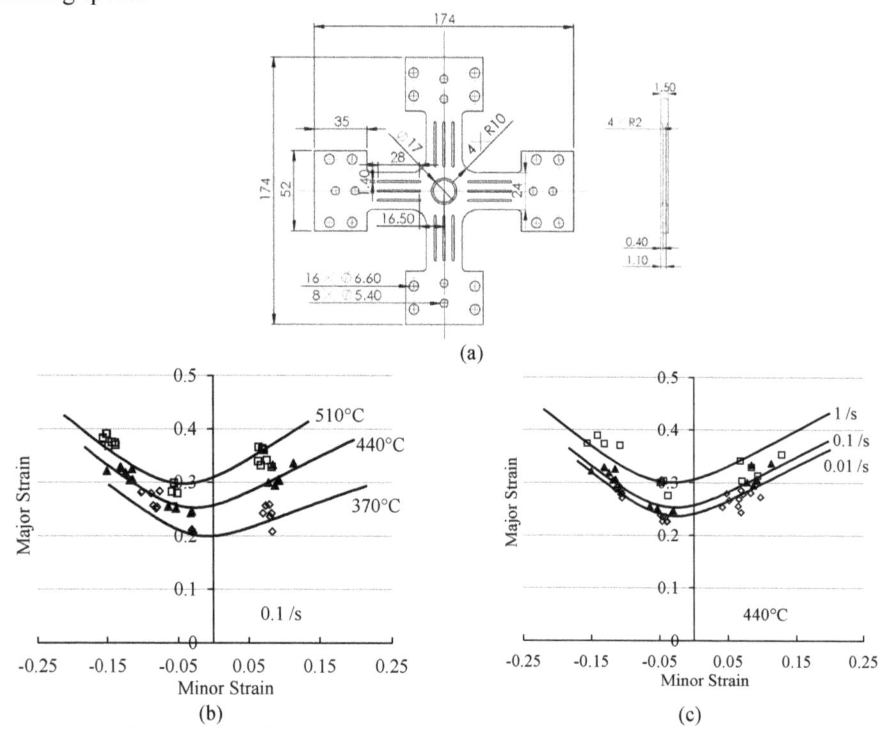

Fig. 4. (a) Cruciform specimen dimensions for biaxial tension (mm); (b) FLD for different temperatures and (c) FLD for different strain rates.

3. Materials Model

Based the previous study [3], a set of unified viscoplastic constitutive equations are developed and formulated as below:

$$\dot{\varepsilon}_e = \left\{\left[\left|\sigma_e/(1-\omega)\right| - R - k\right]/K\right\}^{n_1} \tag{1}$$

$$\dot{\varepsilon}_{ij}^{P} = 3S_{ij}\dot{\varepsilon}_e/(2\sigma_e) \tag{2}$$

$$\dot{R} = 0.5B\bar{\rho}^{-0.5}\dot{\bar{\rho}} \tag{3}$$

$$\dot{\bar{\rho}} = A(1-\bar{\rho})\left|\dot{\varepsilon}_e\right| - C\bar{\rho}^{n_2} \tag{4}$$

$$\sigma_{ij} = (1-\omega)\,D_{ijkl}\left(\varepsilon_{ij} - \varepsilon_{ij}^{P}\right) \tag{5}$$

$$\dot{\omega} = \left[\Delta^*/(\mu_1 - 0.5\mu_2)^{\phi}\right]\left[(\mu_1\varepsilon_1 + \mu_2\varepsilon_2)/(\gamma + \varepsilon_P)\right]^{\phi}\left[\eta_1(\dot{\varepsilon}_e)^{\eta_2}/(1-\omega)^{\eta_3}\right] \tag{6}$$

Where the temperature dependent parameters in Eqn. (1)-(6) are defined as the form of Arrhenius equation:

$$K = K_0 \exp\left[Q_K/(R_g T)\right] \tag{7}$$

$$k = k_0 \exp\left[Q_k/(R_g T)\right] \tag{8}$$

$$n_1 = n_{1_0} \exp\left[Q_{n_1}/(R_g T)\right] \tag{9}$$

$$B = B_0 \exp\left[Q_B/(R_g T)\right] \tag{10}$$

$$C = C_0 \exp\left[-Q_C/(R_g T)\right] \tag{11}$$

$$\eta_1 = \eta_{1_0} \exp\left[-Q_{\eta_1}/(R_g T)\right] \tag{12}$$

$$\eta_2 = \eta_{2_0} \exp\left[Q_{\eta_2}/(R_g T)\right] \tag{13}$$

$$\eta_3 = \eta_{3_0} \exp\left[Q_{\eta_3}/(R_g T)\right] \tag{14}$$

$$E = E_0 \exp\left[Q_E/(R_g T)\right] \tag{15}$$

$$\mu_1 = \mu_{11} \exp(-\mu_{12}/T) \tag{16}$$

$$\phi = \phi_{11} \exp(-\phi_{12}/T) \tag{17}$$

$$\Delta^* = \Delta_{11}^* \exp\left(\Delta_{12}^*/T\right) + \Delta_{21}^* \exp\left(\Delta_{22}^*\dot{\varepsilon}_e\right) \tag{18}$$

Where R_g is the universal gas constant, T is the absolute Kelvin temperature. A, n_2, K_0, k_0, n_{1_0}, B_0, C_0 η_{1_0}, η_{2_0}, η_{3_0}, E_0, Q_K, Q_k, Q_{n_1}, Q_B, Q_C, Q_{η_1}, Q_{η_2}, Q_{η_3}, Q_E, μ_{11}, μ_{12}, ϕ_{11}, ϕ_{12}, γ, Δ_{11}^{*}, Δ_{12}^{*}, Δ_{21}^{*} and Δ_{22}^{*} are material constants to be determined from experimental data of formability tests.

Eqn. (1) is the flow rule, in which effective strain rate $\dot{\varepsilon}_e$ is formulated according to the traditional power law with damage ω taken into account. k is the initial yield point, σ_e is the von-Mises stress and R represents the isotropic hardening. S_{ij} in Eqn. (2) is deviatoric stress components and $\dot{\varepsilon}_{ij}^{P}$ represents plastic strain rate. R in Eqn. (3) is a function of the normalized dislocation density $\bar{\rho}$. Eqn. (4) represents dynamic and static recovery and the accumulation of dislocations due to plastic flow. The flow stress equation was modified to include the effect of damage ω in Eqn. (5), where D_{ijkl} is the elastic matrix of a material. The damage evolution $\dot{\omega}$ in Eqn. (6) was based on the nucleation and growth of voids around particles. The first two terms in Eqn. (6) are proposed to describe the effects of major strain and minor strain on the damage evolution in the material. ϕ is introduced in Eqn. (6) to control the intensity of the effects of principal strains on the damage evolution, thus to control the predicted FLD of the material. μ_1 and ϕ are defined as temperature dependent and Δ^* is defined as both temperature and strain rate dependent, given in Eqns. (16)-(18).

Material constants in the equations (1-18) can be determined by fitting computed stress-strain curves and FLDs to corresponding testing results. The material constants are listed in Table 1. Good agreement between experimental and computed results for each forming limit curve can be seen in Figs.4 (b) and (c).

Table 1. Material Constants for Eqn. (1)-(18) for AA6082.

K_0 (MPa)	k_0 (MPa)	n_{0_1}	B_0 (MPa)	C_0	η_{1_0}	η_{2_0}	η_{3_0}
0.36	0.59	0.10	4.83	7.31	2.46	0.84	0.24
Q_K (J/mol)	Q_k (J/mol)	Q_{n_1} (J/mol)	Q_B (J/mol)	Q_C (J/mol)	Q_{η_1} (J/mol)	Q_{η_2} (J/mol)	Q_{η_3} (J/mol)
25409.5	21693.0	17637.1	15698.9	2112.6	16273.2	837.8	22107.9
Q_E (J/mol)	R (J/molK)	E_0 (MPa)	A	n_2	μ_{11}	μ_{12}	μ_2
27987.4	8.31	249.7	0.19	1.83	0.85	451.9	0.15
ϕ_{11}	ϕ_{12}	Δ_{11}^{*}	Δ_{12}^{*}	Δ_{21}^{*}	Δ_{22}^{*}	γ	
19.09	874.6	79.8	26.1	-81.1	3.52E-3	3.5E-3	

4. Boundary Conditions

In addition to work-piece characteristics, a process is influenced by boundary conditions, which in the case of sheet forming comprise tool/work-piece friction and heat transfer from work-piece to tool. The values of these variables will not be uniform over the surface of a tool and it is necessary to relate quantitatively both friction and heat transfer to pressure and work-piece sliding.

4.1. *Tool/Work-piece friction*

A device in which a test-piece in the form of a strip is pulled through a clamp consisting of two pads representing tool surfaces has been made using which tests can be performed on a tensile testing machine either within or without a furnace. The clamping force divided by the pulling force was taken as the coefficient of friction. Fig. 5 is an example of a result obtained using a lubricated test-piece. Static friction, that at the start of sliding, is seen to be greater than that during sliding, which is known to be so for most sliding situations. Friction coefficient remains fairly constant as temperature increases but at about 300°C it breaks down. Therefore, in choosing the best operating temperature for a process lubrication must be considered in addition to work-piece characteristics.

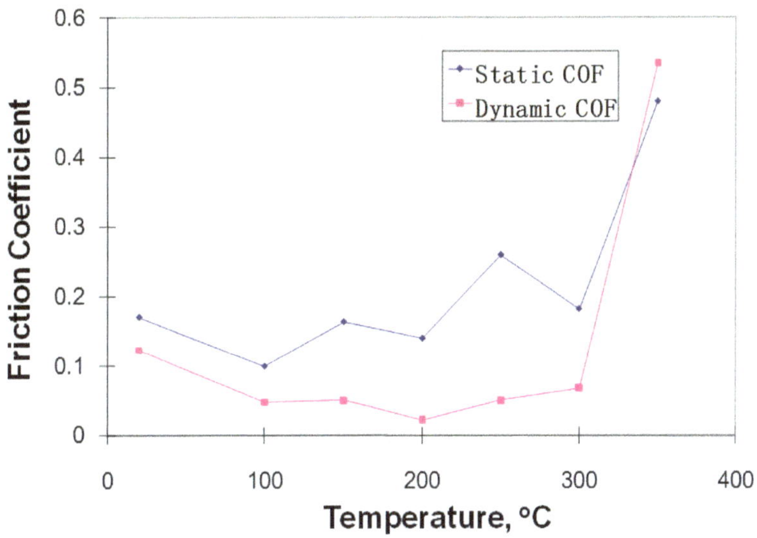

Fig. 5. Friction Measurement for a Lubricated Aluminium Alloy.

4.2. *Interfacial heat transfer coefficient (IHTC)*

In an HFQ® operation, heat is transferred from work-piece to tools both during deformation and during the period when a work-piece is clamped under pressure immediately deformation is concluded. Success in restoring mechanical properties during subsequent ageing depends on the rate of heat loss from the work-piece being sufficiently high to inhibit significant precipitation within its microstructure. To determine the effects of process conditions on heat transfer coefficient, dedicated equipment, the 'IHTC-Mate', was developed and used, as shown in Fig. 6. Essentially they consisted of two flat platens made from die materials, and a strip, made from workpiece material, having thermocouples embedded within the workpiece. The flat platens are changeable enabling die material effects on the IHTC to be studied. The temperature evolutions detected by the thermocouples when a heated test-piece is pressed between the flat surfaces is considered analogous to that arising during an HFQ® operation, either during deformation, or during the post-deformation holding period, when the platens press directly on the work-piece. By reverse engineering using FE simulation, a value of heat transfer coefficient can be related to the measured temperature field and therefore to a particular process condition of tool/work-piece contact pressure or tool/work-piece gap.

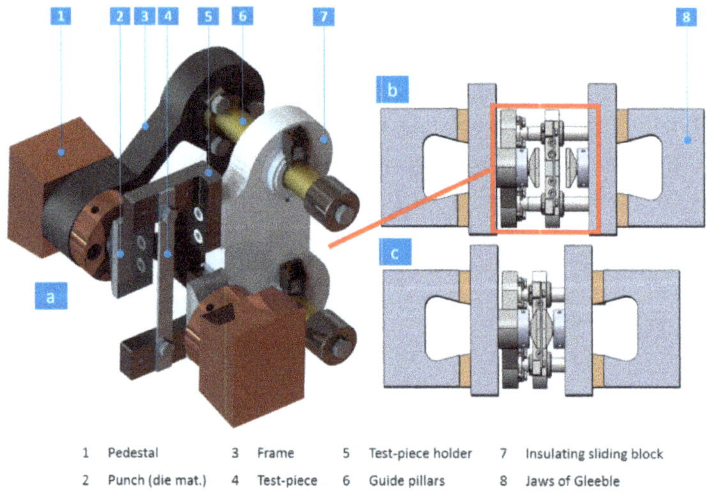

Fig. 6. (a) Schematic structure of the IHTC-Mate (www.smartforming.com) under (b) test-piece loading condition and (c) testing condition.

5. CAD of Tools for HFQ®

5.1. *Shape complexity*

Compared with steel components, greater shape complexity is obtainable in parts made by HFQ®. Reasons for this are illustrated in Fig. 7. First, HFQ® aluminium alloy in general has grater ductility than both conventional DD and boron steels, therefore deeper sections are possible. In addition, tighter radii can be formed. These two attributes allow sections with enhanced second moments of area to be formed which compensate for reduced stiffness resulting from the lower Young's modulus of aluminum alloy.

5.2. *Shape accuracy*

As an HFQ® part is held firmly between the tools until it has reached near room temperature and regained a relatively high yield strength, the degree of spring-back experienced when it is ejected, is very small. Thus consistent accuracy is obtainable in production and in contrast to cold forming, trial and error testing with modification of tool shape is unnecessary. An example relating to a simple top-hat shape is shown in Fig. 8 where the cold formed part is much distorted from the orthogonal design [4].

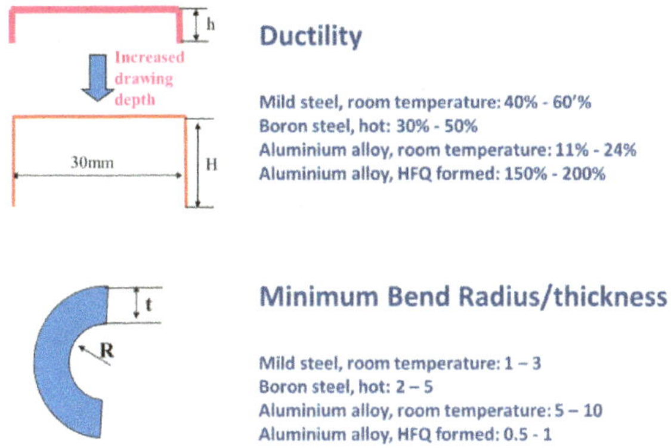

Fig. 7. Geometric Advantages in Shapes Feasible by HFQ®.

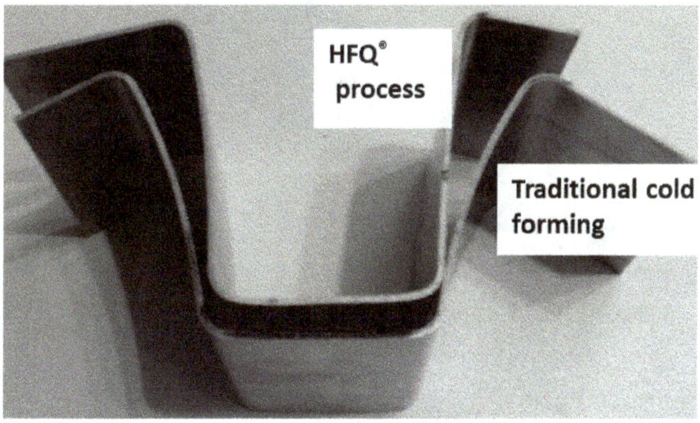

Fig. 8. Comparison of Spring-back in a Shape Cold Formed With One Made by HFQ®.

5.3. *Expert system*

Data consisting of work-piece alloy constitutive properties, plus values of friction and heat transfer coefficients enables accurate process simulation to be undertaken. An expert system is being constructed which will be embedded in commercial software, currently PAM-STAMP, using which process simulation is undertaken and through iteration, component design and process conditions can be tailored to suit manufacture by HFQ®.

Using the system, design consists of several steps:

i. The first step requires producing a digital version of the suggested part shape together with mechanical properties specification.

ii. Using a knowledge-base comprising data from previous tests, two-part or three-part tooling is designed to suit the part shape, including a cooling facility if productivity will be high.

iii. A blank shape, preheat temperature, lubricant and surface treatment of the tool, also press speed, dwell time and load are chosen.

iv. The process is simulated and the output data; comprising product soundness relating to necking, failure, wrinkling and completeness of shape are examined.

v. If defects are observed, such as the thinning shown in the part pictured in Fig. 9, reference will be made to the knowledge base to identify suitable corrective amendments to the process.

vi. Iterations of process variables will be undertaken until a satisfactory part is produced in a manner suitable for production.

vii. Trimming and punching operations will be identified and a suitable ageing treatment chosen.

The system will enable part design and processing to be optimized for production by HFQ®.

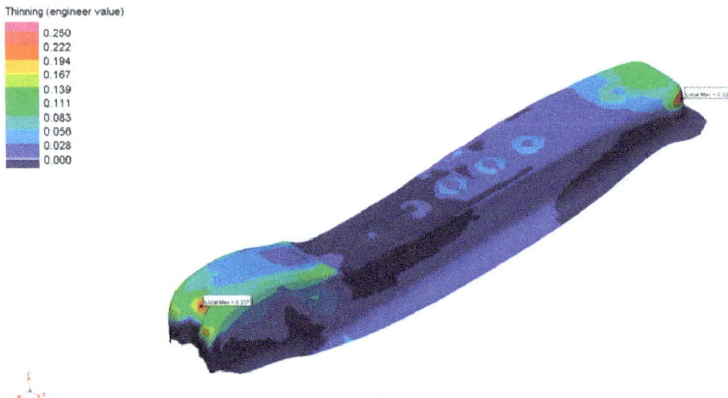

Fig. 9. Simulated Formed Part Showing Thickness Variations.

6. Exemplar Formed Parts

To date several parts have been formed by HFQ® both for feasibility studies and small batch production for use in current automobile on the road.

Fig. 10 shows the lower section of an automobile B-Pillar which demonstrates not only that the shape can be made but also that good mechanical properties are achievable by ageing the formed part.

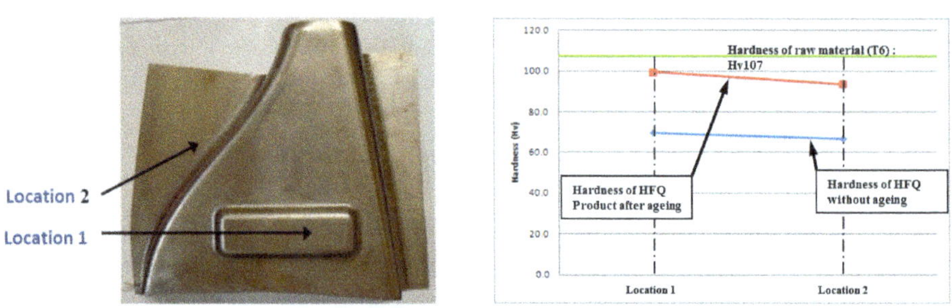

Fig. 10. B-Pillar Part Formed and Aged.

Fig. 11 illustrates the cost savings that may be achieved by using HFQ® to form parts that by tradition have been machined. The material yield is vastly greater and production time very much lower for this part shape which currently is machined from billet. It is obvious that cold forming the part is not possible.

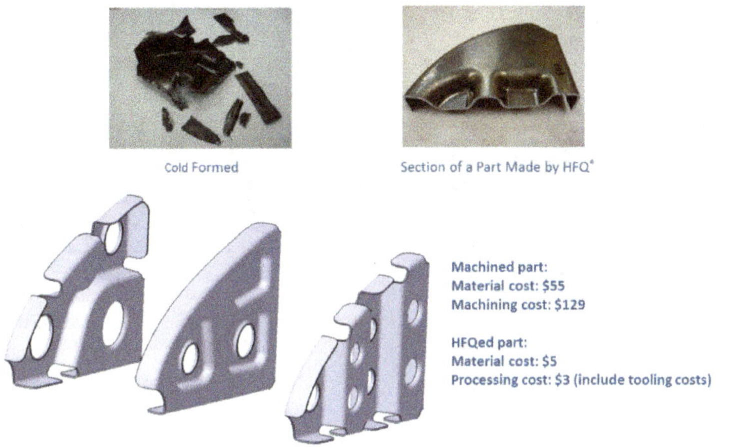

Fig. 11. Aircraft Wing Stiffener – A Feasibility Study in AA2024.

Fig. 12 shows a breakdown of estimated weight and cost of a car door inner panel made by three processes. Note that, due to lack of formability, it is necessary to make the cold formed aluminum version currently used for some premium vehicles, with five individual components joined together.

Fig. 12. Potential Weight and Cost Savings in Using a Automobile Door Inner Panel Made by HFQ®.

A panel made by HFQ® as one part is shown in Fig. 13. This was for a high performance vehicle and had a much deeper draw than is usual.

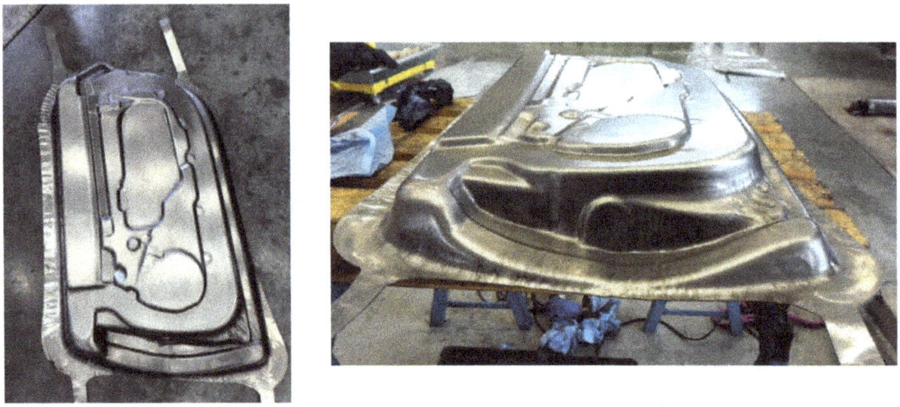

Fig. 13. AA6082 door inner panel With a Maximum Draw Depth of About 200 mm Made by HFQ®.

7. Conclusion

The development stages leading to a novel process named HFQ®, for the manufacture of complex shaped, high strength panel parts, is described in this paper. The main features of the process and the R&D work undertaken to bring the process to a state where it is ready for series production, are explained. Demonstrator parts have been presented and the benefits of the process to enhanced product quality, cost and productivity are highlighted.

References

1. H. Kabasian and A. E. Tekkaya, A Review on Hot Stamping, *J. Mater. Process Tech.* **210**, 2102 (2010).
2. Z. Shao, N. Li, J. Lin, and T.A. Dean, Development of a New Biaxial Testing System for Generating Forming Limit Diagrams for Sheet Metals Under Hot Stamping Conditions, *Exp. Mech.* 1 (2016).
3. J. Lin, M. Mohamed, D. Balint and T.A. Dean, The development of continuum damage mechanics-based theories for predicting forming limit diagrams for hot stamping applications, *Int. J. of Damage Mechanics* **23**, 684 (2014).
4. A.L. Wang, K. Zhong, O. El Fakir, J. Liu, C.Y. Sun, L.L. Wang, J.G. Lin and T.A. Dean, Springback analysis of AA5754 after hot stamping: experiments and FE modelling, *Int. J. Adv. Manuf. Technol.* 1 (2016).

Advanced Manufacturing of Locally Graded Components for Lightweight Car Body Design

O. Kragt[†], C. Koroschetz and L. O. Jönsson
AP&T Sweden AB
[†]*E-mail: Otto.Kragt@aptgroup.com*

J. Eman and E. Sandlund
SWEREA SICOMP AB

A large number of todays manufactured components for automotive car body structures are so called monolithic components. This means that these parts have the same mechanical and technological properties over the entire cross section. Especially for safety relevant components, the monolithic design can have a big effect on the crash performance of the car, since those components can only absorb as much crash energy as plastic deformation remains in the used material after the forming operation. This phenomenon applies not only for steel components, but also for components produced out of Aluminum. Hence there is a need for tailoring the properties of such components. Production solutions which are currently available are limited in design freedom and add a significant cost. Varying from case to case, these do not always provide an optimal solution. Therefore AP&T is developing various production solutions for locally adjusting the mechanical properties of car body components. These solutions are competitive in different aspects. One very promising solution is based on press hardening technology and offers cost effective tailored tempering with high flexibility in part design. This is reached by adding a second heating station to an existing press hardening lay-out. After the blank is homogeneously heated and austenized in the conventional press hardening furnace, the blank is transferred to a second heating station. In this station it is locally heated by radiation, which is directed to specific areas of the blank by means of placing a shield of a specific shape between the blank and the radiation source. The areas which are targeted by the radiation are being heated, in order to keep them in the austenitic phase. The other areas are meant to cool down, transform to ferrite/perlite and lose their hardening potential. This results in partially hardened (martensitic) and partially soft (ferritic/pearlitic) areas after the forming and quenching operation. Another approach to tailored part properties is often addressed with the credo: 'The right material in the right place.' This is of course a very true credo and as such it is not a surprise that AP&T is developing matching production techniques. In this paper a sneak preview is given on the promising combination of aluminum and carbon fiber reinforced plastic.

Keywords: Steel; Aluminum; Carbon fiber reinforced plastic; Press hardening; Tailored tempering; Production solutions.

1. Introduction

Modern automobile manufacturers are challenged by demanding requirements. Increasingly tightened emission regulations are pushing for lightweight car body structures. Whilst increasing safety requirements are pushing for stronger car body structures with build-in ductile zones, absorbing the crash energy. The current mainstream solution is to use high strength steels with tailored properties. The tailored properties can be obtained by welding two steel grades to each other (Figure 1), or by tailored tempering of a single grade [5-7]. Tailored Welded Blank (TWB) and Tailored Tempered Blank (TTB) respectively.

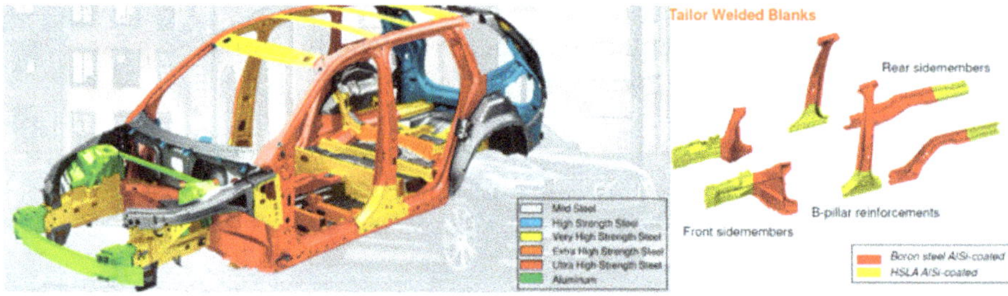

Fig. 1. Tailored welded blanks in the Volvo XC90 as presented at the EuroCarBody 2014.

One approach for tailored tempering is to control the cooling rate of the blank after austenizing. A faster cooling rate, above the critical cooling rate leads to a transformation to martensite. A a slower cooling rate which lies significantly below the critical cooling rate may lead to Ferrite/pearlite. A different cooling rate within one heated blank can be obtained by selectively heating areas of the blank, whilst the other areas cool down on air. [1,2,3,4].

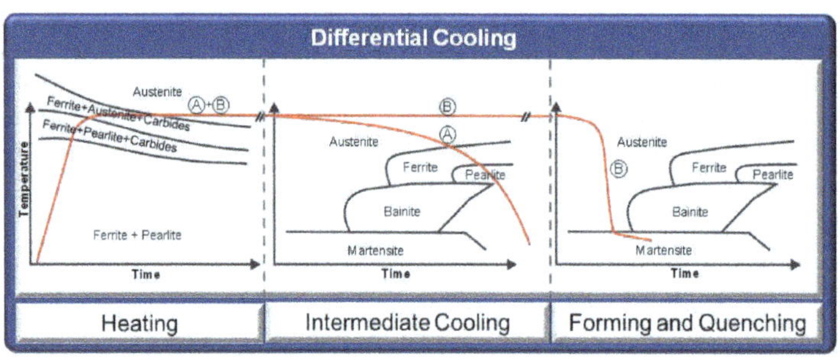

Fig. 2. Tailored tempering by differential cooling with a local intermediate cooling in ambient air, A represents the soft zone, B the hardened zone.

The differential cooling approach (Figure 2) is very suitable for a material which needs to be uniformly heated because of its coating properties. This applies to a heat treatable steel, such as the 22MnB5, with AlSi coating for example.

If the uniform heating of the whole blank is not needed, an alternative to differential cooling is to use differential heating instead (Figure 3). This can be achieved by heating selected parts of the blank so that they form austenite whilst the other areas are kept in ferritic/pearlitic phase. When the blank is quenched only the austenitic areas will transform to martensite.

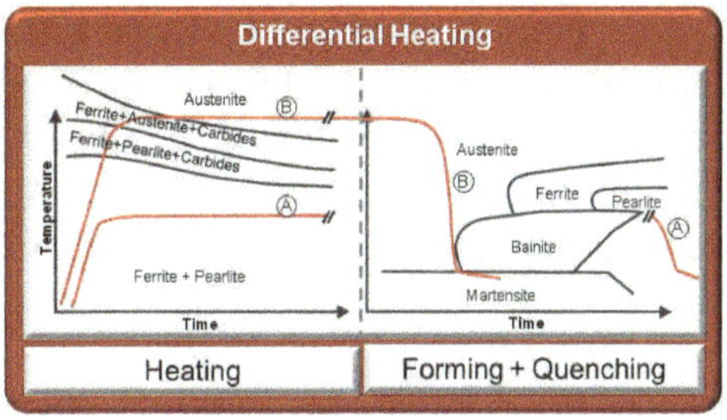

Fig. 3. Tailored-tempering by differential heating, A represents the soft zone, B the hardened zone.

The advantage of the TTB is the in-situ tailoring of the properties in the hot stamping line. This means that no different materials have to be ordered. No extra cost is added for a more complex versus a simple tailored design. This implies more design freedom, less room for errors and a reduction in costs.

In the future it seems that there may be a need for even lighter solutions. One possibility is to use a combination of CFRP and aluminium. This combination creates a cost efficient material with a high light weighting potential. It poses however several challenges as well. One challenge is to find a cost efficient solution to join the two materials together. In this paper the idea of introducing the CFRP in the aluminium hot forming process is discussed.

2. Tailored Properties in Press Hardening

2.1. *Flexibility in tailored properties*

One of the major advantages of tailored tempering lies in its flexibility. The technology allows the creation of multiple zones with different properties. These zones are created simultaneously and can be varied in their size and position.

The size and shape of the zones are created by directing the radiation from an infrared source to the part (Figure 4). Directing the radiation is done by using a mask in-between the IR source and the substrate. The geometry of the mask corresponds to the geometry of the zones to be created. Additionally it is possible to switch of parts of the IR source to increase the energy efficiency.

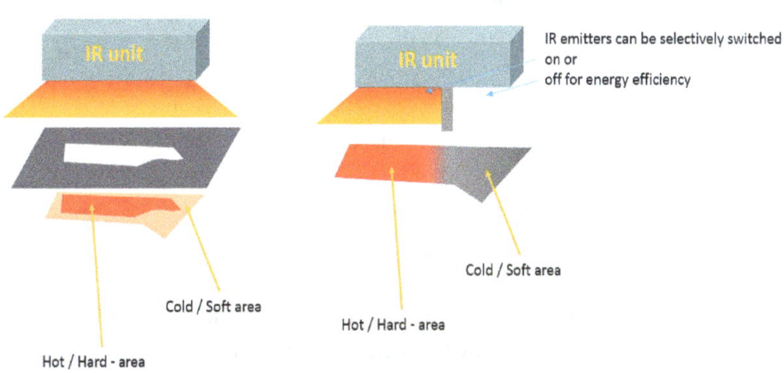

Fig. 4. Partial heating by IR radiation.

With this technology it is possible to create TTB by the differential cooling method and by the differential heating method. For the differential cooling method (Figure 5), a blank will be heated in a conventional furnace such as a multilayer furnace (MLF). After the part is sufficiently heated it will be moved to the partial heating (IR) station. There, selected areas will be kept hot whilst the other areas cool down. Subsequently the blank is transferred to the press where it is formed and quenched.

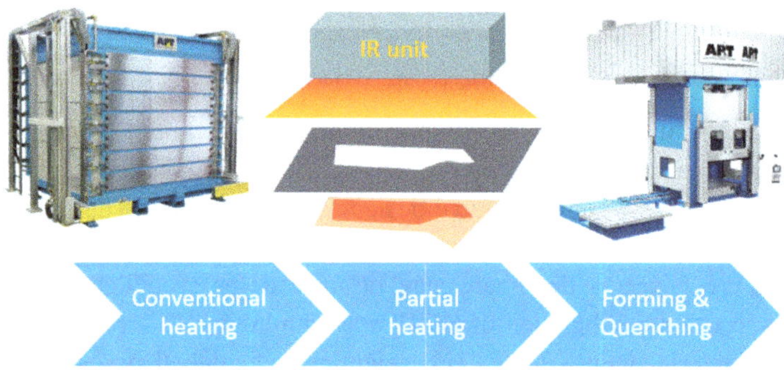

Fig. 5. Differential cooling method.

For the differential heating method (Figure 6) a TTB can be created by placing the blank in the partial heating station. Where selected areas are heated whilst the other areas stay cold. Subsequently the blank is transferred to the press for the forming and quenching operation.

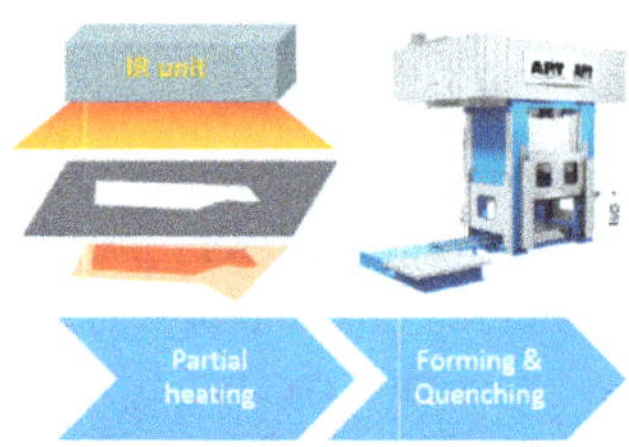

Fig. 6. Differential heating method.

2.2. *Basic experiment*

The basic investigations were carried out at the AP&T R&D and Try-Out centre in Ulricehamn, Sweden. This centre consists of:

- Blank loading station
- Furnace Feeder for blank in and out feeding of the furnace
- 7 cavity Multilayer chamber feeder furnace
- Blank centring table
- 2 speed feeders for press feeding
- 12,000 KN hydraulic high speed press hardening press
- Process control/monitoring system

The partial heating station is located on the opposite side of the multi-layer furnace above the blank loading station. After austenizing the blank, it is transported with the furnace feeder from the furnace to the partial heating station and kept underneath, until the required temperature targets are reached (see Figure 7 left).

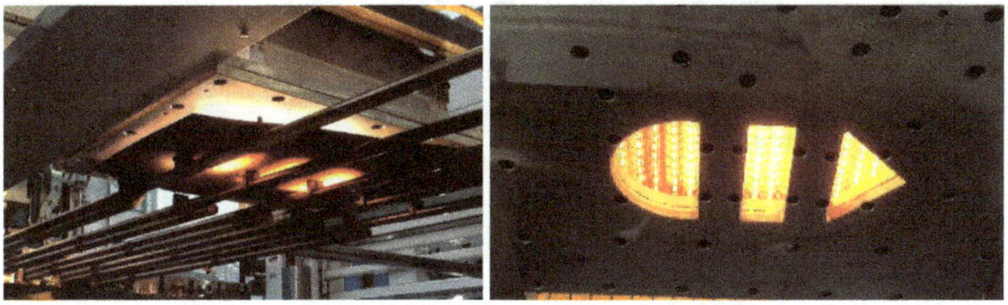

Fig. 7. Left: Experimental setup at AP&T's R&D facility, local heated blank under test device; Right: Complex pattern for soft/hard zones, cooled mask under radiation unit.

For this basic investigation a complex shaped pattern of soft/hard zones (see Figure 7 right) was chosen, in order to analyse:

- max./min. transition zone,
- identifiability between the different hot zones (hardness drop) and
- geometrical boarders.

Applied test parameters:

- Blank material: 22MnB5, AlSi (150 g/m²)
- Sheet thickness: 1,5 mm
- Time in furnace: 300 sec.
- Press force: 4,000 KN
- Holding time under the radiation station: 20-90 sec. (in steps of 5 sec.)
- Standard forming and quenching

Figure 8 shows a formed component. The soft/hard zones can be clearly seen through the different colours of the AlSi coating, due to the different diffusion time of these zones. All parts, even the one with 90 seconds under the heating station, were forming without any failures. 90 seconds under the heating station means 90 seconds of cooling time for the soft zones. This long time is not recommended for serial production, since the coating cools off too much and will cause abrasion in the tool.

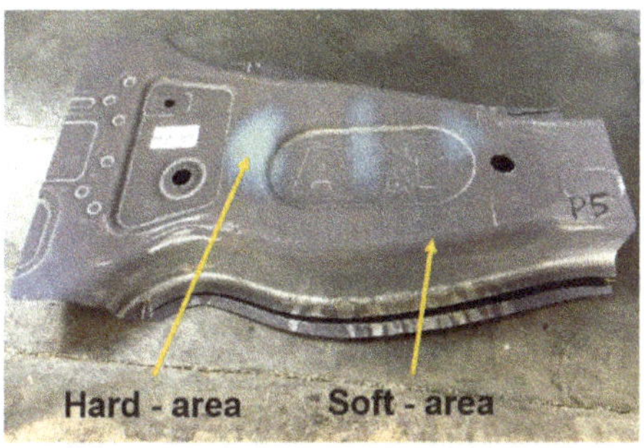

Fig. 8. Press hardened component with locally adjusted soft/hard zones.

Figure 9 shows the time-temperature curve for the partial heating process underneath the heating element. The recommended time for this test setup under the heating station in order to reach fully hardened and soft zones was determined between 20 and 30 seconds. This time is depending on when the phase transformation from Austenite into Ferrite/Perlite occurs. This can be clearly seen through the latent heat occurrence after 20-30 seconds of cooling on air.

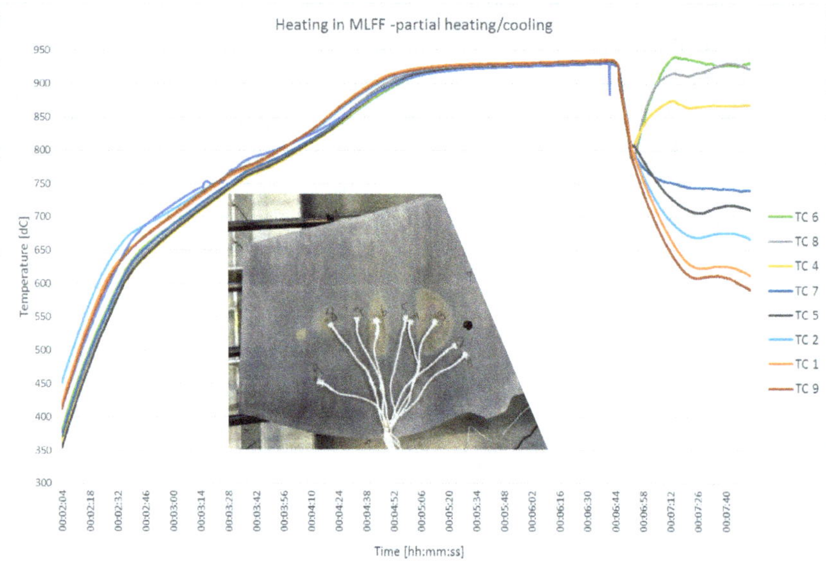

Fig. 9. Time/temperature curves on various thermocouple positions (hot and cold zones).

Figure 9 clearly shows that hot and cold areas can be separated very precisely which leads to the conclusion that quite small transition zones, e.g. 20mm, can be achieved. A variation of the transition zone can be achieved by either varying the distance between the part and the cooled mask or by varying the mask opening design.

2.3. *Results*

After performing the trails the initial hardness measurements were carried out with the non-destructive measurement system 3MA. Figure 10 shows the hardness distribution over a hardened component that was tailored heated for 20 seconds under the partial heating station. The hardness values clearly show the distinction between the soft and the hard zones.

Figure 11 left shows, that the soft zone is fulfilling the OEM target requirements for tailor-tempered zones in terms of maximum strain values. The hardened zones reach the standard values of hardened 22MnB5. Figure 11 right shows the hardness measurements of the part cross section. For this case a transition zone of 40mm was achieved, but as stated earlier it is possible to vary the transition zone to smaller and bigger values by changing various parameters.

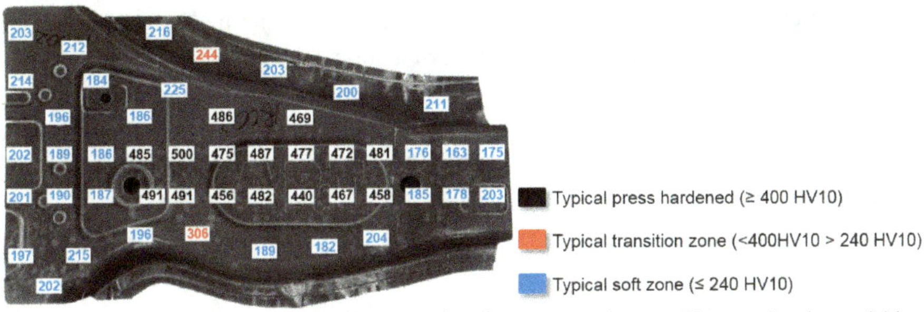

Fig. 10. 3MA measurement results [HV10] of a hardened component that was 20sec. under the partial heating zone.

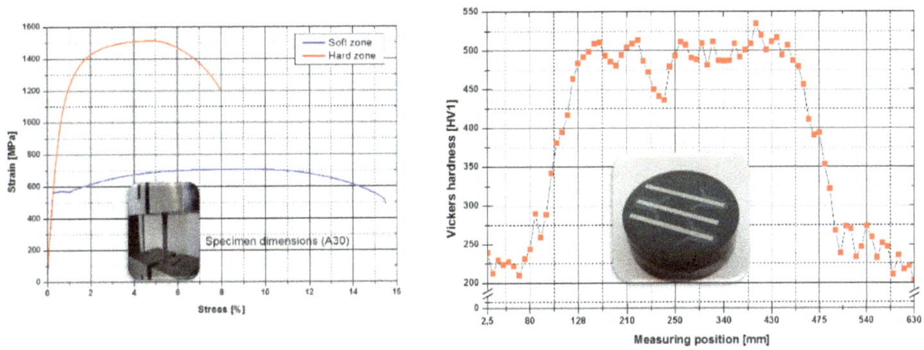

Fig. 11. Left: Stress/strain curve of soft/hard zone, Right: Hardness measurement over the part cross section.

Furthermore the coating composition was measured in terms of total coating thickness, interdiffusion layer thickness and surface roughness.

Hard zone:

- total coating thickness: 40,5 – 44,9 μm
- Interdiffusion layer: 10,1 – 10,7 μm
- Surface roughness:Ra=3,42 μm, Rz=24,4 μm

Soft zone:

- total coating thickness: 39,3 – 44,7 μm
- Interdiffusion layer: 9,1 – 9,8 μm
- Surface roughness:Ra=3,32 μm, Rz=23,8 μm

All measured values, as well as in the hard and the soft zone, fulfil the OEM requirements.

3. Combined Benefits of Aluminium and Carbon Fibre Reinforced Plastic

The positive strength to weight ratio and crash energy absorbing properties for both Aluminium and CFRP leads to a significant weight saving potential compared to a steel solution. The weight savings seem to lie in the range of 40% for aluminium and 100% for CFRP [8]. A drawback of these materials is the added cost compared to the steel solution. The additional costs are reported to lie in the range of 500% for aluminium and 2000% for CFRP [8].

The high cost of a CFRP part is currently mainly due to the high raw material prise which is contributing to about 70% of the costs of a composite part [9].

These high material costs quickly lead to the idea to optimize the material usage. One solution could be to use it only in areas where the additional strength is required. Other areas and functionalities could be fulfilled by a more cost efficient light weight alternative.

Another aspect of discussion is the joining of CFRP to other parts. Current applications of CFRP are joined by gluing [10]. Whilst car body shops are fitted for metal joining techniques like spot welding and riveting. The light weight alternative material should thus favourably combine lightweight properties with metal-like processing requirements.

For the curing of CFRP's, heat and pressure is often required. Hence combined heat and pressure has to be introduced somewhere in the process line. Exactly this combination is already present in the Aluminium hot forming quenching process (HFQ®) (Figure 12). Hence there seems to be a possibility to develop a highly integrated process for both materials.

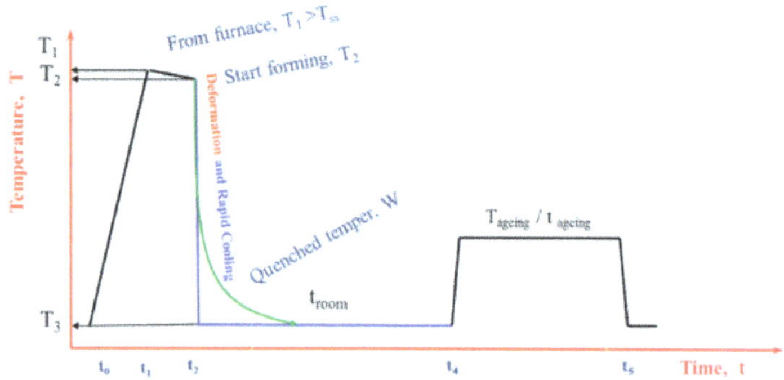

Fig. 12. HFQ® Process.

This has raised the idea of introducing the CFRP in the hot forming step of the HFQ® process (Figure 13). This leads to the following hypothesis: The combined heat and pressure is sufficient to cure the CFRP patch on to the hot aluminium substrate to at least the extend that it is stable for further processing and results in a good adhesion of the patch on the aluminium.

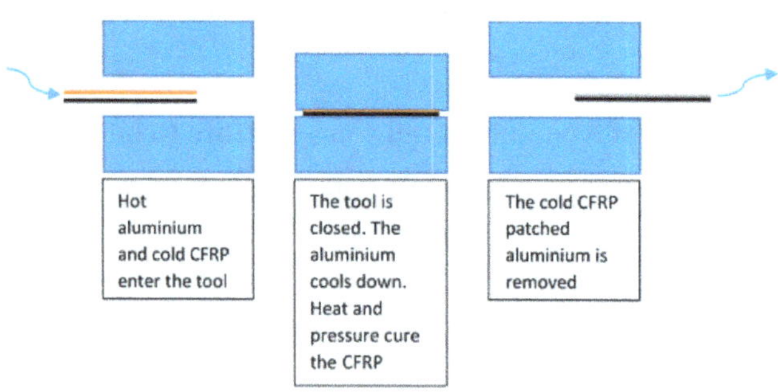

Fig. 13. Basic process flow description of the idea.

3.1. *Feasibility study*

This initial study was kept quit limited and serves as a quick screening test. Most of the equipment is already available and deemed suitable for the study. The materials used were chosen based on expert opinions. Two parameters were identified as critical for the success of this test:

- Adhesion and
- the extent to which the laminate can be cured in the HFQ® process.

For the scope of this test it can be assumed that the adhesion depends on the surface conditions of the substrate and the laminate quality. The laminate quality is assumed to be dependent on the process window: pressure, time and temperature. This leads to the simplification that both parameters (Adhesion and curing) are, at least partially, dependent on the process (time, temperature and pressure). The optimal process window is given and known. What is achievable within the HFQ® process is unknown. This leads to two tests: An adhesion test with the optimal process. And a process test to see whether a suitable time and temperature window can be achieved.

3.2. *Adhesion*

To ensure the adhesion, the tests were executed with optimal laminate quality. The samples were made in the laboratory of SWEREA SICOMP. In this environment the pressure, time and temperature process was exactly matched to the suppliers recommendations:

- Curing time: 5h
- Curing temperature: 120°C
- Pressure: 0,2Mpa

The epoxy resin in the prepreg is supposed to act as an adhesive and should bond to the aluminium substrate. This bonding can however possibly be enhanced by a special adhesive which was developed to match the properties of the epoxy resin. Therefore samples were prepared where this adhesive was used in addition (Figure 14).

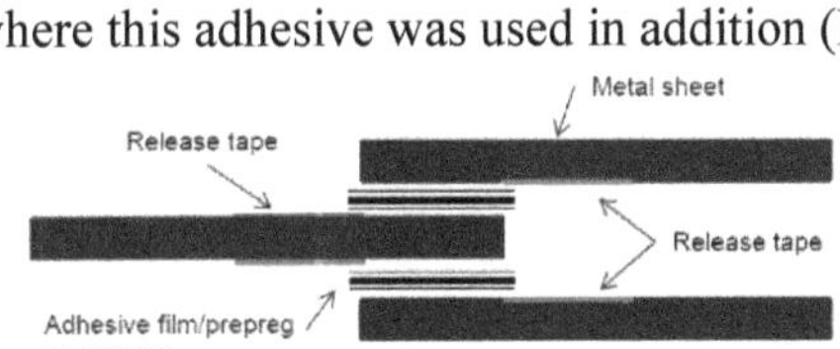

Fig. 14. Schematic drawing of a double lap shear test sample.

To get an idea of the bonding potential of the epoxy resin without the adhesives, a sample was made where the epoxy was bonded to a substrate of the same material: epoxy to epoxy. This sample served as a reference.

Several aluminium samples were produced according to the HFQ® process, in order to be able to simulate the process typical surface conditions on the sheet surfaces.

The adhesion test is executed by means of a double lap shear test (Figure 15) with the following parameters:

- Standard ASTM-D-3528-96

- Specimen length: 160-175 mm
- Nominal width: 25.4 mm
- Nominal joint length: 25 mm
- Grip distance: 110 mm
- Pulling speed: 1.2 mm/min

Fig. 15. Double lap shear test sample in a tensile test machine at SWEREA SICOMP in Piteå, Sweden.

The results from the test are shown in Figure 16. From the results it seems that the adhesive improves the results. When compared to the reference sample it seems that the adhesion to the aluminium substrate is quite low. Several hypothesis can be formulated which explain the root cause, such as: poor surface conditions of the substrate, different mechanical behaviour of the CFRP and the aluminium substrate, poor joint design, etc. For the purpose of this test however, it is clear that the bonding potential in the HFQ® process is expected to be low.

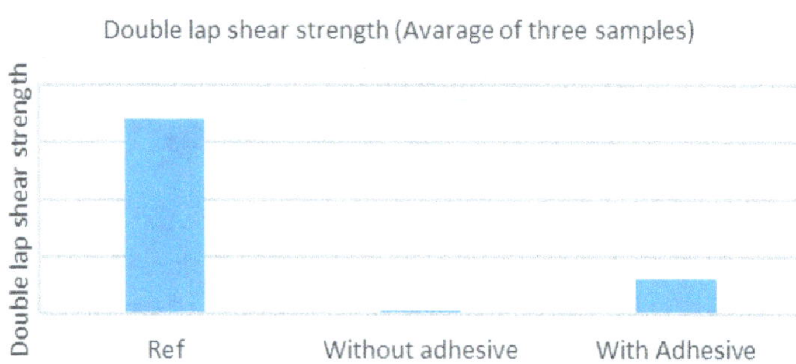

Fig. 16. Double lap shear strength (normalised).

3.3. *Process*

The goal of the process test was to see whether or not a suitable time-temperature window can be achieved by introducing CFRP into the HFQ® process. Therefore samples with different time temperature exposures were processed. Two strategies were used to alter the time temperature exposure:

- Altering the temperature of the incoming hot aluminium and
- changing the cooling rate of the hot aluminium in the quenching dies.

With these two strategies three samples were processed with different time and temperature exposures. The test was executed in AP&T's R&D centre in Sweden. This centre consists of a furnace, Press and automation. A flat quenching die was used for the first test (Figure 17 left). After successful flat die quenching tests, additional tests were carried out with a simple forming and quenching tool (Figure 17 right).

Fig. 17. Left, Flat quenching die, Right, Simple forming and quenching tool.

When the epoxy resin will be exposed to too high temperatures, it will appear in a more brownish colour. This effect can be used to visually judge the results of the test. However, due to the black carbon fibres this effect is quite difficult to see on a CFRP. Therefore glass fibres, which are more transparent are used instead.

Three samples were processed with the following strategies:

- Pre cooling of the aluminium sheet
- Low cooling rate in the quenching dies
- High cooling rate in the quenching dies

The results are shown in Figure 18.

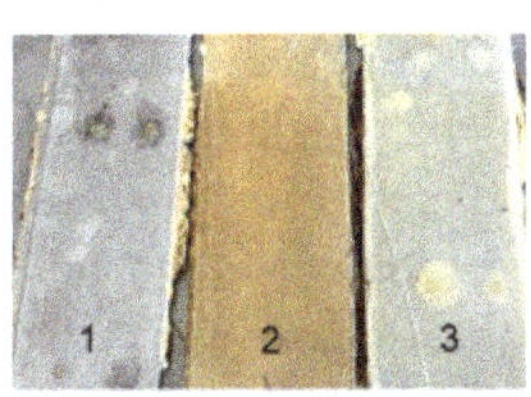

Fig. 18. Parts processed with different exposures to time and temperature.

Interpretation of the results shown in Figure 18:

- The epoxy resin is hard all over the part and thus seems to be cured quite well. The colour of the epoxy resin is quite clear and shows no sign of overheating on the part. On the edges of the part a more brownish colour was seen. This indicates that the temperature has been too high there.
- A much darker colour than the one on the left. Here the epoxy resin has clearly been too hot. What is not seen on the picture, but what is seen by the visual inspection, is that the laminate is detached from the substrate. Especially in the centre of the part.
- The sample on the far right shows results which lie a bit in-between the other two samples. The epoxy resin seems to be fully cured, it is not detached from the substrate. The colour is a bit brownish. This leads to the hypothesis that these settings are on the edge, it is almost too hot.

The conclusion from this small test is that there seems to be a possibility to make a suitable time temperature window with the HFQ® process in order to cure the epoxy resin. The results from sample 1 are deemed to be the most promising. The time temperature window from this test is therefore used to make a part with the forming and quenching tool.

The results are shown in Figure 19. The CFRP on the entire part is hard and the adhesion seems at least enough to hold the part together. Even after the heat treatment. This test seems to be quite successful.

Fig. 19. A part formed and joined in one step.

4. Conclusion

Although the presented technologies are different in the way that they work, they both show the potential of creating parts with tailored properties. In the case of the tailored tempered press hardened parts the advantages are the flexibility of the material properties within a one piece part. Within its own physical limitations, the technology allows for the creation of sizable soft and hard zones where needed. This can be used to optimize the crash performance but also to optimize the joining performance by creating a soft flange to aid self-piercing riveting technologies. Such an optimization seems not necessary for an aluminium part which despite its strength is suitable for several riveting technologies. CFRP on the other hand seems not suitable for such joining technologies. Here there seems to be room for improvement which the proposed combination of CFRP and aluminium tries to fulfil in a simple and cost effective manner. The direct application of CFRP into the HFQ® process seems to lead to poor results on the adhesion of such a patch. The positive results of the process test have shown that it is physically possible to cure the CFRP directly on to the hot aluminium in the HFQ® process.

References

1. K. Steinhoff, Press Hardening - From Ultra-High-Strength Steel to Body-In-White Application, in *Proceedings of the 1st International Seminar on Hot Sheet Metal Forming of High-Performance Steel*, (Hannover, Germany, 2010).
2. M. Geiger, M. Merklein and C Hoff, Basic Investigations on the Hot Stamping Steel 22MnB5, *Adv. Mater. Res.* **6**, 795 (2005).
3. M. Maikranz-Valentin, K. Bergmann, U. Weidig and K. Steinhoff, Components with Optimized Properties due to Modification of Thermo-Mechanical Process Design for Press Hardening, in *Proceedings of the 2nd International Conference on New Forming Technology*, (Bremen, BIAS-Verlag, 2007).
4. K. Steinhoff, New Process Strategies and Resulting Product Structures by Locally Induced Thermo-mechanical Interaction, in *Neuere Entwicklungen in der Massivumformung*, (Frankfurt, America,2005).
5. T. Stoehr, J. Lechler and M. Merklein, Investigations on different strategies for influencing the microstructural properties with respect to partial hot stamping, in *Proceedings of the 2nd International Conference on Hot Sheet Metal Forming of High-Performance Steel*, (Auerbach, Germany, 2009).
6. U. Weidig, K. Bergmann, B. Scholtes and K. Steinhoff, Functionally Graded Properties by Controlled Thermo-Mechanical Interaction in Metal Forming Processes, in *Proc. 8th International Conference on Technology of Plasticity ICTP*, (Verona, Italy, 2005).
7. G. Deinzer, A. Stich, K. Lamprecht, G. Schmid, Rauscher, M. Merklein and J. Lechler, Presshärten von Tailor Welded Blanks: Werkstoffauswahl, Eigenschaften und Verbindungstechnik, in *Erlanger Workshop Warmblechumformung*, (Bambsberg, Meisenbach Verlag, 2008).
8. O. Täger, Einsatz von Faserverbunden für den automobilen Leichtbaus – Chancen und Hearuasforderungen, in *Automotive Circle, Conference Proceedings, Materials in Car Body Engineering*, (Bad Nauheim, Germany, 2016).
9. P. Mårtensson, The challenges of carbon fibre structures, the cost and weight balance, in *Automotive Circle, Conference Proceedings, Materials in Car Body Engineering*, (Bad Nauheim, Germany, 2016).
10. Z. Peterson, Building by Bonding: BMW, the i3 and Carbon Fiber, in *Automotive Design & Production*, 2013.

Hot Forming of an Automotive Skin Panel in Aluminum and in Magnesium

F. Melotti, D. Szegda and A. D. Foster[†]
Impression Technologies Ltd, Lyons Park
Sayer Drive, Coventry, CV5 9PF, UK
[†]*E-mail: a.foster@impression-technologies.com*

Aluminum alloys, such as the 6082 alloy, is now becoming established for future lightweight vehicle structures, while Magnesium alloys remain a specialist material. By developing a low-cost, mass-production route for sheet magnesium, designers can start to explore the use of pressed magnesium panels into future vehicles. In this work, it is presented the feasibility study for an application and development of a hot forming technology to the forming an automotive external body (skin) panel. The reported forming trials of the front cowl panel were conducted using AA6082 and AZ31b alloys.

Keywords: Hot forming; Aluminum; Magnesium; HFQ™.

1. Introduction

The use of aluminum alloy in automotive and other transport industries has been growing rapidly lately, thanks to new manufacturing processes which allow replacement of heavier materials, predominantly steels, at commercially viable costs. One of the breakthroughs in aluminum forming is the solution Heat treatment, Forming and in-die Quenching (HFQ™) process. This patented technology has developed rapidly and it is well regarded for forming complex, lightweight structural panels such as those forming the crash structure.

HFQ™ process, as reported in the literature (e.g. [1, 2]), exploits the high ductility, high formability, of aluminium alloys in their solutionized condition. The rapid forming and holding under load both forms and quenches the light alloy components in the die, which leads to very low levels of spring-back. Solution hardening and precipitation hardening alloys, such as 6xxx and 7xxx aluminium series, are the natural choice for the HFQ™ process since the major benefits gained. Previously the development of HFQ™ has been for structural components where surface finish is not critical. The technology is suited to skin panels after appropriate design refinement of both the panel and the tool surface has been made, although this ability has not been demonstrated until now.

However, the HFQ™ process is also applicable to other light alloy metals. Magnesium is the lightest of all structural metals used in the automotive industry and it offers substantial weight saving in sheet form over both steel (less than ¼ of the density) and aluminum (⅔ of the density) solutions. However, until recently its use was limited to cast components due to very low formability at room temperatures that on a microstructural level is due to the limited slip systems. Recent research of room temperature ductility adds further insight into the active mechanisms leading to low ductility [3] which may lead to improvements in energy absorption for crash structure applications. At elevated temperatures, more slip systems are active and this extends substantially the ductility of the material. The research community has demonstrated the ability to superplastically form magnesium sheet [4], but the process is slow and expensive. The introduction of Quick Plastic Forming (QPF) has led to shorter forming times [5]. However, QPF remains a low volume manufacturing route. By contrast, the HFQ™ method offers the possibility of high-temperature forming coupled with high productivity.

Laboratory scale trials on HFQ™ forming of Mg sheet have shown good promise for a mass-production capable means to form shapes in an otherwise brittle material [6]. One of the key to hot form Mg alloy is the pre-determining the critical forming temperature, several researchers showed that Mg alloys show high plastic formability between 200 and 450°C. Although, in separate work, it was observed that above 450°C phase transformations occur with a consequent decrease in plasticity, experiencing a brittle fracture [7].

This paper reports the result of a project to evaluate the use of HFQ™ to form an automotive skin panel in both Aluminum (AA6082) and Magnesium (AZ31b) sheet materials. The project, named MagHot, took an existing skin panel geometry formed using a low-volume manufacturing route, simulated the formability under HFQ™ conditions, manufactured prototype tooling and produced prototype parts. Here, an overview of the work undertaken is presented and the most recent results are discussed.

2. Experimental Procedure

In order to generate a basic material data set for forming simulation the tensile response of Magnesium over a range of temperatures and strain rates was needed.

Simulation requires a full suite of thermo-mechanical flow stress curves in order to calibrate the material model. Details of such methods have been reported in the literature, e.g. [8]. In the section below, recent advances in the gathering of such data are discussed.

2.1. *Generating high temperature materials data*

High-temperature materials flow data has been captured using a purpose-built materials test rig with infrared heating and air-blast quench facilities, together with a maximum 5000mm/s crosshead speed. The rig, shown in Fig. 1, has been designed to capture the thermo-mechanical response of material that is subject to the same thermo-mechanical cycle as the industrial hot forming process. The rig simulates: i) heating material to the oven temperature; ii) fast quench to test temperature; iii) constant true strain rate testing at constant temperature. The rig has facilities to quench samples after a set strain has been induced to study post-forming material properties. However, in this study the samples were pulled to failure.

Fig. 1. View of the purpose-built tensile testing machine for physical simulation of metal heating, forming and in-tool quenching processes (pictured with chamber doors open). Not shown: High-speed camera for strain measurement; air-blast cooling nozzles.

2.2. *Strain measurements*

In this work, an innovative strain evaluation system is presented for flat specimen, which includes a high speed camera coupled with an image analysis routine. Tensile specimens are painted white, with three black stripes in the gauge length. This novel painting system firstly, enhances the uniformity of the emissivity value (having accurate temperature measurements) and secondly, it enhances the contrast for the subsequent image analysis. A pyrometer, mounted on the tensile machine, monitor the temperature on the black surface of the specimen.

Six strain values are measured in a post-process operation by tracking the movement of the edges of the black stripes from high speed video footage. This method allows to having an understanding of strain localization within the gauge length, whilst setting the minimum tracked gauge length at an appropriate level for subsequent simulation work. The strain measuring process can be further expanded by painting numerous stripes on the specimen, allowing localized strain to be captured in more detail. However, the usefulness of fine resolution data in highly strain localized samples for the simulation of hot forming processes should be questioned before use. The calculated strain is then correlated to the force measurement to output the stress-strain curve.

2.3. *Simulation of a pressed skin panel*

A front cowl panel geometry was selected as a demonstration panel. A similar panel geometry had been formed using a low-volume production route and the geometry features were sufficiently complex to challenge the forming of magnesium sheet.

Forming simulation was carried out in dedicated forming simulation software PamStamp [9]. The process of simulating and optimizing a panel for HFQ™ forming is shown in Fig. 2.

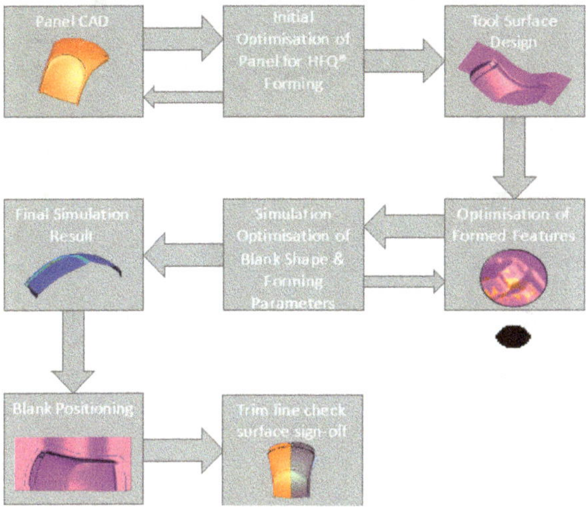

Fig. 2. Process flow for development of HFQ® tool surfaces and forming parameters (blank shape and positioning, forming speed, blank temperature, lubrication, etc.).

The panel geometry was firstly assessed based on known HFQ™ formability criteria and the design modified to be suitable for hot forming. Next, the tool surface was created; the process is similar to that used in developing cold forming tools, although the optimization of addendum geometry is tailored for the flow response of hot aluminum.

Simulation was performed using a fully-coupled thermo-mechanical solver with element refinement to capture wrinkling. A lubricant-free contact condition was assumed. The CAD geometry of the panel and tool were updated based on thinning map results to optimize the final part feasibility. Several iterations of simulation were then performed to optimize the blank design and positioning.

Once the tool surface was finalized, this was sent to the tool maker to construct the physical tooling.

2.4. *Tool try-outs and panel production*

Tools were manufactured from cast iron. In this instance, cooling of the tools was not considered since at trial volumes the cooling was deemed unnecessary.

Figure 3 shows the tool mounted to the press. Blanks were transported manually from the solutionising oven to the tool. For AA6082 alloy, no lubricant was used, while for AZ31b, a small amount of graphite-based lubricant was used.

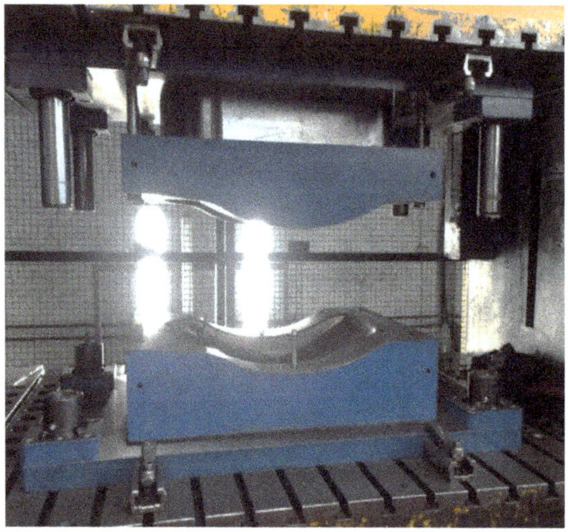

Fig. 3. Cowl tool mounted to the press.

3. Results and Discussions

3.1. *Flow stress curves*

A sample of flow-stress data for AZ31b is shown in Fig. 4. The data given shows the wide range in flow stress that is observed between hot, mid-strain rate (approx. 60MPa flow stress with very little work hardening) and high strain rate at lower temperatures (approx. 190 MPa with pronounced work hardening). The ductility of the material is sufficient for forming complex parts (in excess of 20% elongation), although it is markedly less than is seen in the HFQ™ hot forming of Aluminum alloys where strains at failure can exceed unity. The tensile data was combined with further data from previous studies [3], which was scaled to reflect the strains observed from the more recent hot tensile work.

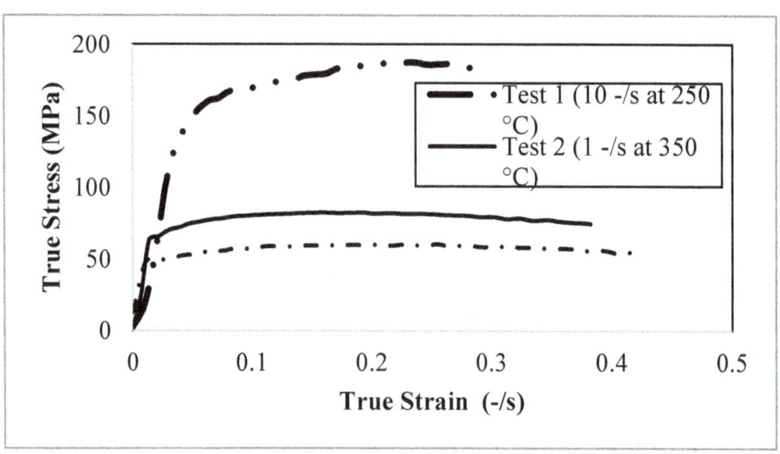

Fig. 4. Flow stress curves from high-temperature testing of commercial Mg AZ31b sheet material.

3.2. *Thinning maps comparison*

Simulation was carried out on the forming process, both for Al and Mg alloys, with a high friction coefficient to represent a non-lubricated pressing. The simulated thinning map for the panel pressed in AA6082 alloy is shown in Fig. 5(a). Some localized thinning was predicted near the grill's edge, indicated by the white arrow. There, AZ31b showed 20% less thinning compared to the AA6082 alloy, as reported in Fig. 5(b). Thus, the panel complexity was set to challenge the formability of Magnesium whilst being suitable for demonstrating a HFQ™ formed skin panel in Aluminium.

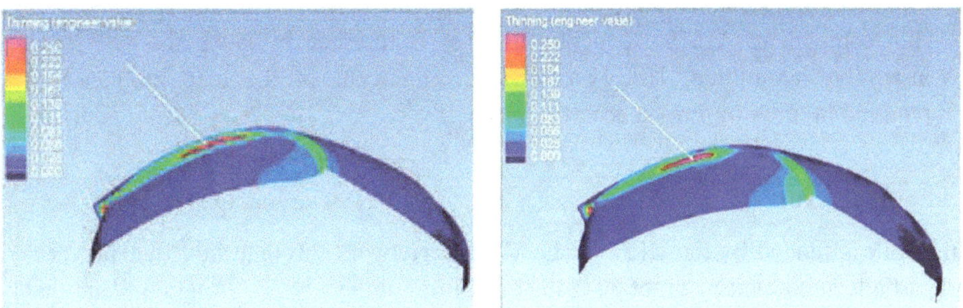

(a) (b)

Fig. 5. PamStamp simulation data for thinning maps of AA6082 (a) and AZ31b (b).

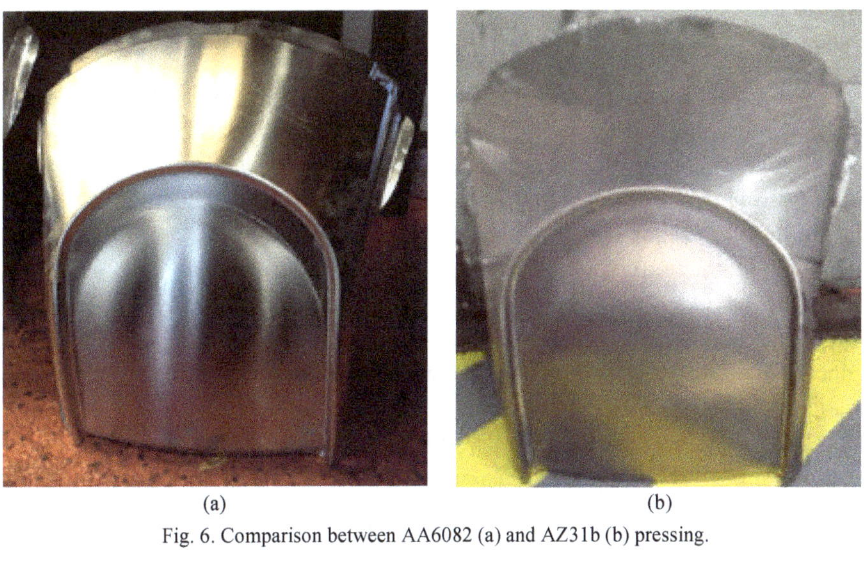

(a) (b)

Fig. 6. Comparison between AA6082 (a) and AZ31b (b) pressing.

Considering the thinning map, precautions were taken and the Front Cowl panel was pressed. Fig. 6 showed the final pressed components for both the Al (a) and Mg (b) alloy. The aluminum formed component shows good surface finish which indicates the HFQ™ process is suitable for skin panels with minimal surface finishing. As suggested by simulation, wrinkles were visible in the Mg panels. Further, some brittle-type splitting was observed. This was traced back to a greater than expected drop in temperature between the oven and the tool due to lower density and higher emissivity than the aluminum blank. However, the output was positive, as the potential for HFQ™ with Mg alloy was demonstrated and further trials will be conducted on an HFQ™ optimized line having automated fast blank transfer between the oven and the furnace.

4. Conclusion

1) HFQ™ Aluminum forming is suitable for skin panels with minimal surface finishing. Further work on the tool surface will improve further the part finish.
2) This study has demonstrated potential towards the feasibility of an Mg pressing for a skin component. A 35% reduction in mass is foreseen by using the AZ31b alloy instead of aluminum. However, further refinement of the material handling is required in order to form successful panels.

Acknowledgments

This work is funded by the UK's Niche Vehicle Network through the 'MagHot' project and partially supported by grant 604240 (LoCoLite) of the EU FP7 Program. With thanks to the Morgan Motor Company for supplying panel CAD data and to PAB (Coventry) for their support in pressing prototype panels.

References

1. D.J. Politis, A.D. Foster, N. Li, L. Wang and J. Lin, Prediction of thinning behaviour for complex-shaped, lightweight alloy panels forming through a hot stamping process, in *Advanced High Strength Steel and Press Hardening: Proceedings of the 2nd International Conference (ICHSU2015)*, (Changsha, China, 2015).
2. A. Foster, D. Szegda and J. Sellors, Design considerations for HFQ® hot stamped aluminium structural panels, in *Proc. of Int. Conf. New Forming Tech.* (Glasgow, UK, 2015).
3. M. Lentz, M. Risse, N. Schaefer, W. Reimers and I.J. Beyerlein, Strength and ductility with {1011} – {1012} double twinning in a magnesium alloy, *Nat. Commun.* Article number: 11068 (2016).
4. A. El-Morsy, K. Manabe and H. Nishimura, Superplastic Forming of AZ31 Magnesium Alloy Sheet into a Rectangular Pan, *Mater. Trans.* **43**, 2443 (2002).
5. R. Verma, L.G. Hector, P.E. Krajewski and E.M. Taleff, The finite element simulation of high-temperature magnesium AZ31 sheet forming, *JOM* **61**, 29 (2009).
6. O. El Fakir, S. Das, I. Stone, G. Scamans, Z. Fan, L. Wang, D. Balint, JP. Dear and J. Lin, Solution heat treatment, forming and in-die quenching of a commercial sheet magnesium alloy into a complex-shaped component: experimentation and FE simulation, *Key Eng. Mater.* **622-623**, 596 (2014).

7. H. Yang, L. Huang and M. Zhan, Hot forming characteristics of magnesium alloy AZ31 and three-dimensional FE modeling and simulation of the hot splitting spinning process, *Magnesium Alloys: Des. Process. Prop.* (2011).
8. M.S. Mohamed, A.D. Foster, J.G. Lin, D.S. Balint and T.A. Dean, Investigation of deformation and failure features in hot stamping of AA6082: Experimentation and Modelling, *Int. J. Mach. Tools Manuf.* **53**, 27 (2012).
9. ESI PamStamp (version 2015.1).

New Research Progresses on Hot Stamping Technology of High-Performance Sheet Metal

Y. L. Wang†, B. Zhu and Y. S. Zhang

State Key Laboratory of Materials Processing and Die and Mould Technology, Huazhong University of Science & Technology, Wuhan, 430074, China

†Email: wangyilin@mail.hust.edu.cn

This paper introduces the research work progresses of "High Performance Sheet Metal Forming Team" from Huazhong University of Science & Technology, in hot stamping technology and its application. Aiming at the basic theory and engineering application of hot stamping, the team has carried out lots of innovative researches in materials, process and equipment for hot stamping of high-performance sheet metal since 2008. The representative achievements in many aspects of the team are illustrated, including high strength steel hot stamping process strategies with tailored properties, conductive heating technology and equipment, constitutive and fracture modeling of multiphase, tribological behavior in different process conditions, and non-destructive testing method for hot stamped parts. In addition, the hot stamping of aluminum alloy and high strength steel tube is still developed to meet the industrial requirements.

Keywords: Hot Stamping; Tailored properties; Conductive heating; Aluminum; Hot tube forming; Non-destructive testing.

1. Introduction

In 2015, there are more than 25 million vehicles were produced in China. Meanwhile, 543 million tons oil was consumed in China, dependence of foreign imports over 60 percent. Oil consumption in automobile industry was about 65 percent of Chinese petroleum oil consumption, and over 10 million tons CO_2 was produced from vehicle fuel every year, so energy saving and emission reduction of automotive industry is urgent. A series of result from studies and tests show that there was a linear relationship between the automotive weight and fuel consumption, take passenger car as example, fuel consumption and emissions will reduce 6-8 percent under the conditions of 10 percent of car weight reduction. Lightweight is a direct and effective means of energy saving [1].

Under this condition, lightweight materials were applied, which not only can ensure the vehicle safety performance, but also can meet the fuel consumption regulations. High-performance sheet metals such as high strength steel and aluminum alloy exist forming difficulty because of the spring-back and tools wearing problems during the cold forming processes. Hot stamping was a kind of technology of getting ultra-high strength components without the preceding problems, which has been used to produce high strength components presently in automotive industry. This paper introduces some research work of HUST team in materials, process and equipment of hot stamping in recent years.

2. Research on Innovative Processes

With the increasing requirements of the automotive industry for reduced weight and safety improvement, high strength steel hot stamping parts have been increasingly manufactured as structure components of the car body-in-white. The hot stamping technology is widely used because of the good formability, lower forming force, less spring back and high strength parts obtained. The simultaneous forming and quenching in cooled dies within the direct hot stamping process offers the chance to gain a fully martensitic structure assuring tensile strength up to 1500 MPa. However, some components manufactured through the conventional hot stamping process would benefit of regions characterized by a reduced strength but higher ductility in order to enhance their crashworthiness [2].

The compatibility of mechanical properties and safety performance of automotive components has been proposed that different mechanical properties are needed in different regions of the same components, namely tailored tempering properties (TTP). The B-pillar automotive component (Fig. 1) is an example of such components in which the bottom part should have higher energy absorption capacity to enhance the side crash performance, whereas the upper part must retain the martensitic structure to show high intrusion resistance. To achieve this, variants of the conventional hot stamping have been recently introduced, with the aim of tailoring the microstructure of the hot stamped components and their mechanical properties. In this paper, several different process strategies to obtain tailored properties are presented.

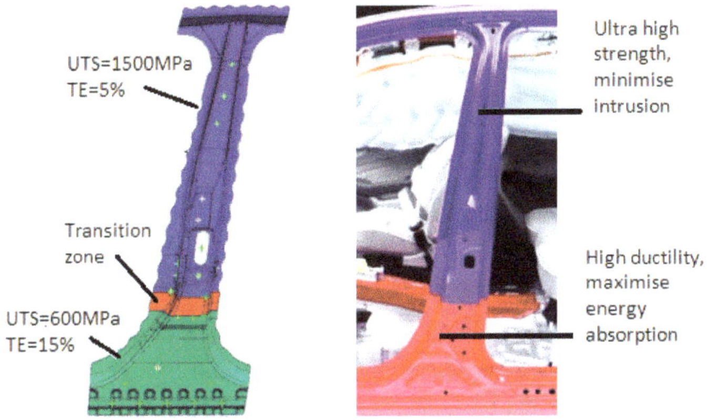

Fig. 1. B-pillar automotive component with tailored properties.

2.1. *Partial heating*

The implementation of the partial heating strategy is that certain regions of the blank is heated above its specific Ac3-temperature in where martensitic transformation is requested, while other regions are kept below Ac3-temperature to prevent the transformation into austenite, which will result in lower tensile strength but higher ductility during the heating step of the hot stamping process. Afterwards, the blank with the local temperature distribution is transferred into the die then formed and quenched simultaneously by the water-cooled die for 5–10s, depending on the part thickness. This heating strategy will gain a transition zone with a partial austenitization, therefore, the local temperature distribution develops mixed grain structure, which leads to a continuous reduction of hardness from the martensitic to the ferritic-pearlitic region, and the development of local mechanical properties after the forming and cooling steps.

The strategy of thermal control of the blank during the heat treatment can be solved in different ways such as using separated chambers, absorption masses or thermal shielding in a furnace. Most of those methods are realized local temperature distribution by improving the furnace, but only the simple combination of heating region and non-heating region is easy to implement in production line, it is very difficult to customize the complex temperature distribution according to design requirement with tailored properties. As a new heating strategy, conductive heating has the capability to select the heating regions by heating electrode arrangement, and, therefore, obtain customization of temperature distribution.

A movable conductive heating device on die was developed in terms of the Joule's Law (Fig. 2), including electric controlling system, temperature control & detection system, electrode & insulation system and pneumatic system [3-4]. Some experiments are carried out from blank heating to forming and quenching, and obtain formed specimen, then test the local mechanical properties distribution and observe the microstructure.

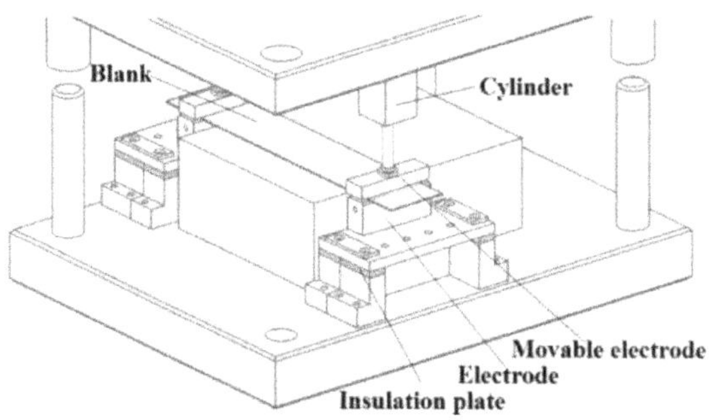

Fig. 2. A conductive heating device with movable electrodes.

Fig. 3 shows the temperature and hardness distribution along length direction. The results indicates that the average temperature of the central zone is about 940°C and hardness is over 470HV as the microstructure of central zone is fully transformed from austenite into martensite, whereas the temperature on two side of blank decreases gradually and hardness also decreases.

The SEM images in different zones are displayed as shown in Fig. 4. The hardening zone is a complete martensitic structure, transition zone is mainly a mixture of martensite and bainite, with a small amount of ferrite/pearlite structure, whereas the softening zone is the ferrite/pearlite structure as supply state.

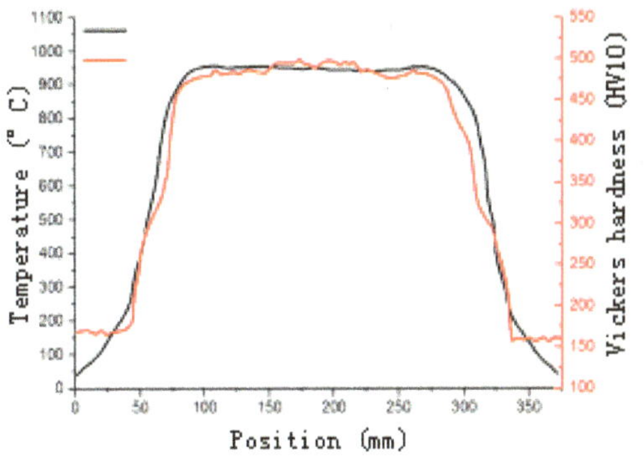

Fig. 3. Temperature and hardness distribution along length direction.

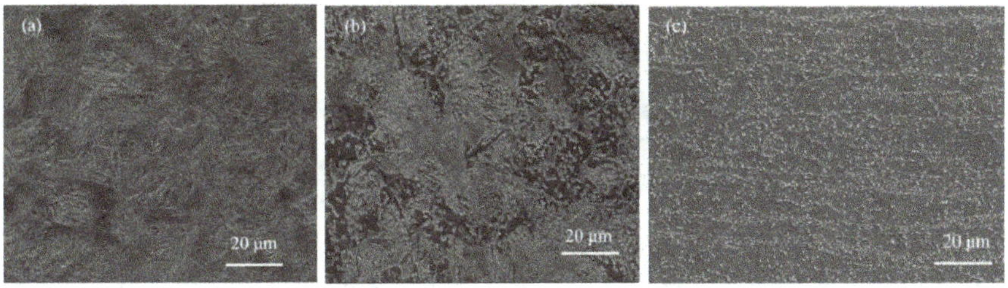

Fig. 4. SEM images in different zones along length direction: a) 150mm (hardening zone); b) 75mm (transition zone); c) 0mm (softening zone).

2.2. *Differential cooling*

The differential cooling is another strategy for producing components with tailored properties, which is accomplished by using the effect of different cooling conditions during the quenching process inside the tools. Fast cooling rates above 27 K/s lead to the formation of a martensitic structure, while lower cooling rates force the formation of a more ductile microstructure with lower strength, such as bainite and ferrite–pearlite. The method to control cooling rate includes heated tool tailoring, using tool steels which differ in thermal conductivity, adjusting cooling channel distribution and different contact condition.

Firstly, the heated tool tailoring is studied through the die partition experiment (Fig. 5). The die is divided into a heating and a cooling zone. With the different die temperatures, the cooling rates differ in different regions [5]. When the hot blank is quenched in this die, its thermal resistance mainly relies on temperature difference during the heat exchange process between the blank and the die, thus affecting the cooling rate of the blank. When a hot blank is contacted with a cold die, the cooling rate will be so large that austenite will be completely converted into high-strength martensite, resulting in high strength parts. However, when a hot blank is in contact with a hot die, the thermal contact resistance increases, the cooling rate of the component will be reduced or even lower than the critical cooling rate (27°C/s). Therefore, multi-phase microstructure comprising bainite and martensite is obtained with the reduced strength and enhanced plasticity.

Fig. 6 shows the hardness distribution of the heating zone, the cooling zone and the transition zone at different temperatures. Each hardness data represents an average of three through-width measurements. When the die temperature in the heating zone was below 400°C, the minimum hardness value emerged in the transition zone. When the die temperature reached 400°C, the higher hardness in cooling zone gradually transited to the lower hardness in heating zone, which coincides with the results of others [12].

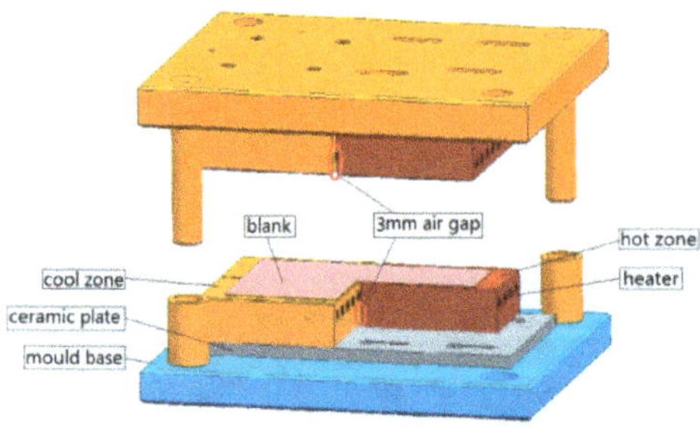

Fig. 5. Die partition for tailored properties.

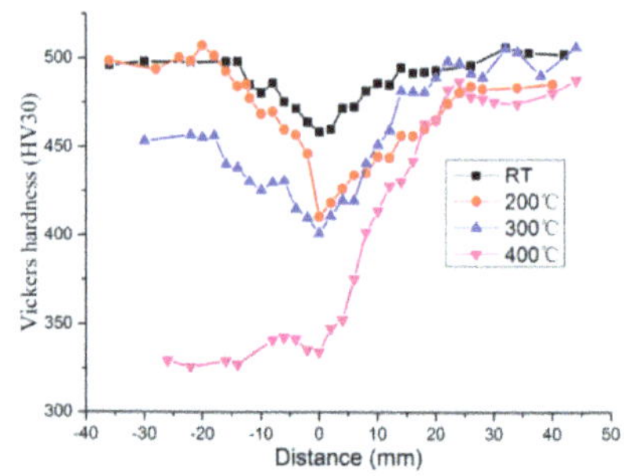

Fig. 6. Hardness results across the transition zone at different die temperatures.

Then, another method being called air cooling for testing different contact condition is studied by the experiment in Fig. 7. Air cooling method is aimed to the absorption region, a large clearance is used between blank and tools to get a mixed microstructure by reducing the cool rate. Two solutions for the new hot stamping process (shown in Fig. 10) are proposed according to the new idea to make tailored properties [6].

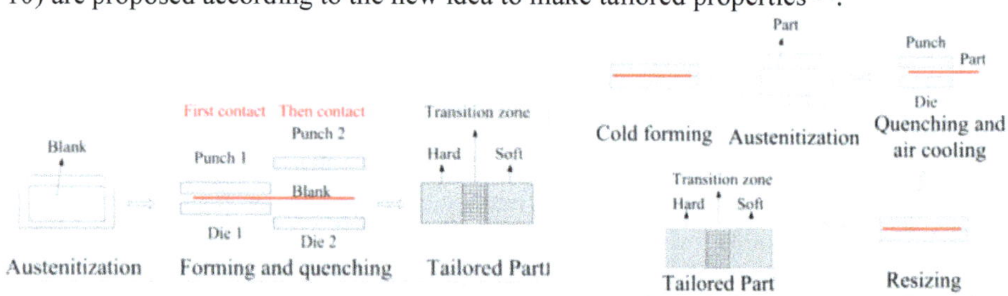

(a) Solution 1.

(b) Solution 2.

Fig. 7. Two solutions to make tailored properties by air cooling.

By the premise of guarantee of full austenization, the total elongation after air cooling is larger, with the decrease of heating temperature and heating time. And the heating parameters have little effect on ultimate tensile strength. With the increase of heating temperature and heating time, the grain size of original austenite increased, and the volume fraction of ferrite decreased, the volume fraction of bainite and martensite increased. As a result, the total elongation decreased. The photographic measurement results show that the formability of solution 2 is better than solution 1.

2.3. *Annealing*

Besides the adjustment of the component mechanical properties within the heat treatment or forming and quenching steps, an additional annealing step after hot stamping can be used. Because the annealing step is independent of the conventional process route, its application is very agile and tailored properties can be adjusted easily in a wide range regardless of their position and the component dimensions. In addition, the annealing is used as the way of reduction of tensile strength in favor of local trimming performance and welding property. The disadvantage of annealing is lower production efficiency and distortion of component after annealing treatment.

The experiments are conducted using tempering temperature factor with 6 levels (200, 300, 400, 500, 600 and 700 °C) and tempering time with 6 levels (20, 40, 60, 80, 100 and 120 min) and 36 sets of factorial experiments were obtained [15]. The experiment results show that the tempered specimen hardness mainly are affected by tempering temperature (Fig. 8).

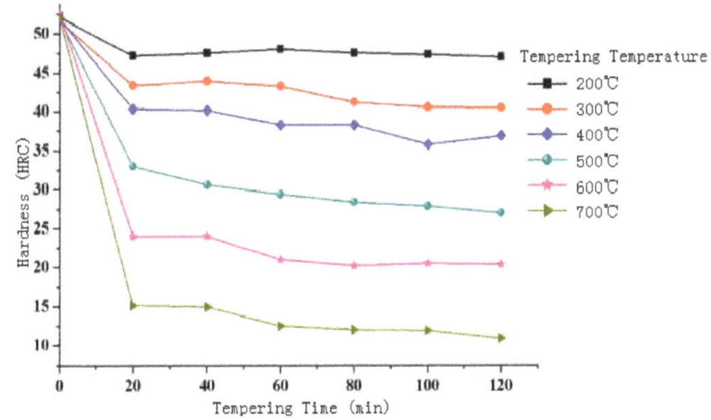

Fig. 8. Variation of speciman hardness with time under the different tempering temperature.

A U-shaped fully-martensitic part is tested through electromagnetic induction heating (Fig. 9). The experiment results show that when the tempering temperature is 880°C, the hardness of softening region is the lowest, and the length of transition region being about 20mm, which has greater transition gradient, is obviously smaller than the length obtained by die partition.

2.4. *Q&P processing*

The above three different strategies can improve the elongation of the energy absorbing zone, but greatly reduced the strength of this zone, meanwhile the elongation of the high strength zone cannot improved. By applying the Quenching and Partitioning (Q&P), processing combined with PHS, the ductility of PHS parts is expected to improve significantly without degradation of the strength.

Q&P processing is a method of stabilizing austenite in steel by quenching the steel to produce a specific volume fraction of martensite and subsequently partitioning the C from martensite to austenite during the partitioning stage. Fig. 10 shows a schematic diagram of the heat treatment cycle modified for the hot forming and Q&P processing, which is based on the direct hot stamping method. The blank is heated above the Ac3 temperature in a furnace, and then transferred to the die-quenching press and subsequently formed. The cooling in the dies is interrupted at a quench temperature in the range of Ms – Mf temperature. The partitioning treatment at a temperature higher than quench temperature follows the die-quenching stage in another furnace. In this stage, the carbide-free bainite is formed, and, simultaneously, C diffuses from both the primary lath martensite and the bainitic ferrite into the untransformed austenite. Finally the remaining austenite get stabilized upon cooling to room temperature.

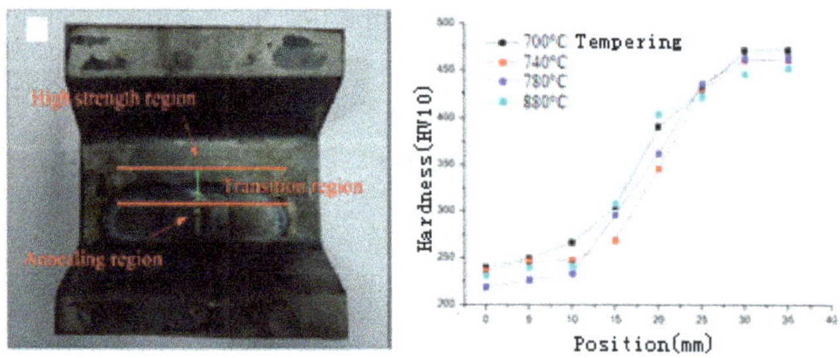

(a) the part (b) hardness distribution

Fig. 9. Tailored properties by tempering.

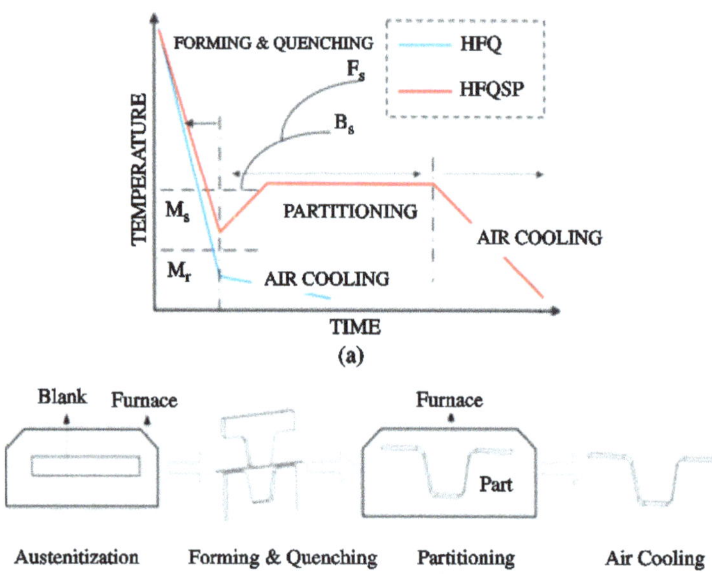

(b)

Fig. 10. Hot forming and Q&P processing: (a) schematic diagram (b) process diagram.

Different quenching temperatures affect the final residual austenite content, and then affect the mechanical properties of the formed parts (Fig. 11). The mechanical property of the formed part is greatly improved by elongation over 10%. With the increase of quenching temperature, the product of strength and elongation (PSE) first increases and then decreases. When the quenching temperature is about 300°C, the PSE reaches the maximum.

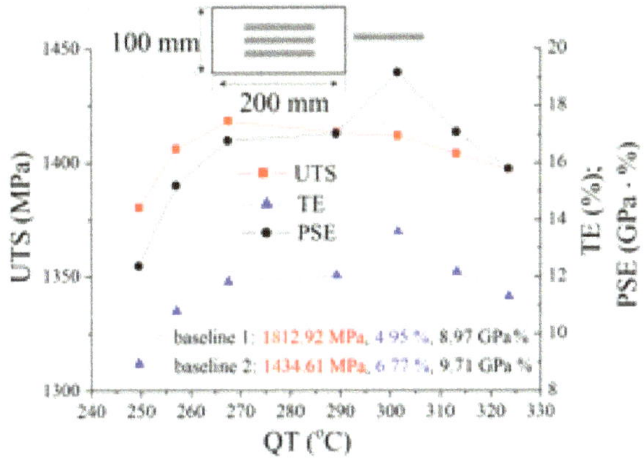

Fig. 11. Effect of quenching temperature on mechanical properties.

After partial quenching, the blank is transferred to the furnace to carry out the partitioning treatment, and the partitioning time affects the final residual austenite content, and, therefore, their mechanical properties (Fig. 12). With the extension of partitioning time, the tensile strength of the material decreased gradually. This is because the carbon in martensite will be diffused into the austenite, so the content of martensite is decreased, the internal dislocation is reduced and the intensity is reduced. When the partitioning time reached 150s, the change of tensile strength was no longer significant. The elongation of the material will gradually increase with the increase of the partitioning time, and then slowly decrease. This is due to the extension of treatment time, some carbon will form carbide and precipitate. With the extension of partitioning time, PSE first decrease slowly, then increase, and finally decrease slowly, which reflects the synthesized effect of the tensile strength and elongation.

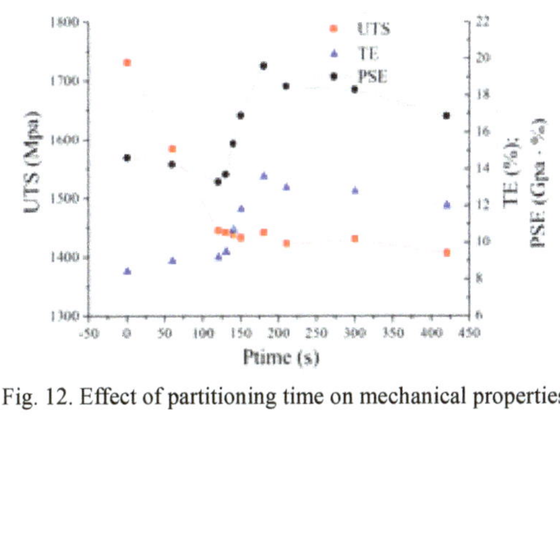

Fig. 12. Effect of partitioning time on mechanical properties.

3. Research on Application Basis

3.1. *Numerical modeling*

In order to study the performance of hot stamped parts, the modified Mohr-Coulomb model is used to predict the fracture behavior. As the equation (1) shows is the traditional Mohr-Coulomb model. When the combined value of normal stress and shear stress reaches a certain value, the parts will fracture.

$$(\tau + c_1\sigma_n) = c_2 \tag{1}$$

In equation (1), c1 and c2 are respectively friction coefficient and shear resistance, which are related to the material. As equation (2) shows is the modified Mohr-Coulomb model. kf, nf, c1, c2 and c3 are parameters related to the material. η is the stress triaxiality and $\bar{\theta}$ is the lode angle parameter.

$$\bar{\varepsilon}_f(\eta,\bar{\theta}) = \left\{ \frac{k_f}{c_2}\left[c_3 + (1-c_3\frac{\sqrt{3}}{1-\sqrt{3}})(\sec(\frac{\pi}{6}\bar{\theta})-1) \right] * \left[\sqrt{\frac{1+c_1^2}{3}}\cos(\frac{\pi}{6}\bar{\theta}) + c_1(\eta + \frac{1}{3}\sin(\frac{\pi}{6}\bar{\theta})) \right] \right\}^{-\frac{1}{n_f}} \tag{2}$$

A second integral part of the fracture prediction model is the so-called damage indicator D. As equation (3) shows is the proposed linear relationship between the damage indicator and the equivalent plastic strain. In the simulation, the damage indicator of one of the candidate elements reaches the critical value of 1 exactly identified the fracture. Elements that reach the fracture criterion D>=1 are eliminated from the simulation.

$$D(\bar{\varepsilon}_p) = \int_0^{\bar{\varepsilon}_p} \frac{1}{\bar{\varepsilon}_f(\eta,\bar{\theta})} d\bar{\varepsilon}_p \tag{3}$$

The parameters are obtained through physical experiments. The process of plane strain and uni-axial tensile are simulated using the modified Mohr-Coulomb model. The changes of the parameters are shown in Figs. 13(a) and (b). Figs. 14(a) and (b) show the simulation results of plane strain and uni-axial process. The simulation results fit well with the experiments one.

3.2. *Tribological properties*

The tribological behavior during the hot stamping affect the cost, product quality and producing efficiency directly, which cannot be neglected. The self-developed friction testing machine is used to measure the friction factor and wear between die and blank. Considering the process to get tailored tempering properties, the machine has been improved to investigate the tribological behavior in the condition of different initial blank temperature, different die temperature and different cooling rate as shown in Fig. 15.

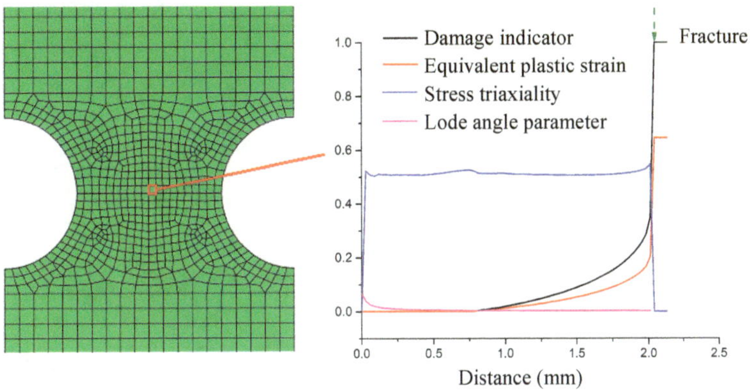

(a)

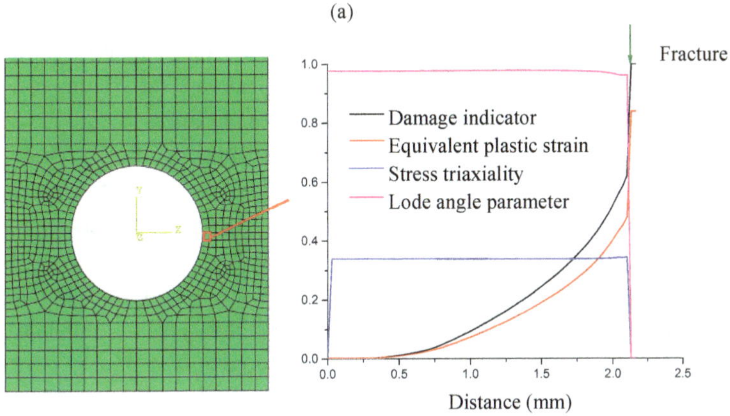

(b)

Fig. 13. Parameters changing in a) plane strain and b) uni-axial process.

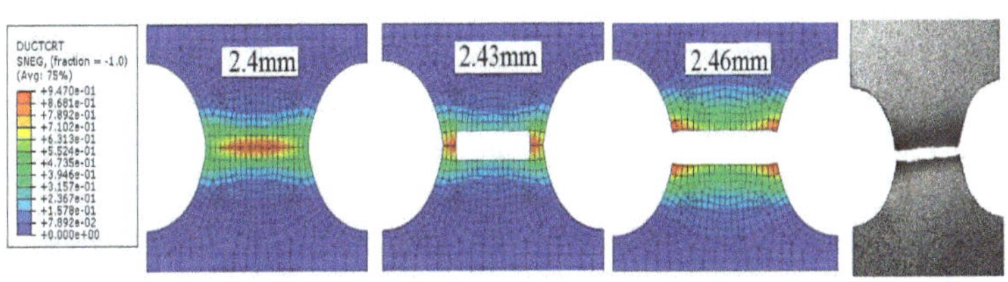

(a)

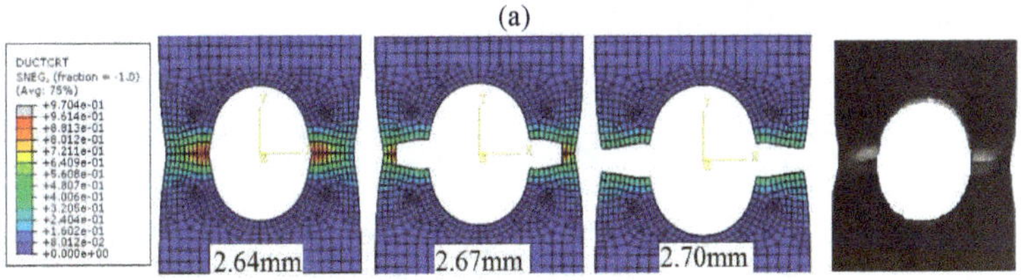

(b)

Fig. 14. Fracture process in a) plane strain and b) uni-axial strain.

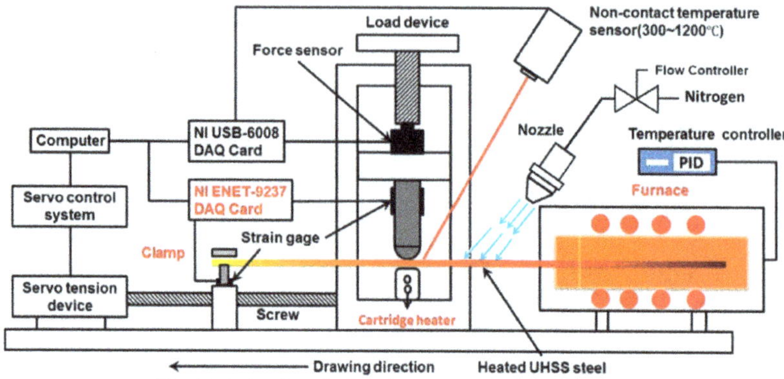

Fig. 15. A developed tribo-simulator for friction testing.

The surface roughness of the blank increases with the increase of the temperature, and brings the increase of viscous wear. Under hot forming, the frictional factor is much more than the frictional factor under the condition of cold forming.

With the initial gradient temperature distribution of strip, the change of frictional factor can be divided into three stages: lower temperature section, transition section and higher temperature section as shown in Figs. 16(a) and 16(b). From lower to higher temperature section, the surface transition from hardening to softening occurs the peak of frictional factor. On lower temperature section, the frictional factor of coated HSS strip is higher than that of uncoated HSS strip, whereas on higher temperature section the frictional factor of coated strip is lower than that of the uncoated one. Low temperature section is mainly abrasive wear, the transition section is heavy viscous wear and high temperature section includes viscous wear and abrasive wear [7, 8].

During cooling step, when the temperature is about 900°C, microstructure of the strip is soft high-temperature austenite and the coefficient of friction is about 0.35. As the temperature decreases, the coefficient of friction increases to 0.6 - 0.7. Considering only the cooling stage from 800°C to 400 °C, the behavior of the increasing friction with decreasing temperature is quite evident. Due to the fact that the friction tool makes contact with the substrate of the strip, when the temperature is high, the existence of the oxidation particle reduces adhesion. As temperature decreases, the oxidation particles decrease, therefore the coefficient of friction increases.

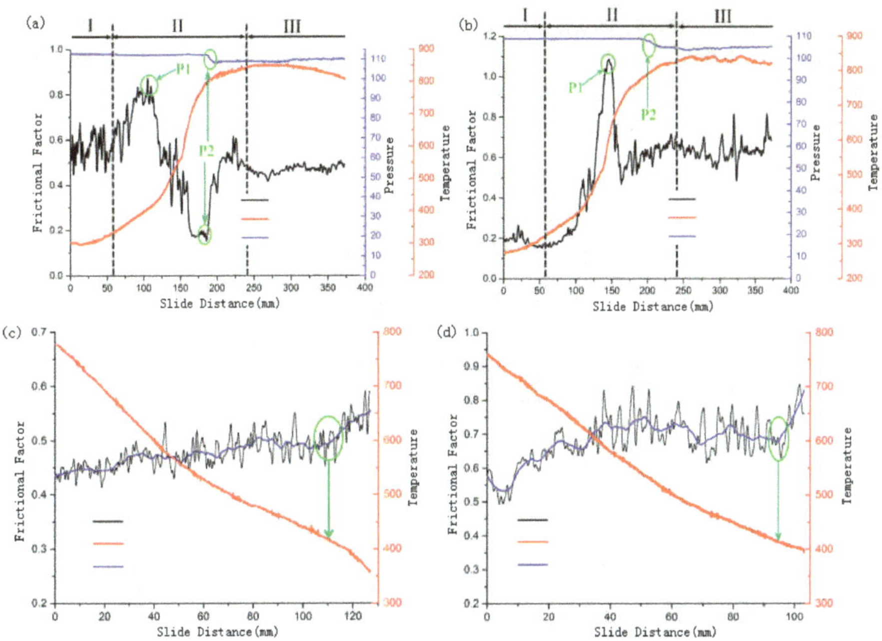

Fig. 16. Variation of frictional factor of (a)(c) coated steel; (b)(d) uncoated steel.

3.3. *Non-destructive testing of hot stamped parts*

To ensure mechanical properties of the formed high strength steel parts, the on-line testing is need. However, the traditional detection methods, such as metallographic observation and hardness measurement, are not only time-consuming, but also cause the irreversible damage of parts. Therefore, in order to realize the rapid and efficient measurement of mechanical properties of hot stamping high-strength steel automobile structure and the nondestructive testing of intensification of advanced high strength steel in forming process, a nondestructive detection method base on Barkhausen noise (Magnetic Barkhausen Noise, MBN) was put forward [13-14].

For the detection of mechanical properties of hot stamping high strength steel, the relation of hardness and time-frequency characteristic value was studied. The research indicates that the MBN signal including peak value and RMS gradually decreased with the increase of hardness. Meanwhile, the peak value of low frequency component decreased and the high frequency increased. The essential reason is that the revolving of domain became more difficult and the release of MBN is harder. Moreover, it is found that the increase of heating rate and the reduction of holding time and heating temperature weaken the MBN with the same microstructures in the study of the influence of rapid electric heating parameters on MBN signal. The fitting function relation between the hardness, heating parameters and the time-domain characteristic value could provide the quantitative basis for the on-line detection. And the analysis of frequency components brings the foundation for the qualitative analysis of the microstructures.

Finally, the hardness testing method base MBN was designed and the portable MBN detector was developed for the on-site production [9, 10, 11]. The hardness routing inspection system integrated hardness measurement system base MBN hardness and industrial robot can effectively improve the efficiency quality control and extend the application scenarios.

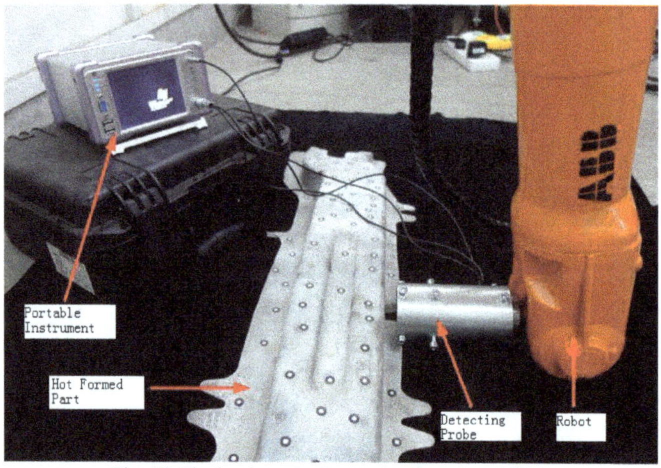

Fig. 17. The hardness testing system based on MBN.

4. New Technology of Hot Stamping

4.1. *Aluminum hot forming*

Aluminum, as a lightweight material, possesses many advantages such as low specific weight, high corrosion resistance, high thermal conductivity, recyclability. The need of weight reduction leads to an increasing trend of using different Aluminum alloys (6xxx, 7xxx) for producing car body components, but high strength aluminum cold forming is very difficult because of low formability and bigger spring back. Therefore, aluminum forming at elevated temperature becomes an inevitable choice for high strength aluminum alloys.

In the aluminum hot forming process, the blank is heated to solution heat treatment, and kept there some minutes for soaking. It is then formed and quenched in one step in cooled dies. The followed artificial aging process gives the part its final strength. On one side, the maximum formability at high temperature is obtained, on the other hand, the spring back is reduced significantly.

There are still some challenges when aluminum hot forming goes into industrial application for high serial production. One main challenge is that the mismatching of cycle time between solution heat treatment, press hardening and artificial aging make configuration of production equipment difficulty. Another challenge is to prevent some defects such as fracture and galling from occurring during aluminum hot forming.

4.2. *Hot tube forming*

The conventional process to form some products with tubular shape such as exhaust and chassis is adopted hydroforming. With increasing application of UHSS, the strength of the applied material is limited in hydroforming process. One solution is hot tube forming, which combine the advantages of profiled shape components with high strength.

The hot tube forming process is similar to current hot sheet stamping process except for using tubular blank. Sometimes a moderate inner pressure is applied by sealing both ends of the tube in order to keep close contact between the blank and the die. A new strategy to forming high strength tubular components with tailored properties is presented. The conductive heating will be used to control the temperature distribution during heat step, and the process planning is finished (Fig. 18).

Fig. 18. The process planning of the tubular part

5. Conclusion

The conductive heating strategy is suitable for manufacturing components with tailored properties in current production line without adjusting of processes.

The Q&P processing has great potential as a trend of hot stamping technology, but need to face some challenges in material, process and equipment.

There is a growing demand for aluminum hot forming and tube hot forming.

References

1. M. T. Ma, Y. S. Zhang, L. F. Song and Y. L. Wang, Research and Progress of Hot Stamping, in *International Conference on Hot Stamping of UHSS*, (Chongqing, China, 2014).
2. M. Merklein, M. Wieland, M. Lechner and S. Bruschi, Hot stamping of boron steel sheets with tailored properties: A review, J. Mater. Proc. Tech. **228**, 11 (2016).
3. W. K. Liang, L. Wang, Y. Liu, Y. L. Wang, Y.S. Zhang. Hot stamping parts with tailored properties by local resistance heating, in *11th International Conference on Technology of Plasticity (ICTP 2014)*, (Nagoya, Japan, 2014).
4. W. K. Liang, Y. Liu, B. Zhu, M. L. Zhou and Y. S. Zhang, Conduction heating of boron alloyed steel in application for hot stamping, *International Journal of Precision Engineering and Manufacturing* **16**, 1983 (2015).
5. Z. J. Wang, Y. Xu, W. T. He and Y. S. Zhang, Hot stamping of two different high strength steel with tailored properties, *Mech. Mater. Trans. Tech.* **395**, 909 (2013).

6. P. X. Liu, K. Wang, Z. J. Wang, B. Zhu and Y. S. Zhang, A new hot stamping process to make tailored properties based on air cooling, in *5th International Conference on Hot Sheet Metal Forming of High-performance Steel*, (Toronto, Canada 2015).
7. X. Tian, Y. Zhang and J. Li, Investigation on tribological behavior of advanced high strength steels: influence of hot stamping process parameters, *Tribol. Lett.* **45**, 489 (2012).
8. B. Zhu, Y. S. Zhang and C.H. Wang, Modeling of the Austenitization of Ultra-high Strength Steel with Cellular Automation Method, *Metall. Mater. Trans. A* **45**, 3161 (2014).
9. Y. Xu, Z. J. Wang and Y. S. Zhang, Experimental analysis of hardness distribution of the transition region in tailored tempering stamping of ultra-high strength steel, in *The 2013 International* Forum *on Mechanical and Material Engineering*, (Guangzhou, China, 2013).
10. R. George, A. Bardelcik and M. Worswick, Hot forming of boron steels using heated and cooled tooling for tailored properties, *J. Mater. Process. Tech.* **212**, 2386 (2012).
11. X. Y. Luo, Y. L. Wang, L. Wang, J. Xie and Y. S. Zhang, Non-destructive hardness measurement of hot-stamped high strength steel sheets based on magnetic barkhausen noise, in *11th International Conference on Technology of Plasticity, (ICTP 2014)*, (Nagoya, Japan, 2014).
12. M. Merklein and T. Sevec, Hot stamping: manufacturing functional optimized components, *Production Engineering* **7**, 141 (2013).
13. X. Y. Luo, Y. S. Zhang and Z. J. Wang, Non-destructive Testing Device for Hot Forming High Strength Steel Parts Based on Barkhausen Noise, *Mechanics Mater.* **423-426**, 2555 (2013).
14. X. Y. Luo, Y. Zhang, Y. L. Wang and Y. S. Zhang, Application of Magnetic Barkhausen Noise Test System in Hot Stamping, in *The 1st International Conference on Hot Stamping of UHSS(ICHSU2014)*, (Chongqing, China, 2014).
15. Z. J. Wang, K. Wang, P. X. Liu, Y. L. Wang and Y. S. Zhang, Prediction of thermal softening of hardened high strength steel, in *5th International Conference on Hot Sheet Metal Forming of High-performance Steel*, (Toronto, Canada 2015).

Forming and Application of Auto Body Panels Made of Tailor Welded Blanks and Ultra-High Strength Steel Sheets

L. Hua†, Y. L. Song, J. N. Liu, J. Liu, Y. Han and Y. H. Shen
Hubei Key Laboratory of Advanced Technology for Automotive Components, Wuhan University of Technology, Wuhan, 430070, China
Hubei Collaborative Innovation Center for Automotive Components Technology, Wuhan, 430070, China
†E-mail: hualin@whut.edu.cn

To satisfy the lightweight demands of auto body panels, high and ultra-high strength sheet metal becomes widely used nowadays. Auto body panels with large dimension, thin wall and complex shape, however, are difficult to be formed by using high and ultra-high strength sheet metal due to increased deformation resistance, decreased formability and forming precision. This paper presents the state of the art of cold/hot stamping of high strength sheets (e.g., tailor welded blanks and ultra-high strength steel) and its applications in manufacturing auto body panels.

Keywords: Lightweight; Plastic forming; High strength sheet metal; Auto body panels.

1. Introduction

Lightweight of automobiles is an important measure to deal with global issues such as energy exhaustion and environment pollution. To satisfy the lightweight demands of auto body panels, high and ultra-high strength sheet metal, e.g., tailor welded blanks, high and ultra-high strength steel and aluminum alloy, becomes widely used nowadays [1]. Since auto body panels commonly own large dimension, thin wall and complex shape, they are difficult to be formed by using high strength sheet metal due to increased deformation resistance, decreased formability and forming precision [2-4].

2. Tailor Welded Blanks

Tailor welded blanks (TWBs) have been widely applied in auto body panels due to their competitive advantages over conventional sheet metal forming, e.g., reduced weight, recollected offal and cost saving [2,5]. Some research work on yield locus of base material, constitutive modeling of inhomogeneous weld and process optimization of TWB auto parts are carried out in the Lab.

2.1. *Yield locus of base materials*

A yield criterion is a hypothesis defining the onset of plastic deformation for a material under certain combination of stresses. More recently, several yield criteria have been developed to account for the intrinsic nature of metals. The predicted yield behavior of a given metal may quite differ according to different yield criteria and therefore distinctly affect the accuracy of computer simulation. In order to find the influence of yield model on different metals, four base materials of TWBs including DX56D+Z, DP600m, 6111-T4 and 5182-O were investigated with four different yield models, i.e., Von Mises, Hill'48, Yld89 and Yld2000 (Fig. 1). It appears that the four yield loci differ distinctly in the equibiaxial area and also the Yld2000 model produces a yield locus with strongest anisotropy for all the steel and aluminum alloy sheets [6].

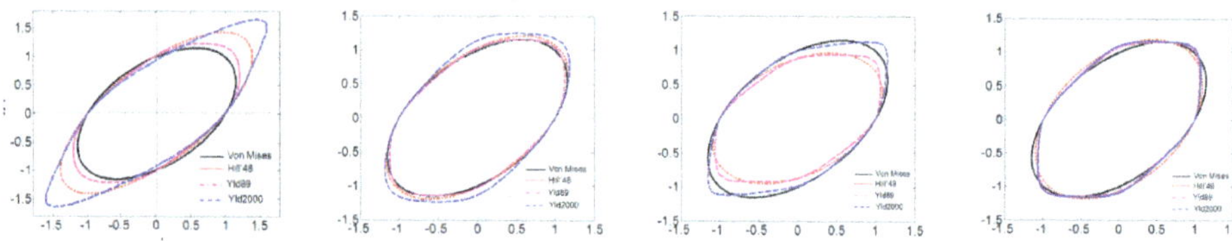

(a) DX56D+Z (b) DP600 (c) 6111-T4 (d) 5182-O

Fig. 1. Normalized yield surfaces for four materials with different yield model [6].

The comparison between experimental and simulated Erichsen cupping test results demonstrated that simulations with the Yld2000 models always exhibit excellent agreement with the experimental ones for all the investigated sheets; the Yld89 models are nearly accurate for the aluminum alloy sheets, but in lower precision for the steel sheets; as for the Von Mises and Hill'48 models, these predicted results slightly differ from the experimental ones For the steel sheets, while the difference are more distinctly for the aluminum sheets (Fig. 2) [6].

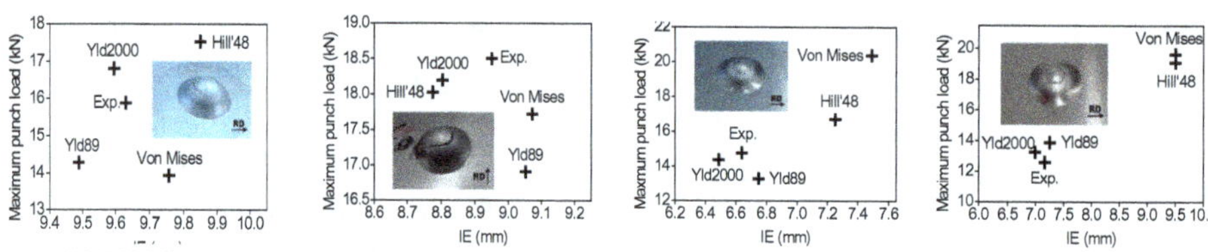

(a) DX56D+Z (b) DP600 (c) 6111-T4 (d) 5182-O

Fig. 2. Erichsen cupping test results for four materials with different yield model [6].

2.2. *Constitutive modeling of weld material*

In previous FE simulations for TWB forming, the weld was either directly neglected or simply represented by a row of connected nodes or elements [2]. Significant loss in simulation accuracy may occur in case that the welds are located in large deformation zones. To solve this problem, the inhomogeneous property traversing the weld line was taken into account. A constructive model was build via a Nano Indenter G200 system. The mechanical properties of investigated materials can be estimated from the measured load-depth (P-h) response (Fig. 3) [7].

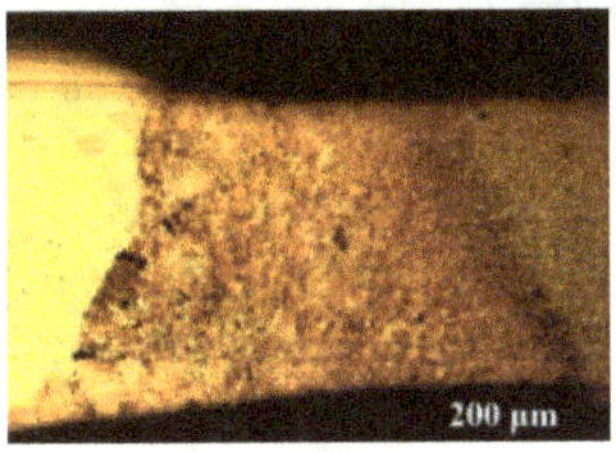

(a) Metallographic section of weld

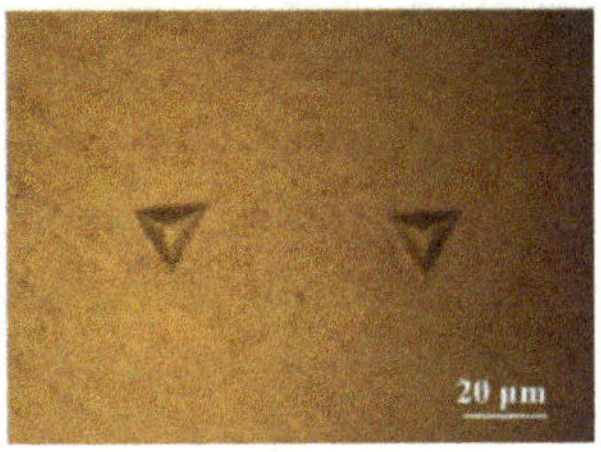

(b) Nano indention imprints

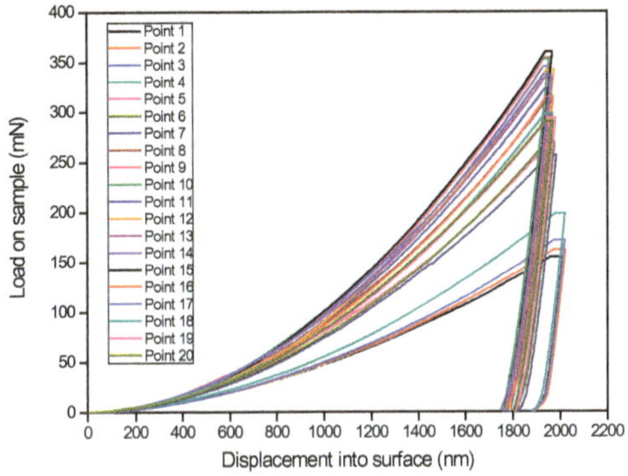

(c) Measured load-depth response

Fig. 3. Etched metallographic section of weld zone [7].

By calculating the material parameters including Young's modulus E, the initial yield stress σ_s, the strain hardening exponent n and the strength coefficient K of the weld metal, the constructive model of the weld accounting for the performance inhomogeneity can be precisely built, as shown in Eq. (1) [7].

$$\sigma(\varepsilon,x)=\begin{cases}(-6.116\times10^{-5}x^2+0.008772x+257.1)\varepsilon, & \sigma\le -0.001441x^2+0.453x+907.6\\ \dfrac{1777}{e^{(0.0012x-0.2372)^2}}\varepsilon^{-1.217\times10^{-10}x^3+1.294\times10^{-7}x^2-3.445\times10^{-6}x+0.112}, & \sigma> -0.001441x^2+0.453x+907.6\end{cases} \tag{1}$$

Here σ, ε are the true stress and true strain respectively; x denotes the distance from the centerline of the weld, and e (≈2.718) is the base of the natural logarithm.

The constitutive model of welds was also built in the oretical-experimental approach, namely micro-Vickers hardness test and the rule of mixture. The relationship between the Vickers hardness number HV and the material parameters can be deduced as Eq. (2) [8].

$$n=\ln HV-\ln K+1.1151/\ln(3.05HV/E+0.29) \tag{2}$$

The load equilibrium equation, as shown in Eq. (3) was obtained to analyse the virtual uniaxial tension process for the entire "HAZ and WN-only" specimen based on the rule of mixture (Fig. 4).

$$K_{hw}\varepsilon^{n_{hw}}A_{hw}=K_h\varepsilon^{n_h}A_h+K_w\varepsilon^{n_w}A_w \tag{3}$$

where subscripts *h*, *w* and *hw* refer to the HAZ, WN and mean value of the "HAZ and WN-only" specimen, respectively. By using a series of the statistical data processing method, the material parameters were determined, as listed in Table 1. The whole flow chart for inhomogeneous constitutive modeling method for weld metals is listed as Fig. 5. Further study has proved that the results calculated by the proposed theoretical-experimental method were in good agreement with the nano-indentation testones.

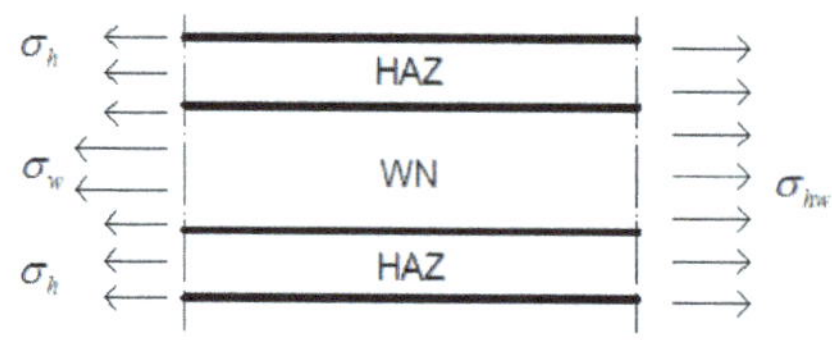

Fig. 4. Load equilibrium of TWB during uniaxial tensile test [8].

Table 1. Determined hardening parameters for the HAZ and WN materials [8].

TWB	Material	*K*(MPa)	*n*	Tol_{over}(N)	*STDEV* (N)
TWB 1 and TWB 2	HAZ	625	0.1664	0.189	3.823
	WN	900	0.1584		
TWB 3	HAZ1	700	0.2022	−50.539	1.055
	HAZ2	1100	0.1754		
	WN1	1500	0.1676		
	WN2	1700	0.1598		

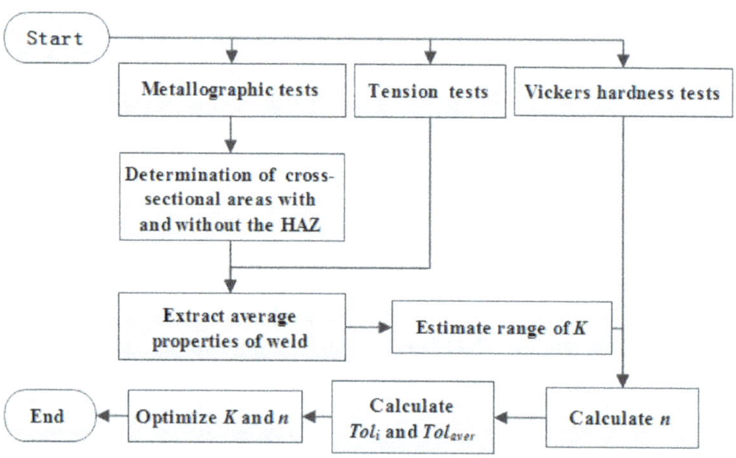

Fig. 5. Flow chart for inhomogeneous constitutive modeling method for weld metals.

2.3. *Process optimization of TWB auto parts*

Because of the existence of welds, the formability of a TWB is significantly lower than that of base metals. Moreover, the springback of a TWB also become more complicated due to great difference in mechanical properties of base metals and weld [5]. Therefore, the process optimization is a must for the forming of TWB auto parts. Therefore, the formability of the B-pillar reinforcement was predicted. BP neural network and genetic algorithm were used to optimize the forming process and the crack and springback were then well controlled, being consistent with the experimental samples in Fig. 6 [9]. Similarly, the process optimization of some other TWB auto parts was also conducted (Fig. 7).

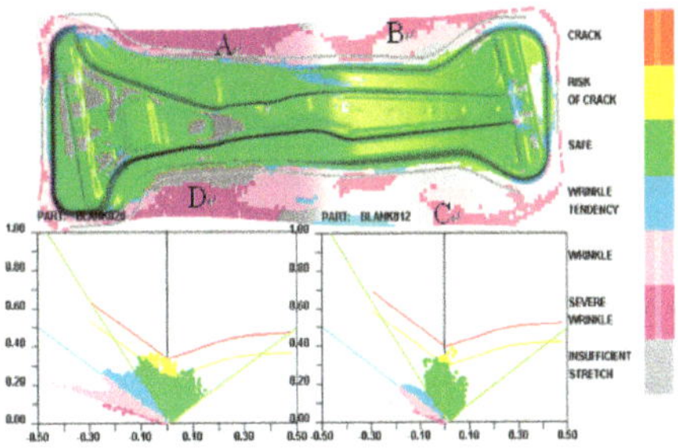

(a) Formability

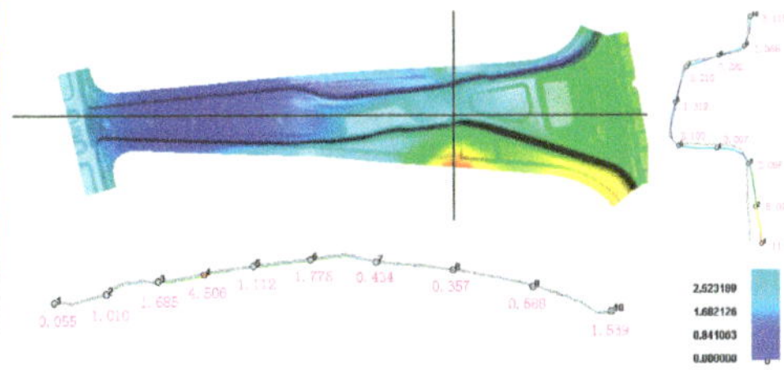

(b) Springback

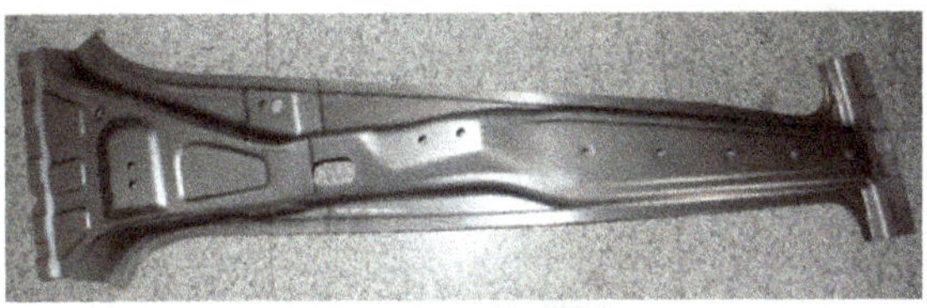

(c) Experimental sample

Fig. 6. FE simulated and experimental B-pillar reinforcement [9].

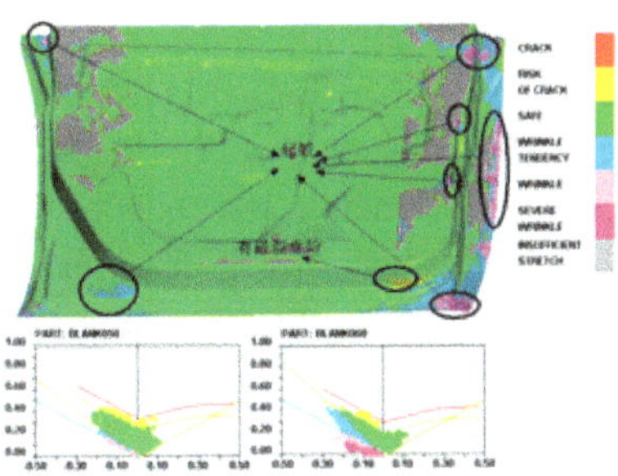

(a) TWB door inner

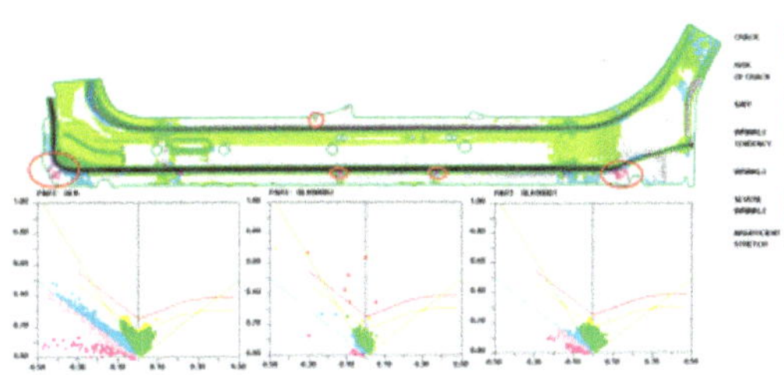

(b) Side rails

Fig. 7. FE simulated for other TWB auto parts.

3. Hot Forming of Boron Steel Sheets

Hot stamping of high-strength boron steels is another advanced metal widely used for automotive lightweight requests. In addition to greatly reducing weight, it can also improve safety performance without affecting the performance of the final products [3].

3.1. *Material microstructure and texture evolution*

High strength boron steel sheets are mainly used for hot stamping and their thermomechanical behavior is important to understand the deformation mechanism of hot stamping. Hence, a series of isothermal unaxial tensile tests were conducted for B1500HS boron steel specimens. The true stress–strain curves at different temperatures from 25℃

to 900℃ under a strain rate of 0.1 s^{-1} are shown in Fig. 8 [10]. The effects of deformation temperature and strain rate on the investigated material were analyzed. The material constitutive model was then built to describe the deformation behavior at elevated temperature, as shown in Eq. (4).

$$\dot{\varepsilon} = 2.115\times10^{20}[\sinh(7.57\times10^{-3}\sigma)]^{16.365}\times\exp\left(-\frac{4.26\times10^{5}}{RT}\right) \tag{4}$$

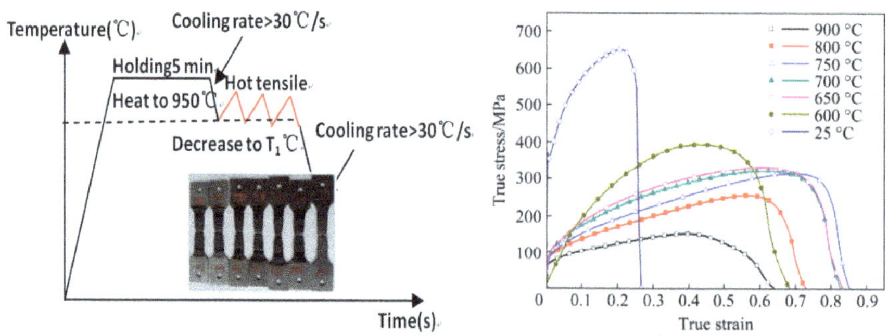

(a) Scheme of isothermal uniaxial tensile tests (b) Ture stress-true strain curves

Fig. 8. Uniaxial tensile test scheme and results for B1500HS boron steel [10].

To estimate the flow behavior of the steel, the fracture morphologies of the specimens were observed, as shown in Fig. 9 [10]. It was found that the centre of the ductile fracture exhibits parabolic shear dimples and fracture mechanisms is the nucleation and growth of microvoids. Texture evolution of B1500HS at various temperatures was observed by EBSD method and the results are shown in Fig. 10. It can be seen that the grains are significantly refined by the dynamic recrystallization under large deformation combined with high temperature. The intensity of γ-fiber texture in the rolling direction is higher than that in the vertical direction and therefore plasticity in the rolling direction is better than that in the other direction.

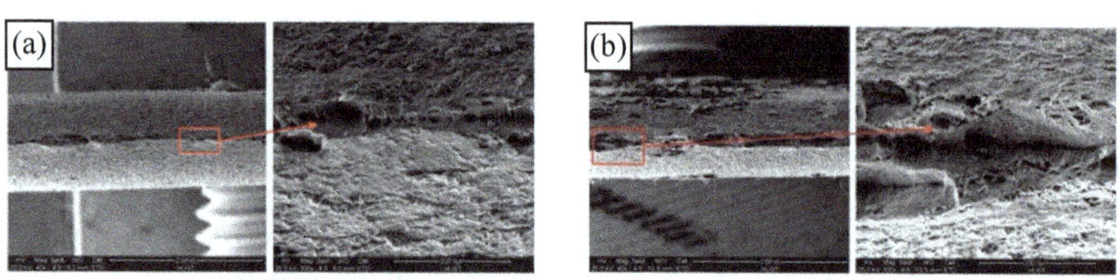

(a) 800 °C (b) 600 °C

Fig. 9. Fracture morphologies on the center of samples [10].

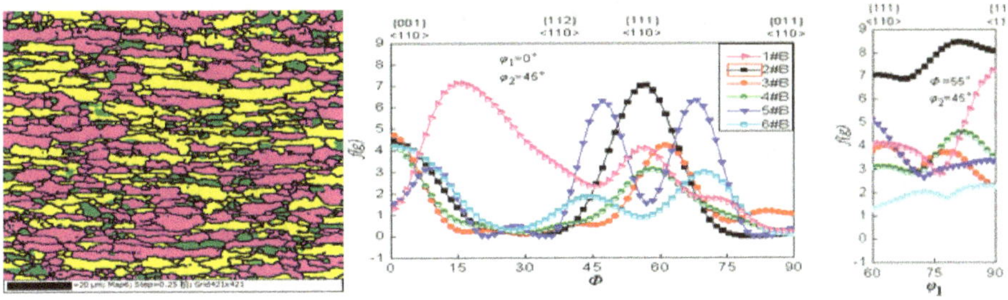

(a) Texture of blanks (b) α-fiber (c) γ-fiber

Fig. 10. Texture evolution by EBSD method.

3.2. *Forming prediction and defect control of auto parts*

In the hot forming of high-strength boron steel sheets, local cracking and loss of precision are the main problems. The forming process of the B-pillar reinforcement were optimized by FE simulation based on orthogonal tests and the results were in good agreement with the experimental ones [Figs. 11(a) and (b)]. Precision analysis proved that a certain degree of loss in the forming precision occurred at the small end of the hot formed part [Fig. 11(c)]. Similarly, the process optimization of some other hot formed auto parts was also performed as shown in Fig. 12. The function relationship between structural characteristics of automotive beam parts and wrinkling in hot stamping [11].

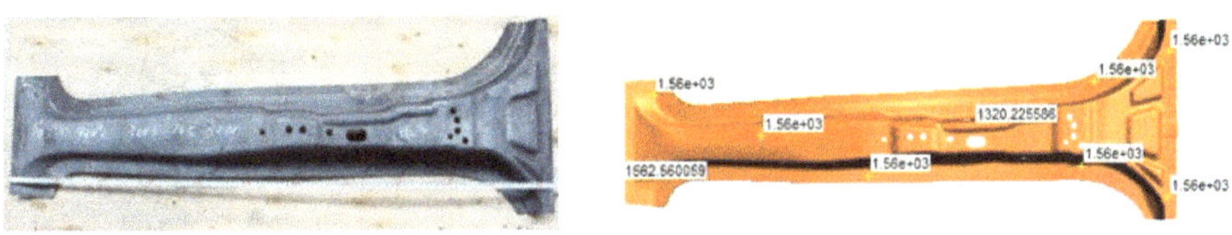

(a) Experimental sample. (b) Thickness contour.

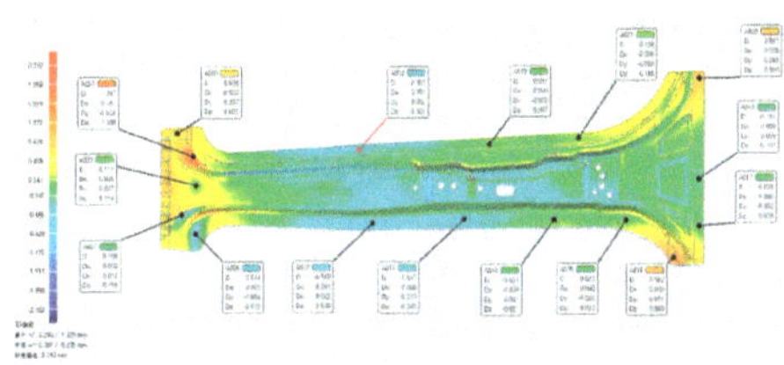

(c) Precision analysis

Fig. 11. Experimental and simulated B-pillar reinforcement.

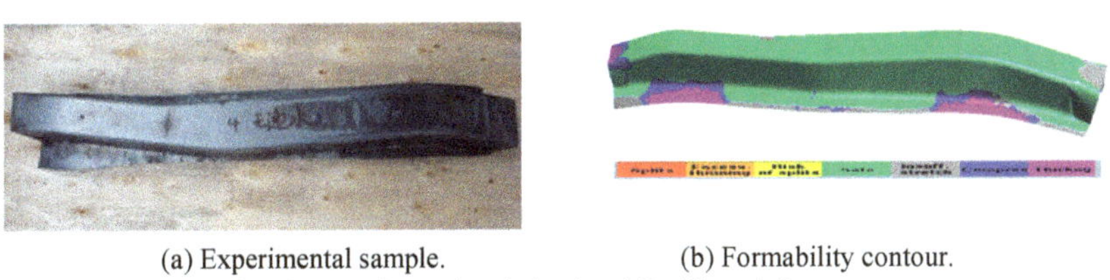

(a) Experimental sample. (b) Formability contour.

Fig. 12. Experimental and simulated B-pillar reinforcement.

3.3. *Variable strength design of hot formed auto parts*

The variable strength design and manufacturing of hot formed parts are important due to the consideration to both strength and energy absorption. Optimal design of B-pillar reinforcement based on side impact has also been carried out (Fig. 13). The adaptive respond surface method (ARSM) was used to obtain the reasonable strength distribution under the environment of Hyperstudy software [12]. It is demonstrated that the auto body using the B-pillar with variable strength can not only own better side impact performance, but also reduce the B-pillar assembly weight by 12.1%.

(a) FE model of vehicle side impact

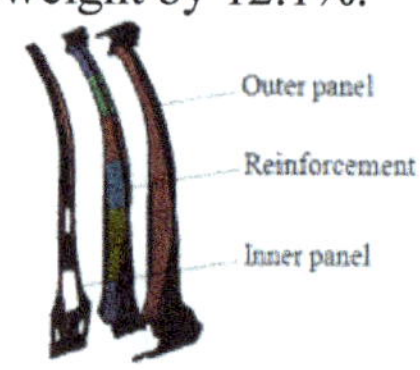

(b) Optimized B-pillar

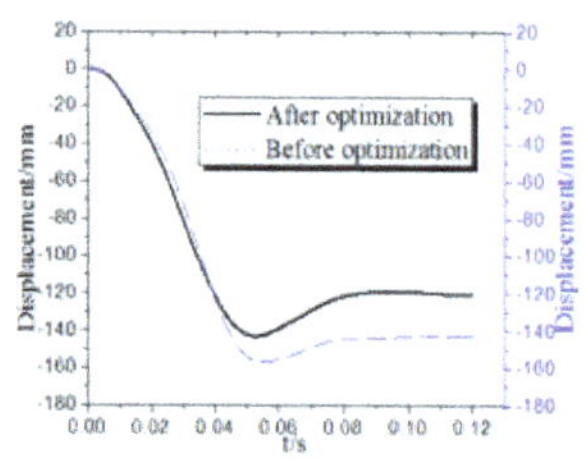

(c) Intrusion displacement at chest

Fig. 13. Variable strength design of hot formed B-pillar reinforcement [12].

4. Conclusion

The usage of tailor welded blanks and hot stamping of ultra-high strength steel can obviously meet the rigid requirement of auto body lightweight. By studying material microstructures, mechanical properties and process optimization, the deformation regularity and mechanism of the high strength but lightweight sheet metals are understood and the forming defects are effectively controlled.

Acknowledgments

This project funded by National Natural Science Foundation of China (Grant No. 51305317) and China Automobile Industry Innovation and Development Fund (Grant No. U1564202).

References

1. P.K. Mallick, Materials, design and manufacturing for lightweight vehicles, (Elsevier, 2010).
2. W. Lee, K.H. Chung, D. Kim, J. Kim, C. Kim, K. Okamoto, R.H. Wagoner and K. Chung, Experimental and numerical study on formability of friction stir welded TWB sheets based on hemispherical dome stretch tests, *Int. J. Plasticity* **25**, 1626 (2009).

3. R. George, A. Bardelcik and M.J. Worswick, Hot forming of boron steels using heated and cooled tooling for tailored properties, *J. Mater. Process. Tech.* **212**, 2386 (2012).
4. T. Dursun and C. Soutis, Recent developments in advanced aircraft aluminum alloys, *Mater. Des.* **56**, 862 (2014).
5. V.S. Suresh, S.P. Regalla, A.K. Gupta and G. Padmanabham, Weld Line Shift in the Case of Tailor Welded Blanks Subjected to Differential Strengths with Respect to TIG and LaserWelding, *Materials Today: Proceedings* **2**, 3501 (2015).
6. Y.L. Song and L. Hua, Formability evaluation of steel and aluminium alloy sheets with different plane stress yield criteria, in *The 14th Metal Forming International Conference*, (Kraków, Poland, 2012).
7. Y.L. Song, L. Hua and F.Z. Meng, Inhomogeneous constructive modeling of laser welded bead based on nanoindentation test, *Ironmaking and Steelmaking* **39**, 95 (2012).
8. Y.L. Song, L. Hua, D.N. Chu and J. Lan, Characterization of the inhomogeneous constitutive properties of laser welding beams by the micro-Vickers hardness test and the rule of mixture, *Mater. Des.* **7**, 19 (2012).
9. J. Liu, H.J. Mao, F. Yin, R.C. Chen and W. Cao, BP-GA-based multi-objective optimization method for drawbead design of B-pillar, *Journal of Hunan University Natural Sciences* **37**, 16 (2010).
10. J.N. Liu, Y.L. Song, J. Lu and W. Guo, Effect laws and mechanisms of different temperatures on isothermal tensile fracture morphologies of high-strength boron steel, *Journal of Central South University* **22**, 1 (2015).
11. Y.H. Shen, Y.L. Song, L. Hua, L. Yang and Z.X. Lv, Function relationship between structural characteristics of automotive beam parts and wrinkling in hot stamping, in *Proceedings of the Second International Conference on Advanced High Strength Steel and Press Hardening (ICHSU 2015)*, (Changsha, China, 2015).
12. Y.L. Song, Y. Han, L. Hua, J. Lu and C. Yu, Optimal design and hot stamping of B-pillar reinforcement panel with variable strength Based on Side Impact, in *Proceedings of the Second International Conference on Advanced High Strength Steel and Press Hardening (ICHSU 2015)*, (Changsha, China, 2015).

Local Patchworking of Sheet Properties Combining Carburization and Hot Stamping

B. F. Rolfe*, A. Abdollahpoor, Y. Wang and M. Pereira
School of Engineering, Deakin University, Geelong, VIC 3216, Australia
**E-mail: brolfe@deakin.edu.au*

H. Kong and D. Fabijanic
Institute for Frontier Materials, Deakin University, Geelong, VIC 3216, Australia

K. Hu, R. Han, K. Gao, A. Liu, L. Pan and Y. Bi
Research and Development Center, Wuhan Iron and Steel (Group) Corp., 430080, Wuhan, China

Z. J. Wang and Y. S. Zhang[†]
State Key Laboratory of Materials Processing and Die and Mould Technology, Huazhong University of Science & Technology, Wuhan, China
[†]E-mail: zhangys@mail.hust.edu.cn

Hot stamping is now commonplace in the automotive industry. The continuing need by automotive manufacturers to reduce weight while increasing crashworthiness has driven the industry to seek new hot stamping solutions. Currently tailored hot stamping is being introduced to hot stamp panels for automotive body structures. A tailored hot stamped part has at least two regions of different material properties to satisfy engineering design targets. For example, the B-pillar can be separated into the incredibly strong upper region to reduce intrusion, while the lower region is softer and tailored for energy absorption. We can then consider the tailored hot stamped part as a patchwork of hard and soft regions. Patchwork in this context means there is a relational organization (topology) to the network of hard and soft regions. The next generation of tailored hot stamping sheets may see extremely small regions of the blank have tailored properties.

We have investigated local carburization of a low-carbon blank sheet to locally tailor the properties of a hot stamped component. After hot stamping and in-die quenching, regions with a high carbon concentration (and high hardenability) transform to martensite while regions with low carbon concentration (and low hardenability) transform to ferrite, pearlite, or bainite. The location of carburizing can be controlled by using a commercially available stop-off coating (or mask) for carburizing. When this coating is used, the gradient in carbon concentration across the boundary between the masked and non-masked (that is, carburized and non-carburized) region of the sheet blank is on the order of the sheet thickness. The results of this process are reported in this paper.

Keywords: Tailored hot stamping; Tailor Welded Blanks; Carburization.

1. Introduction

Automotive structural parts with locally tailored properties have the potential to improve the crash behavior [1]. The zones with higher strength are beneficial where the aim is to protect the passenger compartment against intrusion, while the higher ductility allows higher energy absorption [2]. The common methods to achieve parts with locally tailored properties via hot stamping include: local austenitization of the blank in the furnace; pure air cooling of selected regions of the hot stamped part; post-forming heat treatment; and using a partially heated tool which allows controlling the cooling rate between the hot blank and the heated tool. The tailored heating of the hot stamping process was found to be less robust compared to standard hot stamping by Abdollahpoor et al. [3]. As a result, their work suggests that a material solution may be more appropriate than a tailored process solution.

The tailored part can be thought of as a part with a patchwork of material properties. In this context, patchwork means that there is a relational organization (topology) to the network of material properties. Our previous work (Figure 1) studied the effect of changing the patchwork of a simulated tailored blank on the crash performance of the deformed part. The results indicated that if genuine local control over the material properties is possible, then the part designer has a greater influence on the final performance of the part [2].

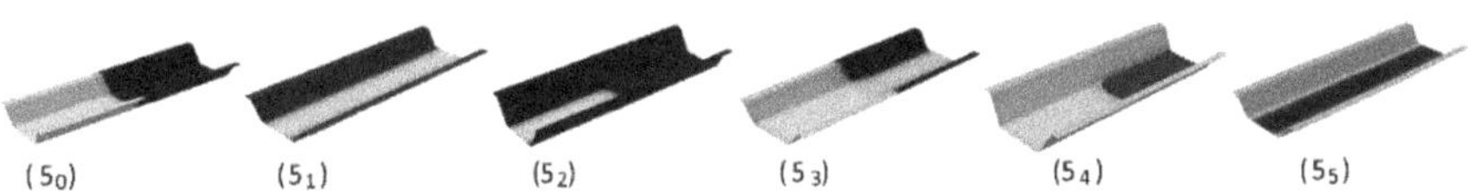

Fig. 1. Different alternatives of patchworking the soft and hardened zones in a tailored hot stamping part (black: hardened, light gray: soft).

The question remains of how to achieve such fine control over the patchwork blank material properties. In this study, we investigate carburization of a low-carbon steel to provide genuine local control over a blank's properties.

2. Local Carburization for Tailored Properties

We use local carburization of a low-carbon blank sheet to locally tailor the properties of a hot stamped component. After hot stamping and in-die quenching, regions with a high carbon concentration (and high hardenability) transform to martensite while regions with low carbon concentration (and low hardenability) transform to ferrite, pearlite, or bainite. The location of carburizing can be controlled by using a commercially available stop-off coating (or mask) for carburizing, see Figure 2. When this coating is used, the gradient in carbon concentration across the boundary between the masked and non-masked (that is, carburized and non-carburized) region of the sheet blank is of the order of the sheet thickness.

Table 1 lists the chemical composition of the two low-carbon steels that are used for carburization experiments. The sheet thickness was 1 or 2 mm. Carburization was performed in a fluidized bed furnace at 925°C with an atmosphere of propane and nitrogen. The propane concentration was varied between 5 to 10% during carburising. After carburizing, the steel was quenched in water or oil. Water quenching resulted in the formation of cracks at the mid-thickness of the sheet.

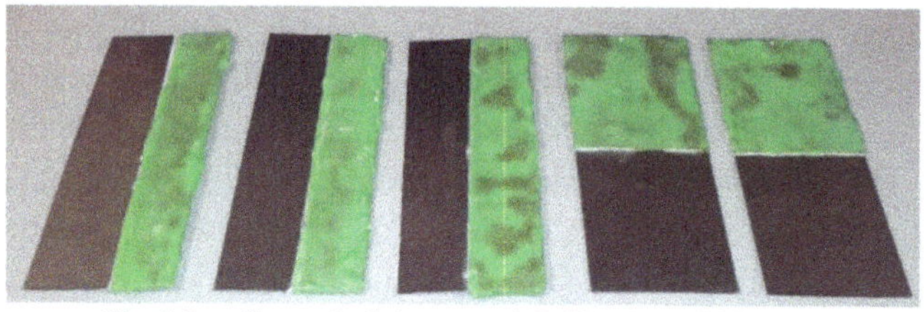

Fig. 2. Locally masked sheet metal (DC01, 1mm thickness).

Table 1. Chemical composition of the low-carbon base steels used for carburising experiments (wt. %).

Steel	C	Si	Mn
HA3	0.05	0.005	0.20
DC01	0.04	0.004	0.23

The details of the carburizing can be found in the previous paper by the authors [4].

3. Hot Forming

The carburized samples were prepared to be formed using the conventional hot stamping process. The advantage of the carburization in tailoring the properties is that there is no need to heat the tool to achieve tailored properties. This means that the tool at room temperature can be used for whole sample and it is the carbon content that makes differences in localized zones. For this paper, a bumper hot stamping tool at room temperature was used, see Figure 3. The carburized blanks were heated and kept for 5 minutes in an electric furnace at 930°C. The heated blank was transferred to the tool within approximately 7 seconds. The in-die quenching time was chosen as 4 seconds.

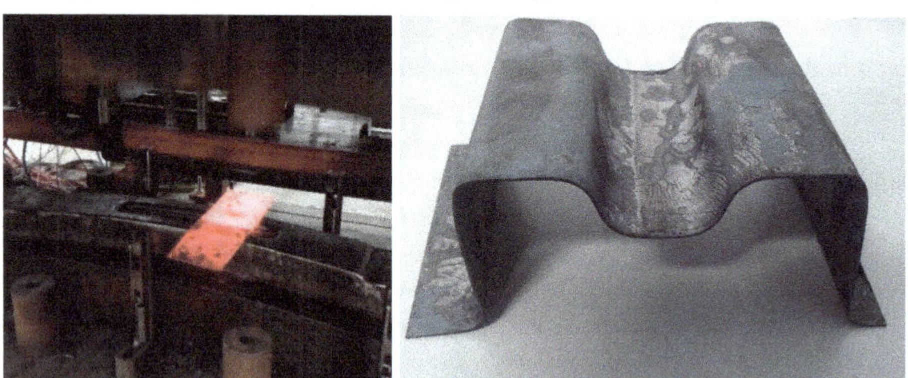

Fig. 3. (left) the hot stamping tool used for forming, (right) the formed samples.

4. Results

4.1. *Microstructure-hardness relationship*

The surface hardness of the hot stamped blank was measured using a load of 2kg and a dwell time of 15s. A continuous decrease from ~255HV in the carburized region to ~125HV in the non-carburized region was indicated in Fig. 4. The transition zone was approximately 3 mm. The hardness values in the current carburized region after hot stamping were much lower than that of the conventional hot stamping steel (i.e. ~500HV for the 22MnB5 steel). This might be associated with the pearlite-ferrite microstructure produced after hot stamping in the present steel (Fig. 5a). A microstructure consisting of ferrite and pearlite was also observed in the transition zone (Fig. 5b). But the fraction and grain size of the ferrite enhanced, which contributed to the decreased hardness compared to the carburized region. A full ferritic structure was revealed in the non-carburized region after the hot stamping process, resulting in the much lower hardness values.

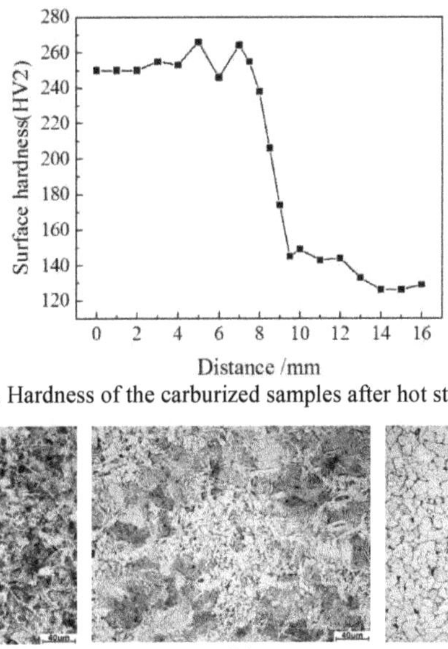

Fig. 4. Hardness of the carburized samples after hot stamping.

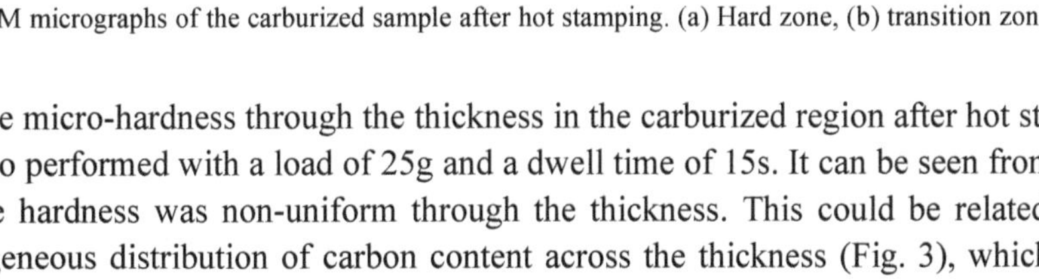

Fig. 5. OM micrographs of the carburized sample after hot stamping. (a) Hard zone, (b) transition zone, (c) soft zone.

The micro-hardness through the thickness in the carburized region after hot stamping was also performed with a load of 25g and a dwell time of 15s. It can be seen from Fig. 6 that the hardness was non-uniform through the thickness. This could be related to the heterogeneous distribution of carbon content across the thickness (Fig. 3), which might lead to the difference in transformation behavior during the hot stamping heat treatment. A high hardness of up to 400HV was observed in the near surface region while the value decreased gradually as the distance to the surface increased. The lowest hardness of ~270HV was found in the middle of the thickness. This might be explained by the coarser grain size observed in the middle zone in comparison to the finer microstructure in the near surface region.

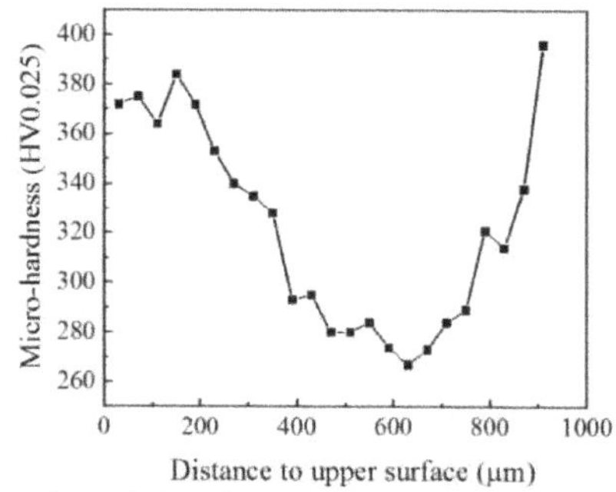

Fig. 6. Micro-hardness through the thickness in the carburized region after hot stamping.

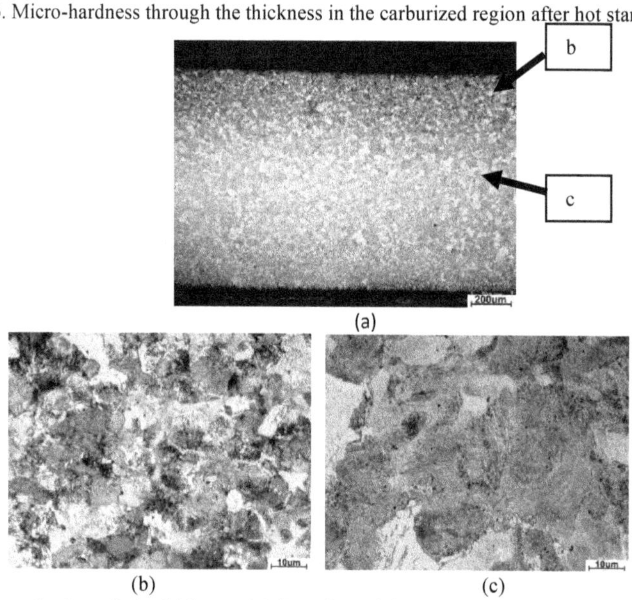

Fig. 7. OM micrographs through the thickness. (a) Overview of the cross-section, b and c in (a) indicate (b) near surface region and (c) middle zone, respectively.

4.2. *Tensile properties*

Tensile testing was carried out in the carburized and non-carburized samples after hot stamping. The tensile stress-strain curves were shown in Fig. 8. An ultimate tensile strength (UTS) of ~780 MPa and a total elongation (TE) of ~18% were obtained in the carburized sample, while a UTS of ~420 MPa and a TE of ~23% were observed in the non-carburized sample. Thus, a tailored property was produced after hot stamping a blank with carburized and non-carburized regions. However, the strength in the carburized sample was much lower than that of the typical hot stamping steel (i.e. ~1500 MPa for the 22MnB5 steel), which might be due to the pearlite-ferrite microstructure observed in the current study compared with the full martensitic structure produced in hot stamped 22MnB5 steel. There is promise in this method; however, it obviously needs some development to ensure a fully martensitic structure is produced via hot stamping. Analysis to improve the process is being currently undertaken.

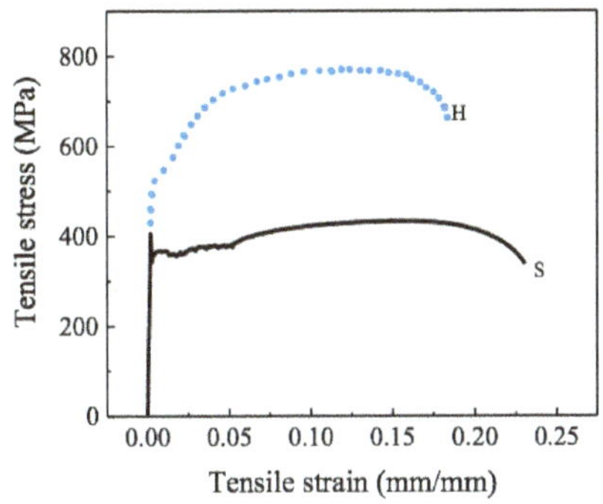

Fig. 8. Tensile stress-strain curves of the hot stamped samples. H and S represent hard (i.e. carburized) and soft (i.e. non-carburized) samples, respectively.

5. Conclusion

This paper explored a method for obtaining tailored properties (or a patchwork microstructure) in hot stamped components using locally carburizing low-carbon steel in order to locally modify the hardenability. As a result, carbon-rich regions transform to martensite during in-die quenching while carbon-lean regions transform to ferrite, pearlite, or bainite during in-die quenching. The effects of carburization parameters on the carbon concentration are currently being investigated. This process results in a very thin transition zone (on the order of the sheet thickness) between regions that are carburized and regions that are not carburized. Initial results suggest that a material solution (that is, local modification of chemical composition) by may be more appropriate to obtain a patchwork microstructure than a tailored process solution.

References

1. H. Karbasian and A.E. Tekkaya, A review on hot stamping, *J. Mater. Process. Tech.* **210**, 2103 (2010).
2. A. Abdollahpoor, X. Chen, M.P. Pereira, N. Xiao and B.F. Rolfe, The crash behaviour of hot stamped components – the effect of tailoring conditions, in *5th Int. Conf. Hot Sheet Metal Form. High Performance Steels*, (Toronto, Canada, 2015).
3. A. Abdollahpoor, X. Chen, M.P. Pereira, N. Xiao and B.F. Rolfe, Sensitivity of the final properties of tailored hot stamping components to the process and material parameters, *J. Mater. Process. Tech.* **228**, 125 (2016).
4. B.F. Rolfe, A. Abdollahpoor, M.P. Pereira, H. Kong, E. Pavlina, M. Cai, D. Fabijanic, K. Hu, R. Han, L. Pan, Y. Bi, Z.J. Wang and Y.S. Zhang, Advances in tailored hot stamping – innovations in material and local patchwork topology, in *2nd Int. Conf. Advanced High Strength Steel and Press Hardening*, (Changsha, China, 2015).

Recent Studies of Q&P Hot Stamping Process

X. H. Han†, Y. N. Ding, S. L. Tan, S. S. Chen and Y. Y. Zhong
Institute of Forming Technology & Equipment, Shanghai Jiao Tong University
Shanghai, 200030, P.R. China
†*E-mail: hanxh@sjtu.edu.cn*

Q&P hot stamping is a novel process to produce high strength steel parts with better elongation than the conventional ones. In this paper, recent studies of Q&P hot stamping process are presented. Firstly, the process of the Q&P hot stamping is discussed and a fractional cooling steps strategy is introduced to optimize the procedure. Secondly, several different materials are introduced to test the positive effects of Q&P process, and the suitability of these steels are also discussed. Thirdly, the formability of the hot stamping process is studied, as well as the springback phenomenon and its relationship with process parameters. Finally, microstructure evolution during the Q&P hot stamping process is modelled and discussed.

Keywords: High strength steel; Hot stamping; Quenching & Partitioning; Product of strength and plasticity.

1. Introduction

Hot stamping technology has been widely used to produce high strength steel parts. the tensile strength of the final part reaches 1500MPa or even more, but the elongation is only about 5% [1]. Product of strength and plasticity $\sigma_b \cdot \delta_t$ (abbreviated as PSP) is regarded as an index of comprehensive mechanical properties and is used to measure the energy absorption during the car crash. Q&P hot stamping process has been developed in the last few years to increase PSP of ultra-high strength steels [2-4], in which the Q&P heat treatment is integrated into the conventional hot stamping process, and the dual-phase microstructures of martensite and residual austenite can be achieved through such treatment, which induces notable improvement of elongation and PSP as well.

The research of Q&P hot stamping has been in progress, but there are still some problems need to be solved before it can be used in industry, such as, the process optimization and the corresponding stamping tool, the suitable materials for the new process, the formability especially the springback of the final part, and the microstructure evolution during the process and its modeling method. In this work, the above issues are discussed successively.

2. Q&P Hot Stamping Process

Fig. 1 (a) shows the schematic diagram of the conventional hot stamping process. When the hot blank is moved into the cool tools, it undergoes forming and quenching process, and the temperature drops continuously to room temperature. Fig. 1 (b) and (c) illustrate the procedure of Q&P hot stamping, when the elevated blank is moved into the tool, the forming and quenching processes begin simultaneously as the conventional hot stamping process does. But the blank is not quenched to room temperature directly, it is interrupted at a certain temperature called as quenching temperature (T_Q) which is between the martensitic transformation start temperature (Ms) and the martensitic transformation finish temperature (Mf). Then the blank is held at another temperature called as partitioning temperature (T_P) for a certain duration named as partitioning time (t_P). T_P can be larger than T_Q as shown in Fig. 1(b), or be equal to T_Q as shown in Fig. 1(c). The former is called as the two-step Q&P hot stamping process and the latter is called as the one-step Q&P hot stamping process.

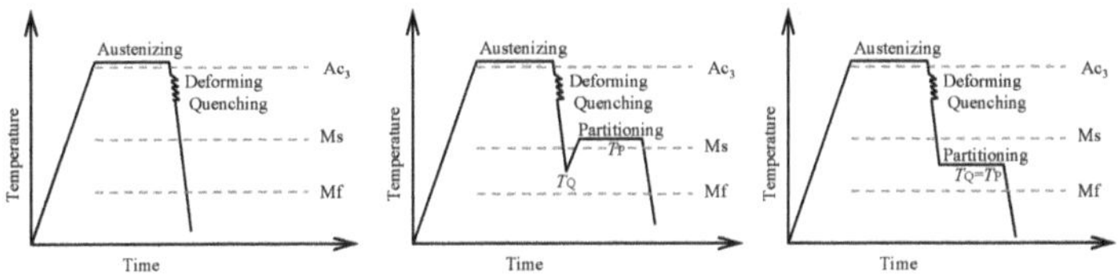

(a) (b) (c)

Fig. 1. Scheme for (a) conventional hot stamping process, (b) two-step Q&P hot stamping process, (c) one-step Q&P hot stamping process.

The study Zhong [5] shows that the steel's plasticity after the two-step Q&P process is higher than that of the one-step process. However, the temperature changes complicatedly during the two-step process and can be hardly realized with the presented tools. So only the one-step Q&P hot stamping process is studied in this work.

Generally, there are two different ways to perform the Q&P process. One is the "in-mold" method, which means that the whole procedure including forming & first quenching-partitioning-second quenching are all completed in the mold. The other one is "out-mold" method, in which, the forming & first quenching process are still carried out in the mold, but the partitioning and the second quenching process are done out of the mold. For the "in-mold" method, the tool design is the key problem [6], while the temperature monitor and deformation control are important to the "out-mold" method.

In order to shorten the cycling time of the Q&P hot stamping process, the so-called fractional cooling method can be integrated with the Q&P process [7]. Therefore, the rapid cooling technology is used during the transfer stage, so the cooling time is reduced. And a bonus of such method is that controlling of the forming temperature can lead to a better formability than the traditional hot stamping process, the optimized forming temperature for B1500HS is about 650°C.

3. Material for Q&P Hot Stamping

According to the previous studies of hot stamping process and Q&P treatment, the following elements are commonly mentioned, including carbon (C), manganese (Mn), silicon (Si), aluminum (Al), and boron (B). Si promotes carbon diffusion from as-quenched martensite to austenite, and retards cementite precipitation from these phases, which is important during the partitioning process. The content of Si is suggested more than 1% for such purpose [8]. Element B is very commonly used in the existed hot stamping steels, it pushes the TTT curve to the right side and increases the hardenability effectively. However, recent research [9] shows that B may lead to the formation of intermetallic Fe23(C,B)6 particles that precipitate on the grain boundary at lower austenite temperature, which is harmful to the final properties.

Three steels including B1500HS, 27SiMn and TRIP780 are introduced to study their performance during the Q&P process by using Gleeble 3500, all of these steels have been mass produced and can be purchased in the market.

After Q&P treatment with different parameters, the mechanical properties including strength, elongation and PSP are shown in Fig. 2. It can be seen that the strength range of 27SiMn is higher than TRIP780 and B1500HS, while the elongation range of TRIP780 is higher than 27SiMn and B1500HS. Overall, PSP values of TRIP780 are generally better than others.

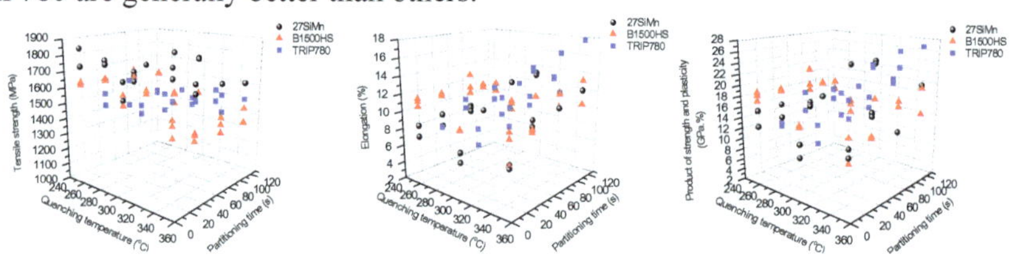

(a) (b) (c)

Fig. 2. Mechanical properties with different Q&P parameters (a) strength, (b) elongation, (c) PSP.

4. Formability and Springback

Since the forming process happens during the first quenching process, the steel formability of the Q&P hot stamping is the same as that of the conventional hot stamping process. But when the fractional cooling strategy is applied as discussed above, the effect of forming temperature is worth to be discussed.

As shown in Table 1, an unequal-high & unequal-wide U-shaped mold is applied, it can be found that the formability is highly related to the forming temperature. Failures easily happen for this part because of the complex geometry, but it can be avoided when the forming process is carried out near 650°C.

On the other hand, the cooling speed from the austenitizing temperature to the forming temperature also affects the forming result. For example, failure happens for the air cooling case but can be avoided for the water cooling case although both of the forming temperatures are 700°C.

Table 1. Hot stamping parts by integrating fractional cooling process.

Temp. / Process	550°C	600°C	650°C	700°C	750°C
Air cooling					
Spray cooling					

An asymmetrical U-shaped mold is adopted to test the springback of the Q&P hot stamping as shown in Fig. 3. It can be found that the fillet angle (angle 2&3) is positive and the round angle (angle 1&4) is negative; the springback angle decreases as the forming velocity increases and the maximum forming velocity is 825mm/s; and the springback angle increases with increasing of the forming temperature; while the partitioning time doesn't have an obvious influence on the springback.

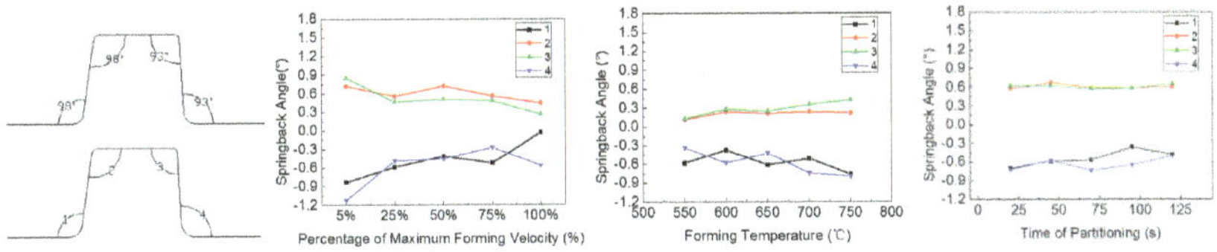

(a) (b) (c) (d)

Fig. 3. Springback study (a) part geometry, (b) forming velocity, (c) forming temperature, (d) partitioning time.

5. Microstructure Evolution

High strength ferrous alloys containing significant fractions of retained austenite can effectively improve the elongation with a little decrease of strength. As a result, the ideal microstructure of the final part to get high PSP is a dual-phase mixture of martensite and residual austenite. When the austenite fraction is 1.5%, the material strength is 1558MPa with a small elongation about 6.1%, and the PSP is about 9.5GPa%. However, when the austenite fraction is 6.3%, the material strength is about 1476MPa, but the elongation reaches to 10.7%, and the PSP is about 15.8GPa%. It can be seen that with the increase of the austenite amount, the elongation and the PSP have a great improvement, with only a little decrease of the material strength.

Fig. 4 presents the metallographic images of B1500HS for both full martensitic transformation and Q&P treatment. The lath shape martensite is clear as shown in Fig. 4 (a), which reflects the characteristic of full martensitic transformation. For Q&P process, the sizes of martensite remain almost unchanged as illustrated in Fig. 4 (b), but more white austenite can be observed among the martensitic structure. The change of austenite content also can be approved by XRD measurement [10].

(a) (b)

Fig. 4. Microstructure of B1500HS after (a) hot stamping process, (b) Q&P process (T_Q = 350°C, t_P = 80s).

6. Conclusion

Q&P hot stamping is a novel process to produce high strength steel parts with better elongation than the conventional ones. In this paper, building on research works in recent studies of Q&P hot stamping process, we look into several different materials to study the positive effects of Q&P process, and the suitability of these steels. Finally, we investigate the formability of the hot stamping process, and the springback phenomenon, and theirs impacts on the Q&P hot stamping process.

Li model, Lee model and CCE models are investigated for diffusion transformation, non-diffusion transformation and carbon partitioning process, respectively. The predicted value of retained austenite by Q&P treatment (the partitioning temperature is 300°C) is about 5.2%, which is comparable with the experiment result that is 6.3%.

References

1. H. Karbasian and A. E. Tekkaya, A review on hot stamping, *J. Mater. Process Tech.* **210**, 2103 (2010).
2. J. G. Speer, D. K. Matlock, B. C. Decoonman, J. G. Schroth and B. C. DeCooman, *et al.* Comments on "On the definitions of paraequilibrium and orthoequilibrium" by M. Hillert and J. Agren, scripta materialia, 50, 697-9 (2004), *Scripta Mater.* **52**, 83 (2005).
3. H. Liu, X. Jin, H. Dong and J. Shi, Martensitic microstructural transformations from the hot stamping, quenching and partitioning process, *Mater. Charact.* **62**, 223 (2011).
4. X. Han, Y. Zhong, P. Xin, Z. Cui and J. Chen, Research on one-step quenching & partitioning treatment and its application in hot stamping process, *P. I. Mech. Eng. B-J. Eng.* **84**,163 (2015).
5. N. Zhong, Research on high strength Q&P and Q-P-T steels, Ph.D thesis, Shanghai Jiaotong University, (2009).
6. X. Han, Z. Cui and P. Xin, A set of U-cap bending tool for hot stamping with one-step Q&P heat treatment: China Patent. ZL201310335789.1 (2015).
7. X. Han, S. Chen and Y. Zhong, A hot stamping process integrated with Q&P treatment and fractional cooling procedure: China Patent. 201510019472.6 (2015).
8. M. H. Saleh and R. Priestner, Retained austenite in dual-phase silicon steels and its effect on mechanical properties, *J. Mater. Process Tech.* **113**, 587 (2001).
9. H. Mohrbacher, Martensitic automotive steel sheet-Fundamentals and metallurgical optimization strategies, *Adv. Mater. Res.* **1063**, 130 (2015).
10. F. Xu, S. Hu and Y. Chen, Quenching and Partitioning Process of Hot Stamping Steel 22MnB5, *J. Mater. Sci. Eng.* **31**, 365 (2013).

Extended High-Strength-Ductility by Advanced Hot Stamping Treatment

H. Li*

College of Materials Science & Engineering, Beijing University of Technology, 100124/Beijing, China
E-mail: hongli@bjut.edu.cn

G. Y. Li

Central Iron & Steel Research Institute, CISRI, 100081/Beijing, China
E-mail: cisrilgy@sina.cn

M. T. Ma

China Automotive Engineering Research Institute, CAERI, 401122/Chongqing, China
E-mail: mingtuma@126.com

Y. S. Zhang

Huazhong University of Technology, HUST, 430074/Wuhan, China
E-mail: zhangys@mail.hust.edu.cn

Hot stamping technology is facing new challenges and opportunities for auto-part products, not only in the ultra-high-strength TS>1500MPa grades with balanced ductility for intrusion-resistant components (e.g. B pillars), but also the potential demands of future steel vehicle FSV car body on cold rolled and continuous annealed high-strength-ductility 9.5% DP800, 9.5% TRIP780, 10% DP1000 and 9.3% CP1000-1470 for energy-absorbing components. For such a large mass fraction (~35%) of TS800-1000MPa grade sheet steels, the conventional cold-stamping are facing serious spring-back and forming defects. Therefore hot stamping has been further developed and extended in combination with different austempering AT and Q-P treatments for TS800-1000MPa grade auto-parts through lab-experimental investigation by the authors. As the results of first stage experiments, the high-strength-ductility steels of TS700-1000MPa grades are obtained with good ductility El_t=18-34% and PSE≥18-24GPa•% by advanced hot stamping treatment AHST processing with the same one low alloyed SiMnNb sheet steel. The extended AHST method deals with the different processing routes, lab-experiment device, chemical composition design, mechanical properties and microstructure analysis. At the same time, the new progresses in ultra-high-strength-ductility steel of TS1500 grades (TS=1200-1700MPa and balanced ductility El_t=12-15% and PSE≥15-20GPa•%) by AHST of Q-P treatment are compared with TS800-1000 grades by AHST of AT or Q-P treatment. It is promising to develop a series of high strength steel TS700-1700MPa auto-part with improved ductility El_t=12-34% based on different F-B-TRIP and M-TRIP (Q-P) multi-phase steels by advanced hot stamping treatment AHST.

Keywords: High-strength-ductility; Auto-parts; Advanced Hot stamping treatment (AHST); Austempering treatment (AT); Quenching-Partitioning treatment (Q-P); Ultra-high-strength-ductility.

* Corresponding author.

1. Introduction

Hot stamped auto-part products are facing two challenges for the further development. First, it is necessary to improve the ductility of ultra-high-strength TS≥ 1500MPa grade auto-parts with balanced elongation for further enhancement of passive safety for the car and vehicle. [1-6] The conventional hot stamping process has to be modified and innovated for ultra-high-strength-ductility auto-parts. One method is tailored hot stamping.[2] Another method is advanced hot stamping treatment AHST. [3-6] The quenching–partitioning Q-P process suggested by Speer et al. [7-12] and the further proposed Q-P-T treatment by Hsu T.Y [13-20] have been combined with hot stamping to produce certain amount of retained austensite RA in the quenching martensite matrix to improve the ductility of ultra-high-strength TS≥ 1500MPa grade auto-parts. [3-6, 21-30]

Secondly, the car body structure of future steel vehicle FSV is designed with 9.5% DP800, 9.5% TRIP780, 10% DP1000 and 9.3% CP1000-1470 of cold rolled - continuous annealed sheet steels, as shown in Fig. 1. [31] The mass fraction of AHSS in TS800-1000MPa grades will be remarkably increased to about 35% besides 11.1% hot stamped HF1500. To meet the requirement on geometry accuracy and quality control, such high strength TS800-1000MPa grade sheet steels are facing serious spring-back and forming defects (e.g. wrinkling, splits, angular change, sidewall curl etc.) during conventional cold-stamping.

For the improvement of the ductility of hot stamped TS1500 grade auto parts, the new developed AHST process is effective to improve the total elongation from El_t=6-8% to El_t=12-15%, and to increase the product of strength and elongation PSE from PSE=9-11 GPa•% to PSE≥15-20GPa•% for ultra-high-strength TS=1.4-1.5GPa steels. [5,6, 13-20]

As for the high-strength-ductility TS800-1000 grade auto-parts for FSV car body components, the advanced hot stamping treatment AHST process can be adopted and extended to substitute conventional cold stamping to avoid serious spring back and various forming defects for high geometry accuracy and high quality.

Up to now the cold stamping technology of DP600 has been developed for more than 20 years to meet the requirement of auto parts (e.g. typical B-pillars) on geometry accuracy and defect control, especially on spring-back, wrinkling and splits. It is a systematic and time-consuming work based on multi-pass stamping process design, FEA formability analysis, the try-out cold stamping, measurements and adjustments, the negotiation between part-producer and part-designer on try-out auto-part quality, the experience data optimization.

Some of the important experiences have been summarized for the control of spring-back, wrinkling, splits and other defects in cold stamping of DP600 parts. For example: [31]

- Multi-passes of cold stamping process, e.g. from first die operation to fifth die operation with 5 stamping passes;
- Sidewalls should be designed with open angles to facilitate over-bending for spring back compensation;
- Parts should be designed with opened end to minimize cold-forming problems;

- 1 degree over-bending compensation can be used from a common neutral line based on experience;
- Die cut to compensation file with 1 degree over-bending;
- FEA spring-back modeling: predicted results must be correlated well with actual spring-back;
- Acute angles tend to cause splits, the feature radii should be larger than 3×thickness for DP590;
- Draw die and punch material should be at least one grade higher in strength than blank steel DP600, and more frequently maintenance for die radii and draw beads.
- Press load prediction accuracy: based on engineering formula, bead length, punch opening, part feature-length, blank thickness and strength.
- Post-stretching is one of the important techniques for the reduction of spring-back, and it is effective for both angular change and sidewall curl.

When the tensile strength is further increased to TS800-1000MPa, cold stamping process will be much more complex to compensate stronger spring back and to control more sensitive forming defects on components to meet required geometry accuracy and consistent quality.

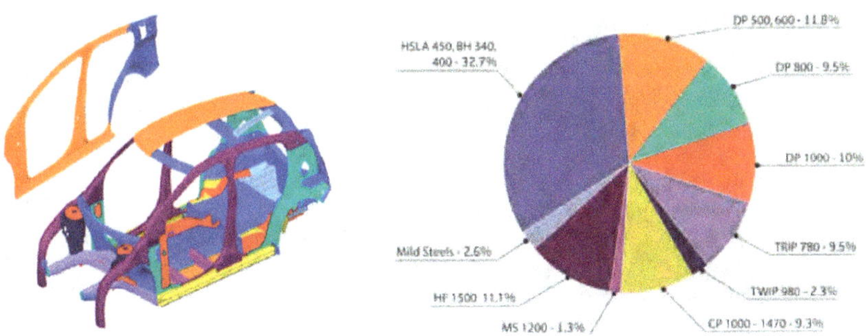

Fig. 1. The car body structure design of future steel vehicle FSV. [31]

Due to the advantages of hot stamping in high geometry accuracy without spring back and forming defects, the advanced hot stamping treatment AHST process has been further developed in combination with different austempering AT and/or Q-P treatments for the high-strength-ductility TS800-1000MPa grade auto-parts in FSV car body. Through the lab-experimental investigation in the present paper, it can be seen that the high-strength TS700-1000MPa grades with better ductility El_t=18-34% and PSE≥18-24GPa•% have been obtained by different type of AHST processes from one low alloyed SiMnNb sheet steel. The microstructure in TS800-1000 steels is featured by F-B matrix and B-M matrix with small amount retained austenite RA, which leads to high-strength-ductility with good elongation. While the microstructure in TS1500 steel is the lath martensite matrix and the inter-lath film of retained austenite in nano-scale, which leads to ultra-high-strength with improved ductility.

It is promising to develop a series of high strength steel auto-part in a wide range TS700-1700MPa with improved ductility El_t=12-34% based on different F-B-TRIP and

M-TRIP (Q-P) multi-phase steels by different type of advanced hot stamping treatment AHST process. The further research work is necessary for the enhancement and control of blank-steel hardenability, tailored hot stamping in AHST process, improved spot weldability and better impact toughness.

2. AHST Lab-Experiment Procedure for TS800-1000 Grade Steels

The low alloyed SiMnNb sheet steel was designed for the AHST experiment. The blank specimen with a chemical composition of 0.15C-1.44Si-1.50Mn-0.048Nb (wt%) in thickness of 1.0mm has been prepared with 50kg vacuum furnace and by forging-hot rolling-cold rolling.

The simulation experiments for hot stamping – quenching – austempering or partitioning treatment of the AHST process are performed on Gleeble 3500 thermo-mechanical simulator, as shown in Fig. 2.

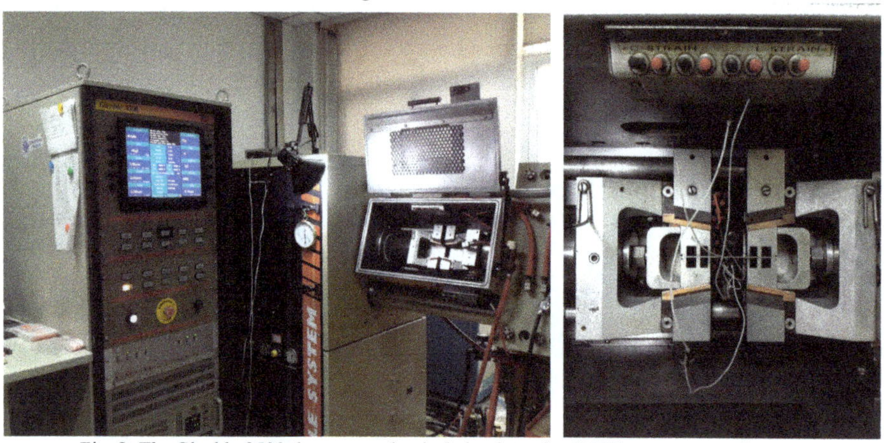

Fig. 2. The Gleeble 3500 thermo-mechanical simulator for AHST process experiments.

The simulation schedule for different AHST processes are preset on the control panel before the experiments according to the process models as shown in Fig. 3, and the process parameters are automatically controlled and recorded during the experiment, including the reheating rate, the austenization temperature and holding time, hot deformation strain and strain rate, cooling rate and so on. The process temperature is monitored with a thermocouple welded on the centre of specimen surface. The tests were conducted under a vacuum condition of 1.23×10^{-4} Pa.

The transformation temperature A_{C1}, A_{C3}, M_S and continuous cooling transformation curves for CCT diagram are measured with DIL805A dilatometer, as shown in Fig. 4. The fully austenization temperature A_{C3} and the starting temperature of martensite transformation M_S are also calculated with different formulae for the tested SiMnNb TRIP assisted steel and compared with the dilatometer measurement. The results show that Park equation (1996) for A_{C3} [32] and Andrews equation (1965) for M_S are more accurate for the tested SiMnNb steel.

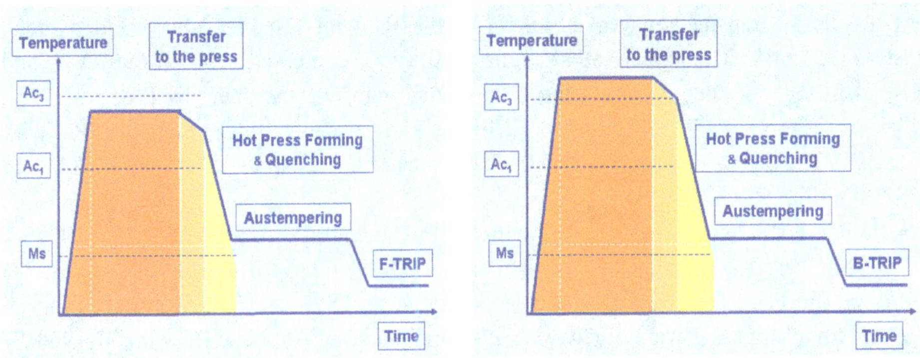

(a) (b)

Fig. 3. The process model for TS800-1000 grade high-strength-ductility auto-part steels,(a) Intercritical (F+A) Annealing: Hot Press Forming + Quenching + Austempering Treatment, Ferrite Matrix + Reined Austenite. F-TRIP Steel Auto-Part, (b) Austenite Region (A) Annealing: Hot Press Forming + Quenching + Austempering Treatment, Ferrite Matrix + Retained Austenite. B-TRIP Steel Auto-Part.

The specimens were reheated at a heating rate of 10°C/s to fully austenization temperature T_A=960°C holding for 5 min to get uniform prior austenite. The hot deformation was simulated with 10% tension strain at a strain rate of $1s^{-1}$. Then the specimens were immediately quenched to T_B in Bainite transformation zoon for austempering treatment, or quenched to T_Q below martensite starting temperature M_S and heating to T_P for partitioning treatment by holding certain time before finally quenching to room temperature.

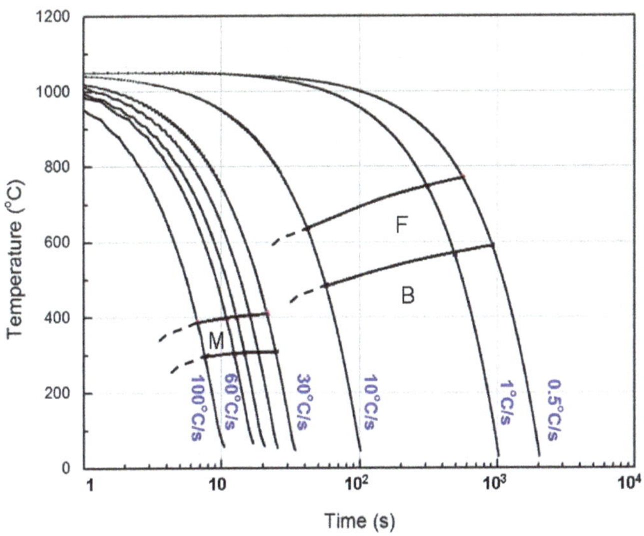

Fig. 4. The measured CCT diagram of 15SiMnNb steel.

3. AHST Lab-Experiment Results for TS800-1000 Grade Steels

Lab-experiment results of AHST process for TS800-1000 grade auto-part steels are listed in Table 1. When hot stamping is combined with austempering treatment on T_B by

Process I, the tensile strength is TS=715-749MPa with El_t=32.5-34.0% and average PSE=24.2GPa%. This grade of strength-ductility is expressed with TS690 A32 and PSE24 in Fig. 5. When hot stamping is combined with Q-P treatment by Process II with T_A=880°C, T_Q=50-250°C, T_P=400°C, the tensile strength is TS=828-843MPa with El_t=22.0-26.0% and average PSE=20.0GPa%. This grade is expressed as TS 780 A24 and PSE20 in Fig. 5.

When hot stamping is combined with Q-P treatment by Process III with T_A=960°C, T_Q= 50-250°C, T_P=300°C, the tensile strength is TS=987-1050MPa with El_t=18.0-20.7% and average PSE=18.6 GPa%, which is expressed as TS980 A18 and PSE18 in Fig. 5. When hot stamping is combined with Q-P treatment by Process IV with T_A=960°C, T_Q=50°C, the tensile strength is TS=1191-1210 MPa with El_t=11-12% and average PSE=13.6GPa%. This grade is expressed as TS1180 A11 and PSE13 in Fig. 5.

The strength could be further increased to TS1500MPa matched with low elongation of El_t=7% according to Koh-ichi Sugimoto's work on isothermal transformation (IT) process at temperatures below martensite-finish temperature (M_F) with a SiMnNb steel of 0.20C, 1.50Si, 1.51Mn, 0.044Al, 0.050Nb (wt%) [33], as shown by dotted line zone in Fig. 5 for comparison. It can be seen that the quenching hardenability of this steel is good and stable in the tested range of T_Q=25-200°C.

Table 1. Lawwsb-experiment results of AHST process for TS800-1000 grade auto-part steels.

AHST Types	Steel Composition	AHST Parameters	Tensile Strength TS (MPa)	Total Elongation El_t (%)	Product of TS & Elt PSE (GPa%)
Process I Hot Stamping + Austempering	15SiMnNb	T_I=800°C T_{AUS}=400°C	715-749	32.5-34.0	24.2
Process II Hot Stamping + Partitioning	15SiMnNb	T_A=880°C T_Q=50-250°C T_P=400°C	828-843	22.0-26.0	20.0
Process III Hot Stamping + Partitioning	15SiMnNb	T_A=960°C T_Q= 50-250°C T_P=300°C	987 -1050	18.0-20.7	18.6
Process IV Hot Stamping	15SiMnNb	T_A=960°C T_Q=50°C t=60s	1191-1210	11.0-12.0	13.6
Process IT Isothermal Transformation[33]	20SiMnNb	T_A=900°C T_Q=25-200°C t=1000s	1500-1600	7.0-7.5	10.8

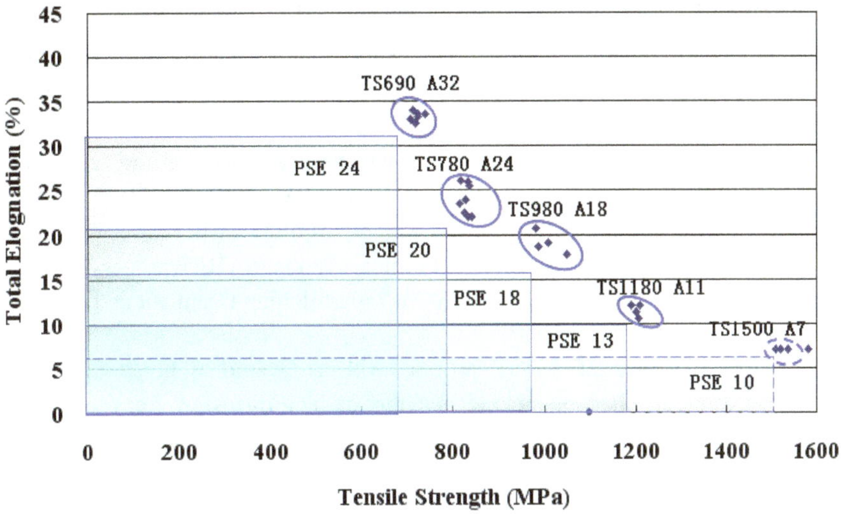

Fig. 5. A series of high-strength grade with balanced ductility of SiMnNb steel by different AHST processes.

The microstructure of above AHST processed specimens are observed with Zeiss Ultra 55 field-emission scanning electron microscope FE-SEM, as shown in Fig. 6. The wide lath of martensite in micrometer size are mixed with ferrite, bainite and very few amount (around 3%) of retained austenite including some martensite-austenite islands.

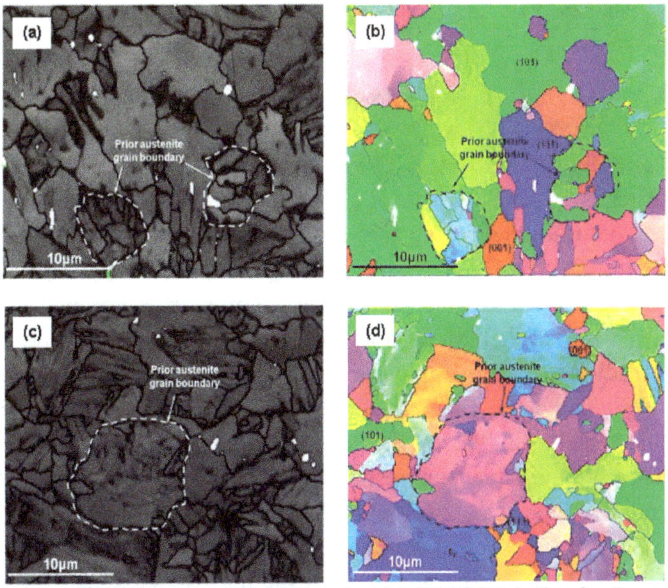

Fig. 6. Microstructure of AHST processed SiMnNb steel with SEM EBSD, (a)(b) after Process III, (c)(d) after Process IV, (a) (c) Kikuchi–line image quality maps, (b)(d) orientation maps (inverse pole figure IPF map): Blue (111), Red (001), Green (101).

4. Discussion: Comparison of Different Type of AHST Processes

For different grade of high-strength-ductility TS800-1000 auto-parts, hot stamping can be combined with different austempering treatment AT and/or Q-P treatment to obtain corresponding multi-phase steels with different F-B TRIP and B-M-TRIP matrix. Therefore the optimal quenching temperature will be decided according to the required strength-ductility grade, as introduced by Li and his co-workers. [3-6]

For the ultra-high-strength TS1500 grade auto-part of M-TRIP (i.e. Q-P) matrix steel with improved ductility, the advanced hot forming treatment AHFT process is developed by hot press forming–quenching to $M_S>T_Q>M_F$ – partitioning treatment at $T_{P\text{-}T}>M_S$ (or at $T_{P\text{-}T}=T_Q<M_S$).

The main difference of AHST process with conventional hot stamping is the partitioning treatment after quenching for the carbon diffusion from martensite to retained-austenite and to get the mixture of martensite and retained austenite.

At the same time, another important difference of AHST process with conventional hot stamping is the quenching temperature T_Q. In conventional hot stamping, the hot pressed parts are immediately quenched to room temperature T_Q=20°C to attain full martensite microstructure.

In AHST process, the hot pressed parts are quenched to the temperature between $M_S>T_Q>M_F$ to control the volume fraction of martensite and retained austenite.

The quenching temperature T_Q can be considered according to its influence on the martensite fraction based on the Koistinen-Marburger kinetics equation for martensite transformation as follows: [34]

$$f_m = 1 - \exp[-\alpha(M_s - T_q)] \tag{1}$$

where $\alpha=1.10\times10^{-2}$ for carbon content $0.37\leq C\leq1.10$（wt%）.

In the research work of A. Arlazarov et al,[28] the optimal quenching temperature was estimated with the Q-P theory of Speer,[7] the formula for starting temperature of martensite transformation M_S by Steven & Haynes,[35] and the kinetics equation for martensite transformation of Kiostinen-Marburger[34] before the experiments.

The normal blank steel 22MnB5 for conventional hot stamping has been directly adopted for the advanced hot stamping in combination with Q-P treatment by several researchers. [21,22,23] The better ductility of AHST processed 22MnB5 has been obtained with total elongation El_t =13-14% for the ultra-high-strength TS=1440-1600MPa by one-step quenching-partitioning treatment respectively on the temperature $T_Q=T_P$=250-300°C with different holding time in the range of 60-300s, as shown in Table 2.

When the SiMn TRIP-assisted steels are selected for the advanced hot stamping in combination with Q-P treatment in the lab experiments, the ductility of ultra-high-strength steel can be further improved by different researchers[24-30] as shown in Table 2. It can be seen that the total elongation are increased to El_t=13-21% for the ultra-high-strength steel in the range of TS=1215-1730MPa, and the product of strength and elongation are increased to PSE=15.8-32.5GPa%.

The main feature of the microstructure in advanced hot stamping treatment AHST processed steel is the lath martensite matrix and the inter-lath film of retained austenite, as shown in Fig. 7.

Table 2. The composition and mechanical properties of AHST processed steels. [21-30]

Chemical Composition	AHST Process	TS. (MPa)	El_t (%)	PSE (GPa·%)	Researcher & Reference
0.22C-0.28Si-1.28Mn-0.21Cr-0.003B	T_A=910°C, T_Q=300°C T_P=300°C, t=60 s	1440	14.2	20.4	Xu F.Y. Hu S.B. [21]
0.26C-0.82Si-1.72Mn-0.038Ti-0.0044B	T_A=900°C, T_D=750°C, ε=60% T_Q=T_P=300°C, t=300 s	1600	13.4	21.4	Chen M.M. Shen Y. [22]
0.23C-0.25Si-1.35Mn-0.19Cr-0.0032B	T_A=920°C, T_Q=250°C T_P=250°C, t=80 s	1604	13.2	21.2	Han X.H. [23]
0.19C-1.52Si-1.53Mn - 0.048Nb	T_A=910°C T_Q=260°C T_P=400°C t=10 s	1220	13.0	15.8	J. Zhang H. Ding. [24]
0.19C-1.57Si-1.52Mn-0.029Nb	T_A=960°C. T_Q=300°C T_P=450°C, t=30 s	1215	15.2	18.5	K. Zhang. Y. Rong [25]
0.25C-0.89Si-1.71Mn-0.026B	T_A=900°C T_Q=T_P=260°C t=300 s	1520	13.4	20.3	Wang C.L. Jin X.J. [26]
0.28C-2Si-2.5Mn-0.15Cr	T_A=850°C, T_Q=240°C T_P=450°C, t=500 s	1425	15.8	22.5	I. e Diego-Calderon I. Sabirov [27]
0.29C-1.4Si-3.0Mn-	T_A=820°C T_Q=200°C, T_P=400°C, t=100 s	1433	15.9	22.8	A. Arlazarov. O. Bouaziz. [28]
0.27C-1.40Si-1.28Mn-0.060Al	T_A=900°C T_Q=T_P=300°C t=120 s	1730	13.1	22.7	Zhong Y.Y Han X.H. [29]
0.20C-0.93 Si - 1.53Mn-0.85Cr, 0.021Ti- 0.024B.	T_A=900°C, T_D>A_{e3}, ε=0.2, 0.4, 0.6 T_P=300°C, t=30 s	1700	15.0	25.5	Heping. Liu Huer.Sun Bin Liu [30]
0.42C-1.58Si-1.46Mn-0.028Nb	T_A=850°C, T_Q=200°C T_P=450°C, t=30 s	1549	20.3	31.4	K. Zhang. Y. Rong [25]
0.20C-0.93 Si - 1.53Mn-0.85Cr-0.021Ti-0.024B	T_A=900°C, T_D>A_{e3}, ε=0.2, 0.4, 0.6 T_P=300°C, t=300 s	1550	21.0	32.5	Heping. Liu Huer.Sun Bin Liu [30]

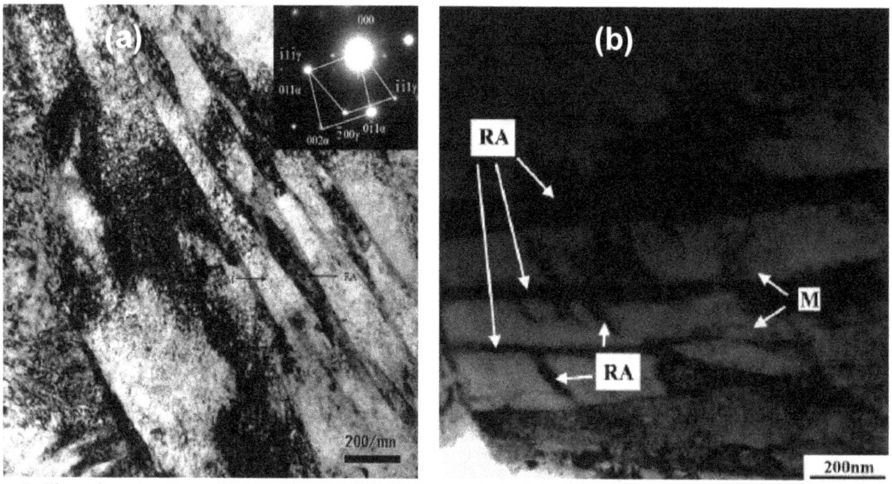

Fig. 7. The microstructure of quenching-partitioning treated steels, (a) the fine lath martensite and needle-like retained austenite between martensite lath in 0.20C-0.93Si-1.53Mn-0.85Cr, 0.021Ti-0.024B steel by DIFT-Q-P treatment, [30] (b) retained austenite after hot stamping and one-step Q&P treatment: T_D=900°C, ε=0.5, T_P=300°C, Δt_P=30s in 0.25C–0.89Si–1.71Mn– 0.026B steel. [26]

The retained austenite film could be the most important and crucial factor for the improvement of strain hardening ability and ductility during the plastic deformation of the M-TRIP (i.e. Q-P treated) steel with lath martensite matrix in sub-micrometer sized layers, even though the retained austenite films are in nano-scale dimensions and small volume fraction.

In the strengthening of ultra-high-strength-steels, the high dislocation density in lath martensite matrix is very important, especially when carbon pining effect occurs in the form of Cottrell atmosphere. Therefore the moderately increase of carbon content is not only favorite for retained austenite, but also prompting for carbon pinning effect. For micro-alloyed steels, the very fine precipitation of carbides (~5nm) is another important strengthening factor.

5. Conclusion

In the car body structure design of future steel vehicle FSV, the high-strength-ductility steels are selected with 9.5% DP800, 9.5% TRIP780, 10.0% DP1000 and 9.3% CP1000-1470 of cold rolled and continuous annealed sheet steels. The mass fraction of AHSS in TS800-1000MPa grades will be remarkably increased to about 35%. However, such high-strength-ductility-steels are facing serious spring-back and forming defects (e.g. wrinkling, splits, angular change, sidewall curl etc.) in conventional cold-stamping for the required geometry accuracy and component quality. It will be more difficult for cold stamping to produce complex-shape components in the new structure design with TS800-1000MPa grade sheet steels.

The advanced hot stamping process AHST has been further developed in combination with different austempering AT treatment and Q-P treatment for TS800-1000MPa grade auto-parts through lab-experimental investigation. As the first stage results, the different grades of high-strength-ductility steels in TS700-1000MPa can be obtained with better ductility El_t=18-34% and PSE⩾18-24GPa•% due to the multi-phase steels with different F-B TRIP and B-M-TRIP matrix from one low alloyed SiMnNb blank steel.

Various high-strength-ductility steels in TS700, TS800, TS1000 and TS1200 grade will be obtained by different AHST processes from one composition SiMnNb steel through one shot hot stamping under low press loading without spring back, wrinkles and splits.

To improve the ductility of TS1500 grade ultra-high-strength-steel auto-parts, the quenching-partitioning Q-P treatment is combined with hot stamping to generate retained austenite RA in martensite matrix to obtain M-TRIP or Q-P steels with ultra-high-strength TS=1200-1700MPa and improved ductility El_t=12-15% and PSE≥15-20GPa·%. Different blank steels are selected and developed from 22MnB5, TRIP assisted steels and Q-P steels for above Q-P type AHST process.

The advanced hot stamping treatment process AHST would be promising to deliver an extended series of high strength TS700-1700MPa steels with improved ductility El_t=12-34% based on different F-B-TRIP and M-TRIP (Q-P) steels, for different applications including ultra-high-strength intrusion-resistance components and high-strength energy-absorption crumple auto-parts.

The further research work in advanced hot stamping treatment is necessary for the enhancement and control of blank steel hardenability, for tailored hot stamping by AHST process, for improved spot weldability and better impact toughness, for the development of new production line and equipment of advanced hot stamping treatment AHST processing. [5-6]

References

1. H. Karbasian and A. E. Tekkaya, A review on hot stamping, *J. Mater. Process Tech.* **210**, 2103 (2010).
2. M. Merklein and M. Wieland, Hot stamping of boron steel sheets with tailored properties: A review, *J. Mater. Process Tech.* **228**, 11 (2016).
3. G. Y. Li, M. T. Ma, X. P. Mao and C. X. Zhao, Innovative Research in Hot Stamping Technology, *Adv. Mater. Res.* **1063,** 223 (2015).
4. G. Y. Li, M. T. Ma, Y. S. Zhang and R. Luo, Advanced hot forming treatment AHFT of deep working technology for ultra-high-strength-ductility auto-parts, *Steel Rolling.* **32**, 1 (2015).
5. G. Y. Li, M. T. Ma, Y. S. Zhang and H. Li, Advanced High Strength Steel and Press Hardening, in *Proc. ICHSU 2015*, (ChangSha, China, 2015).
6. H. Li, G. Y. Li, M. T. Ma and Y. S. Zhang, Advanced hot stamping treatment AHST for ultra-high strength-ductility auto-parts, in *IDDRG 2016. Proceedings*, (Stahlwelt, Linz, 2016).
7. J. G. Speer, D. K. Matlock, B. C. De Cooman, and J. G. Schroth, Carbon partitioning into austenite after martensite transformation, *Acta Mater.* **51**, 2611 (2003).

8. D. K. Matlock, V. E. Bräutigam and J. G. Speer, Application of the Quenching and Partitioning (Q&P) Process to a Medium-Carbon, High-Si Microalloyed Bar Steel, *Mater. Sci. Forum* **426**, 1089 (2003).
9. J. G. Speer, D. V. Edmonds, F. C. Rizzo and D. K. Matlock, Current Opinion in Solid State and Materials Science, *Solid State Mater. Sci.* **8**, 219 (2004).
10. J. G. Speer, F. C. R. Assunção, D. K. Matlock and D. V. Edmonds, The "quenching and partitioning" process: background and recent progress, *Materials Research.* **8**, 417 (2005).
11. D. K. Matlock, AHSS Microstructures, Effect on Failure, in *Advanced High Strength Steel Workshop*, (Arlington Virginia, USA, 2006).
12. J. G. Speer, E. D. Moor, K. O. Findley, D. K. Matlock, B. C. De Cooman and D. V. Edmonds, Analysis of Microstructure Evolution in Quenching and Partitioning Automotive Sheet Steel, *Metall. and Mater. Trans. A.* **42**, 3591 (2011).
13. J. G. Speer, E. D. Moor and A. J. Clarke, Critical Assessment 7: Quenching and partitioning, *Mater. Sci. and Technol.* **31**, 3 (2015).
14. T. Y. Hsu (Xu Z Y), New processes for steel heat treatment, *Heat Treat.* **22**, 1 (2007).
15. T. Y. Hsu (Xu Z Y), Quenching - portioning - tempering (precipitation) (Q - P - T) process for Ultra - high strength steel, *Heat Treat.* **23**, 1 (2008).
16. T. Y. Hsu (Xu Z Y), A brief introduction to quenching - partitioning - tempering (Q - P - T) process, *Heat Treat. Met.* **34**, 1 (2009).
17. Z. Y. Xu, X. J. Jin and Y. H. Rong, Strengthening and toughening mechanisms of quenching - partitioning - tempering (Q - P - T) steels, *J. Alloy. Compd.* **577**, S568 (2013).
18. N. Zhong, X. D. Wang and Y. H. Rong, Interface migration between martensite and austenite during quenching and partitioning (Q & P) process, *Mater. Sci. Technol.* **22**, 751 (2006).
19. N. Zhong, X. D. Wang, L. Wang and Y. H. Rong, Enhancement of the mechanical properties of a Nb – micro – alloyed advanced high strength steel treated by quenching - partitioning – tempering process, *Mater. Sci. Eng. A* **506**, 111 (2009).
20. Y. H. Rong, Advanced Q - P - T steels with ultrahigh strength - high ductility, *Acta Metallurgica Sinica.* **47**, 1483 (2011).
21. F. Y. Xu, S. B. Hu and Y. Chen, Quenching and partitioning process of hot stamping steel 22MnB5, *J. Mech. Sci. Technol.***31**, 365 (2013).
22. M. Chen, An Integrated Processing Method for Steel's Hot Deformation and Q&P Heat Treatment, Phd Thesis, Shanghai Jiaotong University (2013).
23. X. H. Han, Y. Zhong, K. Yang, Z. Cui and J Chen, Application of hot stamping process by integrating quenching and partitioning heat treatment to improve mechanical properties, *Procedia Eng.* **81**, 1737 (2014).
24. J. Zhang, H. Ding, R. D. K. Misra and C. Wang, Microstructural evolution and consequent strengthening through niobium-microalloying in a low carbon quenched and partitioned steel, *Mater. Sci. Eng. A.* **641**, 242 (2015).

25. K. Zhang, P. Liu, W. Li, Z. H. Guo and Y. H. Rong, Ultrahigh strength-ductility steel treated by a novel quenching–partitioning–tempering process, *Mater. Sci. Eng. A*. **619**, 205 (2014).
26. C. L. Wang, Master Thesis, Shanghai Jiao Tong University, (2015).
27. D. D.Calderon, I. Sabirov, J. M. M. Aldareguia, C. Föjer, R. Thiessen and R. H. Petrov, Microstructural design in quenched and partitioned (Q&P) steels to improve their fracture properties, *Mater. Sci. Eng. A* **657**, 136 (2016).
28. Arlazarov, O. Bouaziz, J. P. Masse and F. Kegel, Characterization and modeling of mechanical behavior of quenching and partitioning steels, *Mater. Sci. Eng. A* **620**, 293 (2015).
29. Y. Y. Zhong, S. S. Chen and X. H. Han, Study on quenching & partitioning heat treatment for mechanical properties and microstructure of steel 27SiMn, *J. Plasticity Eng.* **22**, 124 (2015).
30. H. Liu. H. Sun, B. Liu, D. Li, F. Sun and X. Jin, An ultrahigh strength steel with ultrafine- grained microstructure produced through intercritical deformation and partitioning process, *Mater. Des.* **83**, 760 (2015).
31. S. Keeler and M. Kimchi, *Advanced High-Strength Steels Application Guidelines*, Version 5 (World Auto Steel, 2014).
32. S. H. Park, Posco Technical Report, 50 (1996).
33. K. I. Sugimoto and A. K. Srivastava, Microstructure and Mechanical Properties of a TRIP-Aided Martensitic Steel, *Microstruct. Anal.* **4**, 344 (2015).
34. D. P. Kiostinen and R. E. Marburger, A general equation prescribing the extent of the austenite-martensite transformation in pure iron-carbon alloys and plain carbon steels, *Acta Metall.* **7**, 59 (1959).
35. W. Steven and A. G. Haynes, The temperature formation of martensite and bainite in low-alloy steels, some effects of chemical composition, *J. Iron & Steel Inst.* **183**, 349 (1956).

Research on Q&P Process and its Springback Property of High Strength Steel

J. Zhu, Y. S. Zhang† and B. Zhu

State Key Laboratory of Materials Processing and Die and Mould Technology, Huazhong University of Science & Technology, Wuhan 430074, China

†E-mail: zhangys@mail.hust.edu.cn

In this paper, the contents mainly include two aspects, one of which is the study of Q&P parameters effects, the main parameters are quenching temperature, partition temperature and partition time. Several groups experiments are carried on under different conditions, the results are determined by the tensile strength and elongation rate which are obtained by tensile test. The second is to carry out experiments on U shaped parts in the same conditions respectively with the traditional hot forming-quenching process and Q&P process, comparing the effect of Q&P and traditional hot forming process on springback, the results of springback are evaluated by the rebound angle. Finally, by analyzing and summarizing the experimental results, finding that Q&P process has little influence on springback.

Keywords: High strength steel; Hot forming; Quenching and partitioning; U shaped part springback.

1. Introduction

For environmental protection and energy saving, to realize the lightweight of vehicles, the high strength steel has been paid more and more attention by the automobile industry. In order to make better use of high strength steel, a new hot forming and quenching-carbon partitioning (Q&P) process has been developed. This paper mainly studies the parameters rules of Q&P and its spring back feature with the traditional process, to find and use the advantages compared with the traditional process.

2. Parameters Affecting Hot Forming and Q&P Process

2.1. *Experimental materials*

In this paper, a high strength steel plate 30CrMnSi2Nb for car with a thickness of 1.5mm is used to carry on the experiment, which has a chemical composition in table 1 and phase-transition temperature in table 2.

Table 1. The chemical composition of 30CrMnSi2Nb.

Element	C	Si	Mn	P	S	Cr	Nb
	0.28	1.60	1.10	0.012	0.005	0.99	0.029

Table 2. The phase-transition temperature of 30CrMnSi2Nb.

Ac_3	Ms	Mf
836	350	235

2.2. *Experiment method*

Fig. 1 shows the temperature curve during the entire process. The experiment process mainly contains several steps. Firstly, the steel is heated to the austenitization temperature and kept a period of time, then the plate is transformed onto the press mould along with mold closing and quenching. Next, it is transformed into another furnace to carry on partition, finally, the plate is taken out and the whole process is finished. Aiming at the quenching temperature, partition temperature and partition time [1,2, 3, 4], the experiments are designed as table 3.

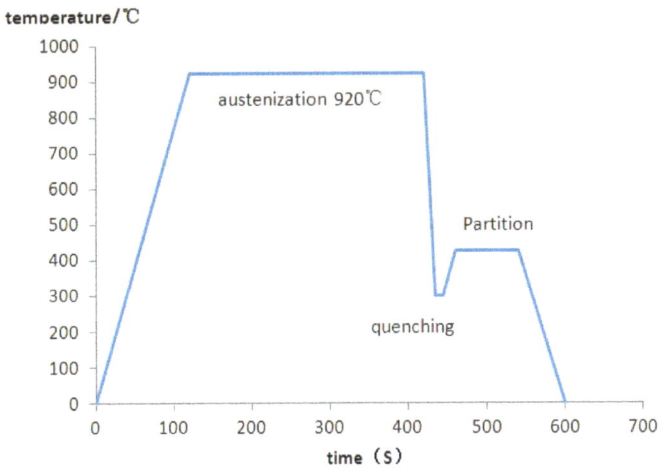

Fig. 1. Temperature curve.

Table 3. Experiments list.

QT / °C (PT=425, Pt=180)	PT / °C (QT=300, Pt=180)	Pt / s (QT=300, PT=425)
265	300	50
290	350	120
300	400	150
350	450	200
		300

2.3. *Discussion*

As to quenching temperature, when the quenching temperature is low, with the increase of quenching temperature, the tensile strength decreases, the elongation increases, but as the quenching temperature continues to raise, the elongation rate decreases. This is because the higher the quenching temperature, the lower the martensite content and the higher the austenite content after the quenching stage, it also makes the final austenite content higher. But when the quenching temperature is too high, the carbon is not stable enough in a very high content austenite, part of the austenite decomposes and makes the final austenite content decreased, along with the plasticity decreases.

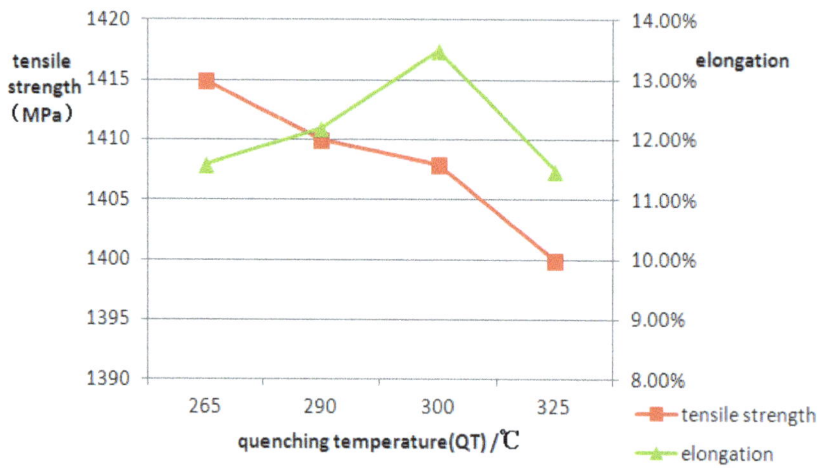

Fig. 2. Quenching temperature.

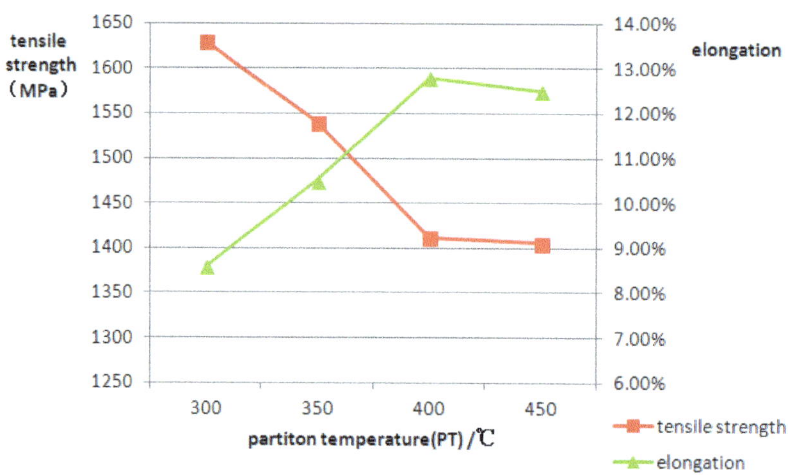

Fig. 3. Partition temperature.

As for the partition temperature and partition time, they have similar impact, but influence each other. In brief, the increment of the partition temperature and partition time has advantages of the increment of the retained austenite, however, too high partition temperature and time will make the carbon in austenite separates out, and reduce the content of retained austenite. Therefore, in order to obtain a better plasticity, it's necessary to configure the partition temperature and partition time reasonably.

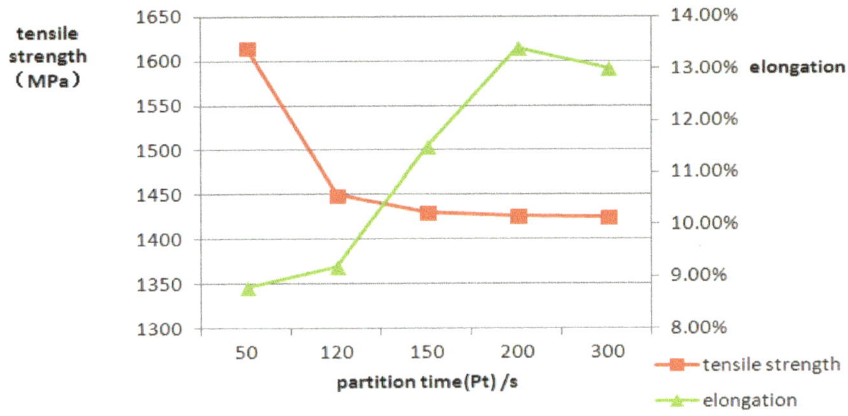

Fig. 4. Partition time.

3. Effect of Hot Forming and Q&P Process on U Shaped Parts' Springback

In U shaped parts' springback experiment, the same steel plate is used, the plate is cut into 280 mm×140 mm, and the experiment is carried on by Q&P one step method, that is, the quenching temperature is equal to the partition temperature. In this experiment, the quenching temperature and partition time are mainly studied, as well as the comparison with conventional hot forming. There are different groups quenching temperature and partition time are selected to carry on the experiment, and the parts are obtained as Fig. 5, the results as table 4 and table 5.

Through the data analysis, first of all, whether the traditional hot forming or hot forming -Q&P process, the parts have small springback angle, the biggest of which is less than 1 degree. This shows that the hot forming process can effectively reduce the springback of the parts, compared with the cold forming, it has a great advantage in the inhibition of springback. Then, compared with the traditional hot forming with the hot forming -Q&P process, it can be found that both of the springback angle is very small, this shows that the Q&P heat treatment process has little effect on the springback of the parts, and the main factors affecting the springback of the parts are the hot forming process.

Fig. 5. U shaped parts.

Table 4. Angle of bend at the bottom.

PT（℃）	Pt（25 s）	Pt（60 s）	Pt（180 s）
250	119	120.2	120.2
300	120	121	120.4
350	120.2	119.5	120.2

Table 5. Angle of springback.

PT（℃）	Pt（25 s）	Pt（60 s）	Pt（180 s）
250	-1	0.2	0.2
300	1	1	0.4
350	0.2	-0.5	0.2

4. Conclusion

1) To the quenching temperature, partition temperature and partition time, the effect of each parameter is similar, as it increases, the elongation increases and the strength decreases. But when their value is too high, the elongation decreases with the increase of them, and the plasticity decreases. Therefore, to get better plasticity, a better quenching temperature must be determined first, and then configuring the partition temperature and partition time reasonably.
2) Hot forming process can effectively reduce the spring back of U shaped parts, on this basis, coupled with the Q&P process influence on springback is very small, all the spring back angles of these parts are less than 1 degree.

Acknowledgments

This research work is financially supported by the National Natural Science Foundation of China (grant no. U1564203) and National Natural Science Foundation of China (grant no. 51405171).

References

1. M. Naderi, M. Ketabchi, M. Abbasi and W. Bleck, Analysis of microstructure and mechanical properties of different high strength carbon steels after hot stamping, *J. Mater. Process. Tech.* **211**, 1117 (2011).
2. J. G. Speer and D. K. Matlock, Developments in the Quenching and Partitioning Process, *World Iron&Steel* **9**, 31 (2009).
3. M. J. Santofimia, L. Zhao, R. Petrovb and J. Sietsma, Characterization of the microstructure obtained by the quenching and partitioning process in a low carbon steel, *Mater. Charact.* **59**, 1758 (2008).
4. J. G. Speer, D. V. Edmonds, F. C. Rizzo and D. K. Matlock, Partitioning of carbon from supersaturated plates of ferrite， with application to steel processing and fundamentals of the bainite transformation, *Curr. Opin. Solid State Mater. Sci.* **8**, 219 (2004).

Effect of Process Parameters on Formability of Aluminum Alloy 7075 in Hot Stamping Process

W. C. Xiao and B. Y. Wang†
School of Mechanical Engineering, University of Science and Technology Beijing, Beijing 100083, China
†*E-mail: bywang@ustb.edu.cn*

The elongation of aluminum alloy 7075 at room temperature is extremely low. To overcome the defects in cold forming methods, aluminum hot stamping process was proposed. This process can effectively improve the stamping formability of aluminum. In this paper, heat treatment experiments and deep drawing tests were used to study the formability of aluminum alloy 7075 in hot stamping process. In the deep drawing tests, the effect of hot stamping process parameters on formability were also studied, which can be used to improve hot stamping formability and reduce forming defects.

Keywords: Aluminum alloy 7075; Hot stamping process; Process parameters; Formability.

1. Introduction

For the requirement of automobile lightweight, there is more and more application of aluminum alloy in the automotive industry [1]. 7xxx series aluminum alloys belong to the Al-Zn-Mg-Cu series aluminum alloys, which have characteristics of low density, high specific strength, good corrosion resistance, etc. As a typical material of 7xxx series aluminum alloys, 7075 possesses high tensile strength, which made it can be used as a structural part in the automotive body [2].

However, the elongation of 7075 aluminum alloy at room temperature is extremely low, making this material hard to be plastically deformed [3]. Nowadays, the plastic forming methods that commonly used for 7075 aluminum alloy at room temperature include superplastic forming method, synchronous hydro-bulging method, etc [4,5]. However, superplastic forming method requires a low strain rate, making it not suitable for automobile industry's assembly line [6]. Hydro-bulging method requires specialized equipment and relatively complex molds, making it not a wise choice to efficiently produce large quantities of auto parts in terms of cost and efficiency.

To overcome the defects in cold forming methods, professor Lin Jianguo in Imperial College London proposed the aluminum hot stamping process [7]. In this process, aluminum blank was first heated to solution temperature, then high temperature blank was quickly transferred to the tool and quenched in cold dies. At last, artificial aging treatment was used to improve the strength of the material. This process can effectively improve the stamping formability of aluminum. Many researchers had already used this technology for research and production. Mohamed [8] studied the hot stamping formability of 6xxx series aluminum alloy by designing a stamping experiment tool.

Wang [9] introduced hot stamping process to the research of AA2024, and studied the effects of blank temperature and stamping speed on stamping formability.

In this paper, heat treatment experiments and deep drawing tests were used to study the formability of aluminum alloy 7075 in hot stamping process. In the deep drawing tests, the effect of hot stamping process parameters on formability were also studied, which can be used to improve hot stamping formability and reduce forming defects.

2. Experimental Details

Material studied in this paper was aluminum alloy 7075 with thickness of 2mm. Solution treatment experiments were done to verify the feasibility of using this material in hot stamping process. The specimen was heat treated at solution temperature for tens of minutes. Then, it was quickly quenched to room temperature, and followed by the artificial ageing treatment (120 °C/24h) to make the material obtain the maximum strength. Solution treatment temperature of 460 ~ 500 °C and solution treatment time of 10 ~ 50 minutes were studied.

Deep drawing experiments were used to study the formability of aluminum alloy 7075 in hot stamping process. Process parameters that have effect on the hot stamping formability were studied, including blank temperature, stamping speed, and blank holder force.

3. Results and Discussion

The effects of solution treatment temperature and solution treatment time on the material's mechanical properties were shown in Figure 1 and Figure 2. It can be found from the experiment results that the aluminum alloy 7075 can be solution treated in a wide range of solution treatment temperature and solution treatment time. In this range, maximum strength the material can be obtained.

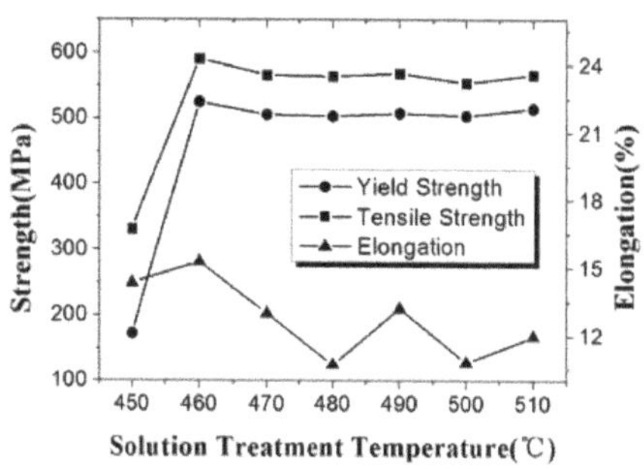

Fig. 1. Effect of solution treatment temperature on material's mechanical properties.

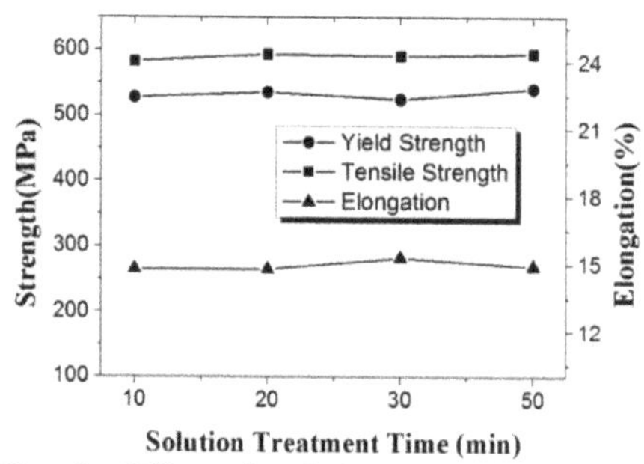

Fig. 2. Effect of solution treatment time on material's mechanical properties.

Our research group's unpublished work showed that in the cooling process, when the material was pressed in the dies and contact pressure was greater than 5MPa, the material can obtain the same mechanical properties as water quenching did. This characteristic of the material make it can be well used in hot stamping process.

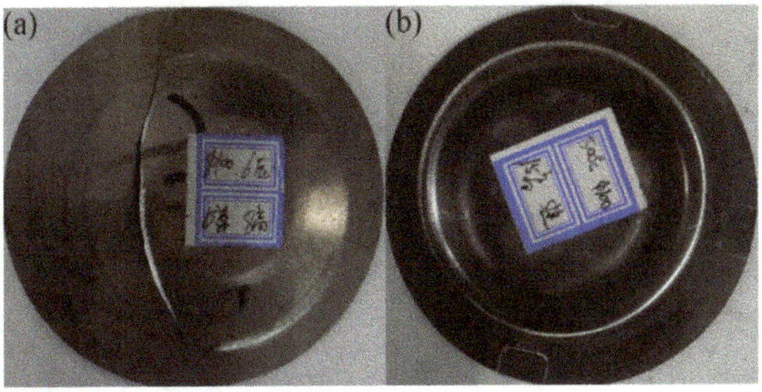

Fig. 3. Comparison of cold and hot stamping process.

Figure 3 shows the comparison of cold and hot stamping process. When the material was cold stamped, serious rupture could be found even when the stamping depth was less than 8 mm (Figure 3a). As a comparison, the stamping depth can reach 15mm when the blank was heated to 500 °C in hot stamping process (Figure 3b).

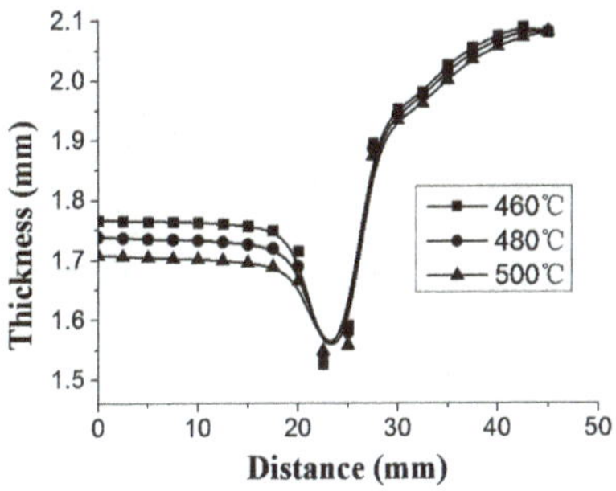

Fig. 4. Effect of blank temperature on formability.

Figure 4 shows the effect of blank temperature on formability. In this case, stamping speed was 100mm/s, blank holder force was 5KN, stamping depth was set as 15mm. It can be found that the blank thickness was evenly distributed in the bottom of the part, and the thickness of side wall between two corners was the smallest. The effect of blank temperature on formability is small, only the thickness in the center zone of the deep drawing part is affected.

Figure 5 shows the effect of stamping speed on formability. In this case, blank temperature was 480 °C, blank holder force was 5KN, stamping depth was set as 15mm, stamping speed was set as 50, 100, 200 mm/s. The part stamped at 50 mm/s was well formed, however, greater rupture was found with the increase of stamping speed. The experiment results shown in Figure 5b, c, and d verified the reliability of the simulation results.

Figure 6 shows the effect of blank holder force on formability. In this case, blank temperature was 480 °C, stamping speed was 100 mm/s, stamping depth was set as 15mm, blank holder force was set as 0, 5, 10 KN. It can be found that the blank thickness decreases with the increase of blank holder force. When the blank holder force was 10 KN, the blank was seriously thinned at the side wall. Although there was no rupture when the blank holder force was 0 KN, the maximum thickness of the blank was greater than 2.1mm, making the part possible to be wrinkling in the flange.

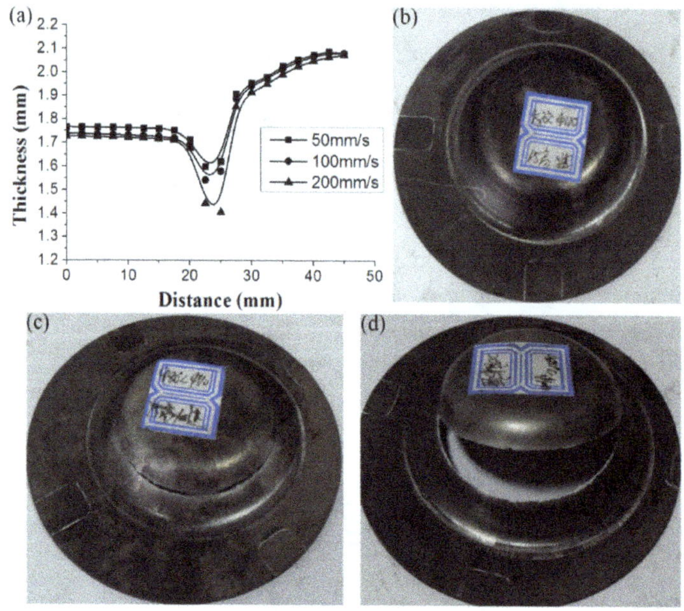

Fig. 5. Effect of stamping speed on formability.

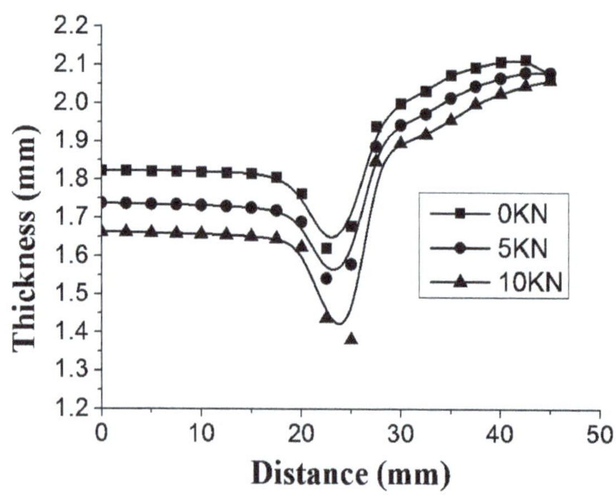

Fig. 6. Effect of blank holder force on formability.

4. Conclusion

This paper studied the feasibility of aluminum alloy 7075 in hot stamping process through solution treatment experiments and the effect of process parameters on formability through deep drawing tests. The following conclusions can be made:

a) Aluminum alloy 7075 can be used in hot stamping process. Stamping depth of hot stamping process is 15mm, more than double of that of cold stamping process.
b) When the stamping speed is greater than 50mm/s, greater rupture can be found with the increase of stamping speed.
c) Blank holder force is needed to prevent wrinkling in the flange, and blank thickness decreases with the increase of blank holder force.

Acknowledgment

This work was funded by the Joint Funds of the National Natural Science Foundation of China (Grant No. U1564202).

References

1. W. S. Miller, L. Zhuang, J. Bottema, A. J. Wittebrood, P. D. Smet and A. Haszler, A Vieregge, Recent development in aluminium alloys for the automotive industry, *Mater. Sci. Eng. A* **280**, 37 (2000).
2. N. R. Harrison and S.G. Luckey, Hot Stamping of a B-Pillar Outer from High Strength Aluminum Sheet AA7075, *SAE Int. J. Mater. Man.* **7**, 567 (2014).
3. J. Zhou, B. Y. Wang, J. G. Lin, L. Fu and W. Y. Ma, Forming defects in aluminum alloy hot stamping of side-door impact beam, *Transactions of Nonferrous Metals Society of China* **24**, 3611 (2014).
4. A. J. Barnes, Superplastic Forming 40 Years and Still Growing, *J. Mater. Eng. Perform* **16**, 440 (2007).
5. S. J.Hashemi, H. M. Naeini, G. Liaghat, R. A. Tafti and F.Rahmani, Forming limit diagram of aluminum AA6063 tubes at high temperatures by bulge tests, *J. Mech. Sci. and Tech.* **28**, 4745 (2014).
6. P. F. Bariani, S. Bruschi, A. Ghiotti and F. Michieletto, Hot stamping of AA5083 aluminium alloy sheets, CIRP *Annals - Manufacturing Technology* **62**, 251 (2013).
7. A. Foster, T. A. Dean and J. Lin, *Process for Forming Aluminium Alloy Sheet Components*. 2009.
8. M. S. Mohamed, A. D. Foster, J. G. Lin, D. S. Balint and T. A. Dean, Investigation of deformation and failure features in hot stamping of AA6082: Experimentation and modelling, *International Journal of Machine Tools and Manufacture* **53**, 27 (2012).
9. L. Wang, M. Strangwood, D. Balint, J. Lin and T. A. Dean, Formability and failure mechanisms of AA2024 under hot forming conditions, *Mater. Sci. Eng. A* **528**, 2648 (2011).

Investigation on Crush Performance of HAT Samples Based on 7075 Alloy HFQ Process

P. Hu, H. Rong, Y. L. Si and L. Ying†

School of Automotive Engineering, Faculty of Vehicle Engineering and Mechanics,
State Key Laboratory of Structural Analysis for Industrial Equipment,
Dalian University of Technology, Dalian, P.R. China
†E-mail: yingliang@dlut.edu.cn

7075-T6 aluminum alloy, with relatively low density and high strength, is one of superior lightweight materials. HFQ hot stamping process improves formability and reduces rebound of 7075-T6 aluminum alloy, which greatly expands its application range. In this paper, quasi-static crush experiment and simulation of HAT samples were conducted to research the crush performance of 7075-T6 alloy with HFQ Process. Forming temperature, solution temperature, solution time and artificial aging were chosen to study their influence on crush energy absorption of 7075-T6 alloy. The result shows that energy absorption significantly improves with forming temperature increasing and the best performance appears at 450°C. Energy absorption is preferable under the condition of solution temperature of 475°C for 30min. Compared with natural aging, artificial aging can effectively increase crush energy absorption of 7075-T6 aluminum alloy.

Keywords: Hot Form Quenching (HFQ); 7075-T6 alloy; HAT sample; Quasi-static crush; Energy absorption.

1. Introduction

7075-T6 aluminum alloy has been applied to automotive lightweight design for its low density and high strength recently. [1] HFQ hot stamping process improves formability and reduces rebound of 7075-T6 aluminum alloy. For HFQ process, the major factors that affect forming performance comprises forming temperature, solution temperature, solution time and artificial aging, etc. [2] In this paper, quasi-static crush experiment was designed to explore effect of the factors on energy absorption of 7075-T6 aluminum alloy through control variable method based on HFQ process. The results will be contributed to understanding the crushing performance of 7075-T6 aluminum alloy with HFQ process and applied to the engineering practices deeply.

2. Quasi-static Crush Experiment and Simulation

2.1. *HFQ process procedures*

The experimental process procedures are shown in Table 1.

Table 1. HFQ Process procedures for 7075-T6 aluminum alloy HAT samples.

Order	Factors	HFQ process			
		Solution temperature/°C	Solution time/min	Forming temperature/ °C	Aging treatment /120°C×24h
1	Forming temperature	475	30	100,300,450	Artificial aging
2	Solution temperature	450,475,500	30	400	Artificial aging
3	Solution time	475	20, 30,40, 60	450	Artificial aging
4	Aging treatment	475	30	450	Artificial aging, Natural aging

2.2. *Experiment platform and FEM model*

Experiment was carried out on Y27Y-40 hydraulic machine, as shown in Figure 1.

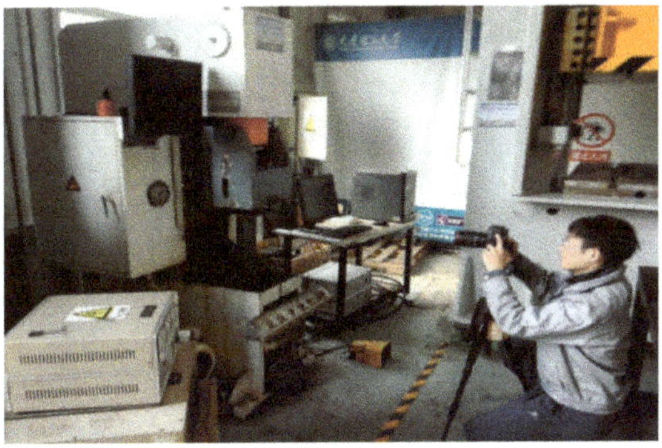

Fig. 1. Experiment platform.

Quasi-static crush speed is 15mm/s and maximum displacement is 100mm [3, 4]. The test force and displacement data in experiment process are acquired by sensors and energy absorption are calculated out through the data. HAT samples were prepared with HFQ process according to Table 1. Fig. 2 (a) gives out the cross section and Geometrical size of HAT samples, whose axial length and thickness are respectively 300mm and 2mm [5, 6]. U-shaped part and plate are riveted with interval of 45mm. Aluminum alloy platen are welded at both ends of HAT samples, as shown in Fig. 2 (b).

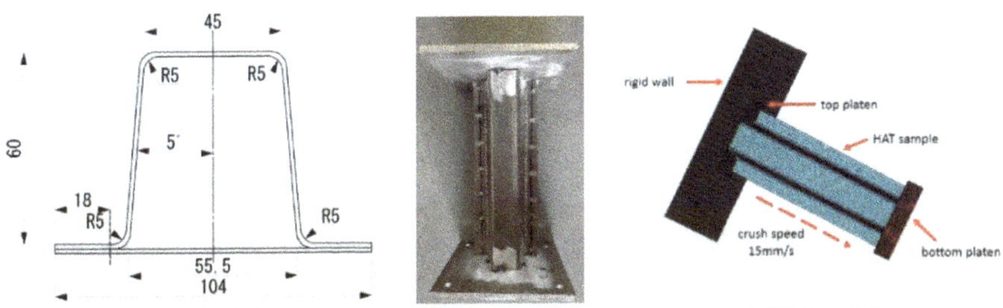

a) Geometrical size. b) HAT sample. c) FEM model.

Fig. 2. HAT sample and finite element model.

Simulation was conducted through LS-DYNA software. Finite element model comprises HAT sample, rigid wall, top platen and bottom platen, as shown in Fig. 2 (c). There are 8441 elements and 8837 nodes. The model are meshed with 4mm quadrilateral Belytschko–Tsay shell elements and warp degree of elements is less than 20°C to ensure the simulation is accurate enough. In order to reduce the amount of calculation, one-dimensional rigid body ais used instead of riveting parts. Top platen is clamped and rigid wall crush HAT sample at the speed of 15mm/s. The contacting type is surface-to-surface.

3. Results and Analysis

3.1. *Forming temperature*

As shown in Fig. 3 and Table 2, with forming temperature increasing, the maximum energy absorption of 7075-T6 aluminum alloy significantly improves from about 4300 J at 100 °C to more than 6600J at 450 °C.

Table 2. Maximum energy absorption with different forming temperature.

Forming temperature / °C		100	300	450
Maximum energy Absorption /J	Experiment	4220	4850	6660
	Simulation	4480	5255	6840

Since the second phase in aluminum substrate can be reduced with high forming temperature, solute atoms maintain high concentration in the solid solution. As a result, the basic mechanical property of 7075-T6 aluminum alloy is improved, which lead to better energy absorption performance.

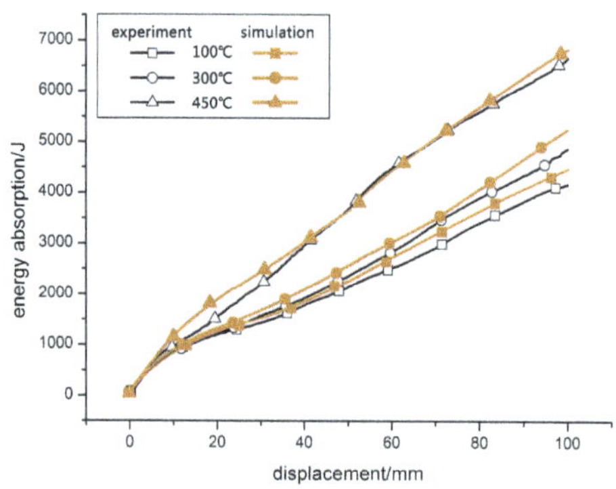

Fig. 3. Energy absorption-displacement curves with different forming temperature.

3.2. *Solution temperature and time*

For 7075-T6 aluminum alloy, when solution temperature reaches to the non-equilibrium eutectic temperature of 475 °C, alloy elements can be fully dissolved in the aluminum substrate. But over burning and grain coarsening will appear if solution temperature is above 475 °C. Therefore, 7075-T6 aluminum alloy has the best energy absorbing property when solution temperature is 475 °C, as shown in Fig. 4 (a) and Table 3.

As shown in Fig. 4 (b) and Table 3, optimal energy absorption emerges with solution time of 30min. This is because the solid solubility of crystalline phase can be improved with appropriate increase of solution time, which makes the coarse second phase fully dissolve into the aluminum substrate.

Table 3. Maximum energy absorption with different solution temperature and time.

Research content	Value	Maximum energy absorption /J	
		Experiment	Simulation
Solution temperature/°C	450	5500	5898
	475	6256	6620
	500	4620	4940
Solution time/min	20	5480	5380
	30	6660	6840
	40	6252	6420
	60	5490	5498

However, if solution treatment time is too long, the second phase will not continue to dissolve. Instead, crystalline grains will gradually grow and coarsen, which will deteriorate energy absorption of 7075-T6 aluminum alloy.

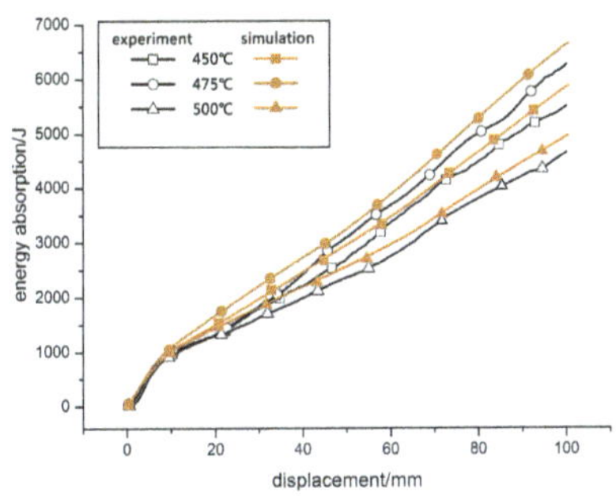

a) Curves with different solution temperature.

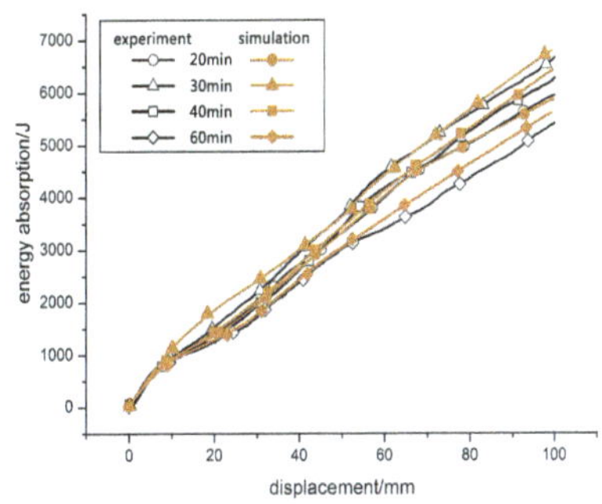

b) Curves with different solution time.

Fig. 4. Energy absorption-displacement curves with different solution temperature and time.

3.3. *Artificial aging*

According to Fig. 5 and Table 4, artificial aging exerts a positive and obvious influence on energy absorption of 7075-T6 aluminum alloy. After heat solution treatment, a kind of supersaturated solid solution in unstable state will form. While, artificial aging can make the supersaturated solid solution decompose and precipitate.

Table 4. Maximum energy absorption with artificial aging or natural aging.

Aging treatment/120 °C×24 h		Artificial aging	Natural aging
Maximum energy absorption /J	Experiment	6660	5187
	Simulation	6840	5940

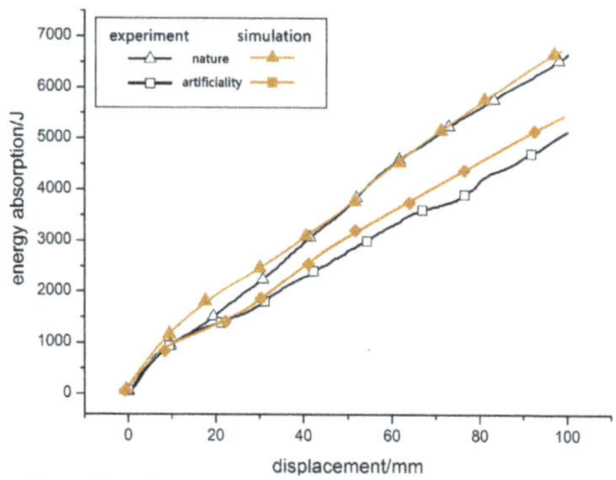

Fig. 5. Energy absorption-displacement curves with artificial aging or natural aging.

4. Conclusion

In this paper, quasi-static crush experiment and simulation of 7075-T6 aluminum alloy HAT samples were conducted to research the relation between energy absorption and HFQ process. The results show that energy absorption significantly improves with forming temperature of 450°C, solution temperature of 475°C, solution time of 30min and artificial aging.

Acknowledgments

This work is funded by the Key Project of the National Natural Science Foundation of China (No. 11272075), China Postdoctoral Science Foundation (2014M561223) and the Fundamental Research Funds for the Central Universities (DUT16RC(4)28).

References

1. N. R. Harrison and S. G. Luckey, Hot Stamping of a B-Pillar Outer from High Strength Aluminum Sheet AA7075, *SAE Int. J. Mater. Manu.* **7**, 567 (2014).
2. J. G. Lin, O. E. Fakira, L. L. Wang, D. Balint, J. P. Dear and T. A. Dean, Numerical study of the solution heat treatment, forming, and in-die quenching (HFQ) process on AA5754, *Int. Mach. Tool. Manu.* **87**, 39 (2014).
3. Y. Ohkubo, T. Akamatsu and K. Shirasawa, Mean crushing strength of closed-hat section members, *SAE Tech. Pap.*740040 (1974).
4. W. Johnson, P. D Soden and S. T. S. Al-Hassani, Inextensional collapse of thin-walled tubes under axial compression, *J. Strain Anal. Eng.* **12**, 317 (1977).
5. Y. Liu and G. Glass, Effects of wall thickness and geometric shape on thin-walled parts structural performance, *Thin Wall. Struct.* **49**, 223 (2011).
6. F. Schneider and N. Jones, Impact of thin-walled high-strength steel structural sections, *P. I. Mech. Eng. D-J. Aut.* **218**, 131 (2004).

Influence of Variables of Solution Heat Treatment-Forming-Quenching-Aging Treatment on the Formability of 6061 Aluminum Alloy

Z. Y. He and Y. L. Wang[†]

State Key Laboratory of Material Processing and Die & Mould Technology,
Huazhong University of Science and Technology,
Wuhan, P.R. China
[†]*Email: wangyilin@mail.hust.edu.cn*

Based on the process of solution heat treatment(SHT)–forming–quenching–aging treatment, some tests of 6061 aluminum alloy were carried out in this paper, in order to guide the hot stamping of 6061 aluminum alloy. To investigate the influence of different temperature and time of SHT on the LDR value of 6061 aluminum alloy, this paper employed the method of SHT-hot drawing and then compared the formability in different conditions. This paper also investigated the effect of different aging time and temperature in mechanical properties, including tensile strength and hardness, by using the uniaxial tensile test and Vickers hardness test after SHT-quenching-artificial aging treatment at certain SHT temperature and time.

Keywords: 6061 aluminum alloy; Solution heat treatment; LDR; Formability; Artificial aging treatment.

1. Introduction

As we all know, with the growing demand for energy conservation and pollution emission reduction, aluminum alloys have drawn more and more attentions in automobile industry due to their low density. [1] 6000-series aluminum alloys possess characteristic properties of good corrosion resistance and high strength, which can well satisfy the requirements on car body. [2] However, poor formability and serious springback at room temperature make it hard to form aluminum alloy at low temperature. [3] By the use of the experience of high strength steel, people have begun to study the formability of aluminum alloy at elevated temperature. Wang M et al. investigated the forming properties of 5182 aluminum alloy sheet for automotive body by deep drawing test [4], while Nia et al. formed an automotive B-pillar outer panel by hot stamping AA7075 aluminum alloy. [5] In 2008, J Lin et al. proposed a novel technology called solution Heat treatment Forming cold die Quenching (HFQ) process in order to form aluminum alloys. [6-7] This technology process is as follows. Firstly, they heat aluminum alloy sheet to SHT temperature and hold for a while. Secondly transfer the sheet to the cold die as soon as possible in order to quench it [see Fig. 1(a)]. [8] It can obviously improve the formability of aluminum alloys by using technology of HFQ. A modified technology by improving die temperature was proposed by He et al. based on HFQ, in order to reduce the temperature drop before forming and make little influence on the effect of quenching. [9]

So far, there are a few systematic investigations on hot forming technology of 6000-series aluminum alloys. To investigate the formability of 6000-series aluminum alloys at high temperature, this paper chooses 6061 aluminum alloys to carry out a series of experiments based on the modified HFQ technology (see Fig. 1(b)), including hot deep drawing test and uniaxial tensile test. The influences on 6061 aluminum alloy deep drawing performance and metallographic structures of different SHT temperature and time were studied by hot deep drawing test. While the influences on 6061 aluminum alloy mechanical properties of artificial aging temperature and time were also studied by uniaxial tensile test and hardness test.

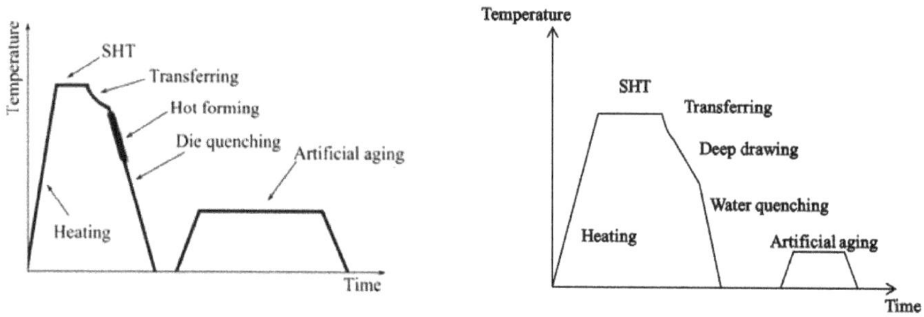

(a) (b)

Fig. 1. Technology of HFQ (a) and modified HFQ (b).

2. Experimental Methods and Material

2.1. *Material*

The as received material was 6061-T4 aluminum alloy with 1 mm in thickness. The main composition is shown in Table 1.

Table 1. Main compositional elements of 6061 aluminum alloy.

element	Si	Mg	Mn	Fe	Ti	Cr	Cu	Al
composition	0.4~0.8	0.8~1.2	0.15	0.7	0.15	0.04~0.35	0.15~0.4	Rem

2.2. *SHT-hot deep drawing test*

To investigate the influence of SHT temperature and time on 6061 aluminum alloy deep drawing formability, this paper decided limit drawing ratio (LDR) as the criterion. The LDR value is defined by Eq. (1).

$$\mathrm{LDR} = \mathrm{D}_{\max} / d \tag{1}$$

Where D_{max} is the maximum diameter of successfully formed blank and d is the diameter of punch.

Device of deep drawing is shown in Fig. 2. The method is as follows. First, a series of circular 6061 aluminum alloy plates were cut from as received materials, whose diameters range from 90 mm to 150 mm at intervals of 2.5 mm. Secondly, deep drawing test was carried out on the die with a 50 mm diameter punch. The die and binder were heated to 300 °C and circular 6061 aluminum alloy plates with coated graphite were put in furnace to solution heat treat at certain temperature for certain time. Then transfer it to die quickly (less than 3 s) and deep draw it at speed of 1 mm/s. At last, quench it by water below 25 °C after drawing.

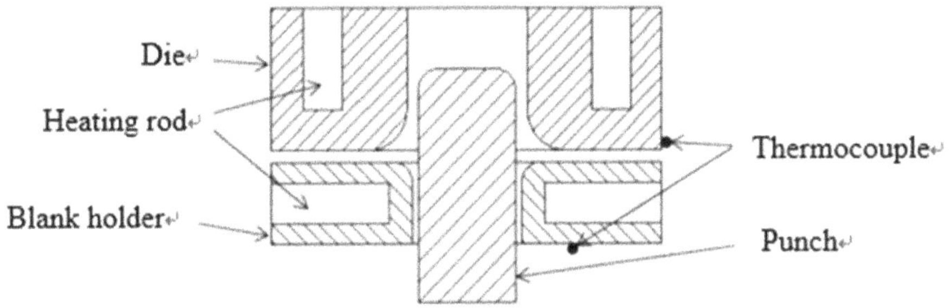

Fig. 2. Device of deep drawing.

2.3. *Artificial aging test*

The tensile specimens were solution heat treated at 530 °C for 15 min and then quenched in water below 25 °C. Then being aging treated at different conditions. Hardness and uniaxial tensile test were carried out after that. The uniaxial tensile strain rate was 0.1 s^{-1}.

3. Results and Discussion

3.1. *Results of SHT- hot deep drawing test*

The results of SHT-hot deep drawing test are shown in Fig. 3. We tested the hardness of the bottom part after deep drawing. It can be found that the LDR value varies as similar as hardness from different SHT conditions. The peak LDR value is 2.15 (blank diameter is 107.5 mm) as the peak hardness is HV 65.2. It indicates that the LDR value related to strength of blank at high temperature. Although higher temperature improves the plasticity of blank, strength is the main factor affected the formability of 6061 aluminum alloy at such high temperature. Fig. 3(a) shows the influence of SHT time on LDR value and hardness. This can be explained by microstructure. At first, phase particles in aluminum alloy gradually dissolve in the Al matrix with the increasement of time, which distorts the lattice and improves the strength of matrix, result in an improvement of LDR value and hardness. While being heated too long time coarsens the grains and decreases the strength of blank, which results are in lower LDR value and hardness. The similar trend is shown in Fig. 3(b) as the temperature changed at identical time of 30 min. The influence of temperature is also similar to the influence of time.

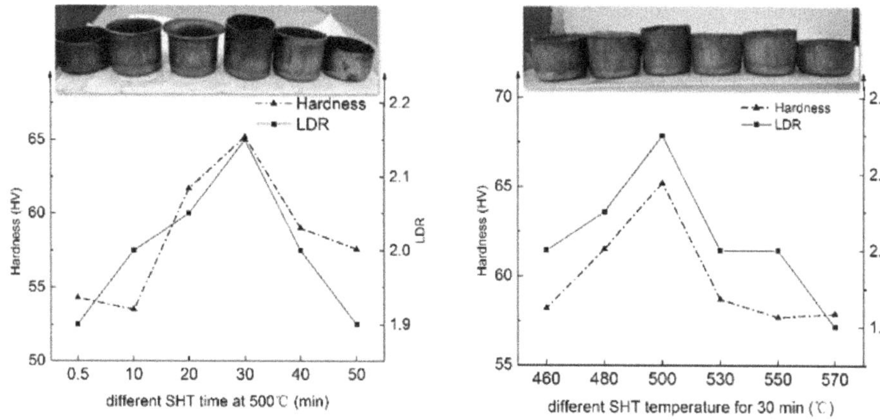

(a) (b)

Fig. 3. Results of SHT-hot drawing test.

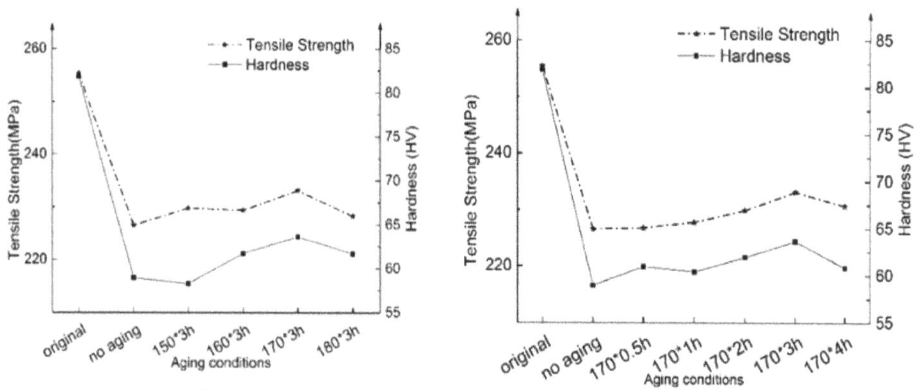

(a) (b)

Fig. 4. Tensile strength and hardness in different conditions.

3.2. *Influence of artificial aging treatment*

As we can see from Fig. 4, the tensile strength and hardness of 6061 aluminum alloy decreases obviously after SHT-quenching compared to original material. They increase after artificial aging treatment though lower than the original. Fig. 4(a) shows the hardness and tensile strength at different aging temperature for 3 h. With the increasement of temperature, the hardness and tensile strength increase at first and decrease after reaching peak at temperature of 170 °C. And the hardness and tensile strength also reach peak at time of 3 h with the increasement of time at identical temperature of 170 °C as shown in Fig. 4(b).

The lower strength of aging treated samples than as received material can be explained for the insufficient time for SHT. Thus second phases do not completely dissolve in the matrix and the driving force of age precipitation is weak, leading to weak reinforcement effect.

4. Conclusion

1) The temperature and time of SHT have a certain effect on the formability of 6061 aluminum alloy. Being kept in 500 °C for 30 min is the optimum condition for 6061 aluminum alloy SHT-hot deep drawing technology above 460 °C and LDR value is 2.15 at the condition.
2) Aging treatment can improve the strength of solution heat-treated 6061 aluminum alloys. With the increase of aging time and temperature, the strength will reach a peak at the optimum condition. In this paper, the optimum aging condition is 170 °C for 3 h. To acquire higher strength by artificial aging treatment, the SHT technology should be moderate to ensure fully solid solution.

Acknowledgments

This research work is funded by the National Natural Science Foundation of China (grant no. U1564203) and National Natural Science Foundation of China (grant no. 51405171).

References

1. M. S. Mohamed, A. D. Foster, J G. Lin, D. S. Balint and T. A. Dean, Investigation of deformation and failure features in hot stamping of AA6082: Experimentation and modelling, *Int. Mach. Tool. Manu.* **53**, 27 (2012).
2. X. B. Fan, Z. B. He and S. J. Yuan, Investigation on strengthening of 6A02 aluminum alloy sheet in hot forming-quenching integrated process with warm forming-dies, *Mater. Sci. Eng. A* **587**, 222 (2013).
3. J. Zhao, L. Zhan and B. Shi, The tensile flow behavior of 6061 aluminum alloy during high-temperature, *J. Plasticity Eng.* **21**, 112 (2014).
4. M. Wang, W. Zhou, J. Ren, C. W. Li, D. Y. Huang and G. Y. Li, Forming properties of 5182 aluminum alloy for automotive body sheet during warm deep drawing processes, *J. Cent. S. U. Sci. Tech.* **41**, 937 (2010).
5. N. R. Harrison and S. G. Luckey, Hot Stamping of a B-Pillar Outer from High Strength Aluminum Sheet AA7075, *SAE Int. J. Mater. Manu.* **7**, 567 (2014).
6. M. S. Mohamed, A. D. Foster and J. Lin, Solution heat treatment in HFQ process, in *Steel Research International*, (Cracow, Poland, 2008).
7. J. Lin, T. A. Dean, R. P. Garrett and A. Dl, Foster, in *Process for forming aluminum alloy sheet components,* (UK, WO2008059242. 2008−09−19).
8. J. Zhou, B. Wang and J. Lin, Forming defects in aluminum alloy hot stamping of side-door impact beam, *T. Nonferr. Metal. S*OC. **24**, 3611 (2014).
9. Z. B. He, X. B. Fan and S. J. Yuan, Review of Hot Forming-quenching Integrated Process of Aluminum Alloy, *J. Netshape Form. Eng.* **6**, 37 (2014).

Experimental Investigation on the Accuracy of Uniaxial Tensile Data of AA6082 Under HFQ® Conditions

X. X. Kong, N. Li[†], R. O'Keeffe and J. G. Lin

Mechanics of Materials, Department of Mechanical Engineering, Imperial College London, London, SW7 2AZ, UK

[†]*E-mail: n.li09@imperial.ac.uk*

Hot stamping of aluminium alloys, named as Solution Heat Treatment, Forming and in-die Quenching (HFQ®), is an innovative sheet metal forming process which allows high-strength complex-shaped aluminium panel components to be formed at one time. In this study, AA6082 was investigated by applying hot tensile tests on Gleeble machine, which simulates the HFQ® process. Newly designed gripping systems were evaluated in comparison with a pair of old grips for the purpose of dealing with the non-uniformly distributed temperature along the specimen. In addition, due to the absence of testing standards under such conditions, test data can be largely scattered through different ways to measure the strain. Through this study the new grips design was found to give great improvement in terms of temperature distribution along the specimen span. Because of this, the stress-strain relations obtained are more accurate than those by the old grips design. Stress-strain data measured based on different gauge length show huge difference; hence the definition of strain measurement is essential for the accuracy of test data.

Keywords: Hot Stamping; Hot uniaxial tensile tests; Temperature distribution; Strain measurement; Viscoplasticity; AA6082.

1. Introduction

Aluminum alloys have ever been limited in applications because of its poor formability properties, especially when forming sheet-like products or with specific design features. A hot stamping technology, Solution Heat Treatment, Forming and in-die Quenching (HFQ®), shows great potential for large complex shaped aluminum panel components to be formed at one time at high efficiency and low cost [1]. In the process, the aluminium alloy sheet is firstly heated to the solution heat treatment temperature (T_{SHT}) and soaked for a certain time when the precipitates dissolve into the matrix and a homogeneous microstructure is achieved, which increases the ductility of the material. Followed by quick transfer from the furnace to the forming tool, the sheet blank is deformed in the stamping press and fast quenched in cold closed-dies [2]. In order to characterise properties of alloy to be subjected to the HFQ® process, rapid heating and cooling of a specimen must be the integral part of a uniaxial tensile test; therefore, conventional hot uniaxial testing which is operated in an oven or furnace is impractical. This study simulates the actual HFQ® process on Gleeble machine to perform hot uniaxial tensile tests of AA6082. During this process, materials exhibit viscoplastic behaviours which are highly affected by the forming temperature and strain rate. From the results of previous tests on Gleeble machine, problems of non-uniform temperature distribution along the specimen span were observed which could bring errors in the obtained data [3]. Existing standards, e.g. EN ISO 6892-2:2011 [4], are applicable to testing only under isothermal conditions, or near isothermal conditions with a small permitted temperature deviation within the gauge region of a specimen. Due to lack of specifically defined standards for Gleeble hot test conditions, when different clamping systems, specimen designs, and different strain measurement methods are applied, the test results could be largely scattered. The objectives of this paper are to demonstrate the significant variations in test results based on uniaxial tensile tests of AA6082 by using two sets of grips and specimens with different strain measurements, and to investigate the factors influencing the accuracy of uniaxial tensile data.

2. Experimental Details

All specimen used for the experiment were machined from the same batch of as-delivered, 1.5mm thick, AA6082-T6 sheet, produced by AMAG Rolling. Two sets of grips were used in this study. One set named 'old grips' is made of stainless steel blocks and cooled directly by the contact with copper jaws during the test, which causes unavoidable heat lost from the specimen; The other set named 'new grips' are bent from copper coils, supported by a nonconductive core inside and warped with thermal and electrical insulation materials, so that excess heat is generated through the copper coil during resistance heating. The concept is to reduce the heat lost from specimen to grips so as to minimise the temperature gradient along the gauge of specimen. Flat dog-bone type specimens with the same total length, and the same gauge width of 8mm, were used for both designs. The parallel lengths were 80 mm and 94mm set on the old grips and new grips, respectively.

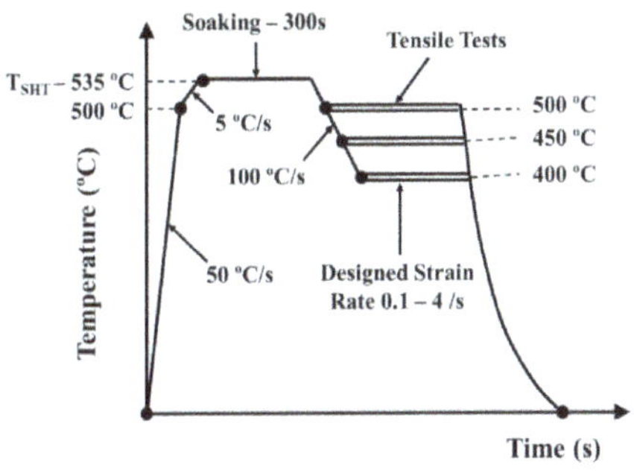

Fig. 1. Temperature profile of the tests.

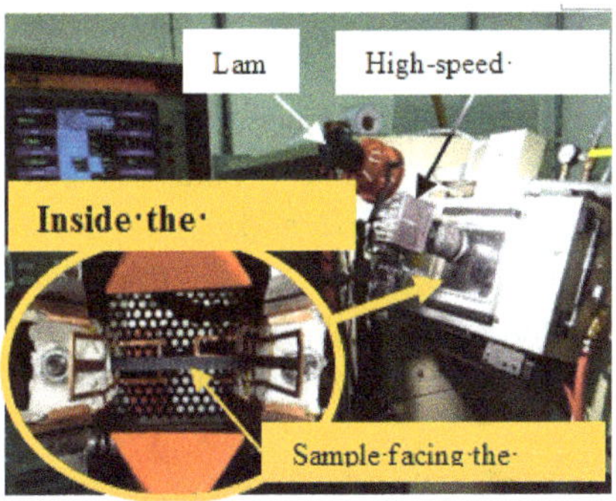

Fig. 2. Set-up of the experimental system.

Fig. 1 shows the specified temperature profile for the tensile testing of AA6082 under HFQ® conditions. It comprises continuous heating, isothermal soaking, quenching, and deformation at constant elevated temperatures. The values of defined conditions are given in the figure. Figure 2 is set-up of the experimental system showing how the sample was placed inside the Gleeble chamber. When deformation tests were conducted, high speed camera was used to capture strain changes. Temperatures distributions along specimens were detected by the welded thermocouples along the gauge of specimens at the positions of 0, 7, 20 and 35 mm away from the centre for both designs. Lateral strain (0mm), and gauge length of 14 and 40 mm were used as the defined gauge lengths to process the test data for new grips design. Gauge length of 14mm was adopted to compare the tensile data of the two designs.

3. Results and Discussion

3.1. *Temperature fields*

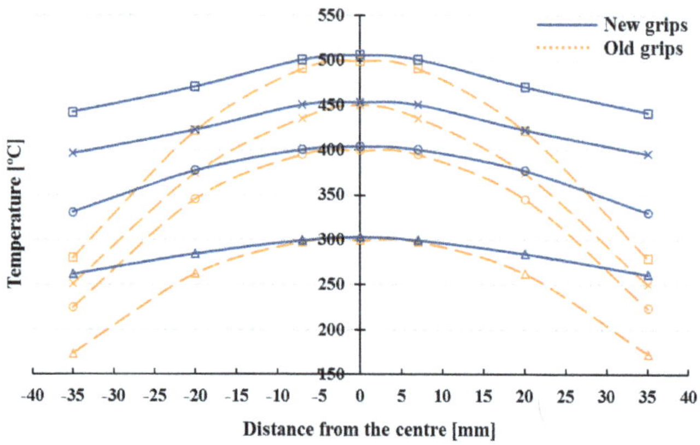

Fig. 3. Temperature distributions along the gauge of specimens.

The temperature fields over specimens obtained by using two sets of grips are compared in Fig. 3, showing the temperature distributions along the gauge of specimens at the start of deformation after soaking at T_{SHT} and quenching. The designed deformation temperatures are controlled at the centre. The solid lines represent temperatures obtained using new grips and the dashed lines correspond to the old grips. At all designed temperatures, the temperature uniformity appears to be much improved with new grips. For the gauge length of 14mm for the designed test temperature of 450°C, the test of new grips give a temperature deviation of 2.7°C from the centre, which is within ISO standard (±3°C); while for that of old grips, this value is 14.4°C. For the gauge length of 40mm, specimens using both types of grips have temperature variation beyond 3°C, while new grips still allowe significant improvement.

3.2. *Stress-strain curves*

The flow stress-strain relations of AA6082 under HFQ® conditions, tested by using the new set of grips and specimen, measured from different designed gauge lengths are presented in Fig. 4. First, significant viscoplastic characteristics the alloy, measured from any gauge length, is revealed. The true stress level decreases with increasing temperature and increases with increasing strain rate; the ductility doesn't change much with either temperature or strain rate. Second, huge difference in stress-strain data from different strain measurements can be seen for all test conditions. Higher levels of flow stress and ductility are obtained with a shorter gauge length and the maximum data are measured from later strain (0 mm). This is due to both the temperature gradients and effect of necking. A representative condition, with the designed strain rate of 1 s^{-1} and deformation temperature of 450 °C, was selected for further analysis. Fig. 5 shows the history of actual true strain rates processed from three gauge lengths. An almost constant strain rate near 1 s^{-1} throughout the deformation could only be obtained by averaging the strain rate along 40mm gauge length; the strain rate is larger at the area closer to the centre of specimen. The small difference at the early stage of deformation attributes to temperature non-uniformity and the dramatically enlarged differences at the late stage are caused by localised necking at the centre.

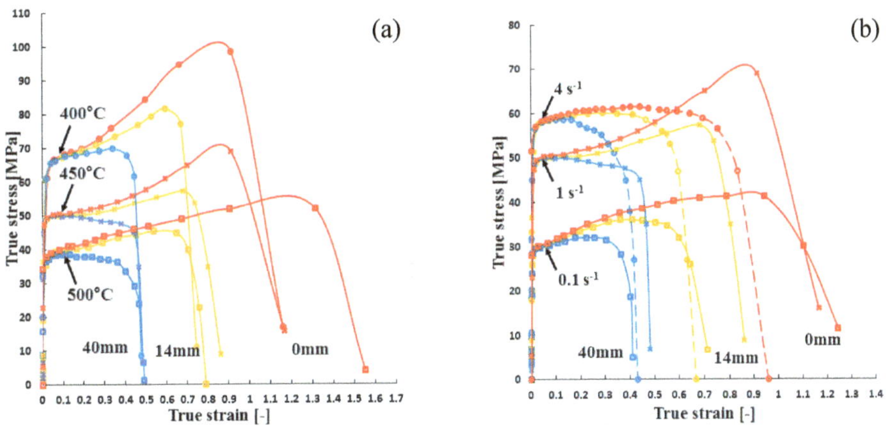

Fig. 4. Stress-strain relations at (a) designed strain rate of 1 s^{-1} and (b) designed forming temperature of 450°C, measured from the gauge length of 0 mm, 14 mm and 40 mm.

Fig. 6 shows that tests with new grips, compared with the old design, result in a lower flow stress level and higher failure strain at the same designed test condition, which is mainly due to the different temperature fields. Applied the same stroke for both gripping conditions, the deformation is more localised around centre for the old design, which causes higher strain rates of material within the 14mm gauge length thus higher stress levels; A shaper temperature gradient for the old design creates a severer triaxle stress states, which could cause reduction in the longitudinal strain to failure.

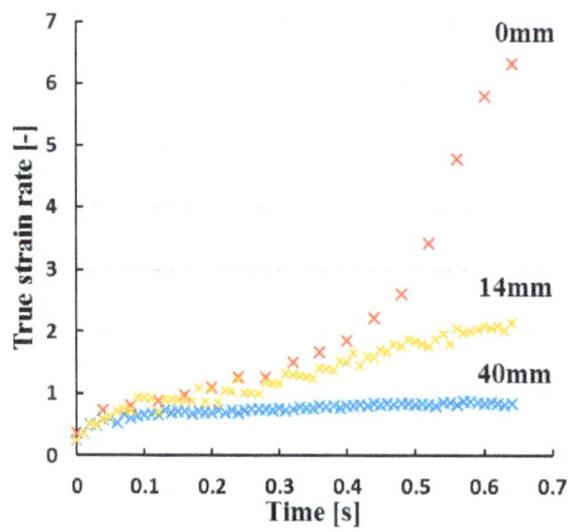

Fig. 5. True strain rates at different gauge length (designed $\dot{\varepsilon} = 1\ s^{-1}$, T = 450 °C).

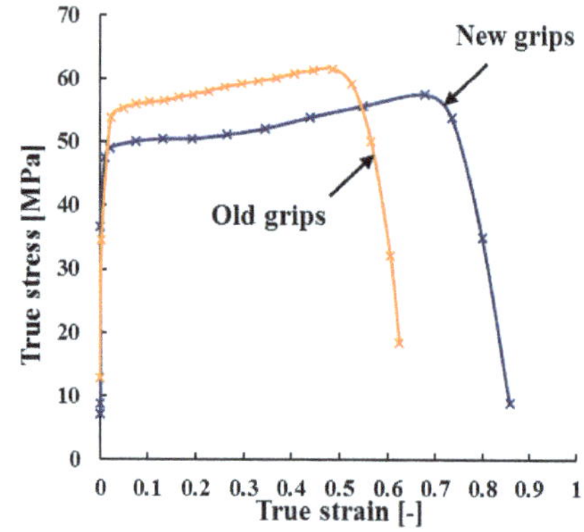

Fig. 6. Stress-strain relations at gauge length of 14mm for old and new grips (designed $\dot{\varepsilon}$ = 1 s-1, T = 450 °C).

4. Conclusion

From the obtained stress-strain relations, AA6082 exhibits viscoplastic behaviors, where material flow stress decreases with raising temperature and decreasing strain rate. These test date are highly affected by the temperature distributions and adopted gauge lengths, therefore a specified standard is required for these tests. The newly designed grips are verified to give great improvements in terms of temperature distribution along the specimen, which in turn improves the accuracy of stress and strain values from the old designs. True strain rates are not constant values throughout deformation as designed, which results in the changing shape of the stress-strain curve and needs to be considered in the analysis.

References

1. M. Raugei, O. E. Fakir, L. Wang, J. Lin and D. Morrey, Life cycle assessment of the potential environmental benefits of a novel hot forming process in automotive manufacturing, *J. Cleaner Production* **83**, 80 (2014).
2. M.S.K. Mohamed, An investigation of hot forming quench process for AA6082 aluminium alloys, *Imperial College London* (2010).
3. J. Cai, Modelling of phase transformation in hot stamping of boron steel, PhD thesis, *Imperial College London* (2011).
4. ISO 6892-2:2011. Metallic materials–Tensile testing–Part 2: Method of test at elevated temperature, International Standard (2011).

Optimization Design for Uniform Cooling in Hot Stamping

C. Wang† and F. X. Xu

Hubei Key Laboratory of Advanced Technology of Automobile Parts, Wuhan University of Technology, Wuhan, Hubei, 430070, China
School of Automotive Engineering, Wuhan University of Technology, Wuhan, Hubei, 430070, China
E-mail: †wchao@whut.edu.cn; xufx@whut.edu.cn

Y. S. Zhang

State Key Laboratory of Material Processing and Die and Die Technology, Huazhong University of Science & Technology, Wuhan, Hubei, 430074, China
E-mail: zhangys@mail.hust.edu.cn

Hot stamping is increasingly used in automobile structure part for the demand of automotive safety and emission reduction. However, due to the non-uniform cooling path of the blank, the mechanical properties are not consistent. This paper expounds that the main reason of non-uniform cooling parts is the non-uniform contact pressure and clearance between parts and sheet metal. Base on the finite element analysis of cooling path during the forming and quenching process, an optimization method that improves contact condition between blank and die is developed by using modified mold block to improve contact condition between die and blank. The result shows that the blank obtains more uniform cooling path.

Keywords: Hot stamping; High strength steel; Optimization design; Uniform cooling.

1. Introduction

In hot stamping, it is generally believed that the component cools with its shape and the cooling system should be designed according to this principle [1]. It is known that the components cool according to its shape. However, this method cannot improve the uneven temperature field of components during the cooling [2].

As we know that the thermal contact resistance (TCR) increases with the pressure decreases. The cooling speed increases with the TCR decreases [3]. The clearance affects the heat transfer between the blank and die. Larger clearance prevents the heat transfer and the smaller clearance makes the heat transfer faster [4]. This is the reason for the uneven temperature field of the blank and die [5].

2. The Effect of the Uneven Cooling of the Components

2.1. *Uneven mechanical properties*

According the forming process of U shape die (Fig. 1, Fig. 2), there are clearances between side wall of the die and blank, between bottom of the die and blank, the temperature is higher.

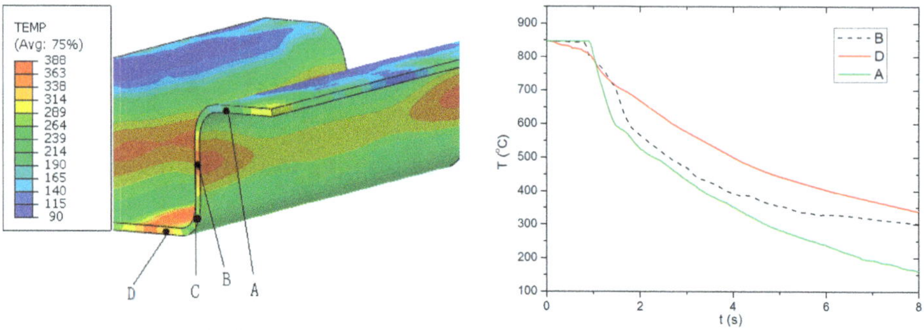

a. Temperature field of U shape die. b. Cooling curves of U shape die.

Fig. 1. Simulation results of U shape die.

However, due to the well contact between blank and die at the corner, the temperature is lower. The hardness of the location A, B, C, D (shown in Fig. 1) is 51, 48, 50.5 and 47.5 respectively.

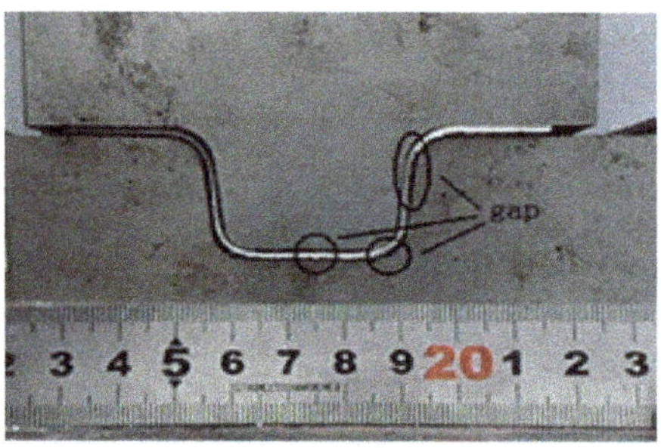

Fig. 2. Schematic of contact U shape die.

Fig. 3. Cracking of the collision beam.

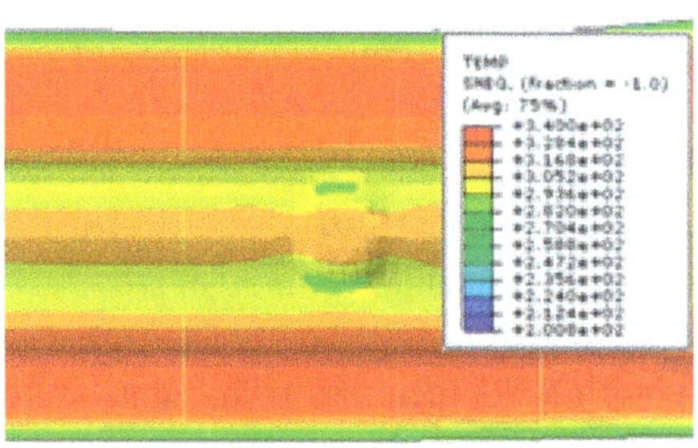

Fig. 4. Temperature field of collision beam.

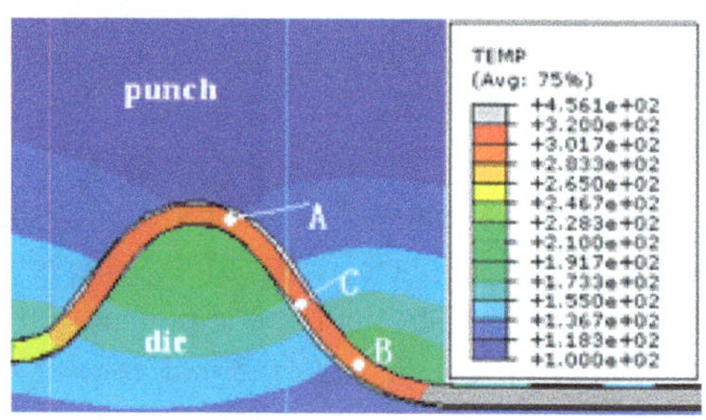

Fig. 5. Temperature field of cross section.

2.2. *Cracking*

Figure 3 shows the cracking of collision beam after hot stamping. Figures 4 and 5 show the temperature field of simulation results. As its shown that the point C have clearance to the die for both sides and point A, B have better contact condition with one side clearance. So the temperature of point C is higher and gets better condition to be elongated compared with point A and B.

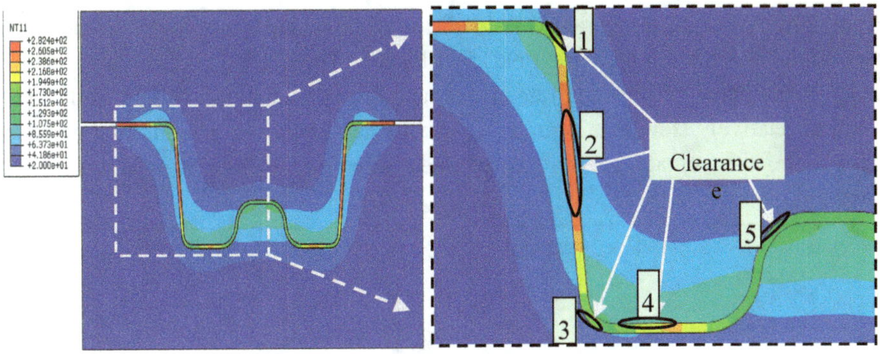

Fig. 6. Temperature field of blank and die.

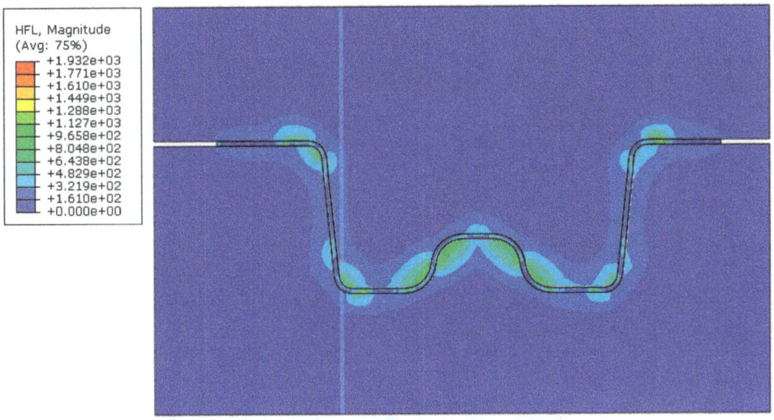

Fig. 7. Heat flux between blank and die.

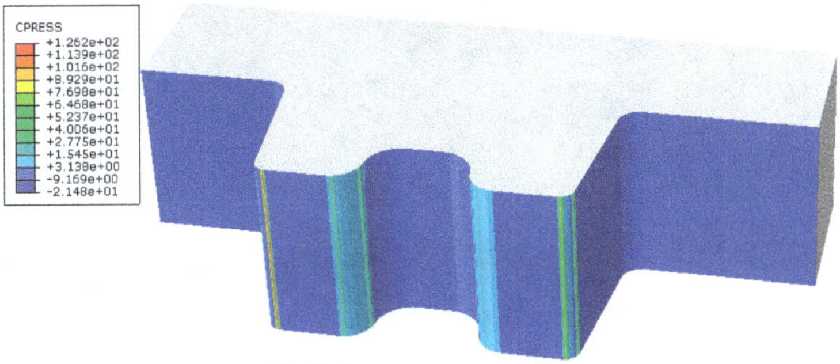

Fig. 8. Contact pressure of die.

3. Cooling Process of Hot Stamping

In the cooling process of hot stamping, the blank partly contacts with the die. There are gaps between blank and die. In figure 6, there are gaps between blank and die at the position 2 and 3. There are gaps between blank and punch at the position 1, 2, 4 and 5. By comparing figure 7 and figure 8, the contact pressure between blank and die is uneven. The heat flux is larger where the contact pressure is larger.

From figure 9, final temperature field distribution in hot stamping is completely different from uniform temperature field distribution. There are good contacts in the rounded part, and has rapid cooling rate. There are poor contact at the bottom of the side wall and has relatively slow cooling rate. That is to say, under the precondition of good cooling system, the contact pressure determines between the heat transfer efficiency blank and die.

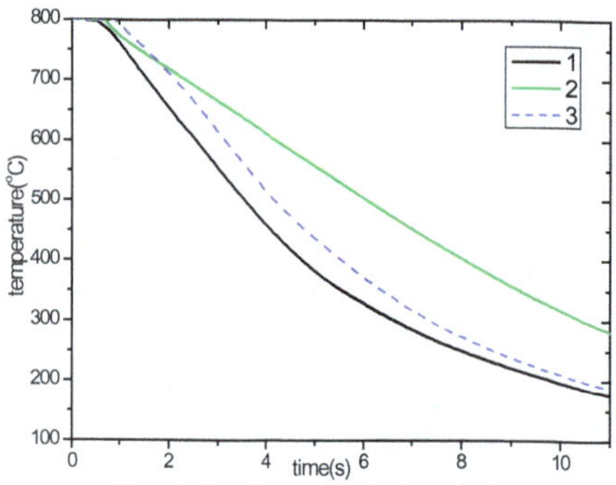

Fig. 9. Cooling curve before optimization.

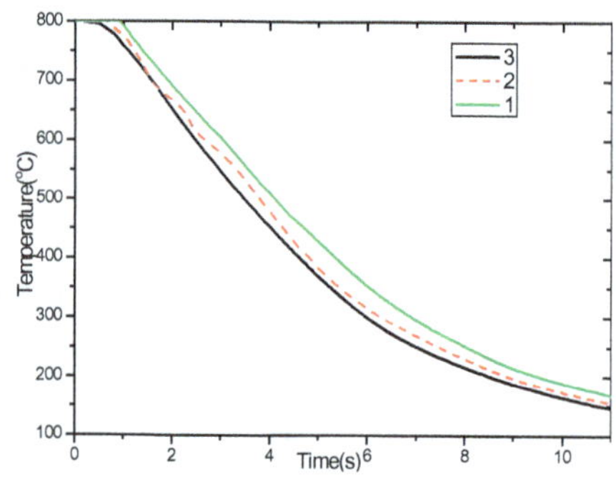

Fig. 10. Cooling curves after optimization.

4. Optimization Design for Uniform Cooling

In order to improve the cooling condition at the quenching stage of hot stamping, the die is optimized and shown in figure 11.

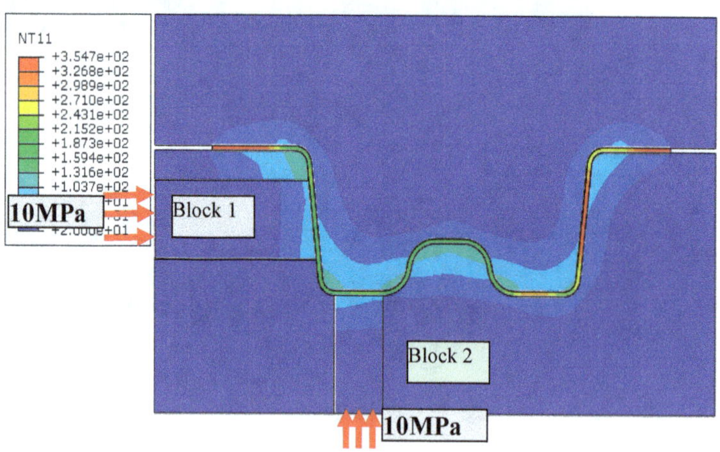

Fig. 11. Temperature field after optimization.

There are two block in the die. There are 10MPa pressures applying on the border. As we can see that the, the cooling rate at sidewall and bottom cooling faster and the temperature is close to the temperature at the fillet. The temperature field becomes more uniform compared with the temperature field in figure 6. By comparing figures 9 and 10, the points noted in figure 6 are more uniform after optimization than before. The reason is that the pressures applied on the blocks improve the contact condition between blank and die. As we can see that, the cooling rate at the sidewall and bottom becomes faster.

Can see from the picture, in the block mould thermal volume increased significantly, its temperature basic consistent with the temperature of the two rounded out. Contrast figure 3 to 11 in the left block blank temperature and the right without blocking in sheet metal temperature, this method can be found to improve mould thermal effect is obvious, has obviously improved the parts cooling uniformity. As shown in figure 3 to 11 in the same location of the point 1, 2, and 3, analysis of the cooling curve, you can see the three locations have almost the same cooling rate.

5. Conclusion

Based on the typical parts a car front bumper as an example, this paper expounds the main reason of non-uniform cooling parts is the uneven contact pressure and clearance between parts and sheet metal. By analyzing the parts under the condition non-uniform cooling parts temperature field distribution and the boundary heat flux density distribution, the method of even cooling is put forward by using modified mold block to improve contact condition between die and blank.

Acknowledgments

The work is partially supported by open fund of State Key Laboratory of Material Processing and Die & Mould Technology of Huazhong University of Science and Technology; Project (2015IVA020, 2016IVA037) is supported by independent innovation foundation of Wuhan University of Technology.

References

1. H. Karbasian and A.E. Tekkaya, A review on hot stamping, *J. Mater. Process. Tech.* **210**, 2103 (2010).
2. K. Mori, T. Maeno and K. Mongkolkaji, Tailored die quenching of steel parts having strength distribution using bypass resistance heating in hot stamping, *J. Mater. Process. Tech.* **213**, 508 (2013).
3. M. Nikravesh, M. Naderi and G.H. Akbari, Influence of hot plastic deformation and cooling rate on martensite and bainite start temperatures in 22MnB5 steel, *Mater. Sci. Eng. A* **540**, 24 (2012).
4. E.J.F.R. Caron, K.J. Daun and M.A. Wells, Experimental heat transfer coefficient measurements during hot forming die quenching of boron steel at high temperatures, *Int. J. Heat Mass Transfer* **71**, 396 (2014).
5. C. Wang, B. Zhu and Y.S. Zhang, Hot-Stamping Process Simulation and Optimize Research for Collision Beam of Automobile Door, *Adv. Mater. Res.* **201-203**, 3 (2011).

Investigation on Uniformity of Temperature Distribution of Blanks in the Process of Conductive Heating

W. J. Tao, W. K. Liang, W. Xiao and Y. S. Zhang†

State Key Laboratory of Materials Processing and Die and Mould Technology, Huazhong University of Science & Technology, Wuhan 430074, China

†E-mail: zhangys@mail.hust.edu.cn

The uniformity of temperature distribution of ultra-high strength steel blanks in the process of conductive heating was investigated on the independent development of conductive heating device. The influence of inhomogeneous heat transfer method on the uniformity of temperature distribution of high temperature blanks was analyzed. Inhomogeneous radiation heat preservation device was developed for the purpose of the uniformity of temperature distribution of blanks in the process of conductive heating. The result revealed that obtained energy of blanks increased with the decrease of the height between blanks and inhomogeneous heat preservation device, so as to obtain higher rising amplitude of temperature on blanks. The uniformity of temperature distribution of blanks which guarantee the subsequent property of parts can be realized through this device. This device can also contribute to the reduction of energy consumption.

Keywords: Conductive heating; Temperature distribution; Inhomogeneous heat transfer; Radiation heat preservation; Uniformity of temperature distribution.

1. Introduction

Heating of blanks was the first step of hot stamping, temperature uniformity of blanks had great influence on property of subsequent parts. Conduction heating was a new heating method, it had good prospect because of rapid heating rate and high production efficiency [1]. But in the process of conduction heating, temperature of blanks nearby copper electrode was lower than that away from copper electrode, it was because more energy transferred to copper electrode form blanks nearby copper electrode that from blanks away from blanks [2]. Owing to the increase of resistivity with the increase of temperature, resistivity of blanks nearby copper electrode was higher than that away from copper electrode. According to Joule's law, heat generation of blanks nearby copper electrode was more, which lead to exacerbation of inhomogeneity of temperature. So in the process of conduction heating, the temperature distribution of blanks was not uniform in natural condition. Based on heat transfer theory, the way to improve temperature inhomogeneity of blanks had been researched in this paper by analyzing the influence of inhomogeneous heat preservation condition on temperature distribution of blanks. Uniform temperature distribution of blanks guaranteed property of subsequent parts [3].

2. Materials and Methodology

The material used in the test was 22MnB5, its chemical composition and mass fraction was shown in Table 1, the sample size was 1160*300*1.8mm. Conduction heating experiments were carried on self-developed conduction heating device, heating rate and heating temperature were respectively set as 20°C /s and 950°C, and then hold on for 60s. Non-heat preservation condition and inhomogeneous heat preservation condition were applied in the process of conduction heating, infrared thermography was used to record the temperature distribution of blanks. The influence of inhomogeneous heat preservation condition on temperature distribution of blanks had been analyzed by comparing temperature distribution of blanks under non-heat preservation condition and inhomogeneous heat preservation condition.

Table 1 chemical composition and mass fraction of 22MnB5.

C	Si	Mn	P	S	Al	Ti	B	Cr	Ni
0.23	0.17	0.95	0.01	0.002	0.04	0.023	0.002	0.24	1.5

3. Theory of Inhomogeneous Heat Preservation Condition

The way of heat loss of conduction heating blanks was convection and radiation under heating and heat preservation condition. Assuming that ambient temperature was 20 °C, convective heat transfer coefficient [4] of blanks was:

$$h_{conv} = [0.14(Gr \times Pr)^{0.33}]\frac{k}{L} \tag{1}$$

Where *Gr* was Grashof number, *Pr* was Prandtl number, *k* was thermal conductivity and *L* is length scale. Radiation heat transfer coefficient [5] of blanks was:

$$h_{rad} = \frac{\sigma\varepsilon(T_1{}^4 - T_2{}^4)}{T_1 - T_2} \tag{2}$$

Where σ was Stefan-Boltzmann constant, ε was emissivity, T_1 and T_2 were respectively blanks temperature and ambient temperature.

It can be learned from Equation 1 and Equation 2 that radiation heat transfer coefficient was one order of magnitude higher than convective heat transfer coefficient for high temperature blanks, this revealed that heat loss of high temperature blanks was mainly in the way of radiation.

Radiation process was phenomenon that electromagnetic transfer energy. When radiation energy was cast to surface of blanks, there were absorption, reflection and penetration phenomenon, that is

$$\alpha + \rho + \tau = 1 \tag{3}$$

Where α, ρ, τ were respectively absorption rate, reflectivity and penetration rate of blanks for radiation energy. Penetration rate was generally considered as 0 for conduction heating blanks in industrial production, so,

$$\alpha + \rho = 1 \tag{4}$$

The absorption and reflectivity property of radiation energy cast on the blanks surface only occurred on blanks surface. Energy loss of conduction heating blanks by radiation would reflect on the inner side of heat preservation device when the inner side heat preservation device had properties of low penetration and high reflectivity, so most energy loss by radiation would reflect to the conduction heating blanks.

For heat preservation device with constant size of width shown in Figure 1, its width was L, the height between it and blanks was h, radiation energy intensity from point source of blanks was E_0, energy radiated in the way of semi-circle in 2-dimension space. Energy intensity, E, accepted in the width of heat preservation is

$$E = E_0 \frac{2arctan\frac{L}{2h}}{\pi} \tag{5}$$

Equation 5 revealed that E increased with the decrease of h. Energy intensity reflected to the blanks increased with the increase of E. So, the amount of energy reflected to the blanks could be appropriate by adjusting the height between heat preservation device and blanks.

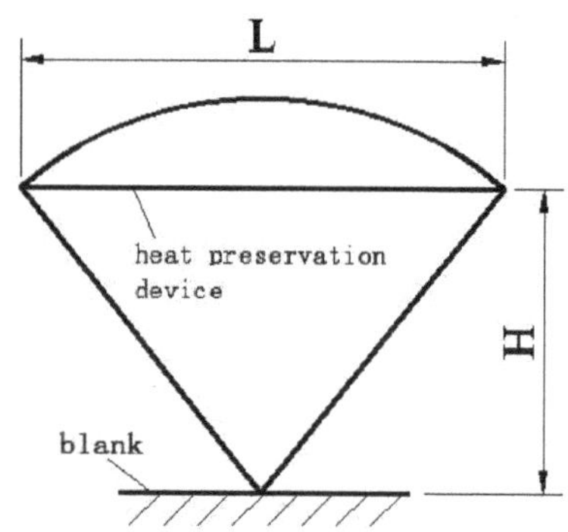

Fig. 1. Relationship between radiation intensity and height.

There was need to add more energy to blanks nearby copper electrode for the sake of temperature uniformity of blanks in the process of conduction heating. So, heating preservation device shown in Figure 2 was applied. The inner side of heat preservation is material with low penetration and high reflectivity, therefore, most energy would be reflected to blanks. The two sections of heat preservation device were respectively located above and below the blanks. The height between heat preservation device and blanks should be close in order to insure more energy could be reflected to the blanks. As a result, temperature of blanks nearby copper electrode rose more. Owing to the increase of resistivity with the increase of temperature, resistivity of blanks nearby copper electrode was higher than that of blanks away from copper electrode, heat generation of blanks nearby copper electrode was more. These two factors contributed to temperature uniformity of blanks. Because the two sections of heat preservation device formed a semi-closed space, which reduce convection between blanks and surroundings, energy loss by radiation also decreased because of the reflection on inner side of heat preservation device, so the final energy consumption decreased.

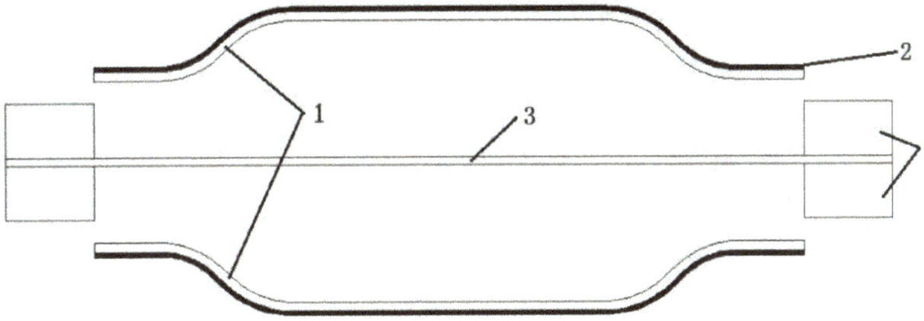

Fig. 2. Schematic diagram of inhomogeneous heat preservation device: 1) heat preservation device; 2) energy storage material; 3) blank; 4) copper electrode.

4. Results and Discussion

Owing to the restriction of the imaging field of infrared thermography, temperature distribution of one half of the blank was taken into consideration. Temperature distribution under non-heat preservation and inhomogeneous heat preservation condition were respectively shown in Figure 3 and Figure 4.

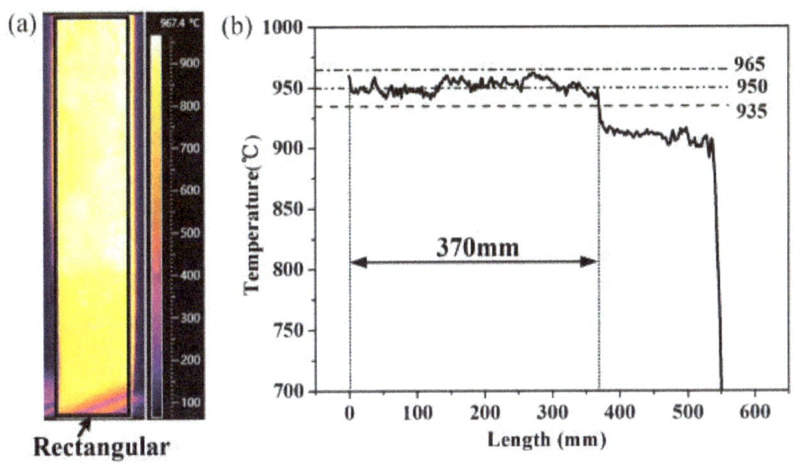

Fig. 3. Temperature distribution under non-heat preservation condition: a) temperature field, b) temperature distribution in rectangular area.

Temperature deviation in area of rectangular was±15°C because of fluctuation of temperature, this area was considered as uniform temperature zone. Experiment results shown that the length of uniform temperature zone of blanks was 370mm under non-heat preservation conditions and the length of uniform temperature zone of blanks was 560mm under inhomogeneous heat preservation conditions, that is to say, material utilization increased from 63.8% to 96.6%. So, temperature uniformity could be improved for the application of inhomogeneous heat preservation device, so as material utilization. Meanwhile, the application of inhomogeneous heat preservation device reduced convection between blanks and surroundings, decreased energy consumption.

Because the reflection of radiation energy utilized by inhomogeneous heat preservation device was physical process, heat preservation efficiency would be better if the height between blanks and device was smaller.

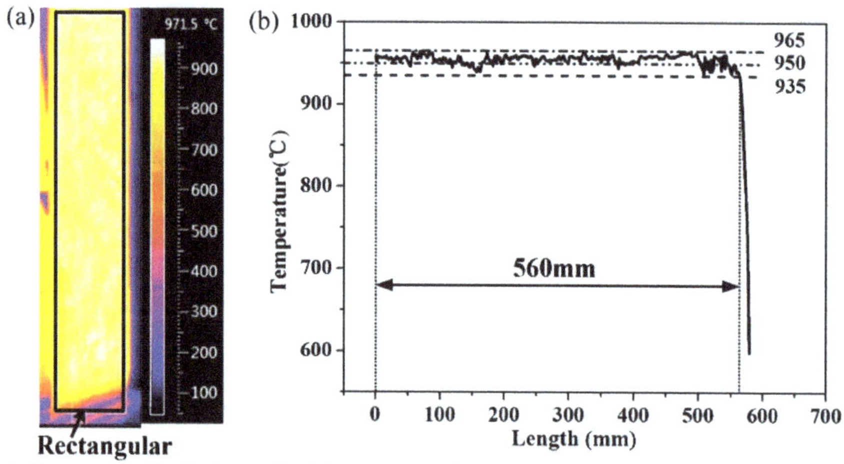

Fig. 4. Temperature distribution under inhomogeneous heat preservation condition: a) temperature field, b) temperature distribution in rectangular area.

5. Conclusion

1) Energy loss of conduction heating blanks was mainly in the way of radiation. Radiation energy would be reflected on the inner side of the device and then cast to the blanks. The application of the device could reduce energy loss of blanks, and decrease convection between blanks and surroundings, which contributed to the reduction of energy consumption.
2) Making sure that the height between the ends of blanks and device was smaller than that between the middle of blanks and device by adjusting the height insured more energy supplement in the ends of the blanks and improved the uniformity of temperature distribution.

References

1. W.K, Liang, Y. Liu, B. Zhu, M.L. Zhou and Y.S. Zhang, Conduction heating of boron alloyed steel in application for hot stamping, *International Journal of Precision Engineering and Manufacturing* **16**, 1983 (2015).
2. W.K. Liang, W.J. Tao, W. Xiao, B. Zhu and Y.S. Zhang, Development and Application of Conduction Heating on Ultra-High Strength Steel, in *advanced high strength steel and press hardening: proceedings of the 2nd international conference (ICHSU2015)*, (Changsha, China, 2015).
3. P. Liu, Y. Zhang, H. Liu and Z. Gui, Modeling for periodic striation and microstructure evolution in active gas melt laser cutting for phase hardened parts, *The International Journal of Advanced Manufacturing Technology* **70**, 1421 (2014).

4. Z.X. Gui, Y.S. Zhang and Z.J. Wang, Simulation of lightweight B-pillar during hot stamping process with thermo-mechanical-metallurgicar model, *Duanya Jishu-Forging & Stamping Technology* **37**, 193 (2012).
5. Z.J. Wang, X.Y. Luo, W.T. He and Y.S. Zhang, Investigation into the tribological behaviors of press hardening steels on the tailored conditions, *Science China Technological Sciences* **58**, 97 (2015).
6. C. Yiran, Z.X.J.X.Q. Tong and L. Derong, Measuring for solar heat gain through glass windows with shading system, *Journal of Southeast University (Natural Science Edition)* **06**, 1169 (2009).
7. J.X. Bao, P.J. Wang and B.F. Wan, Radiant Heat Loss Reduction fo Cathode-Filament Module, *Chinese Jouranl of Vacuum Sciene and Technology* **06**, 522 (2007).

Literature Review on State of the Art in Shot Peening in the Last Decade (2006–2015)

P. Luo[1], Z. Wang[1,2,3,†], G. Wu[1], Y. Wang[1] and M. Cheng[1]

[1]*School of automotive Engineering, Wuhan University of Technology, Wuhan, Hubei, China*

[2]*Hubei Key Laboratory of Advanced Technology for Automotive Components, Wuhan, Hubei, China*

[3]*Hubei Collaborative Innovation Center for Automotive Components Technology, Wuhan, China*

[†]*E-mail: wangzhou@whut.edu.cn*

Shot peening (SP) is a cold working process with the aim of producing a compressive residual stress layer and improving fatigue property of metals and composites, which is widely used in industry due to its own characteristics such as remarkable strengthening effect, low cost and well adaption. This paper provides an extensive literature review and summary on SP process development in the last decade by introducing the development of SP simulation and some new SP methods, such as laser peening, ultrasonic peening and high energy peening and etc., which can give readers a brief overview for the state of the art in shot peeing process in the last decade.

Keywords: Shot peening strengthening; Finite element method (FEM) and experimental method; Laser peening; Ultrasonic peening; High energy peening.

1. Introduction

In the past century, there are various kinds of mechanical surface treatment used to improve fatigue strength of metallic components in industry [1]. So far, SP is still the most widely used in engineering due to its own characteristics that it do not limited by the type of material, material static strength and geometrical shape and dimensions. SP is an established mechanical surface treatment process, well known for its potential to improve fatigue strength of metallic components. The increase of fatigue strength is realized basically by the induction of macroscopic compressive residual stresses and work hardening in the near surface region leading to a suppression and/or deceleration of crack initiation and/or crack propagation during service. SP originated in 1920s to 1930s and independently developed in Germany [2, 3] and the US [4, 5]. Around the world, SP was first used in the strengthening of spring, and then quickly spread throughout the automotive industry. With a tremendous development in the promotion of aviation and other military industry, it backed to the strengthen production in automotive industry and other mechanical parts industry in the 1960s to 1970s. With the purpose of improving the strengthening effect, some modified SP methods, such as stress peening [6], warm peening [7], the secondary peening [8], compound peening [9] and other blasting techniques, were proposed in the following decades. With the development of laser technology, nanotechnology and other related technology, some new-type peening methods were invented, such as laser peening [10], ultrasonic peening [11], high energy peening [12] and etc. in the 21st century, especially in the last ten years.

In order to introduce the development of SP process in the last decade more efficiently, the literature review is subdivided into two main groups with different development directions for this mechanical surface process. The first group is to introduce the development of finite element modeling method for SP simulation, which has already been proved to be an effective method with the aim of minimizing experimental trials and achieving a better understanding of this process. The second group is to introduce some modern SP process, which has originated and developed in the last decade. The aim of this brief introduction is to help readers to know the current development situation of SP process.

2. Finite Element Modeling (FEM)

The great majority of works in the field of SP simulation models is to deal with shot-workpiece interaction with the aim of quantitatively describing the process and predicting the corresponding process results like the residual stress state. Basically, the FE SP models proposed in the literature can be subdivided into two-dimensional (2D) and three-dimensional (3D) approaches.

2.1. *2D simulation models*

2D FE models used to be a very popular approach for the simulation of a single impact on a cylindrical or a semi infinite target body. The very first work on this field was published by Hardy et al. [13], who investigated the flattening of the contact stress distribution in comparison to Hertz's elastic model for ideal elasto-plastic material behavior. Mori et al. [14] then first analyzed the influence of the shots deformation behavior for rigid-plastic. Later, Rouhaud et al. [15] analyzed it for elastic as well as elasto-plastic deformation behavior. The impact of isotropic and kinematic strain hardening on the residual stress state was investigated by Ould et al. [16] in 2006.

Baragetti et al. [17] used X-ray and Evans [18] additionally used neutron diffraction measurements on shot peened test specimens for the determination of the in-depth residual stress profile compared with 2D simulation results. However, a direct comparison of the residual stress depth profile along the axis of symmetry of a 2D model with experimental depth measurements on a shot peened component is questionable, since common measurement techniques like X-ray diffraction or hole drilling only provide macroscopic and averaged stress values, whereas the measurement area is an order of magnitude larger than the diameter of a dimple. Furthermore, the fact that inclined and multiple shot impacts and associated coverage effects can not be taken into account with a 2D FE model leads to the application of 3D models. Therefore, the 3D FE model has become the main choice for SP simulation in modern times.

2.2. *3D simulation models*

In the last decade, 3D FE models are more and more widely used in the field of SP simulation. With different purposes, 3D FE models can be used to investigate the effect of SP process parameters on the process results, such as shot impact angels, SP coverage and etc., or the analysis of SP effect on the target material performance like fatigue property, residual stress state and surface roughness.

In 2011, Kim et al. [19] provided a solution to deal with the evaluation of SP residual stress. They built a practical FE model using plastic shots and integrating some factors like material damping, strain rate, shot diameter, impact and etc., finding that the solution provided was comparable with the 3D multi-impact FE solution and the experimental XRD result. Then in 2012, their group [20] proposed another 3D angled multi-shot FE model for evaluation of residual stress resulting from angled SP. The definition of FE coverages for angled multi-impact analyses they purposed are shown in Fig. 1. After examining the effects of factors such as impact angle, impact pattern and the number of shots using this model and comparing the FE solution with experimental results by X-ray diffraction (XRD), they found that this solution could serve as a base of 3D multi-impact FE model for various incidence angles.

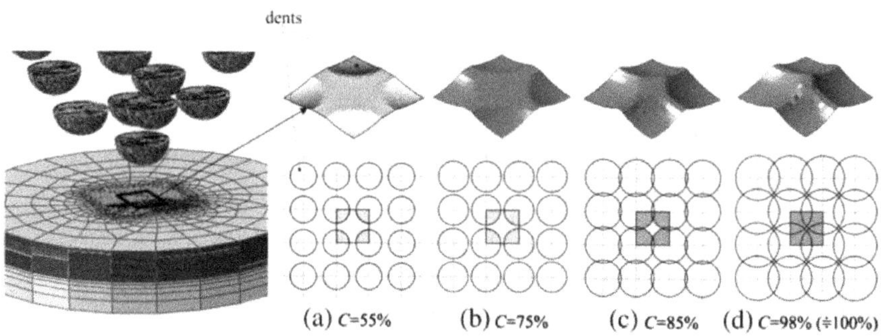

Fig. 1. Definition of FE coverages for angled multi-impact analyses [20].

Since SP simulation deals with shot-workpiece interaction, shot plays an important rolein the SP process. Many investigations focusing on the shot were carried out in different ways. Gangaraj et al. [21, 22] selected two existing FE models that had made an effort to simulate a realistic SP in terms of the shot coverage and made a FE re-simulation of two different symmetry cells (Meguid & Kim [23, 24] and Majzoobi [25]). Figs. 2 (a) and (b) show the comparison between SP models proposed by Kim & Meguid and re-simulated model proposed by Gangaraj. Figs. 2 (c) and (d) show the comparison between SP models proposed by Majzoobi and re-simulated model proposed by Gangaraj. The resulted treated surfaces of re-simulated models were then carefully examined, finding that the existing models had the deficiency of either not capturing a realistic coverage or being very costly to simulate high coverage. On this basis, they proposed that using a variable dimension symmetry cell to acquire full coverage with not increasing the computational cost in 2011 [21] and the application of Avrami equation in a random FE simulation to present an actual coverage evolution in 2014 [22].

Differently, in 2014, Nguyen et al. [26] put forward to use multiphase computational fluid dynamics (CFD) simulations to predict SP coverage. They successfully developed and validated the whole solution procedure in SP process. The advantages of this method was that it could be customized for different geometry configurations and operating conditions for accurate coverage area prediction. In 2013, Gariepy et al. [27] investigated the effect of the peening trajectory in SP. A simple model was proposed to simulate incremental deflections as the shot stream traveled over the samples, correctly predicting the complex radius distribution arising from the peening path and the constraints applied to the sample during peening.

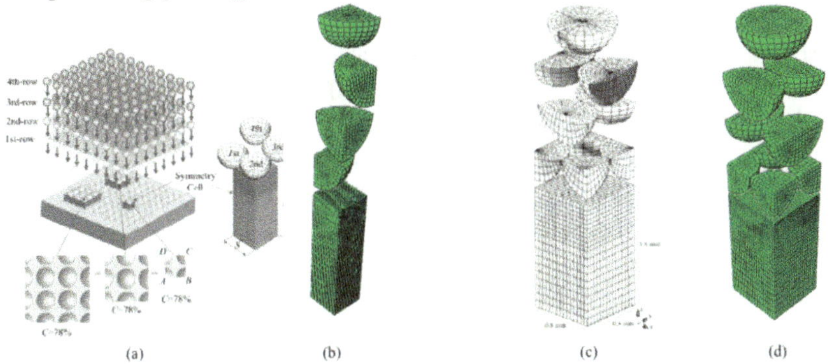

Fig. 2. (a) Symmetry Cell (Kim & Meguid); (b) re-simulated model (Gangaraj); (c) Symmetry Cell (Majzoobi); (d) re-simulated model (Gangaraj) [21].

Apart from the shot, target material is also an important objective that should take into account. In 2014, Sanjurjo et al. [28] investigated the influence of the target material constitutive model on the numerical simulation of a SP process. Three constitutive models were proposed based on the experimental characterization of a duplex stainless steel and incorporated into a FE simulation. By comparing the numerical predictions of residual stress and roughness with the experimental findings, they got a non-linear kinematic-isotropic model as the best target material constitutive model, which could reproduce the behavior of the material subjected to an actual SP treatment, providing a very accurate prediction of both the residual stress field and the surface roughness generated by the process. Another investigation focusing on the material was done by Kim et al. [29] in 2013, they proposed a 3D FE model, which was set with rigid shot, elastic shot, and plastic shot respectively, to study the effects of the Rayleigh damping in the material, dynamic friction and the rate-dependency of the material on 3D residual stress simulation of angled SP, finding that the FE solution for the incidence angle of 60° and plastic shot agreed best with the XRD experimental data.

Besides the SP process parameters, some other investigations focusing on the analysis of SP effect on the target material performance have been done in recent years. In 2012, Song et al. [30] and Xie et al. [31] investigated respectively about the evolution of SP residual stress on GW103 magnesium alloy and the residual stress distribution in deformed surface layer of Ti-6Al-4V after SP. Song' group proposed a simulation of the residual stress evolution in low cycle fatigue (LCF). After observing that under the subsequent compressive loading, the SP plastic strain that caused early yield eventually became useless, they revealed that the effect of SP induced residual stresses on LCF was minimal, due to the fact that during the first loading cycle the eigenstrain within the sample had been modified and flattened by the overwhelming plastic deformation. In Xie's study, a 3D model simulation using four impacts with 36 shot balls was made and compared with the experimental results by XRD, finding that the 3D model of dynamic analysis about shot impacts was appropriate for predicting the distribution of residual stresses introduced by SP. As for the fatigue property, in 2011, Gangaraj et al. [32] analyzed the SP effect on fretting fatigue parameters. During the investigation, a sequential FE simulation was made on normal stress, shear stress, bulk stress and slip amplitude, which were considered to be the controlling parameters of fretting damage, demonstrating that among the modifications related to SP, compressive residual stress had a dominant effect on the fretting parameters. Then in 2013, Liu et al. [33] built an irreversible cohesive zone model based on experimental data for a Nickel-based superalloy to investigate the effect of SP on fatigue crack growth. The specimens were studied for different loading amplitudes and different SP intensities, revealing that the crack initiation position and time depended not only on the SP intensity but also on the cyclic loading amplitude. Retardation of fatigue crack growth was more effective at lower loading amplitude for shot-peened specimens, and higher applied loading would eliminate the beneficial effects of SP. Another important target material performance, so called the surface roughness, was investigated respectively by Bagherifard et al. [34] In 2012 and Taro et al. [35] in 2015. Bagherifard's group built a 3D FE model based on the obtained data to assess the surface roughness evolution induced by SP. After calculating the conventional roughness parameters with the discrete simulation data, they found that the simulation results corresponded well with the roughness values measured experimentally on shot-peened specimens, that is to say, the developed FE model provided an efficient estimation of surface characteristics of the treated components [34]. Similarly, Taro's group proposed to utilize a novel 3D FE model simulating multiple impacts with a deterministic distribution by using a semi analytical method to predict the surface roughness of an initially perfectly flat semi infinite surface [35].

3. The Development of New-type SP Method

With the rapid development of industry and the purpose of improving the strengthening effect, many modified SP methods, such as stress peening [6], warm peening [7], the secondary peening [8], compound peening [9] and etc., were proposed about 30 or 40 years ago, which are not repeated here. Here, three new-type SP methods, which are used in industry gradually or show promising prospect in industry, were chosen as the representative in order to reflect the SP development in recent years.

3.1. *Laser peening*

Laser peening (LP) is a relatively new surface treatment process and very effective in improving the fatigue properties of various kinds of metals and alloys. Usually it is applied in aerospace and automotive industries. Although it may cost more than conventional SP method due to its low production rate, the beneficial effects of LP on static, cyclic, fretting fatigue [36, 37] and stress corrosion performance of aluminum alloys [38], steels [39] and nickel based alloys [40] are obvious. Since LP can introduce a higher and deeper compressive residual stress field [41, 42], it is widely applicable for improving the fatigue properties of metals.

Hatamleh et al. [43, 44] respectively investigated on the effects of LP and conventional SP on fatigue crack growth in friction stir welded joints made by 2195 aluminum alloy and 7075-T7351 aluminum alloy in 2006 and 2008. The crack growth rates were assessed and characterized for different peening conditions, indicating a significant reduction in fatigue crack growth rates using LP compared to SP and untreated welded specimens. In 2010, González et al. [45] investigated the effect of laser shock processing on fatigue crack growth of duplex stainless steel, finding that increasing pulse density, fatigue crack growth rate was reduced and the higher the pulse density was, the greater the compressive residual stress became. At the same year, Zhang et al. [46] studied the effects of multiple LP and pulse intensity used in LP on the residual stresses, micro-hardness and three-point bending fatigue performance of Ti-6Al-4V alloy. The multiple LP was achieved by using the successive laser beams at one spot, concluding that the maximum value of the residual stress increased with increasing the number of laser beams and decreasing the distance from the surface. Later in 2011, Gao et al. [47] investigated the improvement of fatigue property by LP and conventional SP. After experimenting on the machined, laser-peened, and shot-peened specimens made by 1×10^7 cycles of 7050-T7451 aluminum alloy, the results showed that the LP improved fatigue properties more effectively. The compressive residual stress field caused by SP and LP are respectively showed in Figs. 3 (a) and (b).

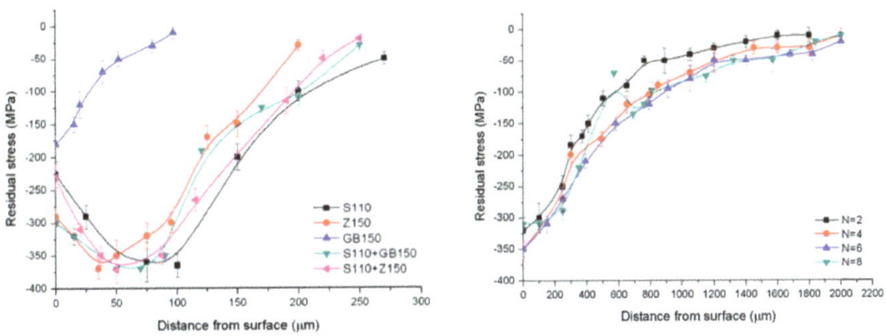

Fig. 3. The compressive residual stress field caused by (a) SP and (b) LP [47].

3.2. *Ultrasonic peening*

Ultrasonic peening (UP) is one of the promising SP methods in the cold treatment of metals, particularly welded structures. It has many industrial applications in aerospace [48], ship and marine [49 - 51], automotive [52], railway [53], and bridge structures [54]. During the process, harmful tensile residual stresses are effectively eliminated [55] and a layer of compressive residual stress is induced into the material surface. The UP process is therefore able to increase surface micro-hardness, corrosion resistance, and enhances fatigue life.

Ultrasonic was firstly invented by Mukhanov et al. [56]. By generating continuous ultrasonic vibrations at a carburized steel tool tip, the treated part surface received direct impacts. Then Krylov et al. [57] proposed a shot between a surface and ultrasonic transducer which imposed high plastic deformation. However, free motion of the shot resulted in negative effects. Some comparative researches proved that cold hammer peening is better than that of weld toe grinding [58 - 60]. Statnikov et al. [61] then invented several advanced different peening methods. They developed an ultrasonic magnetostrictive oscillating system with a peening tool and applied the process to improve the quality and reliability of welded rail [62] and tubular structure joints [63].

Due to its excellent performance on the strengthening of welded structures, more and more UP investigations on welded components have been done in recent years. In 2008, Wang et al. [64] investigated the fatigue design of welded joints by UP. After analyzing the fatigue testing results of ultrasonic-peened parts, it indicated that material strength had effect, to a certain extent, on the fatigue performance of ultrasonic-peened parts. More specifically, in 2012, Abdullah et al. [65] investigated the effect of UP on stainless steel-304 welded parts. A series of experimental fatigue strength tests along with metallography, micro hardness and corrosion resistance tests of welded pieces with and without processing by UP indicated that under the UP treatment, a better mechanical and corrosion resistance was achieved. Table 1 shows the conditions and results of the fatigue tests.

Table 1. Fatigue test conditions and the values obtained for specimens without/with UP [65].

No.	$2\sigma_{Amp}$ (MPa)	F_{Max} (KN)	F_{Min} (KN)	$2F_{Amp}$ (KN)	N (cycle)	Fracture location
Without peening						
1	300	17.325	0.825	16.5	159,552	Weld toe
2					111,544	
3					211,686	
4	316	18.249	0.869	17.38	110,985	Weld toe
5					207,663	
6					84,926	
7	330	19.057	0.907	18.15	60,231	Weld toe
8					103,366	
9					53,336	
With peening						
1	300	17.325	0.825	16.5	329,752	Base metal
2					177,903	Base metal
3	316	18.249	0.869	17.38	228,433	Base metal
4					238,235	Base metal
5	330	19.057	0.907	18.15	106,782	Weld toe
6					118,691	

Apart from the UP strengthening effect of welded structures, a number of investigations on the improvement of different material properties under UP have also been done in the last decade. Mordyuk et al. [66] investigated the effect of structure evolution induced by UP on the corrosion behavior of AISI-321 stainless steel in 2007. This study showed that a nanocrystalline surface layer was produced on the specimens by severe plastic deformation via UP. Through the characterization of the microstructural evolution of the surface layer by means of XRD analysis and transmission electron microscopy (TEM), they found that both the grain refinement and the volume fraction of the strain-induced martensite affected a number of electrochemical quantities of the stainless steel, not only significantly increasing the corrosion potential but also decreasing the passivation-maintaining current density. Then in 2010, their group [67] did another study on the structure, microhardness and damping characteristics of Al matrix composite reinforced with AlCuFe or Ti using UP. This study indicated that composite layers could be formed in aluminum by addition of glycerin suspensions contained either icosahedral quasicrystalline (QC) AlCuFe or hcp Ti fine powders to the zone of severe plastic deformation at the UP processing. According to XRD and TEM, the composite layers formed were characterized by similar level of macroscopic compressive residual stresses, fine grain structure, relatively uniform distribution of reinforcing particulates and their similar volume fraction, which affected significantly the microhardness and damping properties. In 2011, Sanda et al. [68] investigated the effect of processing time, material and quantity of shot balls and distance from radiating surface to sample on the final state of Inconel 718 in precipitated state generated by UP. A change to more compressive residual stresses at the surface of the treated parts had been measured in all cases. For higher UP processing times and/or lower booster sample distances, the degree of plastic deformation in the specimen increased, leading to a change to more compressive surface stresses and a higher density of impact marks in the treated surface. The same occurs when WC/Co balls were used instead of steel balls. The tendency to more compressive stresses reached a saturation level after a certain processing time, when the system was not able to force the material to continue with more plastic deformation. If a higher quantity of balls was used, there would be less impacts of the shots with the surface and their energy would be lower (due to the loss of energy after inelastic collisions). This diminished the effect of the impacts in introducing compressive stresses and leaded to less and shallower impact marks in the treated surface.

3.3. *High energy peening*

The high energy shot peening (HESP) is an effective surface strengthening treatment to change the metallurgical structure and mechanical properties in the surface layer of the components. Different from the conventional mechanical SP, the component is impacted with higher speed and higher energy shots in the HESP treatment and better surface strengthening effects could be obtained [69].

Due to its characteristic, HESP is usually used to fabricate a nanostructured surface layer on material. Nanocrystallization by severe plastic deformation (SPD) in metals has been a subject of many researches [70 - 72] in the last decade. This method offers two significant advantages compared with other techniques such as high energy ball milling, insert gas condensation. First, it is possible to produce large bulk samples. Second, these samples are free from any residual porosity and contamination. Among those methods, HESP is one of the effective treatments for the fabrication of various ultrafine-grained structures by imposing intense plastic strain into the surface of the materials. The microstructure observation at various depth levels provides the clue to the process of structural evolution relevant to various stages of strain. In 2008, Hou et al. [73] investigated the microstructure evolution of AZ91D induced by HESP. A nanostructured surface layer was fabricated on the specimens under the HESP treatment. HESP induced structure along the depth of the treated sample surface layer was characterized by means of XRD, TEM and high resolution transmission electron microscope (HRTEM), finding that a deformed layer of about 50μm has formed after HESP treatment and the average grain size increases from about 40nm in the surface layer to about 200nm at the depth of 40μm. The schematic illustration of refinement mechanism is shown in Fig. 4.

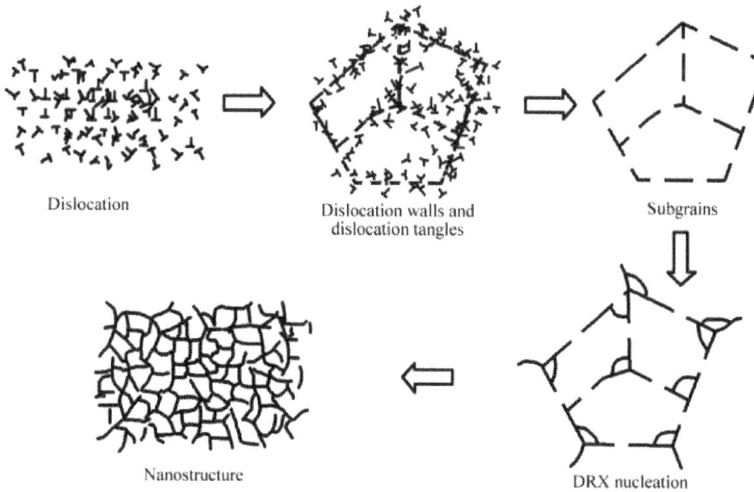

Fig. 4. Schematic illustration showing microstructural evolution process of AZ91D magnesium alloy induced by HESP [73].

Apart from obtaining Nanocrystallization, in 2015, Lu et al. [74] investigated the effect of HESP pressure on the stress corrosion cracking of the weld joint of 304 austenitic stainless steel. The specimens were treated under HESP with different pressures. The grain size and metallographic microstructure of the surfaces layer were analyzed by XRD. Slow strain rate tension tests were then performed to investigate the effect of peening pressure on the stress corrosion sensitivity, showing that in the surface layer of the specimens, the grain refinement, hardness and the plastic deformation all increased with the increasing SP pressure. Besides, Martensitic transformation was observed in the surface layer after HESP, finding that the martensite phase ratio increased with the increasing SP pressure.

4. Conclusion

The paper mainly provided an extensive literature review on SP development in the last decade with the aim of giving readers a brief overview for the state of the art in SP. The first part is to introduce the development of SP simulation models, which help the SP researchers to know the trend of SP simulation recently. The second part is to introduce the development of some new-type SP methods and expect these methods will gain more attention to the SP community in the following years.

Acknowledgments

The financial support of the National Natural Science Foundation of China (NSFC) (No. 51405356) and the Research Fund for the Doctoral Program of Higher Education of China (No. 20130143120015) are gratefully acknowledged.

References

1. V. Schulze, *Modern mechanical surface treatment: states, stability, effects*. (WILEY-VCH Verlag GmbH & Co. KGaA, Weinheim 2006).

2. O. Föppl, G.V. Heydekampf, Dauerfestigkeit and Konstruktion, *Metallwirtschaft Wissenschaft und Technik* **8**, 1087 (1929).
3. A. Thum, H. Wiegand, Die Dauerhaltbarkeit von Schraubenverbindungen and Mittel zu ihrer Steigerung, *Verein Deutscher Ingenieure Zeitschrift* **39**, 1061 (1933).
4. E.G. Herbert, The work-hardening of steel by abrasion, with an appendix on the 'cloudburst' test and superhardening, *J. Iron. Steel* **11**, 265 (1927).
5. J.O. Almen, Peened surfaces improve endurance of machine parts, *Metal program* **43**, 209 (1943).
6. R. Zeller, Shot peening under Prestressing Testing of improved fatigue properties of components, *Materialprufung* **35**, 218 (1993).
7. R. Menig, V. Schulze and O. Vohringer, Optimized warm peening of the quenched and tempered steel AISI 4140, *Mater. Sci. Eng. A* **335**, 198 (2002).
8. P. Fu and C. Jiang, Residual stress relaxation and micro-structural development of the surface layer of 18CrNiMo7-6 steel after shot peening during isothermal annealing, *Mater. Des.* **56**, 1034 (2014).
9. K. Zhan, C.H. Jiang and V. Ji, Uniformity of residual stress distribution on the surface of S30432 austenitic stainless steel by different shot peening processes, *Materials Letters* **99**, 61 (2013).
10. E. Maawad and Y. Sano, Investigation of laser shock peening effects on residual stress state and fatigue performance of titanium alloys, *Mater. Sci. Eng. A* **536**, 82 (2012).
11. M. Malaki and H. Ding, A riview of ultrasonic peening treatment, *Mater. Des.* **87**, 1072 (2015).
12. L.F. Hou and Y.H. Wei, Microstructure evolution of AZ91D induced by high energy shot peening, *Transactions Of Nonferrous Metals Society Of China* **18**, 1053 (2008).
13. C. Hardy, C.N. Baronet and G.V. Tordion, The elasto-plastic indentation of a half-space by a rigid sphere, *In. J. Numerical Methods in Engineering* **3**, 451 (1971).
14. K. Mori, K. Osakada and N. Matsuoka, Rigid-plastic finite element simulation of peening process with plastically deforming shot, *JSME International Journal–Series A–Mechanics and Material Engineering* **39**, 306 (1996).
15. E. Rouhaud and D. Deslaef, Influence of shots' material on shot peening, a finite element model, *Mater. Sci. Forum* **404–407**, 153 (2002).
16. C. Ould, E. Rouhaud, M. François and J.L. Chaboche, A kinematic hardening finite elements model to evaluate residual stresses in shotpeened parts, local measurements by x-ray diffraction, *Mater. Sci. Forum* **524–525**, 161 (2006).
17. S. Baragetti and M. Guagliano, A numerical procedure for shot peening optimization by means of non-dimensional factors, *International Journal Of Materials & Product Technology.* **15**, 91 (2000).
18. R. Evans, Shot peening process: modeling, verification, and optimisation, *Mater. Sci. Technol.* **18**, 831 (2002).
19. T. Kim, H. Lee and H.C. Hyun, A simple but effective FE model with plastic shot for evaluation of peening residual stress and its experimental validation, *Mater. Sci. Eng. A* **528**, 5945 (2011).

20. T. Kim, H. Lee, M. Kim and S. Jung, A 3D FE model for evaluation of peening residual stress under angled multi-shot impacts, *Surf. Coat. Tech.* **206**, 3981 (2012).
21. S.M.H. Gangaraj, M. Gualiano and G.H. Farrahi, Finite Element Simulation of Shot Peening Coverage with the Special Attention on Surface Nanocrystallization, *Procedia Engineering* **10**, 2464 (2011).
22. S.M.H. Gangaraj, M. Gualiano and G.H. Farrahi, An approach to relate shot peening finite element simulation to the actual coverage, *Surf. Coat. Tech.* **243**, 39 (2014).
23. S.A. Meguid, G. Shagal and J.C. Stranart, 3D FE analysis of peening of strain-rate sensitive materials using multiple impingement model, *Int. J. Impact Engineering* **27**, 119 (2002).
24. T. Kim, J.H. Lee, H. Lee and S.K. Cheong, An area-average approach to peening residual stress under multi-impacts using a 3D symmetry-cell FE model with plastic shots, *J. Mater. Des.* **31**, 50 (2010).
25. G.H. Majzoobi, R. Azizi and A.N. Alavi, A three-dimensional simulation of shot peening process using multiple shot impacts, *J. Mater. Process Tech.* **164–165**, 1226 (2005;).
26. V.B. Nguyen, H.J. Poh and Y.W. Zhang, Predicting shot peening coverage using multiphase computational fluid dynamics simulations, *Powder Technology* **256**, 100 (2014).
27. A. Gariepy, S. Larose and C. Perron, On the effect of the peening trajectory in shot peen forming, *Finite Elements in Analysis and Design* **69**, 48 (2013).
28. P. Sanjurjo, C. Rodriguez and I. Penuelas, Influence of the target material constitutive model on the numerical simulation of a shot peening process, *Surf. Coat. Tech.* **258**, 822 (2014).
29. T. Kim, H. Lee, H.C. Hyun and S. Jung, Effects of Rayleigh damping, friction and rate-dependency on 3D residual stress simulation of angled shot peening, *Mater. Des.* **46**, 26 (2013).
30. X. Song, W.C. Liu and J.P. Belnoue, An eigenstrain-based finite element model and the evolution of shot peening residual stresses during fatigue of GW103 magnesium alloy, *Int. J. Fatigue* **42**, 284 (2012).
31. L.C. Xie and J. Zhang, Investigation on experiments and numerical modelling of the residual stress distribution in deformed surface layer of Ti–6Al–4V after shot peening, *Mater. Des.* **41**, 314 (2012).
32. S.M. H-Gangaraj and Y. A-Tabrizi, Finite element analysis of shot-peening effect on fretting fatigue parameters, *Tribology International* **44**, 1583 (2011).
33. J.X. Liu, Numerical analysis for effects of shot peening on fatigue crack growth, *Int. J. Fatigue*. **50**, 101 (2013).
34. S. Bagherifard and R. Ghelichi, Numerical and experimental analysis of surface roughness generated by shot peening, *Applied Surface Science*. **258**, 6831 (2012).
35. M. Taro and T. Chaise, A methodology to predict the roughness of shot peened surfaces, *J. Mater. Process Tech.* **217**, 65 (2015).
36. J.M. Yang, Y.C. Her, N. Han and A. Clauer, Laser shock peening on fatigue behavior of 2024-T6 Al alloy with fastener holes and stop holes, *Mater. Sci. Eng. A* **298**, 296 (2001).

37. L.W. Tsay, M.C. Young and C. Chen, Fatigue crack growth of laser-processed 304 stainless steel in air and gaseous hydrogen, *Corrosion Science* **45**, 1985 (2003).
38. Z. Hong and Y. Chengye, Laser shock processing of 2024-T62 aluminum alloy, *Mater. Sci. Eng. A* **257**, 322 (1998).
39. J.P. Chu and J.M. Rigsbee, Laser shock processing effects on surface microstructure and mechanical properties of low carbon steel, *Mater. Sci. Eng. A* **260**, 260 (1999).
40. Y.Q. Li and W.F. He, Laser shock peening of GH4133 Nickel-based Superallay, *Mater. Sci. Eng. A* **44**, 1517 (2015).
41. P. Peyre, R. Fabbro, P. Merrien and H.P. Lieurad, Laser shock processing of aluminium alloys. Application to high cycle fatigue behavior, *Mater. Sci. Eng. A* **201**, 102 (1996).
42. C.S. Montross, T. Wei, L. Ye, G. Clark and Y.W. Mai, Laser shock processing and its effects on microstructure and properties of metal alloys: a review, *Int. J. Fatigue* **24**, 1021 (2002).
43. O. Hatamleh, A comprehensive investigation on the effects of laser and shot peening on fatigue crack growth in friction stir welded AA 2195 joints, *Int. J. Fatigue* **31**, 974 (2009).
44. O. Hatamleh, J. Lyons and R. Forman, Laser and shot peening effects on fatigue crack growth in friction stir welded 7075-T7351 aluminum alloy joints, *Int. J. Fatigue* **29**, 421 (2007).
45. C.R. González, C.F. Martinez and G.G. Rosas, Effect of laser shock processing on fatigue crack growth of duplex stainless steel, *Mater. Sci. Eng. A* **528**, 914 (2011).
46. X.C. Zhang, Y.K. Zhang and J.Z. Lu, Improvement of fatigue life of Ti-6Al-4V alloy by laser shock peening, *Mater. Sci. Eng. A* **527**, 3411 (2010).
47. Y.K. Gao, Improvement of fatigue property in 7050–T7451 aluminum alloy by laser peening and shot peening, *Mater. Sci. Eng. A* **528**, 3823 (2011).
48. M.P. Nascimento, R.C. Souza, W.L. Pigatin and H.J.C. Voorwald, Effects of surface treatments on the fatigue strength of AISI 4340 aeronautical steel, *Int. J. Fatigue* **23**, 607 (2001).
49. K. Kirkhope, R. Bell, L. Caron, R. Basu and K.T. Ma, Weld detail fatigue life improvement techniques. Part 2: application to ship structures, *Marine Structures* **12**, 477 (1999).
50. T. Deguchi, M. Mouri, J. Hara, D. Kano, T. Shimoda, F. Inamura, T. Fukuoka, and K.Koshio, Fatigue strength improvement for ship structures by Ultrasonic Peening, *J. Mater. Sci. Technol.* **17**, 360 (2012).
51. L.L. Martinez Z. Barsoum and A. Paradowska, State-of-the-art: fatigue life extension of offshore installations, in *ASME 2012 31st International Conference* (Ocean, Off-shore and Arctic Engineering, American, Society of Mechanical Engineers 2012).
52. Y. Kudryavtsev and J. Kleiman, Application of Ultrasonic Peening for Fatigue Life Improvement of Automotive Welded Wheels, International Institute of Welding 2005 (IIW Document XIII - 2075, 9 pp.).

53. S. Maddox, M.J. Doréand and S. Smith, A Case Study Of The Use Of Ultrasonic Peening For Upgrading A Welded Steel Structure, *Welding In The World.* **55**, 56 (2011).
54. D. Yin, D. Wang, H. Jing and L. Huo, The effects of ultrasonic peening treatment on the ultralong life fatigue behavior of welded joints, *Mater. Des.***31**, 3299 (2010).
55. M. Shalvandi, Y. Hojjat, A. Abdullah and H. Asadi, Influence of ultrasonic stress relief on stainless steel 316 specimens: A comparison with thermal stress relief, *Mater. Des.* **46**, 713 (2013).
56. I. Mukhanov and Y.M. Golubev, Strengthening Steel Components by Ultrasonically Vibrating Ball, *Vestnik Mashinostroeniya* **11**, 52 (1966).
57. N. Krylov and A. Polischuk, The use of ultrasonic equipment for metal structure stabilization, *Basic Physics of Industrial Ultrasonic Applications* **1**, 70 (1970).
58. K. Kirkhope, R. Bell, L. Caron, R. Basu and K.T. Ma, Weld detail fatigue life improvement techniques. *Part 1: review, Marine Structures* **12**, 447 (1999).
59. J. Knight, Improving the fatigue strength of fillet welded joints by grinding and peening, *Welding Research International* **8**, 519 (1978).
60. T. Gurney, Effect of peening and grinding on the fatigue strength of fillet welded joints, *British Welding Journal* **15**, 601 (1968).
61. E.S. Statnikov, O.V. Korolkov, V.O. Muktepavel and V.Y. Korostel, Oscillating system and tool for ultrasonic impact treatment, US Patent, Patent No. 20080035627-A1, UITL.L.C., United States, 2007.
62. E. Statnikov, V. Korostel, V. Vityazev and O. Korolkov, Method of improving quality and reliability of welded rail joint properties by ultrasonic impact treatment, US Patent, Patent No.US 20060016858-A1, UIT L.L.C., United States, 2006.
63. P. Mikheev, E. Garf, A. Kuz'menko, V. Zaitsev and E.S. Statnikov, Improvement of fatigue resistance of tubular welded connections by ultrasonic peening, *Int. J. Off shore Polar Engineering* **6**, 1 (1996).
64. T. Wang, D.P. Wang, L.X. Huo and Y.F. Zhang, Discussion on fatigue design of welded joints enhanced by ultrasonic peening treatment (UPT), *Int. J. Fatigue* **31**, 644 (2009).
65. A. Abdullah, M. Malaki and A. Eskandari, Strength enhancement of the welded structures by ultrasonic peening, *Mater. Des.* **38**, 7 (2012).
66. B.N. Mordyuk and G.I. Prokopenko, Effect of structure evolution induced by ultrasonic peening on the corrosion behavior of AISI-321 stainless steel, *Mater. Sci. Eng. A* **458**, 253 (2007).
67. B.N. Mordyuk, M.O. Iefimov and G.I. Prokopenko, Structure, microhardness and damping characteristics of Al matrix composite reinforced with AlCuFe or Ti using ultrasonic impact peening, *Surf. Coat. Tech.* **204**, 1590 (2010).
68. A. Sanda, V.G. Navas and O. Gonzalo, Surface state of Inconel 718 ultrasonic shot peened: Effect of processing time, material and quantity of shot balls and distance from radiating surface to sample, *Mater. Des.* **32**, 2213 (2011).
69. J.C. Sun, G.M. Sheng and Y.T. Wang, Heat Treat. Meterials **5**, 38 (2010).

70. A.V. Korznikov, Y.V. Ivanisenko and D.V. Laptionok, Influence of severe plastic deformation on structure and phase composition of carbon steel, *Nanostructured Materials* **4**, 159 (1994).
71. N.R. Tao, M.l. Sui, J. Ku and L. Ku, Surface nanocrystallization of iron induced by ultrasonic shot peening. *Nanostructured Materials* **11**, 433 (1999).
72. D.H. Shin, B.C. Kim, Y.S. Kim and K.T. Park, Microstructural evolution in a commercial low carbon steel by equal channel angular pressing, *Acta Materialia* **48**, 2247 (2000).
73. L.F. Hou and Y.H. Wei, Nanocrystalline structure of magnesium alloys subjected to high energy shot peening, *J. Alloy. Compd.* **492**, 347 (2010).
74. Z.M. Lu, L.M. Shi and S.J. Zhu, Effect of high energy shot peening pressure on the stress corrosion cracking of the weld joint of 304 austenitic stainless steel, *Mater. Sci. Eng. A* **637**, 170 (2015).

Applied Research on Auto-Body Parts Forming Process with Variable Blank Holder Force Using Nitrogen Cylinders System

C. Y. Wang†, C. Dai, F. F. Guo, Z. R. An, M. Cheng and Y. Y. Yao
School of Materials Science and Engineering, Hefei University of Technology, Hefei, P.R. China
†E-mail: wangchengyong@hfut.edu.cn

Simplistic definition of blank holder force in CAE analyses generally leads to inaccurate results. To improve the reliability of the finite element (FE) simulation, the changing of blank holder force during the stamping process should be taken into consideration. This paper examined the forming process of car fender with the variable blank holder force provided by nitrogen cylinders system. The simulation with constant blank holder force was proved to be inaccurate by conducting the trail test. Three assumptions of variation of blank holder force such as unchanging value, value increasing with linear line and with polynomial fitted curve were carried out to simulate the process. The parameters of the stamping process were verified and optimized with linear approximation and curve fitting plans. An equation for calculating the inflating or deflating of nitrogen was proposed to predict sheet splits in the actual products manufacturing. Further research was carried on to compare the between linear approximation and curve fitting. The results show that the assumption of curve fitting variable blank holder force is superior to the others in terms on obtaining more accurate forming shape.

Keywords: Blank holder force; Nitrogen cylinders system; Curve fitting; Forming simulation.

1. Introduction

Sheet forming simulation is widely used as one of the significant material forming methods in industry such as aviation and automobile manufacture [1-2]. As one significant factor, the blank holder force is often simplified as an unchanging value in simulation of automotive panel stamping, which contributes to remarkable difference between CAE results and real engineering applications. Many researchers took into consideration the optimization of blank holder force in sheet forming simulation [3-4].

In this paper, the fender drawing with blank holder force provided by nitrogen system was selected. Several variable blank holder force optimizations such as average value, linear approximation and curve fitting were introduced to simulate sheet forming process. The simulation results were compared with actual products manufacture and the equation of inflating or deflating nitrogen was derived to solve sheet forming defects like wrinkling and splitting.

2. Theoretical Analysis of Nitrogen System

Nitrogen gas spring is widely used as one of functional components to provide blank holder force in sheet forming. Considering the gas force characteristics, when the spring arrives at certain stroke, the elastic force produced by the single nitrogen gas spring can be found as follows in Eq. (1).

$$F_s = F_{in} \cdot \frac{S_{nom}}{S_{nom} - S_{used} \cdot \left(1 - \dfrac{F_{in}}{F_{end}}\right)} \tag{1}$$

Where S_{used} is the spring used stroke, S_{nom} is the nominal stroke for the spring, F_{in} is the initial force at current charging pressure, F_{end} is the spring force at full nominal stroke, F_s is the spring force at the used stroke S_{used}.

In this paper, 10 nitrogen gas springs named as WDX356204-9512BMP are selected as blank holder force provider and the blank holder force in different strokes is plotted in Figure 1.

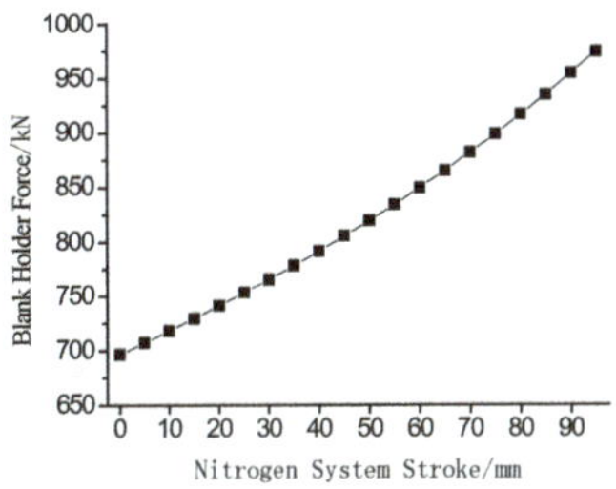

Fig. 1. Load curve of nitrogen system blank holder.

3. Structure and Finite Element Analysis of the Fender

3.1. *Structure analysis of the fender*

Figure 2 shows the structure of the car fenders, the overall dimension of the part is about 695mm×210mm×77mm. To maintain the balance of press and the efficiency of the process, two pieces of part are deformed in one die [5].

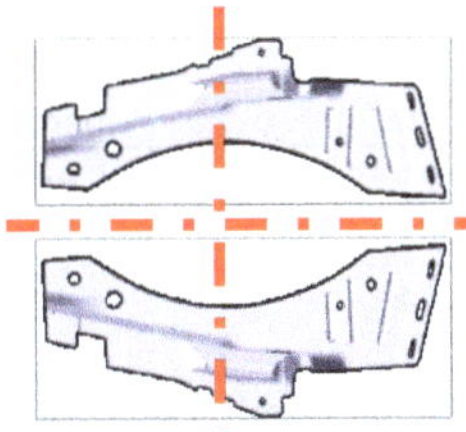

Fig. 2. Structure of the fender.

3.2. *The finite element model*

Under the guide of model structural characteristic, the binder and addendum design of the fender is acquired [6-7]. The material of the sheet is WSS-M1A367, high-strength steel, and the thickness of which is 0.87mm. The material properties used in the simulation are shown in Table 1. The finite element model of drawing process was built and the initial blank holder force is 700kN with 95mm binder stroke.

Table 1. WSS-M1A367 material properties.

Yield Stress σ_s	Work hardening module K	work hardening coefficient n	anisotropy coefficient r_m	Poisson's Ratio	Elastic module
238.4MPa	571.7 MPa	0.26	1.47	0.28	207GPa

4. CAE Results Analysis and Craft Optimization

4.1. *Constant blank holder force result*

Figure 3(a) shows the CAE forming result at initial blank holder force (700kN).The true severe splits in real results are shown in Figure 3(b). From the trail test, it can be concluded that the result obtained from the model with initial constant blank holder force condition is inaccurate.

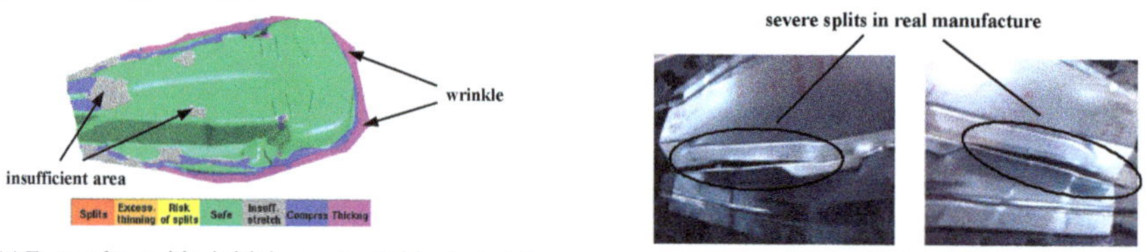

(a) CAE results with initial constant blank holder force.

(b) spilits in trail test.

Fig. 3. Differences between initial CAE and real results.

4.2. *Three variable blank holder force technologies plans*

Based on the characteristics of the loading curve of blank holder force, three plans such as average value, linear approximation and curve fitting were introduced to modify the FEM simulation results and Table 2 shows more details about three plans.

Table 2. Three plans of variable blank holder force technologies.

plans	Plan1 average value	Plan2 linear approximation	Plan3 curve fitting
diagram	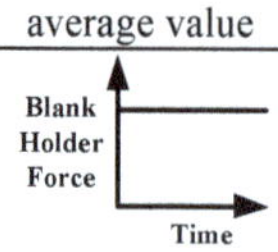	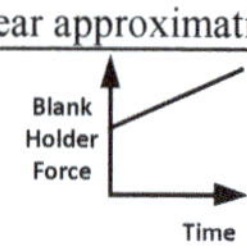	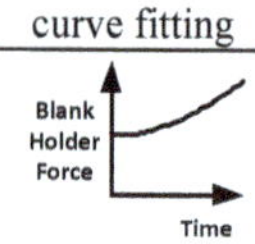
descriptions	constant value at the average: 853kN	connect beginning and end force to a line	fit force at different strokes into cubic function curve

Result of Plan1 is shown in Figure 4. Compared to the initial simulation result, the sheet forms more sufficiently when the upper tools arrive at terminal stroke (95mm). However, wrinkle in some local areas like flange edges still exists.

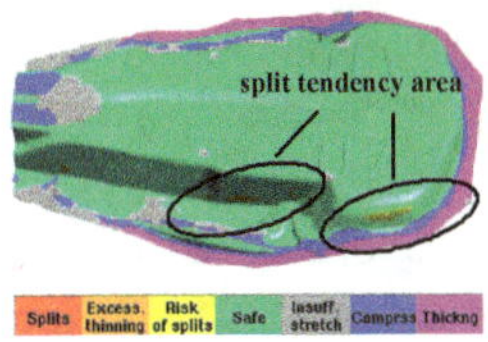

Fig. 4. Plan1 (average value) CAE result.

Result of Plan2 is shown in Figure 5. From Figure 5(a), when the die arrives at 13.5mm behind terminal stroke, deep drawing area starts to emerge with splits and when die continues to move downwards, the splits become more severe shown in Figure 5(b).

(a) 13.5mm behind terminal stroke. (b) terminal stroke.

Fig. 5. Plan2 (linear approximation) CAE result.

Result of Plan3 is shown in Figure 6. From Figure 6(a), when the die arrives at 12mm behind terminal stroke, deep drawing area starts to appear with splits and when die keeps on moving downwards, the splits become increasingly severe shown in Figure 6(b).

(a) 12mm behind terminal stroke. (b) terminal stroke.

Fig. 6. Plan3 (curve fitting) CAE result.

5. Manufacture Solution and 2 Plans Comparison

5.1. *Solutions to find accordance between actual manufacture and CAE analysis*

Due to severe splits with initial constant blank holder force when the die arrives at terminal stroke, blank holder force requires to decrease through deflating gas. The Nitrogen gas spring initial force and pressure is in proportion. The theoretical equation shown in Eq. (2) is established.

$$P_2 = \frac{P_1 \cdot F_{\mathrm{BHF}} \left(S_{\mathrm{nom}} - S_{\mathrm{used}} \cdot \left(1 - \frac{F_{\mathrm{in}}}{F_{\mathrm{end}}}\right) \right)}{N \cdot F_{\mathrm{in}} \cdot S_{\mathrm{nom}}} \tag{2}$$

Where S_{used} is the spring used stroke, S_{nom} is the nominal stroke for the spring, F_{in} is the initial force at current charging pressure, F_{end} is the spring force at full nominal stroke, P_1 is standard pressure, P_2 is required pressure, N is the quantity of the springs, F_{BHF} is the blank holder force provided by the spring at the used stroke S_{used}.

In this study, standard pressure is 11Mpa. According to trial test results, The designed blank holder force changes to 800kN and by using equation (2), the required pressure should be turned to 9Mpa. At the same time, the initial force changes to 57kN and the spring force at full nominal stroke turns to 91.2kN.

5.2. *The results of Plan2 and Plan3 are comprehensively compared*

The FLD results of two plans got from CAE and actual product FLD results are obtained. Detailed parameters are presented in Table 3, it is evident that curve fitting variable blank holder force plan is superior to obtain more accurate results.

Table 3. Comparison between two plans in terms of major and minor strain

Reference point	major strain/%			minor strain /%		
	Plan2	actual product	Plan3	Plan2	actual product	Plan3
1	1.86	2.1686	1.78	-1.1	-4.4388	-1.1
2	15.6	7.2134	14.4	-4.5	-0.7622	-4.3
3	10	16.5623	10	1.91	1.1517	1.98
4	17.3	12.3511	17.1	7.65	6.1389	7.39
5	12.7	11.9553	12.2	0	-0.2923	-0.2

6. Conclusion

This paper numerically and experimentally examined the fender stamping with variable blank holder force provided by Nitrogen cylinder system. It is shown that the CAE result with initial constant blank holder force condition is inaccurate. Based on the mechanism of force supplier, three variable blank holder force plans such as average value, linear approximation and curve fitting were introduced to improve the accuracy. Conclusion can be drawn that the real stamping process was exactly simulated with linear approximation and curve fitting plans. Theoretical formula of inflating or deflating Nitrogen was given to solve sheet splits in the actual products manufacture. Additionally, comparison between linear approximation and curve fitting was drawn. As a result, curve fitting variable blank holder force plan is superior to in terms of obtaining more accurate numerical results.

Acknowledgment

This work is funded by NSFC (Project 51275146).

References

1. L.L. Wang, Y.P. Zheng and G. Yan, Dawbeads force optimum based on orthogonality experiments, *Hot Working Technol.* **43**, 142 (2014).

2. M.A Ahmetoglu, G. Kinzel and T. Altan, Forming of aluminum alloys-application of computer simulations and blank holding force control, *J. Mater. Process. Technol.* **71**, 147 (1997).
3. E. Doege and N. Sommer, Optimie rung derniederhalter krafer beim tiefziehen rechteckig e teile, *Stahl Eisen* **103**, 139 (1983).
4. E.J. Obermeyer and S.A. Majlessi, A review of recent advances in the application of blank holder force towards improving the forming limits of sheet metal parts, *J. Mater. Process. Technol.* **75**, 222 (1998).
5. X.F. Liu, H.Y. Huang and H.R. Zhang, Automotive panels drawing mold design based on CAE technology, *Mach. Des. Manu.* **21**, 242 (2010).
6. W. Chen, Z.J. Liu, B. Hou and R.X. Du, Study on multi-stage sheet metal forming for automobile structure-pieces, *J. Mater. Process. Technol.* **187**, 113 (2007).
7. X. Sun, Front wall inner plate drawing die-face optimum design based on ETA/Dynaform, *New Techonol. Craft* **35**, 79 (2013).

The CSP Process Development of the Press Hardening Steel

J. C. Jin[a], H. R. Gu[b], Q. Q. Lu[c] and W. Zhang[d]

Technology center of Ma'anshan Iron and Steel Company, Ma'anshan, P.R. China

E-mail: [a]jinjiachun1988@163.com, [b]gohero1234@163.com, [c]172451373@qq.com, [d]zwniat@126.com

The press hardening technology is to heat the material to the austenitizing temperature and then stamp it. When finishing pressing, water is injected in the mold. It makes steel quenching into martensite structure which can get a high strength at room temperature, and not easy rebound. So in order to guarantee the martensite structure after quenching, a lot of alloy elements are added to ensure the material's hardenability. Therefore, the production cost is relatively high. In response to the present issue of high cost of press hardening steel, a more economical, more competitive press hardening steel was developed in this paper. This steel was designed based on C-Mn steel addition with Cr and B elements which can improve hardenability and Ti elements which can induce the good comprehensive performance and economy. And the rolling process was done on the CSP production line which is short, high efficiency and more energy efficient. The microstructure and mechanical properties of the steel before and after the hot forming were tested and analyzed. The results showed that the performance of press hardening steel produced by CSP process fully meet the requirements of national standards. Compared with the traditional process of press hardening steel, this steel's production cost is reduced significantly

Keywords: Compact Strip Processing (CSP); Hot stamping; Press hardening steel; Economy.

1. Introduction

Since the late 1970s, the international oil resources are in short supply. As the rise of oil price, the whole society has urgent demand for lightweight of automobile. The application of ultra-high strength steel with incomparable advantages such as high strength, high toughness, high forming precision will be an important way to realize auto lightweight [1-3]. However, the formability of ultra-high strength steel is worse than ordinary steel plate. The higher the intensity is, the more difficult the forming is. The original cold stamping process can no longer satisfy the future automotive industry's requirements of formability of ultra-high strength steel. In recent years, the press hardening technology of ultra-high strength plate had successfully solved the problem of high strength and poor ductility [4]. Therefore, many big steel mills pay more attention to the research of press hardening steel.

While the domestic joint venture has extensively used the press hardening parts, the domestic independent OEM ventures are beginning to design the new vehicles with press hardening parts. Therefore, the press hardening steel will have great profit margins.

At present, most press hardening steel is done through traditional rolling technology, the cost of which is relatively high. In this paper, it is the first time to develop the press hardening steel by the CSP process which is short process, high efficiency and energy

[a] Corresponding author.

saving. The press hardening steel developed by CSP process can greatly reduce the production cost, being a more competitive product at the same time.

2. Alloying Design

In consideration of the domestic and foreign steel mill's design experience of the press hardening steel, this steel was designed not only considering the good hardenability, good impact resistance and high tensile strength, a certain plastic toughness but also reduced cost. So this steel's alloying design was mostly based on the C - Mn steel, adding Cr and B elements for hardenability and economical Ti element for the comprehensive performance at the same time. Eventually the press hardening steel's composition is showed in table 1.

Table 1. Chemical composition (wt%).

Steel	C	Si	Mn	P	S	Cr	Als	B
22MnB5	0.2-0.25	0.1-0.4	1.0-1.5	≤0.05	≤0.05	0.5	≤0.05	≤0.05

3. The CSP Process

The new steel was manufactured by the process: Steelmaking - LF refining - RH refining-CSP continuous casting - CSP hot rolling - coiling – hot-rolled product (or cold-rolled binder) - pickle rolling - continuous annealing -product of continuous annealing. The detail parameters of hot rolling process are shown in table 2.

Table 2. Parameters of hot rolling process.

Parameters	Discharge temperature (°C)	Finishing temperature (°C)	Coiling temperature (°C)
Target	1000-1100	800-900	550-650

4. Results and Discussion

4.1. *Microstructure*

The sample of rolling steel was prepared by the method of linear cutting, its microstructure is shown in the figure below. The microstructure of the press hardening steel by the CSP process is shown in Figure 1.

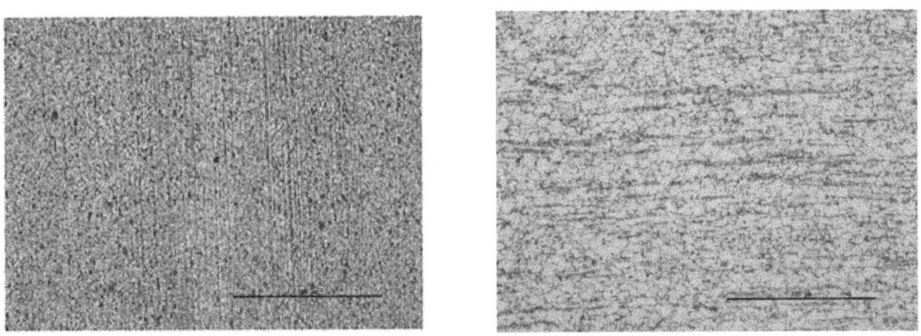

Fig. 1. Microstructure of the press hardening steel by the CSP process.

The rolling microstructures are main ferrite and a small amount of pearlite. And the microstructure distributes dispersively.

The cold rolling products were taken to press hardening line to press, the final parts were shown in the figure below. Select samples at different positions of the parts, observing the microstructure.

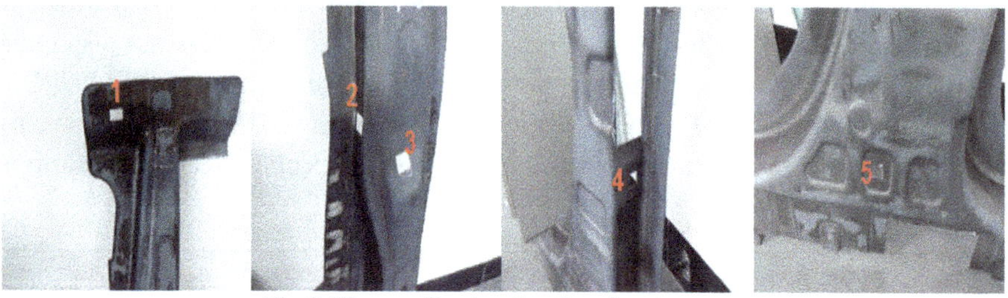

Fig. 2. The sampling location for microstructure.

The sampling locations of the press hardening parts are shown in figure 2. They were chose from the pressure parts, sidewalls and bottom of the press hardening parts. And microstructures of each are shown in the figure below.

Fig. 3. Corresponding microstructure.

Corresponding microstructures are shown in the figure 3. From the figure, it is obvious that microstructure had absolutely changed after the press hardening, from previous ferrite pearlite to lath martensite which reached the desired purpose of over 90%.

4.2. *Mechanical properties*

The sample of rolling steel was prepared by the method of linear cutting. The yield strength, tensile strength, ductility were tested and analyzed, the results are shown in the following table.

Table 3. Mechanical properties before press hardening.

No.	Rp0.2/MPa	Rm/MPa	A80/%
1	481.17	635.92	19.78
2	464.63	629.91	17.56
3	466.28	632.02	17.7
4	462.96	634.87	13.24
Target	320~630	400~800	>13

Mechanical properties before press hardening are shown in table 3. From the table, the yield strength mostly keeps at about 460 MPa, tensile strength keeps at about 630 MPa, but the elongation has a lager rang 13.24% to 19.78%. The minimum ductility has already been over the target 13%. In general, mechanical properties of the press hardening steel all meet the requirements.

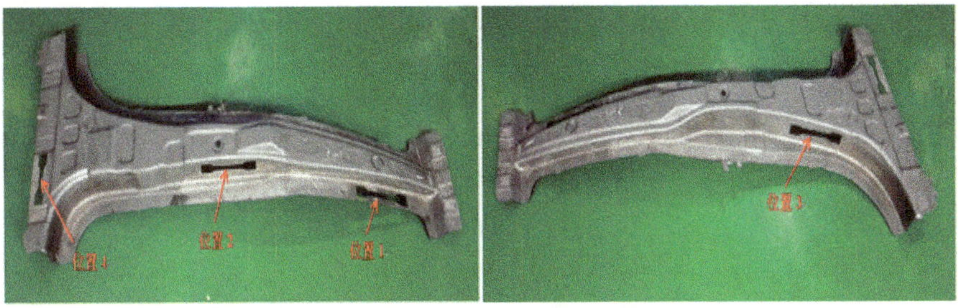

Fig. 4. The sampling location for mechanical properties.

The steel's mechanical properties after press hardening were tested in this paper, the sample locations are shown in figure 4.

Table 4. Mechanical properties after press hardening.

No	Rp0.2/MPa	Rm/MPa	A80/%
1	1014	1512.713	7.64
2	1080	1517.974	7.02
3	1163	1523.023	6.26
4	1085	1538.488	6.95
Target	950～1250	1300～1700	≥4.5

Mechanical properties after press hardening are shown in table 4. From the table, the strength increases obviously after press hardening, the yield strength increases from the previous 460 MPa to1100 MPa, the tensile strength increases from the previous 630 MPa to1500 MPa, the elongation keeps at about 7%. The results just match the change of martensite structure after press hardening. In general, mechanical properties of the steel after press hardening all meet the standards.

5. Conclusion

1) The press hardening steel by CSP process has been successfully developed by Masteel. It was designed mostly based on the C - Mn steel, adding Cr, B and Ti elements.
2) The rolling structures before press hardening are ferrite and pearlite. The yield strength is about 460 MPa, tensile strength is about 630MPa and the elongation is over the target 13%, which meet the requirements of mechanical properties.
3) The parts' structure after press hardening is 90% lath martensite. The yield strength is about 1100 MPa, tensile strength is about 1500MPa and the elongation is about 7%, which meet the standards.

References

1. J.F. Fan and M. Chen, The china early road of auto lightweight, *Casting* **10**, 995 (2006).
2. W.Y. Zhu, The development of the automotive lightweight and high strength steel, *Shanghai metal* **4**, 11 (2003).
3. H.Q. Chen, C.Y. Chen and L.Q. Wei, High strength steel plate and its application in automotive manufacturing, *Mold project* **8**, 88 (2007).
4. R. Kolleck, D. Steinhoefer, J.A. Feindt and P. Bruneau, Manufacturing methods for safety and structural body parts design for lightweight, *IDDRG, Conf. Proc.* **167** (2004).

Tribology & Tools

Wear Observations on Uncoated Tool and Workpiece Surfaces from a Full-Scale Press Hardening Wear Test

L. Deng[†] and M. Oldenburg

Division of Mechanics of Solid Materials, Luleå University of Technology, Luleå, Sweden
[†]E-mail: liang.deng@ltu.se

S. Mozgovoy, J. Hardell and B. Prakash

Division of Machine Elements, Luleå University of Technology, Luleå, Sweden

Tool wear in press hardening has attracted researchers' attention and recently several papers have focused on the wear observations in the laboratory tests mimicking the press hardening conditions. However, the wear observations and quantification in full-scale press hardening tests are rare in view of high costs and longtime requirements. In this work, the wear behavior in a full-scale press hardening wear test has been studied in order to understand the wear mechanisms occurring in the stamping tool. Furthermore, the wear depths in the real stamping tool were measured by using a coordinate measurement machine (CMM) that can provide the quantified wear result and serve as a base to validate the possible wear model. The present study also includes the wear observations in both the counterparts i.e., the blank (workpiece) and the stamping tool with the aim of studying the transfer material/wear particles. Two common tool steels were used in the full-scale press hardening wear test and the differences in the wear severities were observed. Since abrasion and adhesion are major wear mechanisms previously in many laboratory tests, the present study identifies the wear evolution and mechanisms on uncoated tool and workpiece surfaces in the press hardening wear test.

Keywords: Wear mechanisms; Shape deviations; Press hardening wear test.

1. Introduction

Tool wear occurring in the stamping tool not only degrades the production quality but also shortens the service life of the stamping tool because the wear particles generated by the rubbing between the tool and work-piece result in scratches and adhesions in the interface. Furthermore, since the high temperature softens the contacting surfaces and increases the oxide layers in the interface, the wear mechanisms become more complicated in press hardening conditions. There is a variety of papers that present the fundamental studies on the wear in the press hardening conditions, which performs laboratory tests at elevated temperatures to simulate the press hardening process due to the consideration of the high cost and time-consumption. Pin-on-disc test and slider-on-disc test are the main types of the laboratory test. However, setups and process parameters of the laboratory test determine the reproducibility of the press hardening process and the wear mechanisms behind. The effect of high temperatures on friction and wear was earlier studied through a pin-on-disc test [1]. It was found that the formation of compacted wear debris layers consisting of generated wear particles and protective oxide layers prevent the further wear subsequently and decrease the friction coefficient at elevated temperatures. A high temperature tribometer (HTT) was used to study the tribological behaviors in the press hardening conditions [2]. The severe adhesion, three-body abrasive wear and the absence of the protective oxide layers resulted in grooves and cracks in the boron steel strips. The different tribological behaviors in the laboratory tests motivate researchers to study what really happens in the stamping tool of a press hardening production line. The present paper focuses on a specific press hardening wear test designed for an accelerated wear process and presents the wear observations and the measurable wear result. Furthermore, obtaining the measurable wear result in the stamping tool can serve as a validation for the further study of tool wear prediction.

2. Method

2.1. *Full-scale press hardening wear test*

A specific press hardening wear test was performed in a production test line, which was designed to accelerate the wear process resulting in the reduced number of strokes. The setup and process parameters of the test and the blank material compositions can be found in [3]. The stamping tools referring to the upper and lower tools were manufactured by two commonly used pre-hardened tool steels. Their alloying compositions can be found in Table 1. The press hardening processes include heating, forming and cooling and these were consecutively performed in the tools for 200 times.

Table 1. Alloying compositions in mass contents in % and initial hardness (Fe makes up the balance).

Material	C	Si	Mn	Cr	Mo	V	Hardness (HRC)
Tool steel 1	0.38	0.30	0.75	2.6	2.25	0.9	52
Tool steel 2	0.32	0.6-1.1	0.8	1.35	0.8	0.14	45

2.2. *Coordination measurement machine*

A coordinate measuring machine (CMM), ZESIS Contura G2, was used to record the tool shape before and after the strokes and the deviations were regarded as the measurable wear result. There are 32 measured regions distributed evenly in the middle part and curved parts, as seen in Figure 1 (a). In each measured region, 7 measured positions distributed over the radius from 0 to 90 degrees and 4 measured positions evenly located 5 cm away forward and backward radii, as shown in Figure 1 (b).

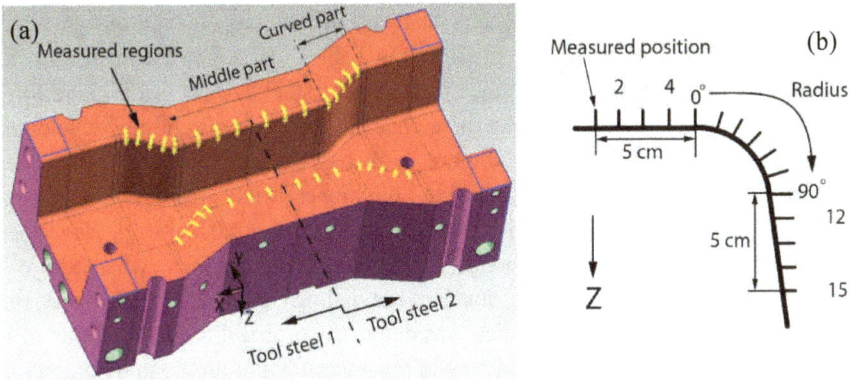

Fig. 1. (a) Illustration of the distribution of the measured regions in the upper tool; (b) distribution of the measured positions in each measured region.

3. Results and Discussion

Figure 2 illustrates a groove formed in the worn blank experienced by tool steel 2 in 192 sequence strokes. The occurrence of the groove was due to the loose wear particles at the interface and the possible adhesion built up in the counterpart. This was in agreement with the observation in the high temperature tribometer test [2]. The tensile straining when the blank was formed into the target shape caused the cracks inside the groove. The worn blank was associated with the evolution of the wear particles in the interface. According to [4], the initial adhesion takes place at asperities due to the formation of bonding junctions and increases in size by plastic deformation as sliding motion continues. During subsequent strokes, the built-up adhered fragment may break away due to plastic shearing and cause three-body abrasive wear to deepen the groove in the surface of the softer body.

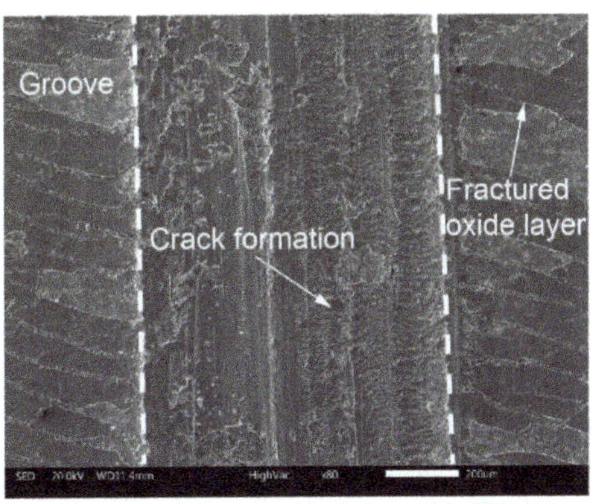

Fig. 2. SEM micrograph of the worn blank formed in 192 sequence stroke.

After 200 strokes, the wear scratches combined by the agglomerated wear particles (adhesion) were observed. These were concentrated in the upper tool radius. In the curved parts of the upper tool, the wear scratches distributed densely and some scratches connected in axial direction. However, the wear scratches located in the middle part of both tool steels were coarse and increased the deviation between the severe wear scars and the measured regions. The statistical analysis of the shape deviations of both the curved parts and the middle parts are presented in Table 2. Based on the mean values, the wear severity in tool steel 2 was higher than in tool steel 1 in both the adhesion and abrasion. This may be attributed to the differences in the hardness and the developing of oxide layer in the tool surfaces. Furthermore, tool steel 2 had a high risk of the adhesion built up in the back part of the tool radius and the highest value was 0.85 mm. A similar height of the adhesion was obtained in a hot stamping tool, 0.07 mm [5]. Nevertheless, there is no evidence to imply the break-away of adherent fragment when a threshold height is reached. Additionally, the standard deviations implied that the wear results varied in different measured radii, which was due to the probability in the motion of wear particles in the interface and the inaccuracy in the measurement. In the middle part of tool steel 2, the occurrence of the adhesion was dependent on the position of the measured region. This indicated a high probability of wear of tool steel 2 in terms of adhesion and abrasion. Figure 3 presents the shape deviations in the curved part, where the value in each measured position was averaged by the measured regions in the curved part. The observed wear scratches concentrated in the tool radius, which started with a deep groove in the front of the radius followed by the probabilistic adhesion. The shape deviations showed the significant adhesion took place in the curved part of tool steel 2. Since the shape deviations obtained from the tool steel 1 were almost negative values, it showed a low possibility of the agglomeration of the wear particles or the agglomerated wear particles built up in the abrasive groove did not exceed the nominally initial tool shape (reference line). According to the observation in the tool radius, an assumption of the wear evolution can be assumed that the wear particles generated in the front part of the radius and then the generated wear particles ploughed and built up in the rest of the radius as the sliding motion continues.

Table 2. Statistical results of CMM in the upper tool.

Unit: mm		Maximum	Mean	Std.
Tool steel 1	Adhesion	0.017	0.0062	0.0042
	Abrasion	0.031	0.0077	0.0070
Tool steel 2	Adhesion	0.085	0.021	0.017
	Abrasion	0.070	0.016	0.017

4. Conclusion

The present study has investigated the wear observations of the stamping tool used in a full-scale press hardening wear test. The groove and cracks in the worn blank was in agreement with the tribological behaviors in the high temperature tribometer. The shape deviations in the upper tool resulted in different wear severities in two commonly used tool steels in terms of presence and absence of the significant adhesion. The shape deviations in tool steel 1 indicated abrasive wear and a low probability of adhesion happened in the tool radius. Meanwhile, significant adhesion was detected in the upper tool radius in tool steel 2. The varying CMM results obtained in the middle part of tool steel 2 implied not only the probability in the amount of the adhesion but also the inaccuracy in the wear measurement. Since the present CMM measurement determined the measured regions before the test, the deviations between the severe wear scratches and the measured regions could lead to the inaccuracy, especially in the middle part of the upper tool where the wear scratches distributed coarsely. Furthermore, the combination of wear mechanisms is beyond the capacity of the present measurement method. The study of the tribological behavior in terms of plastic hardening layer and tribolayer in the worn stamping tool coupled with the present CMM results could deepen the understanding of tool wear.

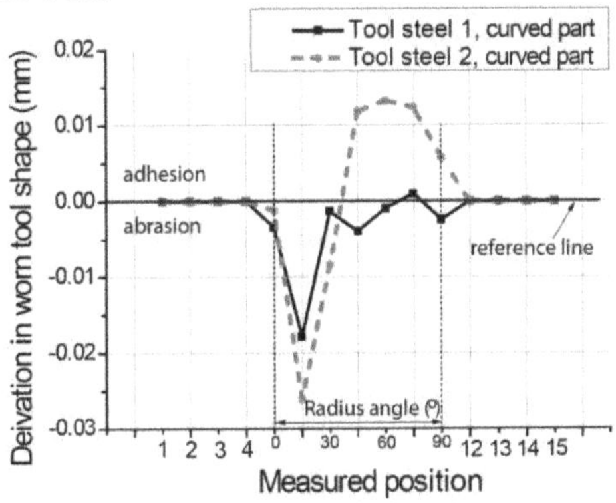

Fig. 3. Mean shape deviations of the measured radii in the curved part of the upper tool.

Acknowledgments

The authors sincerely thank the Swedish Governmental Agency for Innovation Systems VINNOVA for the financial support. The authors gratefully acknowledged SSAB EMEA AB, Oxelösund, Sweden, for providing the tool material and Gestamp HardTech AB, Luleå, Sweden, for performing the press hardening wear test.

References

1. S. Mozgovoy, J. Hardell, B. Prakash, L. Deng and M. Oldenburg, Effect of temperature on friction and wear of prehardened tool steel during sliding against 22MnB5 steel, *Tribol. Mater. Surf. Interfaces* **8**, 65 (2014).
2. S. Hernandez, Friction and wear phenomena in steels at elevated temperatures, PhD thesis, Luleå University of Technology (2016).
3. L. Deng, S. Mozgovoy, J. Hardell, B. Prakash and M. Oldenburg, Validation of tool-wear simulations based on a full-scale press hardening experiment, in *Proceedings of the 5th International Conference on Hot Sheet Metal Forming of High-performance Steel*, (Toronto, Canada, 2015).
4. K.H. Z. Gahr, Microstructure and wear materials, *Tribol. Ser. Elsevier* 1987.
5. J. Pujante, M. Vilaseca and K. Eriksson, Wear mechanism identification on hot stamping tools, in *Proceedings of the 3rd International Conference on Hot Sheet Metal Forming of High-performance Steel*, (Kassel, Germany, 2011).

Investigation of Galling Behaviors Between an Aluminum Alloy and Metal Forming Tool

Y. Hu, Y. Zheng, D. J. Politis and L. Wang[†]
Department of Mechanical Engineering, Imperial College London, London, SW7 2AZ, UK
[†]E-mail: liliang.wang@imperial.ac.uk

M. M. Gharbi
Schuler Pressen GmbH, Bahnhofstrasse 41, 73033 Göppingen, Germany
E-mail: mohanmad.gharbi@schulergroup.com

Hot stamping of Aluminum alloys is a process receiving increased interest from the automotive industry. However, during this process, galling, which has been recognized as one of the major drawbacks, limits the tool life and restricts the complexity and quality of the product. In order to improve the feasibility of this forming technology, an understanding of the mechanism of galling is necessary. The aim of the present study is to investigate the galling behaviours between AA6082 and two advanced anti-wear coatings, CrN and AlCrN. The study was conducted with pin-on-disc experiments at an elevated temperature. The main focus of the study is on the quantitative measurement of galling and identifying the material transfer mechanisms between Aluminum and tool surface.

Keywords: Galling; material transfer; Aluminum alloys; Coating.

1. Introduction

In the automotive industry, one of the main drawbacks of Aluminum sheet metal forming is the adhesion of Aluminum onto the tooling, which subjects the tool and the workpiece to a severe tribological condition. This material transfer gives rise to damage and break down of the functional surface (e.g. coatings) on the tool. Thus, adhesive wear, or galling, is one of the main reasons for reduced quality of the final product, especially on the product surface.

The adhesive wear of Aluminum alloy is generated from both physical and chemical interactions [1, 2]. The welding theory developed by Bowden [3], which indicates that the high transient temperature melt or softening of the material results in welding and transfer, and is the most widely accepted explanation for adhesive wear mechanisms. More detailed studies of material transfer for a single asperity include the plastic shearing model of successive layers of asperity contact [4] and the plastic wave model [5]. Based on detailed micro-scale observations, Janice et al. [6] divided the transfer process under multi-cycle process into two classes, primary and secondary transfer. Schipper et al. [7, 8]

also investigated the multi-cycle material transfer phenomenon and developed a series of bump growth models both on single asperity and macro scale levels.

In the present work, the adhesive wear mechanisms between Aluminium alloy and tool surface under multi-cycle sliding for two types of anti-wear coatings, namely, CrN and AlCrN, at 300°C, were studied. Parameters derived from bearing ratio curves were used to numerically analyse the changes of the coating surface properties after galling occurred and measuring the degree of the changes. The adhesive wear mechanisms of material transfer between Aluminum alloy and coatings were identified and discussed.

2. Experimental Setup

Pin-on-disc tests were conducted to study the adhesive wear behaviour. The Anton Paar high temperature (THT) tribometer was used and experiments were conducted according to the ASTM standards G40 and G99. A semi-spherical pin with radius of 2.5 mm was used. The materials and testing parameters used for this study are shown in Table 1. The surface roughness of the coatings was measured by Veeco white light interferometry (WLI). The adhesive wear phenomenon was studied using WLI, Hitachi scanner electron microscope (SEM) and Zeiss optical microscope.

Table 1. Testing parameters for adhesive wear studies.

Test No.	Pin Material	Disc Material (substrate+coating)	Normal force (N)	Velocity (mm/s)	Radius (mm)	Laps	Temperature (°C)
1	AA6082 T6	GJL+CrN	5	50	15	30	300
2	AA6082 T6	GJL+AlCrN	5	50	15	30	300

3. Results and Discussion

3.1. *Morphology of new coatings*

Micro coating defects were found on tested coatings and assumed to be the main reason for the high surface roughness, which is one of the initial triggers of adhesive wear and thus plays an important role in friction and wear mechanisms [6, 7]. The surfaces of coatings were untreated before testing to simulate the real forming condition. Table 2 lists the average surface roughness (R_a) for the coatings and shows that the CrN coating had higher surface roughness than AlCrN.

Table 2. Average surface roughness (R_a) of coatings before tests.

Test No.	Material	Average R_a(nm)
1	GJL+CrN	214.94
2	GJL+ALCrN	187.17

Two types of micro-defects, bumps and pits (shown in Figure 1), were identified on the coatings, which are evenly distributed on the surfaces. The pit defect was less likely to occur on both coatings, and hence had a small effect on wear. However, the size and amount of bump micro-defects were observed on both CrN and AlCrN coatings.

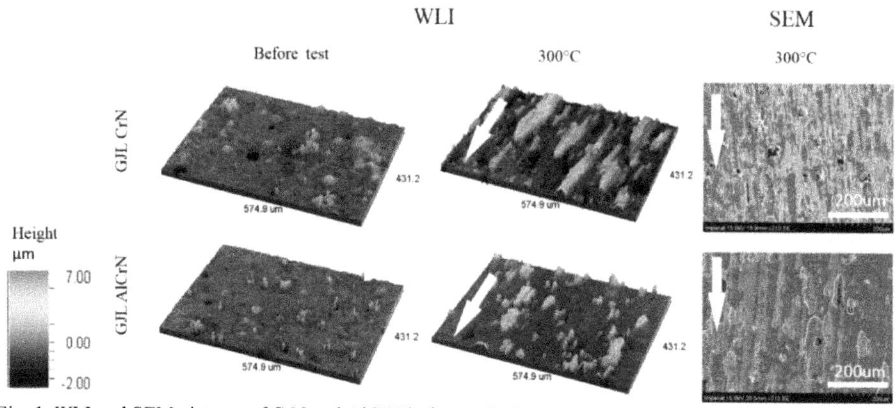

Fig. 1. WLI and SEM pictures of CrN and AlCrN before and after tests. Arrows represent pin sliding direction.

3.2. *Analysis of adhesive wear mechanisms between Aluminium alloy and coatings*

The material transfer on coatings can be clearly observed and measured by WLI due to the height difference between the original surface and bump defects. As shown in Figure 1, the high peaks in WLI pictures indicate Aluminum transfer as it accumulated along the wear track. The material transfer rate under constant operating conditions (transferred Al volume per unit area) was calculated by WLI and shown in Table 3, in which CrN showed higher material transfer rate than AlCrN. In SEM pictures shown in Figure 1, the light area indicates the transfer of Al on the coating surface and the dark area indicates the original coating surface. The larger area of Al transfer shown in SEM pictures compared to WLI indicates that low height Al patches were also widely formed in the sliding process.

Table 3. Aluminum transfer on different coatings.

Test No.	CrN	AlCrN
Material transfer rate (m^3/m^2)	26.3E-9	18.8E-9

Figure 2 shows the surface topography captured from worn AlCrN coating using an optical microscope. An ellipse is used to indicate the contact patch where Al transfer occurs. The full ellipse indicates the adhesive mechanism of bonding. The half ellipse indicates the transfer due to the micro-defect. The micro bump defect is represented by a square, which can be seen obstructing the movement of Aluminum particles or asperities in Figure 3.

Based on the research of Heinrichs et al. [6], two different Al transfer mechanisms can be identified in this figure: 1) Al transfer due to adhesive bonding to the tool material surface, which causes low height Aluminum transfer track; 2) Al transfer due to bump defect on the tool and wear particle entrapment, which leads to high Al bump (3-7μm). As observed in Figure 2, the second transfer phenomenon, which leads to high volume of

Al transfer and more likely activates further transfers, is always triggered by surface defects. Further transfer mechanisms are identified as: 1) overlaying of new layer of material to the existing bump and 2) damage activated transfer. A schematic illustration of the aluminum transfer mechanisms is shown in Figure 3.

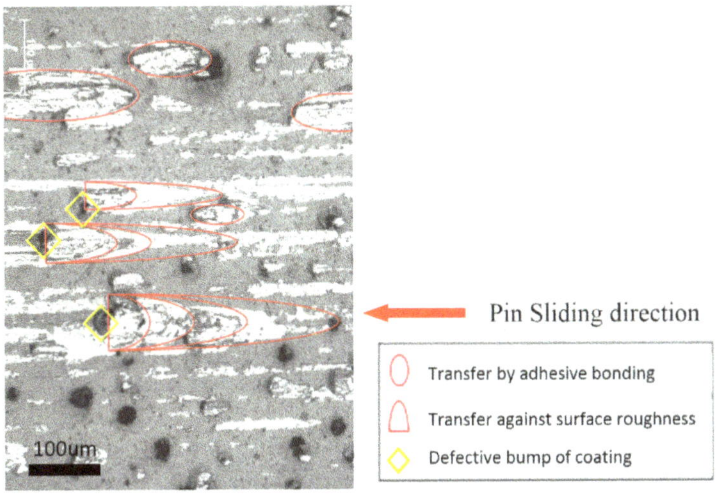

Fig. 2. The determination of main contact patches and mechanisms.

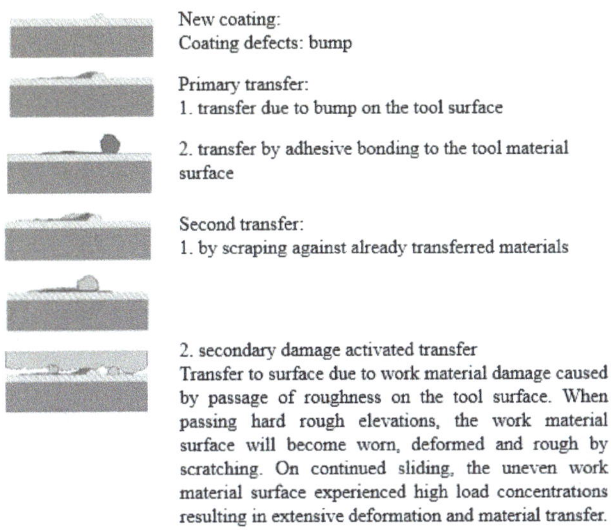

Fig. 3. Schematic illustration and description of the mechanisms of aluminum transfer on coatings.

4. Conclusion

In this study, ball-on-disc tests were conducted at 300°C to study the galling behaviour between an aluminum alloy and coated tool surfaces. It can be concluded that:

1) The Al transfer mechanisms of Al-coating tribo-system can be classified into 2 stages:
   a) Primary transfer due to micro surface defect and adhesive bonding;
   b) Secondary transfer due to scraping against already transferred material and the secondary damage activated transfer.
2) The Aluminum transfer is likely triggered by coating surface defect.
3) GJL+CrN and GJL+AlCrN show similar Al transfer mechanisms. The Al transfer rate on CrN is larger than AlCrN.
4) The wear particles entrapped and adhered on coating surfaces results in large size Al bumps, which may cause high friction, wear and adhesive bonding on the coating surface. Thus a smooth tool surface is recommended to reduce the possibility of large Al particle adhesion, which would lead to a product with good surface quality.

References

1. B. Bhushan, *Introduction to tribology*.(John Wiley & Sons, 2013).
2. J. Pujante, M. Vilaseca, D. Casellas and M. D. Rieral, The Role of Adhesive Forces and Mechanical Interaction on Material Transfer in Hot Forming of Aluminium, *Tribology Letters* **59**, 1 (2015).
3. F. P. Bowden and D. Tabor, *The friction and lubrication of solids*. (Oxford university press, 2001).
4. T. Kayaba and K. Kato, The analysis of adhesive wear mechanism by successive observations of the wear process in SEM, *Wear of Materials*, 45 (1979).
5. E. Schedin, Galling mechanisms in sheet forming operations, *Wear of Materials*, **179**, 123 (1994).
6. J. Heinrichs, M. Olsson and S. Jacobson, Mechanisms of material transfer studied in situ in the SEM: Explanations to the success of DLC coated tools in aluminium forming, *Wear of Materials* **292**, 49 (2012).
7. M. Rooij and D. Schipper, Analysis of material transfer from a soft workpiece to a hard tool: Part I—Lump growth model, *J. tribology* **123**, 469 (2001).
8. M. Rooij, G. van der Linde and D. Schipper, Modelling material transfer on a single asperity scale, *Wear of Materials* **307**, 198 (2013).

Experimental Platforms and Die Design for Hot Stamping of a Patchwork Blank

F. K. Chen[a] and T. H. Hung[b]

Department of Mechanical Engineering, National Taiwan University
Taipei, 10617, Taiwan
E-mail: [a]fkchen@ntu.edu.tw, [b]d99522038@ntu.edu.tw

C. H. Hung

Department of Mechanical Engineering, National Chiao Tung University
Hsinchu, 30010, Taiwan
E-mail: chhung@mail.nctu.edu.tw

T. B. Huang

Department of Mechanical and Computer Aided Engineering, St. John's University
New Taipei City, 25135, Taiwan
E-mail: tbhuang@mail.sju.edu.tw

T. R. Chen

Product Application Research Section, Iron & Steel Research & Development Department,
China Steel Co., Kaohsiung, 806, Taiwan
E-mail: t11c1@mail.csc.com.tw

The hot stamping of a patchwork blank is one of the forming processes being applied to manufacture automobile structural parts. In order to perform the finite element analysis for patchwork blank hot stamping process, the material behavior of blank and interface parameters between blank and tools are required in the finite element model. In this paper, the experimental platforms for obtaining interface parameters including coefficient of friction and interface heat transfer coefficient were established. Then, the finite element analysis model for patchwork blank hot stamping process was constructed. The forming mechanisms, stamping die design, and cooling system design were also examined in this study. Finally, the experimental interface parameters and stamping die design were applied to the finite element analysis for the hot stamping of a patchwork part.

Keywords: Hot stamping; Tailored properties; Patchwork hot stamping; Finite element analysis; Interface heat transfer coefficient; Coefficient of friction; Cooling channel design.

1. Introduction

In order to reach the concept of light weight vehicle, hot-stamping parts have been widely used in automobile structures. The patchwork blank hot stamping process is one of forming processes being developed in recent years. In the patchwork blank hot stamping process, the two blank sheets are first welded to a double layer blank, and then the

---

[a] Corresponding author.

patchwork blank is formed in the hot stamping process. It has the advantages of reducing the number of operation processes and achieving higher accuracy of part shape. However, because of the two welded sheets in the patchwork blank are hot stamped and quenched in the same process, the forming process design and cooling system design in the patchwork blank hot stamping process are more difficult comparing with the single layer blank hot stamping process. In this study, the experimental platforms required for obtaining the interface parameters presented in the patchwork blank hot stamping process were built to measure the coefficient of friction and interface heat transfer coefficient. The measured interface parameters were then applied to the finite element analysis to establish a complete finite element model for the simulations of the patchwork blank hot stamping process. Finally, the forming mechanisms, stamping die design, and cooling system design were studied in the paper.

The flow stress is the required parameter in the finite element analysis for the hot stamping process. Merklein et al. [1] have investigated the flow behavior of 22MnB5 boron steel at various temperatures, strain rates, and cooling rates. The other parameters required in finite element analysis for hot stamping are the coefficient of friction and the interface heat transfer coefficient. Giger et al. [2] have examined the coefficient of friction by the cup deep drawing test with various elevated temperatures of blank. Yanagida and Azushima [3] also investigated the friction behavior of boron steel by the hot flat drawing test. It shows that the coefficient of friction with lubricant is much smaller comparing with the coefficient of friction without lubricant at high temperature. Merklein et al. [4] have studied the interface heat transfer coefficient of 22MnB5 steel at different contact pressures. Hay et al. [5] also investigated the thermal contact resistance at different contact pressures by using the flat surface of the U-shape tool. In recent years, the hot stamping process that could manufacture a part with tailored properties has attracted much attention from the automotive industry. Merklein et al. [6] has classified four types of hot stamping processes that could result in tailored properties of a production part. George et al. [7] also developed a tailored properties hot stamping part by using heating and cooling tools die. However, in order to conduct the finite element analysis for tailored properties hot stamping parts, the flow stress behavior, the interface behavior, and the forming die design are required.

In the present study, the experimental platforms for obtaining interface parameters including coefficient of friction and interface heat transfer coefficient were established, and the finite element analysis model for the patchwork blank hot stamping process was constructed. The forming mechanisms, stamping die design, and cooling system design were then examined, and the results were applied to the finite element analysis for the hot stamping of a patchwork part.

2. The Material Properties

Hot stamping is a complex manufacture process in which the blank sheet is subjected to a thermal-mechanical mechanism in the forming process. In order to perform the finite element analysis for hot stamping, the material properties of 15B22 boron steel used in hot stamping were obtained by calculation from JMat/Pro software. The flow stress of 15B22 boron steel at strain rate $0.1s^{-1}$ with temperature range from 650℃ to 800℃ and at temperature 750℃ with strain rate range from $0.1s^{-1}$ to $10s^{-1}$ are shown in Figure 1(a) and Figure 1(b), respectively. The flow stress behavior of 15B22 boron steel with various temperatures and strain rates were applied to the finite element analysis for hot stamping process performed in the present study.

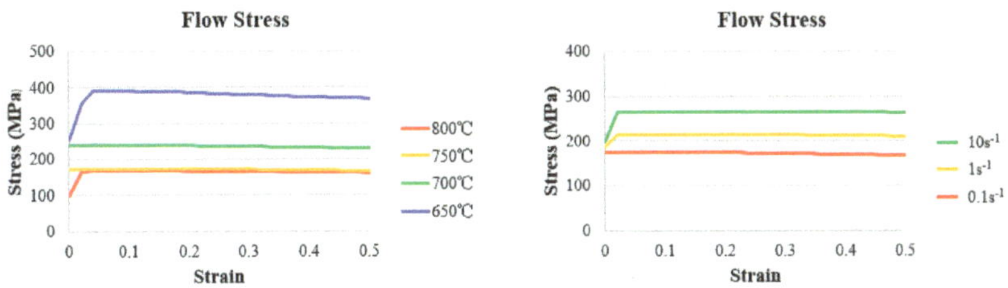

(a) (b)

Fig. 1. The calculated flow stress of 15B22 boron steel from JMat/Pro, (a) Flow stress with various temperature at strain rate $0.1s^{-1}$, (b) Flow stress with various strain rate at temperature 750°C.

3. The Experimental Platform for Measuring Coefficient of Friction

The coefficient of friction is one of the important parameters required in the finite element analysis for hot stamping. In this study, the original experimental platform design of coefficient of friction and modified experimental platform design of coefficient of friction were constructed. These two experimental platforms both have four major systems including sheet drawing system, clamping system, heating system and data acquisition system. The schematic diagram of original press jigs is shown as Figure 2.

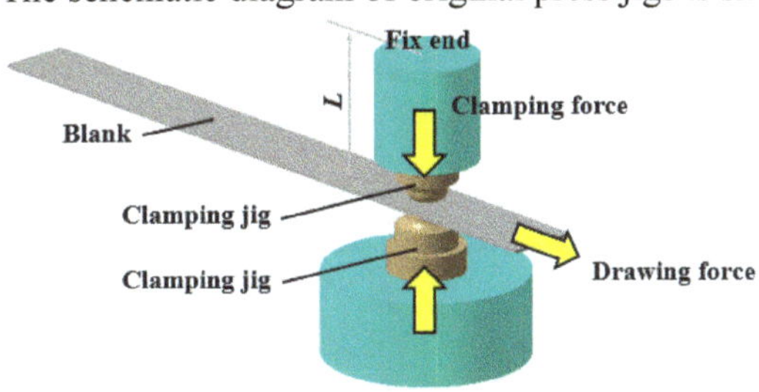

Fig. 2. The original clamping jigs design.

The contact surface between jigs and blank is a circular cross section. However, because the structure rigidity of this original experiment platform is not strong enough due to a large distance between the location of contact surface and the fixed end of the press structure, the force measurement oscillates and a stable result is not achievable during the experiment process. Moreover, due to the contact surface area and its shape not being well designed, the contact condition between blank and jigs is not uniform along the drawing path. For this reasons, the experimental platform was redesigned as the schematic diagram shown in Figure 3. The rigidity of platform structure is increased and jigs are designed for obtaining a uniform contact condition between blank and jigs. The coefficients of friction at various blank temperatures, contact pressures, and blank's surface conditions were also measured in this study.

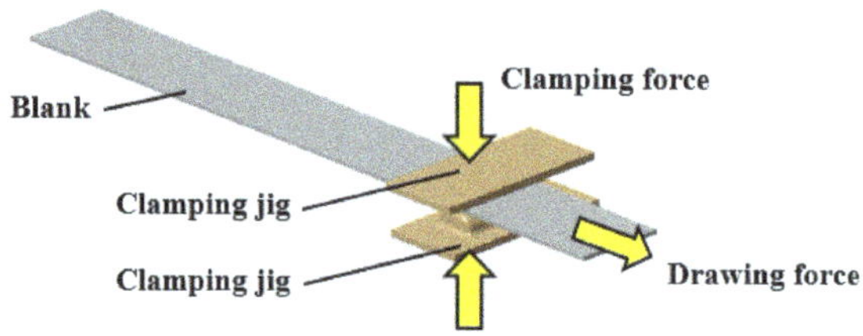

Fig. 3. The modified clamping jig design.

4. The Experimental Platform for Measuring Interface Heat Transfer Coefficient

The interface heat transfer coefficient is also an important interface parameter which is required in the finite element analysis for hot stamping. However, in the patchwork blank hot stamping process, there are two interfaces including blank-blank surface and blank-tool surface. So it requires measuring the interface heat transfer coefficients presented in these two interfaces during the hot stamping process. The schematic diagram of experimental platform of interface heat transfer coefficient is shown in Figure 4.

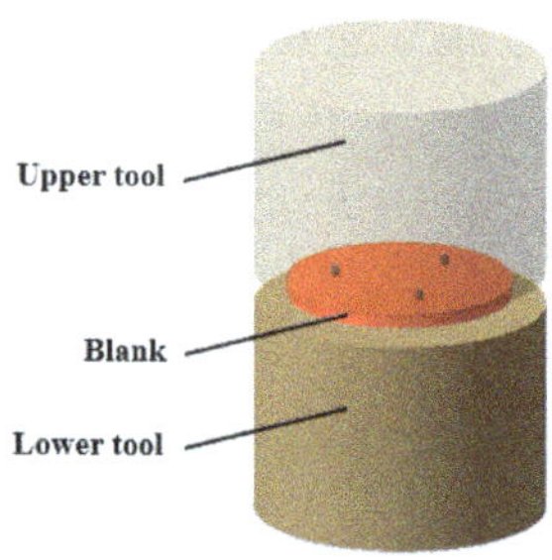

Fig. 4. The experimental platform for interface heat transfer coefficient.

The experiment was conducted in MTS-810 tensile test machine. The temperatures of blank and tools were measured by thermal couples inserted inside the blank and tools and recorded throughout the testing process. The interface heat transfer coefficients on the blank-tool interface at various contact pressure were first measured and calculated from temperature histories. Then the interface heat transfer coefficient on the blank-blank interface could be obtained by simulating the experiment for patchwork blank and comparing to the experiment results. These two interface heat transfer coefficients could be used in the finite element analysis for the patchwork blank hot stamping process.

5. Die Design and Finite Element Analysis for Patchwork Hot Stamping of a B-Pillar

The finite element analysis for patchwork hot stamping of a B-pillar was conducted in the present study. The schematic diagram of patchwork blank and stamping dies is shown in Figure 5.

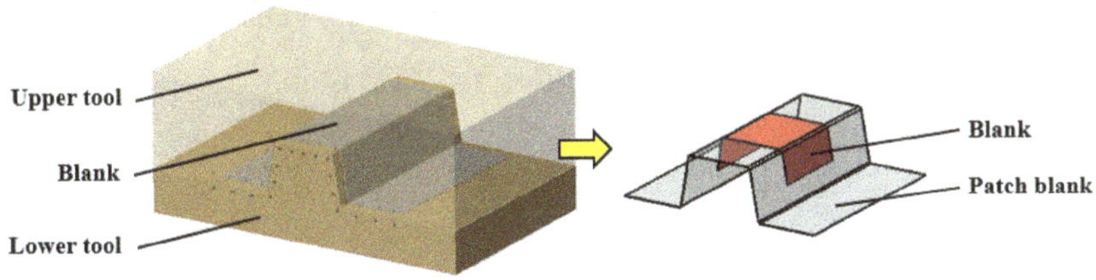

Fig. 5. The patchwork blanks and tools.

A patchwork B-pillar is a typical automotive structural part with tailored properties manufactured by hot stamping process. The forming defects commonly occurred in the hot stamping of a B-pillar are fracture and wrinkle presented at lower side of the B-pillar. The fracture defects are often occurred at the side wall of the B-pillar and the wrinkle defects are often displayed at the top surface of B-pillar, as the schematic diagram shown in Figure 6. In order to eliminate the forming defects, the deformation mechanisms presented in the drawing operation, forming (bending) operation, and forming with pad operation were examined. Moreover, the blank holder position and pad position for each forming mechanisms are also the key factors to eliminate the forming defects of B-pillar.

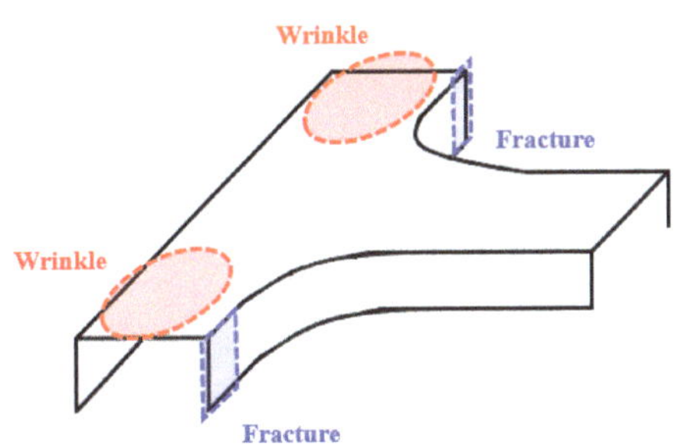

Fig. 6. The schematic diagram of the position of forming defects at T-shaped region of B-pillar.

The other issue encountered in the patchwork hot stamping of a B-pillar is the lower cooling rate due to much more heat to be transferred at the patchwork region. In order to improve the tool's cooling efficiency, the cooling system design at the patchwork region is different from that used in the single sheet area. In the present study, the cooling system design was developed by using a flat plate model as the schematic diagram shown in Figure 7. The design method was then extended to construct the die cooling system used in the patchwork hot stamping.

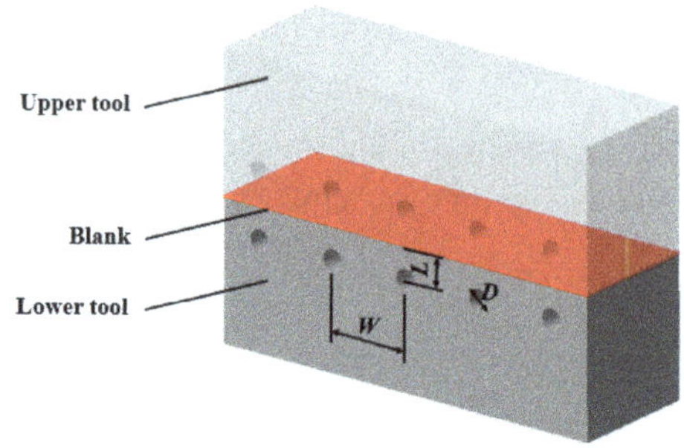

Fig. 7. The flat plate model for cooling system design.

6. Conclusion

In this study, the material properties of 15B22 boron steel has been obtained by calculation from the JMat/Pro software and used in the finite element analysis for hot stamping. The experimental platform of the coefficient of friction was constructed and the coefficients of friction of 15B22 boron steel were measured by the developed experimental platform. The interface heat transfer coefficients of 15B22 steel on blank-tool interface and blank-blank interface were also measured and calculated by the experimental platform and the finite element analysis. The finite element analysis for the patchwork blank hot stamping of a B-pillar was constructed in this study. The formability analysis and quenching analysis for patchwork blank hot stamping of a B-pillar were then conducted. The forming mechanisms and tool design for the patchwork hot stamping were also discussed. The forming defects as fracture and wrinkle were eliminated by stamping die design. Finally, the cooling system design for the patchwork hot stamping was developed and a complete patchwork hot stamping die design for manufacturing a B-pillar with tailored properties was established.

Acknowledgments

The authors wish to thank the Ministry of Science and Technology of Taiwan for the grant under project #MOST 103-2622-E-006-037, which makes this research possible. They are also grateful to ESI-France for the help in running the PAM_STAMP 2G software.

References

1. M. Merklein, J. Lechler and M. Geiger, Characterisation of the flow properties of the quenchenable ultra high strength steel 22MnB5, *CIRP Ann. Manuf. Technol.* **55**, 229 (2006).
2. M. Geiger, M. Merklein and J. Lechler, Determination of tribological conditions within hot stamping, *Production Engineering Research and Development* **2**, 269 (2008).

3. A. Yanagida and A. Azushima, Evaluation of coefficients of friction in hot stamping by hot flat drawing test, *CIRP Ann. Manuf. Technol.* **58**, 247 (2009).
4. M. Merklein, J. Lechler and T. Stoehr, Characterization of tribological and thermal properties of metallic coatings for hot stamping boron-manganese steels, in *Proceedings of the 7th International Conference on Coatings in Manufacturing Engineering*, (Chalkidiki, Greece, 2008).
5. B.A. Hay, B. Bourouga and C. Dessain, Thermal contact resistance estimation at the blank/tool interface: experimental approach to simulate the blank cooling during the hot stamping process, *Int. J. Mater. Form.* **3**, 147 (2010).
6. M. Merklein, M. Johannes, M. Lechner and A. Kuppert, A review on tailored blanks-Production, applications and evaluation, *J. Mater. Process. Tech.* **214**, 151 (2014).
7. R. George, A. Bardelcik and M.J. Worswick, Hot forming of boron steels using heated and cooled tooling for tailored properties, *J. Mater. Process. Tech.* **212**, 2386 (2012).

Modern Tool Steels for Long Life Time in Hot Stamping Application

D. Wang[†], E. Meurisse, R. Rahn and I. Schruff
Kind & Co., Edelstahlwerk, GmbH & Co. KG
Bielsteiner Strasse 124-130
D-51674 Wiehl, Germany
[†]*E-mail: Duan.Wang@Kind-co.de*

Hot stamping, including tailored tempering, is a highly efficient technology to produce high-strength car components. The process therefore requires tools with high mechanical properties such as compression strength, hardness, abrasive wear resistance and thermal conductivity to insure an outstanding performance. In the past, the focus of research was on the thermal conductivity in order to reduce cycling time and to achieve a martensitic microstructure. Considering the distance between cooling channel and tool surface has reduced significantly from approx. 25 mm to 6-10 mm in the last few years, the high thermal conductivity is no longer the key property of tools. As a matter of fact, the requirement of higher hardness of tools increased rapidly. The increasing hardness improves the wear resistance and furthermore increases the tool lifetime consequently. In Europe, instead of standard steels 1.2367, also CR7V-L is been used in wild range due to its high hardness and high wear resistance. Based on the success of CR7V-L and to adjust the market demand, Kind & Co developed recently a special tool steel HS1 especially for hot stamping application which shows higher hardness and higher wear resistance compared with the common tool steels on the market. In this report, the properties of the tool steel HS 1 for hot stamping process will be described firstly. The tool steels, which are suitable for tailored tempering process will be introduced subsequently. Finally risks about stress induced corrosion in the cooling channel will be discussed.

Keywords: Hot stamping; HS1; CR7V-L; Tailored tempering; Stress induced corrosion.

1. Introduction

The world range increasing demand of hot stamped components resulting in large production lots, extended run times and increased loads on the tools. Nowadays, hot stamping production is exposed to harsher using conditions, which force the production process to be more accomplished. As a result of decreased distance between engraving and cooling channels, the loads on the tools increase and consequently the requirements on the tool materials also improve. The selection of tool steels requires a plenary knowledge of various material properties of the tool steels.

2. Increased Loads on Hot Stamping Tools

In the past, the focus of research on hot stamping tools was to improve thermal conductivity. This is based on the demand of shorter cycle time during hot stamping process. More specific, the quenching period of hot stamping process can be reduced by using tool steels with higher thermal conductivity [1-3]. Nowadays, new design and modern tools provide the possibility to reduce the wall thickness – distance between engraving and cooling channel, even until 6-10 mm. With this kind of design, the time of quenching process has been reduced significantly, which means the thermal conductivity of tools is no longer the main factor. In fact, the increased loads on tool steels which coming along with the reduction of the wall thickness is under the spotlight of research now [4].

The increased loads on the remaining material between cooling channels and engraving surfaces make toughness an important property for tool steel. High toughness of steel prevents cracks initiation and growth.

Also abrasive wear resistance of tool steel plays an even more important role. This is because hot stamping products are becoming relative common and are now using everywhere in automobile industry. Up to one million pieces are required on one series and therefore the wear resistance of tool steel must be significantly improved for long lifetime production. Alloy element of tool steels and higher hardness can assistance the tools reach a longer lifetime.

Corrosion has become another critical load in hot stamping application. The possibility of cracks appearance on tool steels increased rapidly since the distance between cooling channels and engraving surfaces decreased as well as the using hardness of tool steels increased. More information about corrosion will be introduced in chapter 4.

Another new trend in automobile industry is to produce hot forming component with different strength and toughness. Tailored tempering is one of the most popular processes for this kind of products. During the process tailored tempering, the single hot stamping components is heated up to 550°C, under this circumstance, a long-time tempering resistance is also required on tool steels [5, 6].

3. Tool Steel Introduction

In this chapter, steels for hot stamping application and steel for tailored tempering application will be introduced. Chemical composites are described in Table 1, tempering properties and mechanical properties of tool steels are given in followings.

Table 1. Chemical compositions and working hardness of recommended tool steels.

Steel Designation		Mass Content in %							Recommended Hardness
Mat.-No.	Brand name	C	Si	Mn	Cr	Mo	Co	V	HRC
1.2344	USD	0.40	1.00	0.40	5.20	1.30	---	1.00	48 - 52
1.2367	RPU	0.38	0.40	0.40	5.00	2.80	---	0.60	50 – 52
---	CR7V-L	0.42	0.50	0.40	6.50	1.30	---	1.00	52 - 54
---	HS1	0.50	0.90	0.80	8.00	1.50	---	1.70	54 - 58
---	HMoD	0.45	0.30	0.40	4.50	3.00	4.50	2.00	50 - 52

3.1. *Steel grad for hot stamping applications*

The general applicable hot-work tool steels for hot stamping application are the steel grades 1.2344 & 1.2367 after international standard [7]. Due to the increasing steel requirement on higher hardness and higher wear resistance, 1.2344 and 1.2367 steels can no longer satisfied the demand. In this circumstance, Kind & Co developed special steel grade CR7V-L. CR7V-L special hot-work tool steel was developed base on the standard chromium-molybdenum-vanadium alloyed hot-work tool steels. Instead 5 % chromium content in 1.2367, CR7V-L contains 6.5 % chromium for higher wear resistance. And instead 0.38 % carbon content in 1.2344, CR7V-L contains 0.42 % carbon for higher hardness. The improved chemical compositions allow CR7V-L tool steel to provide a better hardness and wear resistance at the same time and is especially suitable for the application of hot stamping.

Table 1 shows the chemical components of different tool steel grades for hot stamping application. Fig. 1 compares the tempering property of CR7V-L and international standard tool steels 1.2344 and 1.2367, it is to notice that the maximum secondary hardness of CR7V-L is 56.5 HRC, which is the highest of these 3 different steels. The CR7V-L also reaches generally higher hardness than 1.2344 and 1.2367 under different tempering temperature.

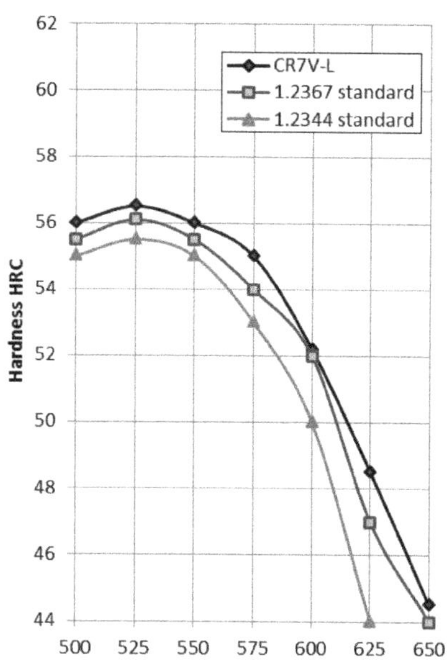

Fig. 1. Tempering diagram of CR7V-L, 1.2367 and 1.2344 tool steels.

Fig. 2 illustrates the high temperature strength of CR7V-L special tool steel. CR7V-L shows a promising high temperature strength, at 500°C for example, the tensile strength Rm still reaches 970 MPa, which at this temperature, 930 MPa for 1.2344 and 950 MPa for 1.2367 [8].

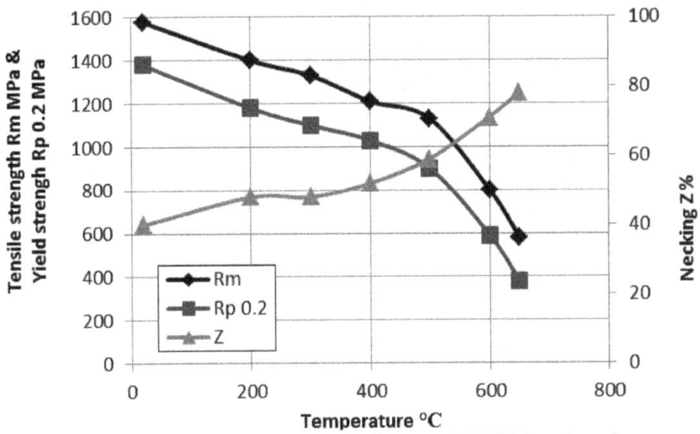

Fig. 2. High temperature strength diagram of CR7V-L tool steel.

HS1, for even higher requirements on hardness and wear resistance especially due to the high lifetime demand, new special hot-work tool steel grade HS1 was been recently developed. These properties are derived from a well-balanced alloy components ration-namely Mo, V and Cr combined with higher carbon content. With the increased carbon content, this steel grade can be used in hardness up to 58 HRC. With the increased vanadium content, the steel provides more fine carbides for better wear resistance. Despite the high hardness, the homogenous microstructure and toughness of HS1 have also been promised by utilizing ESR electro-slag remelting process.

Figure 3 shows the hardness of HS1 after tempering for two different hardening temperatures: 1050°C and 1080°C (quenching in polymer, tempering from 100°C to 650°C, tempering time 2 × 2 hours). In general, there is about 2 HRC difference between these two tempering curves from the two different hardening temperatures. To be more specific, the hardness directly after quenching is 60 HRC and 62 HRC respectively. Also, after tempering at temperature 500°C, HS 1 reached the maximum secondary hardening 58 HRC and 59 HRC separately.

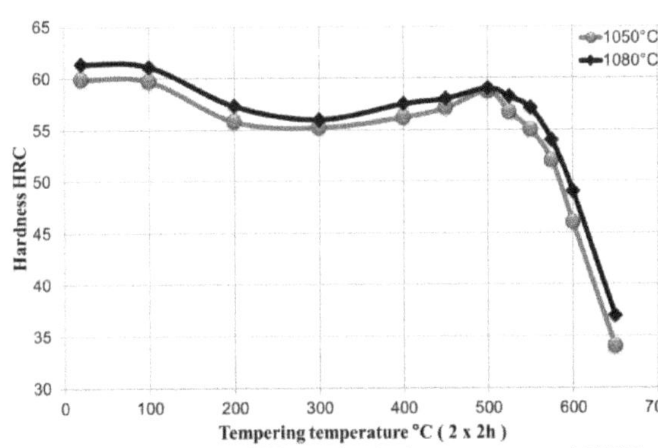

Fig. 3. Tempering diagram of HS1 from two different hardening temperature 1050°C and 1080°C. Quenching in polymer.

Figure 4 shows the microstructure of HS1 at 55 HRC after hardening at 1050°C (in polymer, holding time 45 minutes) and tempering at 550°C (2 × 2 hours) in air. The picture proofs a homogenous martensitic microstructure of HS1, which is as a result of ESR electro-slag remelting process.

Fig. 4. Microstructure of HS1. Grain size Synder Graff 7.

Research from Voss [9] summarized an index to calculate the hot wear resistance Lk. A higher Lk value represents a better wear resistance.

$$Lk = 2 \times (\%Cr) + 5 \times (\%W) + 10 \times (\%Mo) + 40 \times (\% V) \quad (1)$$

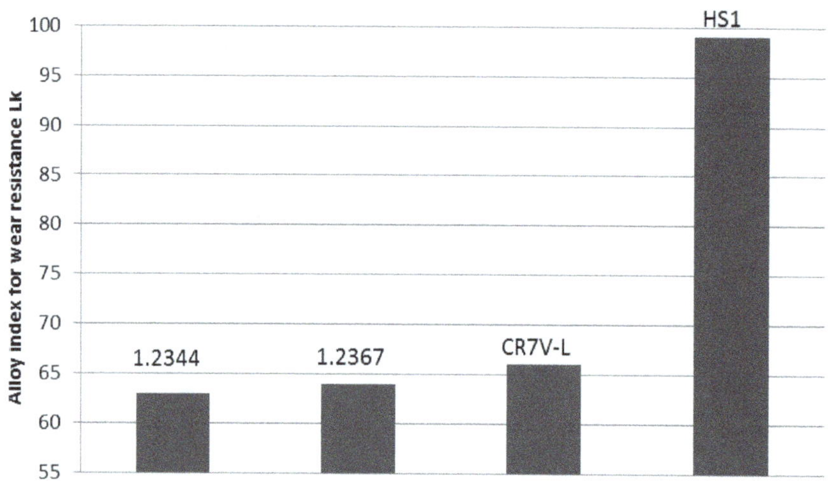

Fig. 5. Alloy index Lk for different tool steels 1.2344, 1.2367, CR7V-L and HS1.

Figure 5 compares the wear resistance based on alloy index Lk of four different steel grades: 1.2344, 1.2367, CR7V-L and HS1. HS1 and CR7V-L special hot-work tool steels exceed standard tool steel 1.2344 and 1.2367 on wear resistance by far.

Table 2 summarized the thermal conductivity of different material at different temperatures. It is to observe that despite the higher wear resistant, HS1 has similar thermal conductivity to standard steels 1.2344 and 1.2367.

Table 2. Thermal conductivity of different material in W/m×K.

Mat.-No.	Brand name	20°C	200°C	400°C	600°C
1.2344	USD	25.5	27.1	27.7	27.6
1.2367	RPU	29.9	29.7	30.1	29.7
---	CR7V-L	23.8	27.3	28.2	28.7
---	HS1	26.6	27.7	30.9	30.5

Based on all the information above, for hot stamping application, both CR7V-L and HS1 bring longer life time than standart steels. HS1, due to its special chemical components and melting process, can fulfill even stricter using demand and lifetime.

3.2. *Steel grad for tailored tempering applications*

Tailored tempering is a process for hot forming components with precisely defined zones and different strength or ductility [10, 11]. In this process, the tool segments for tailored tempering process are heated up until 550°C and are held constantly on this high temperature level for the complete production shifts. Due to the addition of cobalt, HMoD possesses higher long-time tempering resistance and high temperature strength. The high concentration of vanadium also results in an excellent wear resistance. This composition makes HMoD highly suitable for tailored tempering segments.

HMoD has been especially recommended for tailored tempering use due to its extraordinary long-time tempering resistance at tailored tempering applied temperature 550°C.

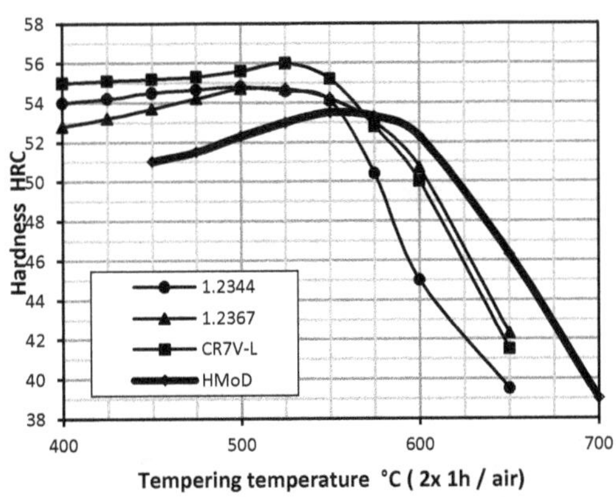

Fig. 6. Tempering diagrams of different tool steels: 1.2344, 1.2367, CR7V-L and HMoD.

Fig. 6 compares the tempering curves of 1.2344, 1.2367, CR7V-L and HMoD steels. Despite the lower maximum secondary hardness of HMoD, the temperature of maximum secondary hardness of HMoD is relative high compared to other steels. This benefits the long-time tempering resistance at 550°C as shown in Fig. 7. Fig. 7 compares the long-time tempering resistance property of CR7V-L and HMoD at 550°C, 53HRC. After 100 hours tempering, there is only 1.5 HRC reduced on HMoD.

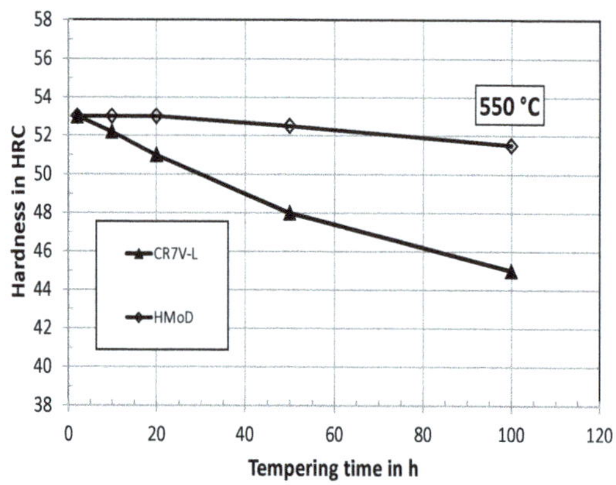

Fig. 7. Long – time tempering resistance of CR7V-L and HMoD tool steels.

4. Corrosion Behavior of Tool Steels for Hot Stamping

It is the trend nowadays in hot stamping industry that the distance between cooling channels and working surface is reduced in order to reduce cycle times [5, 6]. This leads to increased stresses in the remaining material between cooling channel and working surface. In connection with corrosion in the cooling channels due to uncontrolled cooling water, the chance of cracks appearance increased rapidly. It has been observed frequently in these tools corrosion started in the cooling channels und further lead to cracks and water leakage from the cooling channels to the working surface (Fig. 8).

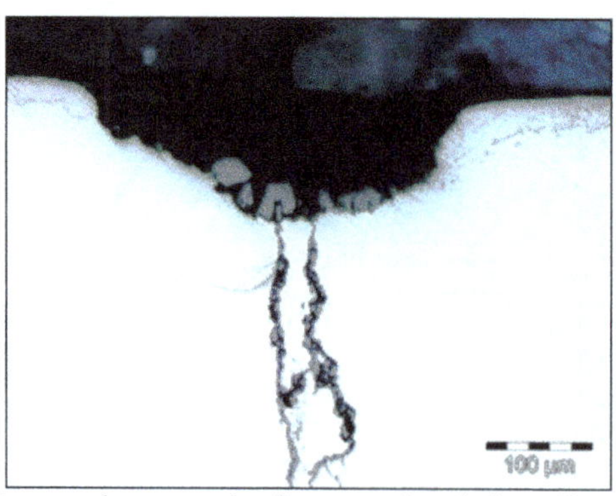

Fig. 8. Corrosion induced stress cracks propagating from corrosion pits in cooling channels to the tool surface.

Three different types of corrosion have been observed and shown in Fig. 9:

(1) General corrosion,
(2) Localized corrosion similar to pitting,
(3) Corrosion induced cracking.

Research from R. Rahn and I. Schruff established several corrosion tests on different hot-work tool steels in simulated cooling water [6]. The results show: the steel 1.2344 appeared to be rather resistant against general corrosion and corrosion induced cracking and the steel 1.2367 did not develop any general corrosion but is sensitive to localized corrosion. On the other hand, special grade CR7V-L is nearly free of general corrosion but react strong with corrosion pitting. With higher hardness and temperature it is even more sensitive to corrosion induced cracking [6].

General corrosion Localized corrosion Corrosion induced cracking

Fig. 9. Microscopic appearance of different corrosion defects.

There is no research on corrosion behavior of steel grade HMoD since it is not intended for use in water cooled tool inserts.

Another factor which influences corrosion is the tempering temperature. If the tempering temperature is approximately 50°C above the temperature of secondary hardness maximum, the carbide nuclei will start to coagulate. This carbides grow-up requires detracting the chromium from their surroundings and further impoverishing this area from chromium. This ended up with a higher local sensitivity to corrosion.

As a result, cooling water controlling, or more specific, closed cooling water system with anti-corrosion additive is a vital issue for the using of hot-work tool steel at high hardness level. If cooling water controlling is not possible, the hardness of tool steel during manufacturing hot stamping components needs to be reduced from secondary hardness maximum to avoid stress induced corrosion and crack.

5. Conclusion

(1) Standard tool steel grades 1.2344 & 1.2367 for hot stamping applications can no longer satisfy the long-lifetime demands in many cases. Thus, we developed CR7V-L and HS1 special tool steels for more specific applications. For hot stamping application, both CR7V-L and HS1 bring longer lifetime than standard steels due to their higher hardness and better wear resistance. HS1, with its special chemical components and melting process, can fulfill even stricter using demand and lifetime.

(2) For tailored tempering it is to recommend using HMoD steel, which has an extraordinary long-time tempering resistance property by contains more cobalt content.

(3) Corrosion or more specifically corrosion induced cracks reduces the lifetime of tools. Water controlling is in this case important. If the water controlling is not possible, it should be avoid to tempering tool steels around the temperature of secondary hardness maximum.

References

1. I. Valls, M. Alsmann and U. Paar, Capital Role of Tool Materials in Hot Stamping of 22MnB5 Sheet, in *Proc. 8th Internat. Tooling Conf.*, (Aachen, RWTH Aachen University, Germany, 2009).
2. S. Gelder and G. Jesner, New high performance hot work tool steel with improved physical properties, in *Proc. 9th Internat. Tooling Conf.*, (Leoben, Montanuniversität Leoben, Austria, 2012).
3. E. Meurisse, C. Ernst and W. Bleck, Improvement of thermal conductivity of hot-work tool steels by alloying design and heat treatment, *Proc. 9th Internat. Tooling Conf.*, (Leoben, Montanuniversität Leoben, Austria, 2012).
4. O. Zöllner, Optimierte Werkzeugtemperierung. Anwendungstechnische Informationen ATI 1104, (Bayer AG, Leverkusen, Germany, January 1999).
5. R. Rahn and I. Schruff, The selection of tool steels for hot-stamping tools with respect to increased loads, in *Proc. 5th Internat. Conf. Hot Sheet Metal Forming of High-Performance Steel*, (Toronto, Canada, May 31st - June 3rd, 2015).
6. R. Rahn and I. Schruff, Modern Tool Steels for Hot Sheet Metal Forming Applications, in *Proc. 4th Internat. Conf. Hot Sheet Metal Forming of High-Performance Steel*, (Lulea, Lulea University of Technology, Sweden, 2013).
7. DIN EN ISO 4957, *Werkzeugstähle, Beuth Verlag*, (Berlin, Germany, 2001)
8. K. Katalog, C.E. GmbH, C. KG, www.kind-co.de/kc_2010/pdf/downloads _prospekte / Werkstoffkatalog_2016.pdf.
9. H.Voss and E.Wetter, Verschleißverhalten von vergütbaren Gesenkstählen, *Archive für das Eisenhüttenwesen* **5**, 379 (1967).
10. S. Sikora, Tailored Tempering–Einstellung lokaler Festigkeitseigenschaften warmumgeformten Bauteilen, *HTM J. Heat Treatm. Mat.* **66**, 323 (2011).
11. B.T. Tang, Numerical modelling of the tailored tempering process applied to 22MnB5 sheets, *Finite Elements in Analysis and Design* **81**, 69 (2014).

Hot Stamping Die Rapid Manufacturing Process Based on Adopted Casting Without Pattern Technology

Z. D. Shan[†], Y. S. Ye, F. Liu and Q. T. Yan

*State Key Laboratory of Advanced Forming Technology and Equipment,
China Academy of Machinery Science & Technology,
Beijing, China*

[†]E-mail: shanzd@cam.com.cn

Hot stamping casting die is a mold manufactured based on patternless casting technology and equipment. In this process, pre-bending conformal cooling pipes need to be fixed in the sand mold before casting a die base. Compared with the drilling process, the conformal cooling pipes in the casting die fits better with the die cavity in the die, resulting in a higher cooling capacity. This paper puts forward a kind of hot stamping die rapid manufacturing process based on adopted casting without pattern technology. In this research, graphite cast steel was used as the casting material. 20 steel pipes were selected as the preset pipes, and zircon sand was filled in the pipes. The casting temperature was set as 1420 °C, which can guarantee the close binding of the preset pipes and the mold base without penetration. Furthermore, the application of patternless casting technology and equipment may help avoid the wood mold manufacturing link, resulting in a shorter casting sand mold development cycle and a lower manufacturing cost. A hot stamping die for motor bracket was trial-manufactured using the above manufacturing process. There were no obvious casting defects in the die and preset cooling pipes were accurately positioned. The sand cores could be discharged through blowing. The mechanical properties of the stamping part could meet the requirements.

Keywords: Hot stamping; Casting die; Without pattern casting technology; Rapid manufacturing.

1. Introduction

The stamping die for ultra-high strength steel hot stamping process is the key factor to form and quench the part. Cooling pipes are designed in the die to transmit the cooling medium and rapidly cool the die. Since the drilling process could only machine straight holes, for the hot stamping die with complex shape, the cooling pipes and the die cavity cannot be fully fit even the block design is adopted. The casting mold used in this paper could pre-bend the conformal cooling pipes according to the mold cavity shape, and set the conformal pipes in the sand mold before casting the die base. This process has the advantages of higher conformal degree of the pipes, lower difficulty for pipe design and machining, and shorter manufacturing cycle [1-3].

There were several problems in the design and manufacturing process of the hot stamping casting mold, such as lacking of conformal pipe design method, pipe deformation and penetration. This paper puts forward a kind of hot stamping die rapid manufacturing process based on adopted casting without pattern technology, and studies the sand mold manufacturing and casting process to solve these problems. The casting sand mold was rapidly manufactured using a digital precision forming machine without pattern casting, which avoided the wood mold manufacturing link in traditional casting process, resulting in a shorter manufacturing cycle and a higher dimensional accuracy of the castings. A hot stamping die for motor bracket was trial-manufactured using the manufacturing process. Fig. 1 shows the design and manufacturing process for the hot stamping casting mold. First, conformal cooling pipes need to be pre-bended according to the mold cavity shape; second, the pipes were set in the sand mold before casting the die base; last, hot stamping casting mold with conformal cooling system was directly casted. The digital without pattern casting technology and equipment can also be used in the rapid development of diesel engines for cars and ships.

(a) Conformal cooling system; (b) Without pattern precision forming; (c) Casting.
Fig. 1. Design and manufacturing process for the hot stamping casting mold.

Even the manufacturing process of hot stamping casting mold can overcome the deficiencies of the drilling process, there are still many difficulties in its implementation. Alloy steel need to be used to cast to guarantee the quality of the hot stamping die. However, the casting temperature of alloy steel is high, making the castability relatively poorer. For the conformal cooling pipes, they were required to be high temperature resistant when casting, and with good thermal conductivity when stamping. Then, the main difficulty for the hot stamping casting mold was the deformation or localized melting of the conformal cooling pipes in the casting process. The pipes may deform when surrounded in high temperature molten steel, having the defects of deviation from the embedded position or pipe wall collapse. Sometimes, localized melting can be found, even penetration. Therefore, one of the key technical problems in the hot stamping casting mold was to closely integrate the pipes and the casting under the premise of guaranteeing the pipe structure.

2. Simulation Analysis of the Casting Process for Hot Stamping Die of Motor Bracket

The manufacture of without pattern casting hot stamping die not only requires adopting advanced and reasonable casting technology and equipment, but also requires the die structure design meeting the requirements of casting. In this paper, the casting process of a die for motor bracket in a car was studied [4].

The structure and the casting process of the hot stamping die of motor bracket were optimized. The die cavity was designed with wall structure of equal thickness. In this way, both the number of hot spots and the use of molding material can be reduced. With fewer hot spots, the cooling of the mold would be more homogeneous, resulting in a fewer casting defects and higher casting quality. With decreased use of molding material, less total heat would be flowed into the mold during the casting process, making the pipes to absorb less heat, resulting in a lower peak temperature in the pipes, which can prevent melt-through and penetration in the pipes [5-6].

The "Visual Cast" Module in software "Visual Environment" was used to conduct the numerical simulation analysis to guarantee the quality of the casting mold and its service life. The temperature field and the stress field of the monoblock die in the casting process were simulated to optimize the casting process.

The monoblock die of motor bracket used graphite cast steel as the casting material, 20# low carbon seamless steel pipes as the preset pipes (Φ 18×4 mm), silica sand for virtual mold box and zircon sand for pipe cores. The heat exchange coefficient between metal and sand mold was set as 500 W/m^2•k, that between metal and metal was 2000W/m^2•k, that between sand mold and ambient air was 10 W / m^2•k, and that between pipe wall and cooling medium under forced convection condition was 100 W/m^2•k. The suitable casting temperature for graphite steel is 1320 ~ 1420 °C, and the paper adopts a casting temperature of 1420 °C for simulation analysis, with casting speed set as 19.2Kg/s.

Temperature field cloud chart of the mold base and the conformal cooling pipes at 3600 s was obtained through the numerical simulation analysis (as shown in Fig. 2).

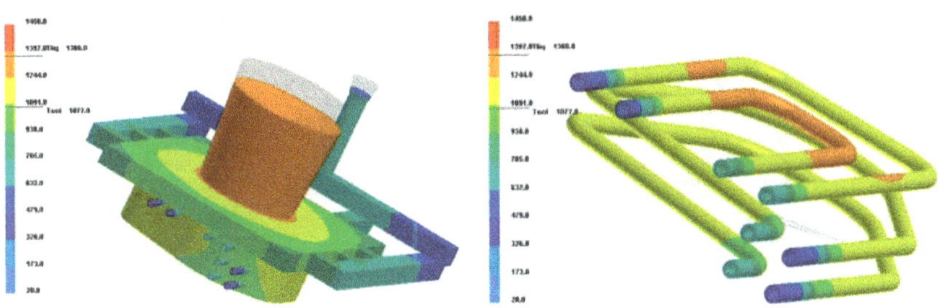

Fig. 2. Temperature field of the mold base and the conformal cooling pipes at 3600 s.

It can be seen from Fig. 2 that partial positions were solidified in the mold at 3600 s, the riser was in the solid-liquid two-phase temperature interval, and temperature of the cooling pipes was further decreased. However, the temperature of the pipes at the bottom of the riser was higher than other places.

The solidification time cloud chart can be obtained through the analysis of solidification time of the monoblock die of motor bracket (as shown in Fig. 3). It can be seen from Fig. 3 that at the time of 7200 s (2 hours), the die base was completely solidified except for the riser. Vertical runner, horizontal runner, and the border of base plate in the die all started to solidify at the time of 600 s.

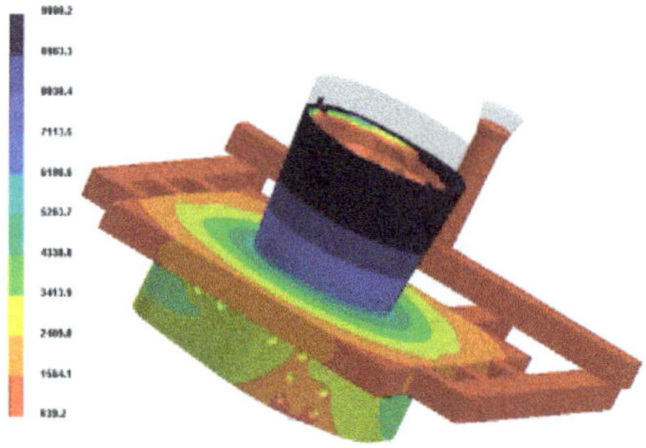

Fig. 3. Solidification time cloud chart of the die.

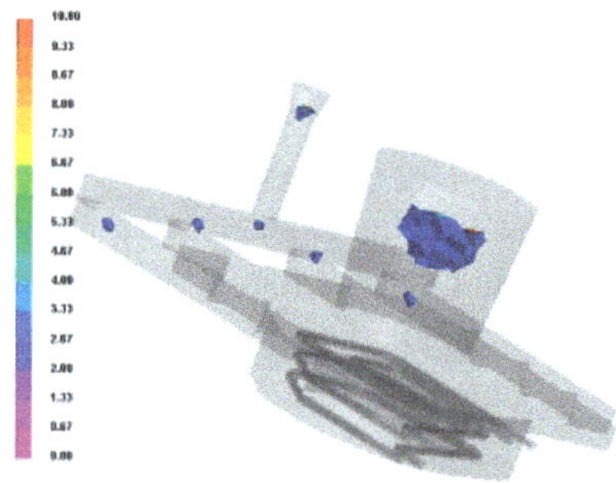

Fig. 4. Shrinkage cavity and porosity cloud chart of the die.

Shrinkage cavity and porosity cloud chart can be obtained through the predictive analysis of the defects in the monoblock die of motor bracket (as shown in Fig. 4). It can be seen from Fig. 4 that the number of shrinkage cavity and porosity in the die was relatively fewer, and all shrinkage cavities appeared in the pouring gate and riser.

3. Digital without Pattern Casting and Experiment for the Hot Stamping Die of Motor Bracket

The Sand mold without pattern casting for the hot stamping die of motor bracket is shown in Fig. 5.

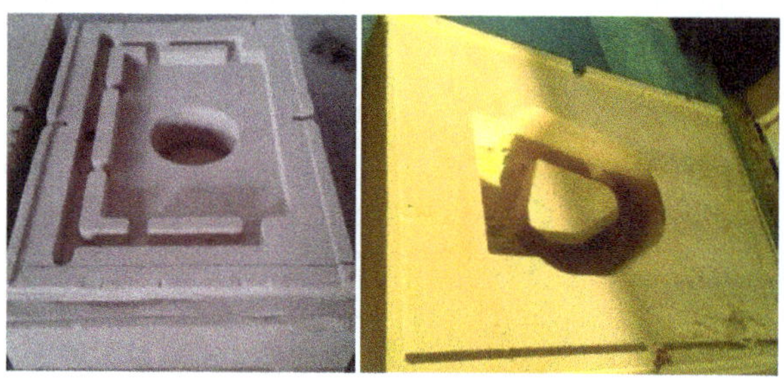

Fig. 5. Sand mold of the hot stamping die.

The sand mold was machined using the patternless digital casting precision forming machine (as shown in Fig. 6) independently developed by the China Academy of Machinery Science & Technology.

Fig. 6. Digital precision forming machine without pattern casting.

There is no need for wood mold or metal mold when the digital without pattern casting technology is used, which can save a lot of time and resources. In the traditional casting process, a casting with complex structure, such as the engine cylinder or the cylinder head, 10～12 sets of molds will be needed at least. What's more, it will also take 2-3 months and spend 0.9-1.2 million RMB. The digital without pattern casting technology developed by the China Academy of Machinery Science & Technology is a typical manufacturing technology for energy saving, material saving, and environmental protection. This technology can also shorten the development process and manufacturing time, realize the digital manufacture of traditional industry, improve production efficiency and reduce cost.

The hot stamping die of motor bracket was obtained after casting, as shown in Fig. 7. It can be seen from the die surface that the quality of the outer surface of the die was good, and there was no obvious defect in the die base. The zircon sand in the pipes was cleared up through blowing and inhaling gas. After cleaning, it was unimpeded in the pipes. The casted hot stamping die billet was machined, and the base plate and guiding device were assembled, then, the machined hot stamping die of motor bracket was obtained (as shown in Fig. 8).

Fig. 7. The casted die billet of motor bracket.

Fig. 8. Monoblock die of motor bracket.

Sample piece was cut from the runner. It was isothermal quenched at 300°C for 45 mins after anneal. Compression and impact toughness performance tests were conducted to the sample, the compressive strength was 3879Mpa, and the impact energy was 4.4J. Hot stamping experiment of motor bracket was also conducted. The blank was heated to 950°C for 5mins, stamping speed was set as 100mm / s, and holding time was 20s. Thus, the ultra-high strength steel motor bracket with excellent mechanical properties was obtained. The hot stamping process and the ultra-high strength motor bracket were shown in Fig. 9.

Fig. 9. Hot stamping experiment and the formed ultra-high strength motor bracket.

4. Conclusion

This paper puts forward a kind of hot stamping die manufacturing process based on adopted casting without pattern technology, and studies the casting process of a hot stamping die for motor bracket in a car. A casting mold for hot stamping die of motor bracket was successfully developed, with graphite cast steel as the casting material, 20# steel pipes as the preset pipes, and zircon sand filled in the pipes, at a casting temperature of 1420°C. The quality of the casting was good, and there was no obvious defect in the die base. The conformal cooling pipes were accurately positioned. The sand cores in the pipes were conveniently cleared up through blowing and inhaling gas. At last, an ultra-high strength motor bracket with excellent mechanical properties was successfully formed using this hot stamping die.

Acknowledgment

This Work was partially funded by National Natural Science Funds for Distinguished Young Scholar (No. 51525503).

References

1. M. Nikravesh, M. Naderi and G. H. Akbari, Influence of hot plastic deformation and cooling rate on martensite and bainite start temperatures in 22MnB5 steel, *Mater. Sci. Eng. A* **540**, 24 (2012).
2. M. S. Mohamed, A. D. Foster, J. G. Lin, D. S. Balint and T. A. Dean, Investigation of deformation and failure features in hot stamping of AA6082: Experimentation and modelling, *Int. J. Mach. Dies Manu.* **53**, 27 (2012).
3. A. Ghiotti, S. Bruschi and F. Borsetto, Tribological characteristics of high strength steel sheets under hot stamping conditions, *J. Mater. Process. Tech.* **211**, 1694 (2011).
4. C. J. Fan, H. F. Wang and T. T. Liu, Simulation of Temperature and Stress Field of Mould in Metal Gravity Casting Mould by ProCAST, *Mach. Build. Autom.* **40**, 84 (2011).

5. Z. D. Shan, Y. S. Ye, B. Y. Wang, C. Jiang and Q. T. Yang, Research on Casting Process for Embedded Internal Flow Passage Mould, *Foundry Tech.* **35**, 531 (2014).
6. W. Q. Jie, Z. Y. Jian, L. Liu, G. L. Liu, J. Shen and H. M. Wang, *Foundry Technology*, (Beijing: Higher Education Press, 2013).

Equipment

Advanced Design for Hot Forming and Press Hardening

B. Dvorak† and J. J. Tawk

Automotive design, BENTELER Mechanical Engineering,
Liberec, Hodkovicka 981/42, 460 06, Czech Republic
†E-mail: borek.dvorak@BENTELER.com

T. Vit

Technical University of Liberec, Department of Energetical Devices
Liberec, Studentska 2, 461 17, Czech Republic
E-mail: tomas.vit@tul.com

This paper presents an advanced innovative design of a continuous roller furnace. These types of furnaces are generally used in hot stamping / press hardening lines. Design is focused on optimal heating layout, modern drives of rollers, new design of furnace's input and output without movable parts and other items respecting the optimal technological and technical aspects. The cycle time of the hot forming process is a function of several technical parameters. Also the technological functions like the dew point temperature regulation, oxygen rate regulation with permanent monitoring function are described. All results are based on the theoretical background of heat and mass transfer, confirmed by numerical FEM (Finite element method) analysis.

Keywords: Hot forming; Press hardening; Roller furnace; Dew point; Heat transfer; FEM analysis; FLOX and automatization.

1. Introduction

Based on the long-time experiences with manufacturing and development of the machinery for automotive and glass making industry, new roller furnaces were designed using modern methods including the FEM (finite element method) analyses for numerical simulations of heating processes and heating power distribution. The numerical solution of many mathematical problems involves some combination of external and internal conditions and different technological processes [2-4].

In this article, we shall cover the development of Furnaces and Automation Equipment (e.g. feeder systems) as well as Robots and Conveyors for the loading and unloading process.

Roller furnaces for hot stamping line (press hardening technology) as shown in Figure 1.

Fig. 1. Example of roller furnace for hot stamping line [7].

2. Design Characteristics

The most important device parameter belonging to the furnace design is the modular concept (Figure 2), which allows building the furnace directly according to customer requirements. Design respects comfortable access for maintenance, smart safety concept which meets all the newest international safety standards, latest technological requirements such as the dew point regulation system and oxygen rate control.

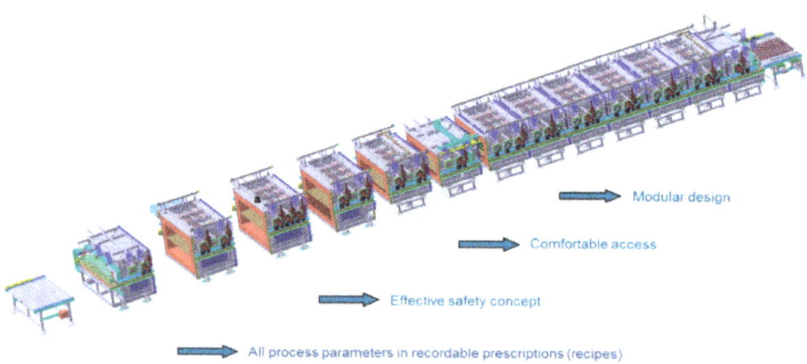

Fig. 2. The modular concept [7].

2.1. *Heating concepts*

Heating concepts include basic models like the recuperative gas burners with the radiant tubes or electric resistance heaters.

The most advanced system is based on highly efficient recuperative burners using the FLOX (Flameless oxidation) technology [1] equipped with the silicon reinforced silicon carbide radiant tubes produced by WS GmbH Company. The main advantages of these burners are the higher efficiency calculated by using equation (1) and reduction of NOx-emissions. The efficiency is up 10% - 15% higher compared to common burners. Design also brings other benefits such as homogenous temperature distribution, reduction of the thermal stress in the burner, noise reduction and lower restrictions on fuels.

$$\eta = 1 - \frac{exhaust\ gas\ losses}{fuel\ input} \tag{1}$$

2.2. *Numerical simulation of the heat transfer*

FEM analysis of heat transfer was used as the main tool for the design of heating system. This calculation takes into account all mechanisms of heat transfer, i.e. conduction, convection and radiation. Table 1 shows an overview of different regimes of heat transfer for different temperature difference between the blank and the furnace. Radiation is the dominant heat transfer mechanism.

Table 1. Heat transfer inside the furnace [7].

Temperature of furnace [°C]	870	900	920	930
Temperature of blank [°C]	25	600	900	920
Emissivity	0.38	0.12	0.54	0.7
Convection coefficient h [W/m^{2}K]	9.0	9.0	9.0	9.0
Heat flux radiation [kW/m^2]	36.6	8.9	4.1	2.7
Heat flux convection [kW/m^2]	7.6	2.7	0.18	0.09
Ratio between	4.8	13.3	22.8	29.4

2.3. *FEM results*

2.3.1. *Results of blank heating simulation*

The example of results is shown on Figure 3. It shows steady temperature field during operation of the furnace and heating of the blanks.

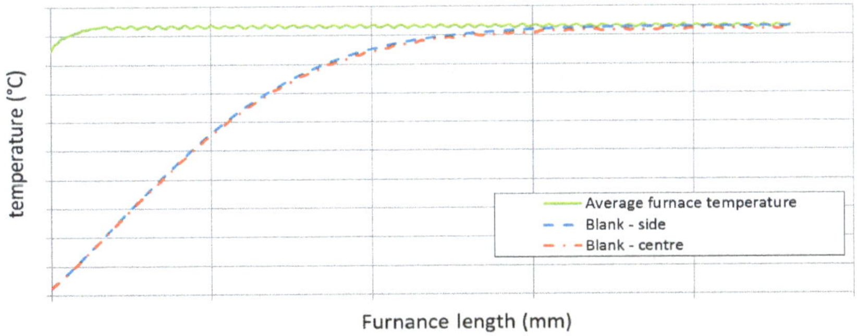

Fig. 3. Temperature in the furnace and on the blank as a function of its position [7 - 9].

The surface temperature of the product at the beginning of the furnace is also visible. Graph in Figure 3 shows temperature inside the furnace and temperature filed on products (Center and the sides) during the transport of the blanks through the furnace. The negligible deviation in temperature across the furnace is shown.

2.4. *Dew point regulation system*

With the knowledge of the problem called the hydrogen embrittlement, we have developed the system for the dew point regulation, which was patented in year 2011. The dew point is the temperature at which the water vapor in a sample of air at constant barometric pressure condenses into liquid water at the same rate at which it evaporates. Basically the dew point temperature is a function of the content of water vapor inside the furnace atmosphere.

Hydrogen embrittlement is the process by which various metals, most importantly high-strength steel, become brittle and fracture following exposure to hydrogen. Hydrogen embrittlement is often the result of unintentional introduction of hydrogen into susceptible metals during heating, forming or finishing operations and increases cracking in the material. The water inside the furnace atmosphere is the source of hydrogen which can cause the embrittlement in processed blanks. Basic chemical reactions during the embrittlement process for the uncoated blanks are:

$$3\mathrm{Fe} + 4\mathrm{H_2O} \rightarrow \mathrm{Fe_3O_4} + 8\mathrm{H} \tag{2}$$

$$\mathrm{Fe} + \mathrm{H_2O} \rightarrow \mathrm{FeO} + 2\mathrm{H} \;\; (\mathrm{Up}\ 570°\mathrm{C}) \tag{3}$$

Those equations show that it is quite important to control the mass of H_2O inside the furnace to prevent the risk of hydrogen embrittlement. The risk is high especially for coated blanks. Basic chemical reactions for Al – Si coated blanks are:

$$2\mathrm{Al} + 6\mathrm{HO} \rightarrow 2\ \mathrm{Al(OH)} + 6\mathrm{H} \tag{4}$$

$$2\mathrm{Al} + 4\mathrm{H_2O} \rightarrow 2\ \mathrm{AlO(OH)} + 6\mathrm{H} \tag{5}$$

$$2\mathrm{Al} + 3\mathrm{H_2O} \rightarrow Al_2O_3 + 6\mathrm{H} \tag{6}$$

$$\mathrm{Si} + 2\mathrm{H_2O} \rightarrow \mathrm{SiO_2} + 4\mathrm{H} \tag{7}$$

When the hydrogen (H) diffuses along the grain boundaries and combines with the carbon (C), which is alloyed with the iron inside the steel, this will form methane gas (CH4). The methane is collected inside the small voids along the grain boundaries where it builds up enormous pressures that initiate crack, see Figure 4. This is quite important during the processing of Al-Si coated blanks; because the hydrogen can be easily collected and locked bellow the coating.

Fig. 4. Example of hydrogen embrittlement crack [5-6].

3. Automatization System Design

One of the most important aspects in the hot forming line is the optimization of the overall cycle time of the hot forming process. We achieved a minimum cycle time of the hot forming process with a compact design of the destacking area of the blanks, unloading area for press hardened parts and a compact feeder system design for transferring the hot formed blanks into the press without interruption of the production process.

There are two different concepts for the general layout which can characterize the layout design of the hot forming line: The straight concept (Press position into the hot forming Furnace: 90°C) and the edge Layout (Press position into the hot forming Furnace: 180°C). The following characteristics show us some of the main convenient points and disadvantages of a straight concept Layout:

Pro:

- Shorter transfer time of hot blanks (1,5s – 2,5s shorter than edge concept) – quite important especially for the thin blanks
- No complicated centering or stoppers on the furnace exit conveyors
- Self-centering grippers are able to run with various types of batch layouts

Contra:

- Bigger opening of the press app. 250 – 300mm

- Worse cycle time of the press (app. + 0,3 s)
- More complication self-centering grippers

The following picture (Figure 5) shows us a general layout overview of a straight concept design:

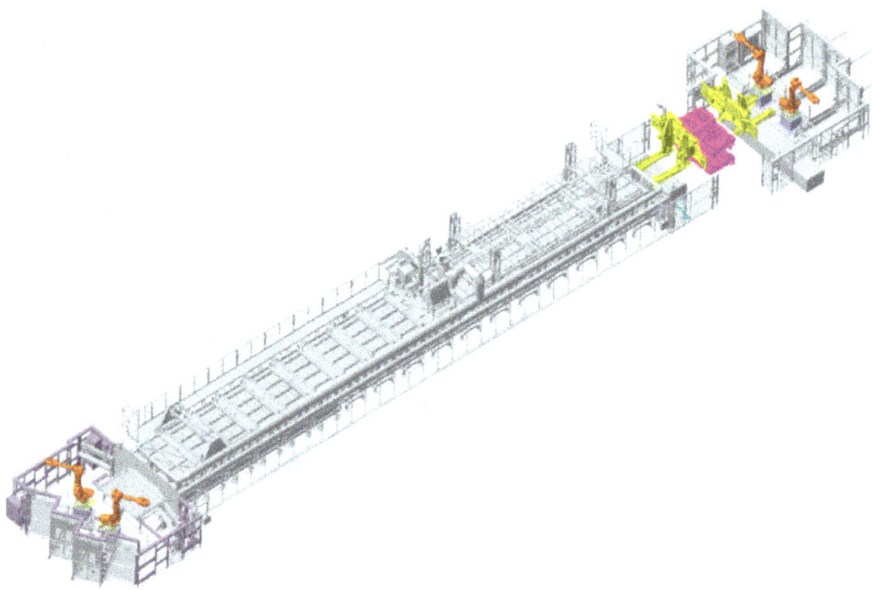

Fig. 5. Hot Forming Line Installation [7].

The general cycle time in the destacking area (Figure 6) depends on the number of destacking devices (for example Robots) and the supplementary equipment in this area. The destacking floor space has additionally a major factor where it has to be compact and to remain as well all the necessary security regulations. Another factor that would complicate the layout design of the destacking area is the production leveling with different blank shapes and sizes.

The amount of robots that are cooperating and alternating within the destacking area can be increased within this area in dependence of the combined functionality of several stations (Batch assembly, destacking and marking devices for blanks identifications / sheet control). As mentioned the challenge of such configurations is to keep the required minimum cycle time without cycle interruption. For determination of the required cycle time it will be necessary to know the detailed information about the dimensions of the processed blanks, which has an impact influence on the desired cycle time. Another major design state of the hot forming line is the transfer feeder (input- and output) to the hot stamping press. This production process has to have perfect synchronized movements with the press ram. This has to be the best cycle time utilization for transferring the hot forming parts in the shortest cycle time and avoiding temperature losses of the austenite tempered blanks. The feeder system works with short cooperation distance between input and output feeders in X-direction. The both feeder closely cooperates with the actual ram position to utilize the maximum of overlay time with ram of press movement.

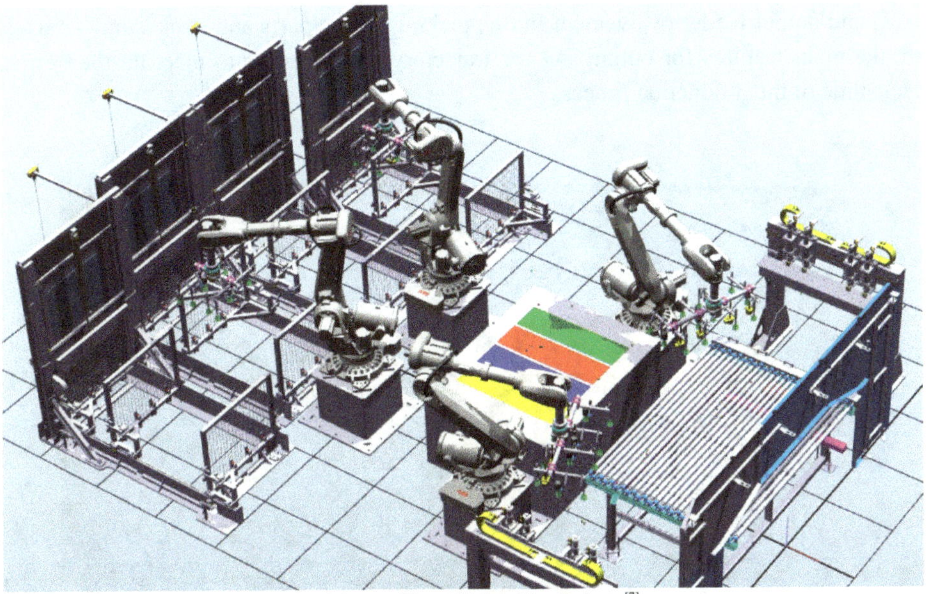

Fig. 6. Destacking with handling Robots [7].

The following picture (Figure 7) demonstrates our dynamic solution of the Input press feeders (4 x 3 Axes) for transferring two hot formed blanks (in one batch size) into the press. The input press feeders are designed with self-centering gripper. Four run without any additional centering on furnace output. This design contributes for a shorten cycle time of the hot forming process.

Fig. 7. The Feeder System for the hot formed blanks [7].

The feeder system contributes for optimizing the complete cycle time of the hot forming process by adjusting the necessary transport time into and out of the press. The movement of the input (movement out of die, putting off the blanks and strokes in X-Z axes) and output feeder (Movement In die, picking up the parts and strokes in X-Z axes) are the main features for optimizing the trajectory of the feeder to meet up the desired cycle time of the production process.

Fig. 8. Unloading area – Design made by BENTELER [7]

The mechanical design of the unloading area (Fig. 8) is based on similar configurations of the destacking area with different supplementary equipment and less complex design of the automatization comparing to the destacking station. Nonetheless the cycle time in that area could be a challenging aspect when the availability of the space floor is limited and the required short cycle time of the customer has to be respected.

The following capture (Figure 9) gives an explanation about all segments of the feeder movement and shows the interface with the other segments of the production line. This simulation contributes for a better visualization of diverse cycle time recording of the hot forming process and simplifies the optimization of the process parameters which is stored inside the process operator panel and can be used for further production.

Fig. 9. All Feeder movement Segments [7].

4. Conclusion

This paper presents a design of a continuous roller furnace incorporated with latest technological development in hot stamping and press hardening lines. Our design goals focused on optimal heating layout, modern drives of rollers, the design of furnace achieve one of our design objective of input and output without movable parts.

Using the theoretical data of heat and mass transfer, all our results are verified and confirmed by numerical simulation based on FEM (Finite element method) analysis.

The design was adopted by BENTELER Mechanical Engineering and commercialize as product of an advanced continuous roller furnaces for hot stamping lines.

References

1. J.A. Wünning and J.G. Wünning, Flameless Oxidation to Reduce Thermal NO-Formation, *Prog.Energy Combust.* **23**, 81 (1997).
2. J.G. Wünning, Energy Saving Potentials for Gas Fired Industrial Furnaces, in *Thermoprocess Symposium 2007*, (Düsseldorf, Germany, 2007).
3. H.C. Hottel and A.F. Saroffim, Radiative Transfer, McGraw hill, 1967.
4. M.F. Cohen and D.P. Greenberg, The hemi-cube: a radiosity solution for complex environments, *ACM SIGGRAPH Computer Graphics* **19,** 31 (1985).
5. Patent, Patenschrift 10 2011 053 634.5, BENTELER Automobiltechnik GmbH, 33102, Paderborn, DE, 15.9.2011.
6. http://www.dr-maiko.com/images/bruchflh2verspr.jpg.
7. BENTELER Mechanical Engineering GmbH, Automotive design, Liberec, Hodkovicka 981/42, 460 06, Czech Republic.
8. W.S. Wärmeprozesstechnik GmbH, Dornierstraße 14, 71272 Renningen.
9. Technical University of Liberec, Department of energetical devices Liberec, Studentska 2, 461 17, Czech Republic.

The Problem with Melted Al-Si in the Hot Stamping Furnace

M. Jönsson

TTE R&D PH Research, Luleå, 971 25 Luleå, Sweden
E-mail: mnjonsson@se.gestamp.com

AlSi coated boron steel is common used in the hot-stamping process. One advantage of applying an AlSi coating (approx. 90 % Al, 10 % Si) on a 22MnB5 steel is that the shot blasting process can be eliminated. In addition, the AlSi coating gives an increased corrosion resistance. And finally the coating hinders decarburization of the boron steel surface. However, the AlSi coating has a melting point of around 600°C. This means that melted AlSi is present in the hot-stamping process. Molten aluminium and its alloys are one of the most aggressive metal melts due to their high chemical activity with nearly all other metals and metal oxides. The melted AlSi sticks to the furnace rolls where it builds up layers. Further, the melted AlSi interacts with the steel in the furnace rolls. The aluminum build-up needs to be removed during regular maintenance stops. During these stops the furnace rollers have been examined visually as well as in the microscope. In addition laboratory trials have been performed in order to better understand the process of the AlSi build-up. In the present work the problem with melted AlSi in the hot-stamping furnaces is investigated and the problem discussed from a process point of view.

Keywords: Hot-stamping; Ultra-high strength steels; AlSi-coating; Process.

1. Introduction

AlSi coated boron steel is commonly used in the hot-stamping process. One advantage of applying an AlSi coating (approx. 90% Al, 10% Si) on a 22MnB5 steel is that the shot blasting process can be eliminated. In addition the AlSi coating gives an increased corrosion resistance. And finally the coating hinders decarburization of the boron steel surface.

However since the melting point of the AlSi coating is around 600°C and the process temperature in the furnace is 900-930°C melted aluminium is present in the process. Molten aluminium, and its alloys, is one of the most aggressive metal melts due to their high chemical activity with nearly all metals and metal oxides [1–3]. The melted aluminium sticks to the furnace rollers where it builds-up layers. In Figure 1 the build-up of the AlSi layers can be seen.

Fig. 1. Al-Si build-up on the furnace rollers.

These AlSi layers need to be removed with even intervals. Consequently, there is a high cost associated with maintenance and production stops due to the problem from the melted aluminium.

2. Experimental

2.1. *Process and material*

As is shown in the figure the blank is transported through the furnace on rollers. The speed of the rollers regulates the total heating time, or the dwell time, for the blanks.
A schematic picture of the direct hot–stamping process is shown in Figure 2.

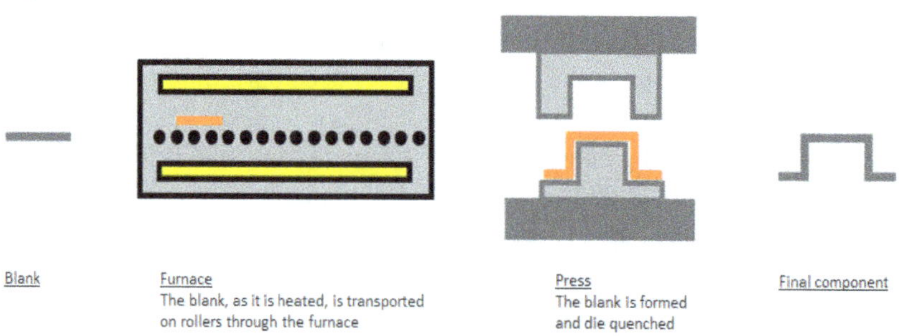

Fig. 2. The direct hot-stamping process.

A furnace located at Gestamp Hardtech in Luleå, Sweden, has been used in the investigation regarding Al-Si build-up. Even though the furnace dwell time is dependent on the thickness of the blank, type of coating etc. some average information about the investigated furnace is given in Table 1.

Table 1. Information about the process conditions.

Furnace length	50 m
Average blank time in the furnace	250 s
Number of rollers	270
Average furnace speed	0.2 m/s
Length furnace /number of rollers	0.19 m/roller

The incoming blanks in the investigated furnace have been 22MnB5 coated with 150 g/m^2 AlSi (AS150).

2.2. *Investigation the Al-Si build-up in the furnaces*

Due to the Al-Si build-up, the furnace rollers is removed from the furnace with even intervals and reconditioned. During the recondition the build-up of Al-Si is removed mechanically. During these maintenance stops the furnace rollers have been visually examined regarding the amount of aluminium build-up.

2.3. *Investigation of the AlSi coating after different dwell times*

In order to better understand the appearance seen in the production furnace, a lab scale test was performed. Small samples of 22MnB5 AS150 were heated at different dwell times in lab furnace and investigated in an electron microscope. The thickness of the samples was 1,2 mm, the size approximately 50x50 mm. The temperature on the furnace was set to 930°C. No specific atmosphere was used. Three different dwell times have been used i.e. 1, 2 and 6 min. Cross sections were investigated with as scanning electron microscope (SEM) model Hitachi S-3400N equipped with EDS model QUANTAX system and a Bruker SDD X-flash Detector 5010.

3. Results

3.1. *Results from the investigation of the Al-Si build-up in the furnaces*

The investigated production furnace has 270 rollers. During maintenance stops the aluminium build-up could be seen mainly on roller no. 35-100. With a maximum build-up found on roller no. 50-70, a schematic image of the aluminium build-up can be seen in Figure 3.

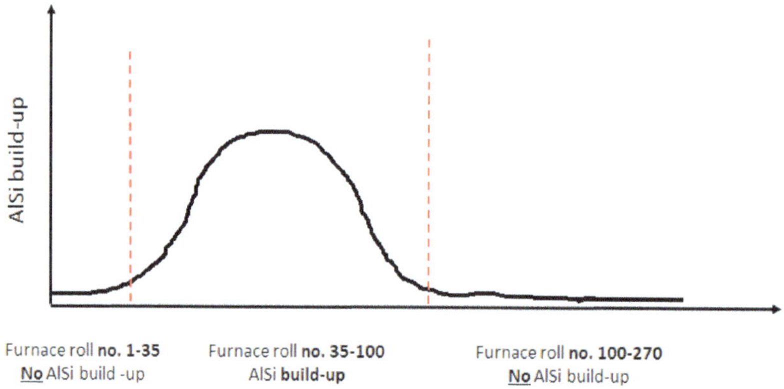

Fig. 3. Build-up of aluminium in furnace with 270 rollers.

From the values given in Table 1 the furnace time for the blanks corresponding to the aluminium build-up could be calculated, the results is given in Table 2.

Table 2. Time in furnace calculated from the values given in Table 1.

Roller (number)	Build-up of AlSi	Corresponding length of furnace (m)	Corresponding time in furnace (s)
1-35	**No**	0 - 6.5	0 - 33
35-100	**Yes**	6.5 - 18.5	33 - 93
100-270	**No**	18.5 - 50	93 - 250

3.2. *Results from the investigation of the AlSi coating after different dwell times*

SEM images of the cross sections can be seen to the left in Figure 4-Figure 7. To the right the corresponding EDS line scan can be seen. The EDS line scan has been measured along the red arrow seen in the figures.

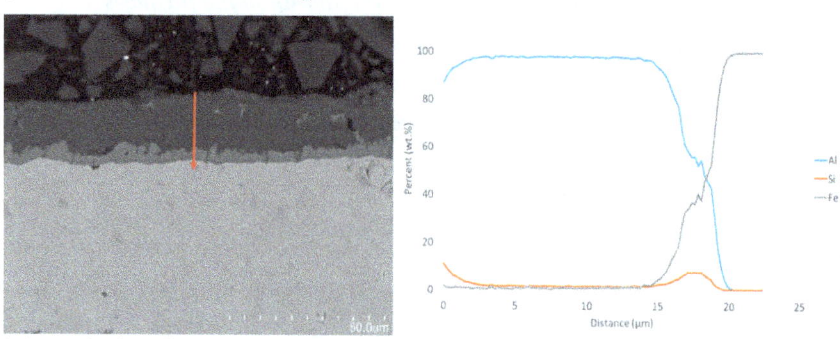

Fig. 4. As received condition Left: SEM image of the cross section of the AlSi coated 22MnB5 steel. Right: EDS line scan along the red arrow.

As can be seen for the "as received" condition the coating consists of an Aluminium Silicon mix. There is only a thin inter-diffusing layer closest to the steel. After 1 minute dwell time the diffusion process has started. The diffusion process is fast creating a mix consisting of approximately 50% Al and 50% Fe. As can be seen the diffusion front is distinct. After 2 minutes dwell time the aluminium and the iron have mixed completely, with a mix of approximately 50% Al and 50% Fe. After 6 minutes the different intermetallics have formed [4].

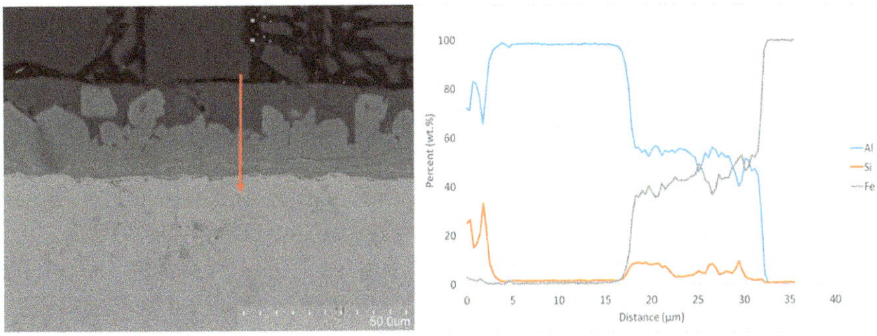

Fig. 5. Dwell time 1 min. Left: SEM image of the cross section of the AlSi coated 22MnB5 steel. Right: EDS line scan along the red arrow.

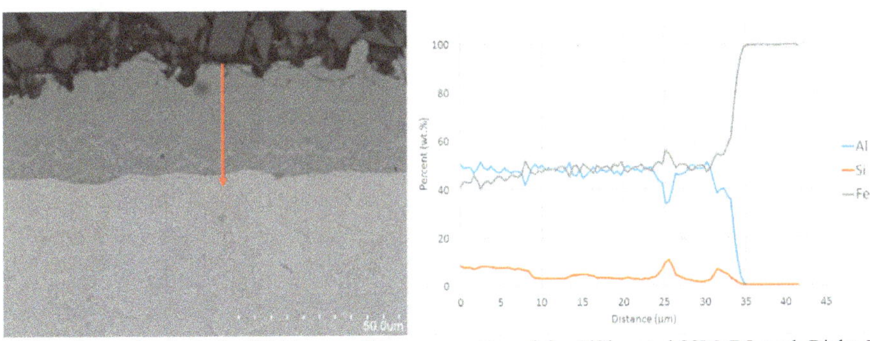

Fig. 6. Dwell time 2 min. Left: SEM image of the cross section of the AlSi coated 22MnB5 steel. Right: EDS line scan along the red arrow.

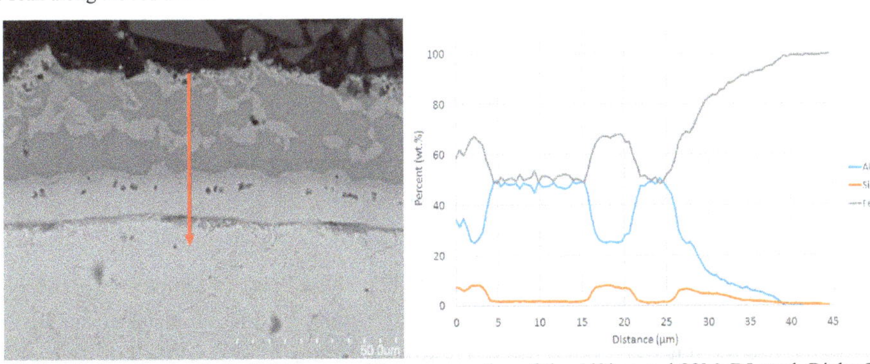

Fig. 7. Dwell time 6 min. Left: SEM image of the cross section of the AlSi coated 22MnB5 steel. Right: EDS line scan along the red arrow.

4. Discussion

Comparing the results obtained from the inspection of the rollers seen in Figure 1 and Table 2 with the microstructure images of the coating after different dwell time, the build-up of AlSi in the furnace can be better understood. On roller no. 1-35 no build-up can be seen. This is due to the fact that the melting point (i.e. ~600°C) of the AlSi coating has not yet been reached in the furnace, hence no AlSi build-up can be seen. With an average speed of 0,2 m/s around 30 seconds have elapsed when the blanks have reached roller no. 35. At this time the coating have reached the melting point and the build-up of AlSi starts on the rollers. This continues between roller no. 35-100, corresponding to approx. 30-90 seconds (Table 2).

The microstructure of the coating during this time interval can thus be assumed to be similar to the coating depicted in Fig. 5. The diffusion between iron and the AlSi layer is rapid. However there is still a layer on the outermost surface that consists of melted Al-Si. It is this outermost AlSi layer that causes the aluminium build-up.

After roll no. 100, when approx. 90 seconds have elapsed, the AlSi build-up have decreased to small levels. It can be assumed that after these 90 second the appearance of the cross-section of the coating is comparable with the appearance seen after 2 minutes dwell time (Figure 6). The AlSi coating and the iron from the base-material have diffused together, forming a mix with around 50% Al and 50% Fe. According to the Fe-Al phase diagram (Figure 8) at a Al level <60wt.% different intermetallic phases starts to form e.g. $FeAl_3$, Fe_2Al_5.

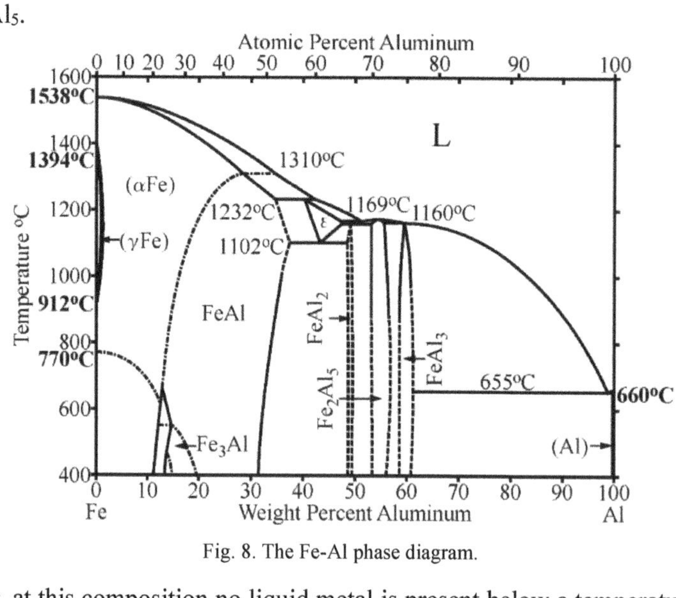

Fig. 8. The Fe-Al phase diagram.

Further, at this composition no liquid metal is present below a temperature of around 1160°C. This temperature is much higher than the temperature in furnaces in the hot-stamping process. Since no liquid metal is present there can be no AlSi build-up on the furnace rollers. However it should be noted that if the AlSi layer is thicker than the stipulated 150 g/m^2 (corresponding to a coating thickness of around 20μm) the diffusion between the AlSi layer and the underlying Fe will take longer time. Consequently the aluminum build-up will affect a larger part of the furnace.

5. Conclusion

The AlSi build-up on the furnace rollers results in a large maintenance costs in the hot-stamping industry. The problem can mainly be seen in the second quarter of the furnace. In this section of the furnace the melting point is reached for the AlSi coating. However the diffusion between the AlSi and the Fe is still ongoing and consequently liquid Al-Si is still present on the surface of the blanks, creating the AlSi build-up.

References

1. M. Yan and Z. Fan, Durability of materials in molten aluminum alloys, *J. Mater. Sci.* **36,** 285 (2001).
2. M. Sundqvist and S. Hogmark, Effects of liquid aluminium on hot-work tool steel, *Tribol. Int.* **26**, 129 (1993).
3. S. Shankar and D. Apelian, Die soldering: Mechanism of the interface reaction between molten aluminum alloy and tool steel, *Metall. Mater. Trans. B* **33**, 465 (2002).
4. M. Windmann, A. Röttger and W. Theisen, Formation of intermetallic phases in Al-coated hot-stamped 22MnB5 sheets in terms of coating thickness and Si content, *Surf. Coatings Technol.* (2014).

Research on Intelligent Process and Production Method of Hot Stamping

L. Wang[1], X. Y. An[2], B. Zhu[1], Y. L. Wang[1] and Y. S. Zhang[1,†]

[1]*State Key Laboratory of Materials Processing and Die and Mould Technology, Huazhong University of Science & Technology, Wuhan 430074, China*

[2]*Dongguan Vision Hot Stamping Technology Co., Ltd., Taozi Industrial Park, Hengli Town, Dongguan City, Guangdong, 523475, China*

[†]*E-mail: zhangys@mail.hust.edu.cn*

The continuously increasing complexity of equipment and processes of hot stamping characterizes production technologies. Hot stamping is a hot sheet metal forming process, by which car body components with a high crash performance at low weight can be produced. The production method is challenging due to: 1) the combination of a forming operation with a heat treatment process, 2) temperature-sensitive properties of products; 3) relatively low productivity compared with traditional stamping. That implies the observation of a large number of parameters and their correlations and corrects any process variations in a very short time in order to ensure a consistent part quality and thus avoid waste. Therefore, the intelligent process and production method was developed. This paper is based on analysis of process-related parameters such as forming temperature, transfer time, press-holding time in terms of their influence on key quality parameters of the final component, and refines intelligent model of equipment and production system. Using the example of forming a B-pillar, the feasibility of this intelligent model of equipment and production system are presented.

Keywords: Hot stamping; Intelligence; Modeling; Production method.

1. Introduction

The vehicle industry has focused on reducing energy consumption and emissions for environment protection for many years, while also repeatedly calling for security improvement. Today, one of the main ways to realize this aim is to use high strength steel instead of traditional steel. Due to the poor deformation capacity of high strength steel, the hot stamping and quenching process [1, 2] of high strength steel has been developed.

The pressure of market competition drives the life cycle of product design and manufacturing system from over 10 years to 3-5 years [3-6]. Correspondingly, the design and manufacturing process of auto parts must also be adapted to the shortened life cycle. In production, the manufacturing mode of auto parts gradually turn into the mode of multiple types and small batch, which requires promotion of the flexibility and configurability [7, 8] of the manufacturing system. The investigation about industry chain shows that the most wanted requirements of hot stamping equipments are production flexibility and floor area reduction.

In hot stamping process, the main contradiction is the mismatches of forming and heating. In order to solve the mismatches, a rising scheme is the multi-chamber furnace [1, 9, 10]. A multi-chamber furnace is a box-type furnace with several chambers piling up. It has independent heating modules; thus it is capable of heating blanks in the parallel mode. Its chambers are able to be controlled whether to participate the production cycle independently. With less chambers involved in, the power consumption of the furnace will be reduced. This is a great degree of flexibility. Like pilot lines with a box-type furnace with single chamber [10-12], a production line with multi-chamber furnaces just takes up more space in high direction and demands a very small floor area compared with ones with roller hearth furnaces. APT reduces the length of line from 70 meters to 25 meters by using a multi-chamber furnace instead of a roller hearth furnace. The SIGMA hot stamping line with multi-chamber furnace is only 18 meters in length. The multi-chamber furnace reduces over 64% of space requirements of hot stamping lines. The box-type furnace has no mechanism inside, good sealing and higher energy efficiency than the roller hearth furnace which helps to reduce the energy cost of parts. Blanks remain motionless inside, which avoids abrasion and surface quality degradation caused by friction.

In terms of control, multi-chamber furnace is much more complex than the traditional scheme, roller hearth furnace. The roller hearth furnace has only one entrance and one exit, and its temperature configured by only a small number of zones. But the multi-chamber furnace always has ten to twenty chamber doors. Every door should be opened and closed at precise time point following proper sequence [10]. In order to feed all the chamber with efficiency, the number of transferring robots are increased from 2 to 6 [10]. The transfer range of robots for furnace covers nearly half the area of the line. On one hand, its motion path must be controlled accurately to avoid collision with the furnace, while on the other hand it must run as fast as possible in order to reduce heat loss of heated blank in air [13-15]. So an intelligent method of hot stamping process and production process are needed to represent and organize hot stamping process and its production, especially on the control of complex production lines with multi-chamber furnaces.

2. Hot Stamping Process and Intelligent Production

Modeling of the hot stamping process procedure should be able to represent and organize hot stamping process properly and suitable for current and future hot stamping production method. A model based on FSM is adopted.

According to the furnace heating method, the core procedure of hot stamping process consists of two manufacturing steps and one transferring steps. The two manufacturing steps are heating (HU) and forming (FU), and the transfer step is transferring heated blanks (THB). As shown in Fig. 1, these three steps are modeled based on FSM, which describe the relationship of the manufacturing units.

This model is independent with production equipment and irrelative with serial or parallel production method. It is capable of theoretically describing pilot lines, lines with roller hearth furnace, and lines with multi-chamber furnace. Pilot lines are generally used for process validation and performance analysis, their production cycles are not limited; they have a simple system structure with one robot for unstacking and transferring, and their parts unloaded by hand. For production lines, THB are handled by robots for efficiency. The structure and operation mode of their furnaces are very different from each other. For roller hearth furnaces, the roller rolls continuously, blank stations are movable and uncertain, thus HU indicates one movable working position inside the furnace. FSMs for roller hearth furnaces are always generating and destroying along as blanks travel in and out. When a cold blank is sent into a chamber, the corresponding FSM transits to the state of Heating from Unoccupied; when the heating reaches the time limit, the FSM transits to Finish-heating; and when a heated blank is removed, the FSM returns to Unoccupied.

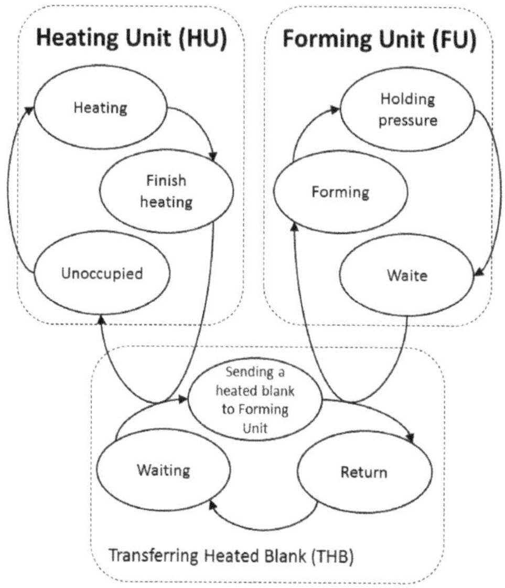

Fig. 1. Modeling of hot stamping process procedures.

In the THB step, an important situation, processing abandoned blanks, should be considered carefully and handled perfectly. There are two robots, furnace unloading robots and press feeding robots, participating into the THB step. And these two robots are reused to handle the abandoned blanks.

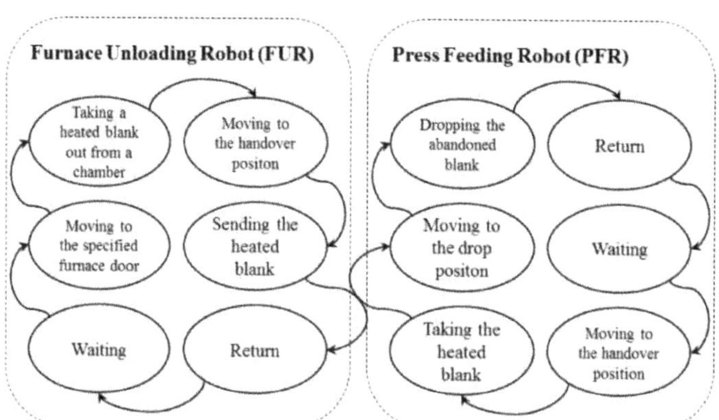

Fig. 2. Modeling of processing abandoned blanks for multi-chamber furnace.

Due to high risk of the heated blanks, when a fault occurs, the heated blanks should be at safety positions. In the hot stamping line with multi-chamber furnace, safety positions for heated blanks are chambers of furnace, inside mold in the press. So an intelligent and automated method of processing faults is important for hot stamping lines. Fig. 2 shows the flow path for two robots to handle abandoned blanks. The path makes use of the space between press and furnaces, and a waste channel is built in this space. The two robots transfer the abandoned blanks to the waste channel. More than half of the flow path is the same as THB steps. The detection of faults is executed at the flow control module which also controls all the equipments in the hot stamping line. When a fault of press is detected, instructions are sent to FUR and PFR and they will turn to fault-handling flow path precisely and rapidly. This method will avoid manual handling of heated blanks, keeps operator and robots away from hazard process, significantly reduce the impact of production by wastes.

3. Experiment

The modeling method has been adopted in a hot stamping line with multi-chamber furnaces for nearly two years. And the fault-handling flow paths also have been executed at this line for dozens of times.

4. Results and Discussion

4.1. *Energy consumption of heating*

With the use of box-type furnace, energy consumption of heating reduces substantially. The power quality analyzer shows that, compared at the same production ability, the power consumption of multi-chamber furnace is less than 10% than roller hearth furnace, the power of which is usually above 1000 kW. And the furnace power consumption can be reduced by quitting chambers from production and power off them separately. The heat preserving power consumption with different number of chambers is shown in Figure 3.

4.2. *Fault response*

For manually handling of abandoned blanks, an operator takes 365 s at least to recover the production with one mold for one part and 683 s at least with one mold for two part. While the intelligent method drops an abandoned blank in less than 6 s. In this case, the next production cycle will not be disturbed and less than 1% of manual time is used. Time-saving also saves much energy, because the furnace still need power to keep temperature inside the chamber while manually handling.

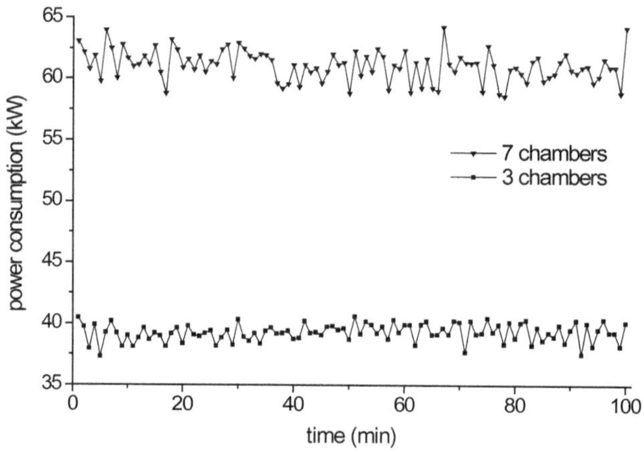

Fig. 3. Heat preserving power consumption with different number of chambers.

5. Conclusion

In this paper, a description model of hot stamping process is built. The model is applied to a hot stamping line with multi-chamber furnaces, a servo press and linear transferring robots. This model helps to schedule multiple equipments and implements the process control of the hot stamping line.

The multi-chamber furnace consumes 90% less power than roller-hearth furnace. And this box-type furnace is able to save further more energy by quitting chambers from production cycle.

The intelligent method to handle abandoned blanks uses less than 6 s and it is less than 1% of manual handling time. The method also saves much energy by shorten fault handling cycle.

The intelligent method now handles part of all kinds of faults. Improvements of the method are desired to reach higher level of automation.

Acknowledgments

This research work is funded by the National Natural Science Foundation of China (grant no. U1564203) and National Natural Science Foundation of China (grant no. 51405171).

References

1. M.T. Ma, Y.S. Zhang, L.F. Song and Y.L. Wang, Research and Progress of Hot Stamping in China, *Adv. Mater. Res.* **1063**, 151 (2015).
2. H. Karbasian and A.E. Tekkaya, A review on hot stamping, *J. Mater. Process Tech.* **210**, 2103 (2010).
3. E.W. Endsley, E.E. Almeida and D.M. Tilbury, Modular finite state machines: Development and application to reconfigurable manufacturing cell controller generation, *Control Engineering Practice* **14**, 1127 (2016).

4. Y. Koren, U. Heisel, F. Jovane, T. Moriwaki, G. Pritschow, G. Ulsoy and H.B. Van, Reconfigurable Manufacturing Systems, *CIRP Ann. Manuf. Technol.* **48**, 527 (1999).
5. C. Renzi, F. Leali, M. Cavazzuti and A.O. Andrisano, A review on artificial intelligence applications to the optimal design of dedicated and reconfigurable manufacturing systems, *The International Journal of Advanced Manufacturing Technology* **72**, 403 (2014).
6. M. Abbasi and M.Houshmand, Production planning and performance optimization of reconfigurable manufacturing systems using genetic algorithm, *The International Journal of Advanced Manufacturing Technology* **54**, 373 (2011).
7. A.J. Niemi, The contribution of pilot systems to the automation of flexible manufacturing, *The International Journal of Advanced Manufacturing Technology* **5**, 275 (1990).
8. A.A. Altumi, A.M. Philipose and S.M.Taboun, Reliability Optimisation of Flexible Manufacturing Systems with Spare Tooling, *The International Journal of Advanced Manufacturing Technology* **17**, 69 (2010).
9. M. Skrikerud and J. Friberg, Press Hardening, Process Design and Tool Systems, in *4th International Conference on Hot Sheet Metal Forming of High-performance Steel*, (Luleå, Sweden, 2013).
10. L. Wang, B. Zhu and Y.S. Zhang, Production Control and Optimization of Hot Stamping Line, in *5th International Conference on Hot Sheet Metal Forming of High-performance Steel*, (Toronto, Canada, 2015).
11. C. Wang, Investigation on Thermal Contact Behavior and Prediction of Mechanical Properties and Process Optimization in Hot-Stamping, Phd thesis, Huazhong University of Science and Technology (2014).
12. D. Landgrebe, N. Piersche, M. Alsmann and J. Schonherr, Intelligent Process Control for Press Hardening, in *5th International Conference on Hot Sheet Metal Forming of High-performance Steel*, (Toronto, Canada, 2015).
13. Z. Shi, K. Liu, M. Wang, J. Shi, H. Dong, J. Pu, B. Chi, Y. Zhang and J. Li, Thermo-mechanical properties of ultra-high strength steel 22SiMn2TiB at elevated temperature, *Mater. Sci. Eng. A* **528**, 3681 (2011).
14. C. Dessain, P. Hein, J. Wilsius, L. Penazzi, C. Boher and J. Weikert, Experimental investigation of friction and wear in hot stamping of usibor 1500P, in *1st International Conference on Hot Sheet Metal Forming of High-Performance Steel*, (Kassel, Germany, 2008).
15. X. Tian, Y. Zhang and J. Li, Investigation on Tribological Behavior of Advanced High Strength Steels: Influence of Hot Stamping Process Parameters, *Tribol. Lett.* **45**, 489 (2012).

Uni-Form: A Pilot Production Line for Hot/Warm Sheet Metal Forming Integrated in a Cloud Based SMARTFORMING Platform

X. Luan, Q. L. Zhang, O. E. Fakir and L. L. Wang†

Department of Mechanical Engineering, Imperial College London
London, SW7 2AZ, UK
†E-mail: liliang.wang@imperial.ac.uk

M. M. Gharbi

Schuler Pressen GmbH, Bahnhofstrasse 41
Goeppingen, 73033, Germany
E-mail: Mohammad.Gharbi@schulergroup.com

Warm/hot stamping is a promising technology for manufacturing high-strength, lightweight, complex shaped components from high strength aluminum alloys. It is well known that the formability and high post-form strength of aluminum alloys are highly temperature sensitive, i.e. manufacturing operations conducted at elevated temperatures could affect their microstructures, leading to variations in the strength of final products. Therefore, the processing routes including forming operations, secondary fabrications, room temperature storage and high temperature curing cycles must be planned strategically to ensure the quality of final products. In the present research, a novel processing route planning system has been finalized and integrated to an online multi-objective platform SMARTFORMING, enabling processing route optimization on a cloud computing environment. The system consists of two major functional modules: '*Tailor*' (processing route optimization) and '*Uni-Form*' (a pilot production line controlled on a cloud based platform for verifying the effectiveness and efficiency of the processing routes).

Keywords: Sheet metal; Warm/Hot forming; SMARTFORMING platform; Uni-Form; Tailor.

1. Introduction

A strong, emerging drive towards curtailing emissions and improving fuel economy has sparked both research interest and industry demand for weight reduction technologies, especially in alternative materials to steels such as aluminium alloys. It is estimated that through the replacement of steel components in vehicles, a weight reduction of up to 40% can be achieved, which in turn yields an approximately 30% decrease in emissions [1]. However, the poor ductility of aluminum alloys at room temperature restricts their widespread adoption. The most common solution to this problem is to form these alloys at elevated temperatures, e.g. using warm or hot stamping processes [2]. Manufacturing processes conducted at elevated temperatures tend to change the microstructures of high strength aluminum alloys. This raises a lot of concerns for engineers because the strength of the final products cannot be determined by individual industry sectors, but instead relies on the entire supply chain, including the material suppliers, tier 1 press shops and OEMs. Therefore, the processing routes for warm/hot stamping of high strength aluminum alloys must be defined strategically. Moreover, as an intrinsic feature of heat treatable aluminum alloys, strong responses to the variations in plastic straining and temperature could also lead to microstructure/strength alterations, thus tailored processing routes must be defined for individual components.

A cloud based multi-objective platform SMARTFORMING has been established by the author's group that can facilitate the definition of such processing routes. Integrated with various functional modules involved in both research and production, SMARTFORMING has been reforming the way scientific research and industrial production operate [3, 4]. Previously, the processing route calculator, '*Tailor*', has been successfully developed and integrated into the platform [5]. In order to finalize the processing route planning system and validate the effectiveness of the processing routes generated by '*Tailor*', a novel pilot production line, '*Uni-Form*' was successfully designed, fabricated and integrated into the SMARFORMING platform. Comprising a dedicated conductive heating furnace, a precisely adjustable conveyor system for hot blank transportation, an alterable forming tool and an incubation chamber simulating multiple curing cycles, *Uni-Form* realizes fully functional and closed-loop sheet metal forming process planning in an intelligent, precise and automated manner.

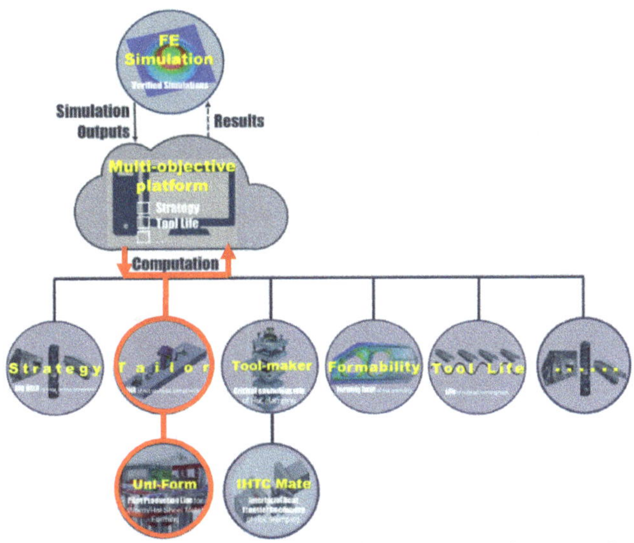

Fig. 1. The structure of the Multi-objective platform (*www.smartforming.com*).

2. The Online Multi-Objective Platform SMARTFORMING

The structure of the SMARTFORMING platform can be seen in Figure 1. Taking advantage of cloud computing, the SMARTFORMING platform enables engineers from each industry sector to develop efficient and cost effective processing routes and to guide the implementation of new forming technologies jointly. Through the online portal *www.smartforming.com*, multiple users can log in to the platform, work simultaneously and remotely. Different functional modules correspond to specific aspects of sheet metal forming research. The modules can be linked together, such as 'Tool Maker' and 'IHTC-Mate', to supervise the material selection and design of stamping tools using the findings of research conducted on interfacial heat transfer coefficients [6]. Modules can also be operated by individual users to achieve specific advanced objectives, such as tool life prediction under multi-cycle loading conditions [7] and formability prediction under hot stamping conditions [8]. Two functional modules of the SMARTFORMING platform, namely '*Tailor*' and '*Uni-Form*', were involved in this study.

3. Tailor: Processing Route Calculator

Integrating multi-disciplinary theories, *Tailor* enables engineers to input a series of inter-related parameters in one software package to calculate each window of the whole processing route, rather than utilizing individual programs. *Tailor* provides a visualized operation interface and customizable selections for users, as shown in Figure 2. The inputs include heating temperatures, hot blank transportation times, blank holding forces, forming rates, die closing pressures, die closing times and curing parameters. According to the capacity of existing facilities in a press shop, the final strength of the component can subsequently be predicted [5]. Alternatively, *Tailor* could be applied in reverse. For example, given the geometry, category and expected post form strength of a formed aluminium alloy sheet metal component, the processing route for such a component including the forming temperature, forming speed, transfer time and artificial ageing parameters, can be optimized according to the facility specifications in a press shop.

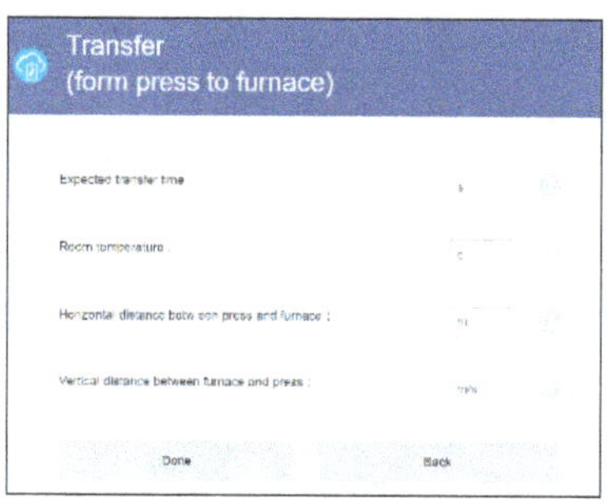

Fig. 2. The user interface of *Tailor*.

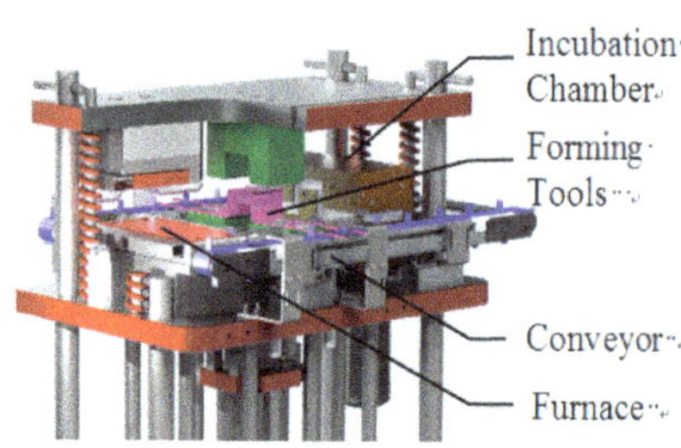

Fig. 3. The design of *Uni-Form*.

4. *Uni-Form*: Pilot Production Line for Verification and Optimization

A dedicated pilot production line, *Uni-Form*, was designed, fabricated and integrated into the SMARFORMING platform, which enables the processing parameters generated in *Tailor* to be realised and verified through the online portal. As shown in Figure 3, *Uni-Form* consists of four sub-systems; the furnace, forming tools, conveyor and incubation chamber. The integration of *Uni-Form* into SMARTFORMING was realized through a telnet connection to a server on the production line using the online portal. A microcontroller programmed in C language was utilized to coordinate the different sub-systems in the production line simultaneously. The dedicated processing route was configured into the server through the online platform and the forming trials were conducted automatically.

Fig. 4. The installation of the pilot production line on the press.

This pilot production line was mounted on a 100 tonne hydraulic press with a gas spring cushion at the bottom, shown in Figure 4. The gas pressure of this cushion can be charged up to 40 bar, achieving an adjustable stamping force. Equipped with a powerful conductive heating facility, a maximum temperature of 600°C and a fast heating rate could be implemented. This was achieved using cartridge heaters embedded in the top and bottom platen after a series of thermal analysis to ensure high temperature homogeneity. The temperature difference between the center and edge of the heater was within approximately 5%. Additionally, a combination of the insulation materials Duratec 750® Machinable Ceramic and PROMALIGHT 1000X-HD was applied, offering an extremely low thermal conductivity of around 0.030 W/m·K. A conservative heat flow scenario of the entire insulation module at the maximum operating temperature was run using PROMAT software and yielded temperature of 45°C on the bottom of the furnace, which was considered safe. The forming tools were designed to be modular, enabling changes to be made to the geometry of the die and the blank holding force. Meanwhile, upon 20 repeated trials of a multi stage process, the lag of the conveyor between full speed and stopping was found to be around 0.23s, and the blank stop position varied by up to 0.5 mm in each trial, confirming the line's repeatability and high positional accuracy. Frame motion analysis tests to confirm no conflict among different components during stamping, evaluations of the capability of the incubation chamber and finally automated hot forming trials of the aluminium alloy 6082-T6 were conducted as well.

5. Forming Trials on *Uni-Form*

A U-shape component with original blank dimensions of 240mm x 86mm x 1.5mm shown in Figure 5 was selected to verify the function of the platform and the capability of *Uni-Form*.

Fig. 5. The formed U-shape component ready for multi-cycle paint/bake.

According to *Tailor*, the suggested set of forming parameters included a forming temperature of 350°C, transfer time within 4-6s, a forming speed of 400 mm/s. This forming route was configured on the server of *Uni-Form* remotely and the forming trials were conducted automatically. Real time monitoring showed that the forming temperature was maintained at 352°C, the transfer time was 5.30s exactly the same as defined, the forming speed was 405mm/s. The formed components ready for artificial ageing is shown in Figure 5.

6. Conclusion

In this work, an intelligent pilot production line, *Uni-Form*, for warm/hot sheet metal forming research was designed, fabricated and integrated into the online multi-objective platform as a functional module. Through this platform, two functional modules, *Tailor* and *Uni-Form*, were unified and the novel processing route planning system has been finalized. Optimized forming parameters obtained through *Tailor* were used to configure the settings for the pilot production line through a telnet connection, and the components were automatically formed. All the operations were accomplished by a computer using the online portal *www.smartforming.com*. Such a scheme not only virtualizes and streamlines the optimization of sheet metal forming processes, but also improves experimental accuracy and reduces workloads dramatically.

The dedicated pilot production line has the capability of reaching a heating temperature of up to 600°C, which realizes the warm/hot forming trials to be adopted not only to aluminium alloys but also to magnesium alloys and steels, enabling novel forming process to be developed. Furthermore, the online multi-objective platform, especially *Tailor* and *Uni-Form*, has provided valuable insight into the scalability of lab-scale research to industrial production for sheet metal components.

References

1. N. P. Lutsey, Review of technical literature and trends related to automobile mass-reduction technology, Reports, UC Davis, (2010).
2. N. Abedrabbo, F. Pourboghrat and J.Carsley, Forming of aluminum alloys at elevated temperatures-Part 1: Material characterization, *Int. J. of Plasticity* **22**, 2 (2006).

3. L. Wang, Knowledge Based Cloud FE Simulation of Sheet Metal Forming Processes, *J. of Visualized Experiments*, 2016
4. A. Wang, Knowledge Based Cloud FE simulation - a multi-objective FEA system for advanced FE simulation of hot stamping process, in *The 10th Int. Conf. and Workshop on Numerical Simulation of 3D Sheet Metal Forming Processes (Numisheet 2016)*, (Bristol, UK, 2016).
5. X. Liu, K. Ji., O. E. Falir, J. Liu and Q. Zhang, Determination of processing windows for the hot stamping of AA7075, in *Proc. 16th International Conference Metal Forming 2016*, (Kraków, Poland, 2016).
6. X. Luan, Characterization of the interfacial heat transfer coefficient for hot stamping processes, in *The 10th Int. Conf. and Workshop on Numerical Simulation of 3D Sheet Metal Forming Processes (Numisheet 2016)*, (Bristol, UK, 2016).
7. Y. Hu, X. Yuan, G. Ma, M. A. Masen and L. Wang, Tool-life prediction under multi-cycle loading during metal forming: a feasibility study, *Manufacturing Review* **2**, 28 (2015).
8. O. El Fakir, Studies on the Solution heat treatment, forming and in-die quenching process in the production of lightweight alloy components, PhD thesis, Imperial College London, (2015).

Kinematics and Dynamics Analysis of Transmission Robot in Hot Stamping Production Line

P. C. Deng, B. Zhu, Y. L. Wang and Y. S. Zhang†

State Key Laboratory of Materials Processing and Die and Mould Technology, Huazhong University of Science & Technology, Wuhan, 430074, China

†E-mail: zhangys@mail.hust.edu.cn

The transmission system of multilayer box-type furnace with high-speed Cartesian robot as the carrier is the key link of automatic production line for high strength steel hot stamping. In order to minimize the loss of heat during transmission and interfere with other equipment, the rapidity and stability of the robot movement must be guaranteed. In this paper, a Cartesian robot was built and its kinematics and dynamics simulation were analyzed in the case of cubic polynomial interpolation and parabolic transition linear interpolation, respectively. The result showed that its movement is more stable when using parabolic transition linear interpolation.

Keywords: Hot stamping; Transmission; Cartesian robot; Kinematics; Dynamics.

1. Introduction

Due to the demand for reduced vehicle weight, improved safety, and crashworthiness qualities, the need to manufacture auto-mobile structural components from ultra high strength steels is apparent [1]. Hot stamping technology combined with advanced forming technology and new material technology, has become one of main methods to improve the automotive light weighting. In the hot stamping process, a blank is heated in a furance, then transferred to the press and subsequently formed and quenched in the closed tool, which obtains the required intensity and plasticity [1]. With the development the hot stamping production line, the new requirements were put forward. Currently, the main equipment in hot stamping production line are as follows: multilayer box-type furnance as a heating system, high-speed Cartesian robot as a transmission system and servo press as a forming system. To minimize the loss of heat and to guarantee the rapidity and stability during transmission, it is meaningful to study the high-speed Cartesian transmission robot [4].

R. Featherstone, D. E. Orin has researched the dynamics of the robot including the equations and algorithms [2]. Claud Vibet has established the model of the kinematics and dynamics of the robot [3]. Y. Sun has analyzed the structure and performance of the transportation fork in hot stamping line, which optimized the performance of the transmission system [4]. Z. Wang has optimized the industrial robot's transmission path in hot stamping line, which guaranteed the rapidity and stability in hot blank transmission system [5]. A. Gasparetto, V. Zanotto has studied the trajectory planning of the robot including cubic splines and fifth-order B-splines interpolation algorithm, and compared them [6].

In this paper, firstly, a three-dimension model of the Cartesian robot was built. Then, the kinematics and dynamics simulation analysis was conducted by cubic polynomial interpolation and parabolic transition linear interpolation, separately. Finally, the movement of the transportation fork and the force of joints were compared in the case of previous two algorithms.

The structure of this paper is as follows: In Section 2, the hot stamping production line is introduced and the simulation model of the hot blank transmission robot is established. In Section 3, the kinematics theory of Cartesian robot and the trajectory planning theory are studied, including the movement equation of Cartesian robot, cubic polynomial interpolation and parabolic transition linear interpolation algorithm. In the end, some conclusions are given in Section 4.

2. Hot Stamping Production Line and the Model of Cartesian Robot

In this paper, the model of production line is composed of seven-layer box-type furance, Cartesian robot and sevo perss, the layout of production line is shown in Fig. 1, the production of the beat of the whole line is 24s.

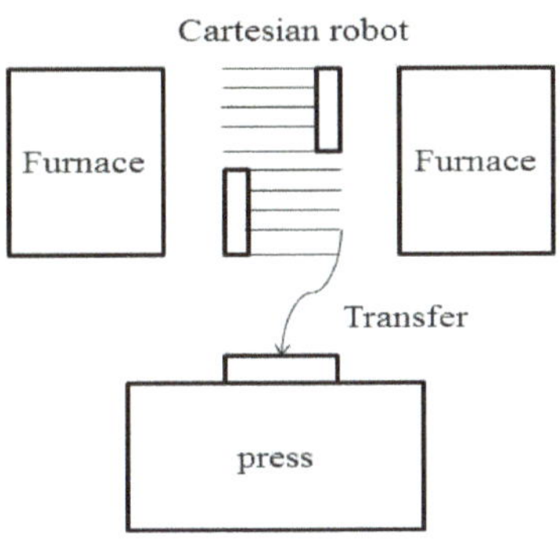

Fig. 1. Layout of the production line.

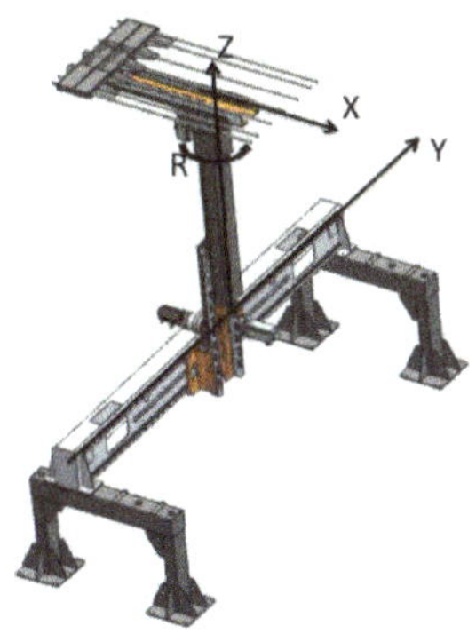

Fig. 2. Simulation model of Cartesian robot.

The three-dimensional simulation model of Cartesian robot is shown in Fig. 2. The Cartesian robot has four degrees of freedom, including three translation joints X,Y and Z, a rotation joints R. The process of hot blank transmission is as follows: Firstly, the transportation fork put into the furnace to get the hot blank along the X direction. Then, after stretching out the furnace, Y joint, Z joint and R joint move at the same time to arrive at position where it is close to the press. Finally, X joint is forward into some distance to hand over with the robot on the press. During the transmission, because it is the most difficult to transfer in the seventh layer of furnace, and poor stability, this paper main studied the seventh layer transmission.

3. Theory of Kinematics and Trajectory Planning

To simplify the research, the components of robot are regarded as rigid parts, and the whole robot is seen as an open loop kinematic chain, its movement equation is built by homogeneous coordinate transformation. Because the transmission robot is composed of three translation joints and one rotation joint, the kinematics equation of transportation fork can be expressed as:

$$T = \text{Trans}(0,y,0)\text{Trans}(0,z,0)\text{Trans}(0,x,0)\text{Rot}(z,\theta) = \begin{bmatrix} \cos\theta & -\sin\theta & 0 & x \\ \sin\theta & \cos\theta & 0 & y \\ 0 & 0 & 1 & z \\ 0 & 0 & 0 & 1 \end{bmatrix} \tag{1}$$

Where: x, y, z are the displacement of X,Y and Z direction, separately; ϴ is the angle of R rotation; Trans(x,0,0),Trans(0,y,0),Trans(0,0,z) are the translational transformation matrix of X,Y,Z direction, separately; Rot(z,ϴ) is the rotational transformation matrix of R direction, T is the displacement matrix of transportation for k.

The trajectory of robot is the kinematics trajectory of transportation fork during working process, namely, the displacement, velocity and acceleration of moving point. specifically, trajectory planning is defined as: according to the requirement of the transportation fork, the suitable interpolation function and constraint conditions are chosen for every joint, and then to solve the function. In this paper, cubic polynomial interpolation and parabolic transition linear interpolation are discussed.

For cubic polynomial interpolation, assuming cubic polynomial as:

$$\text{S(t)} = \alpha_0 + \alpha_1 t + \alpha_2 t^2 + \alpha_3 t^3 \tag{2}$$

In equation (2), 4 constraint conditions are needed: displacement constraints of start and end time, velocity constraints of start and end time, namely,

$$\begin{cases} S(\text{t}_0) = 0 \\ S(\text{t}_f) = \text{S}_f \end{cases} \quad \begin{cases} \dot{S}(t_0) = 0 \\ \dot{S}(t_f) = 0 \end{cases} \tag{3}$$

Then, t derivation of equation (2)

$$\dot{S}(\text{t}) = \alpha_1 + 2\alpha_2 t + 3\alpha_3 t^2 \tag{4}$$

According to equation (2), (3), (4), a_0, a_1, a_2, a_3 can be solved:

$$\alpha_0 = \frac{\left(9t_0^2 t_f + t_0^3\right) S_f}{t_0^3 - t_f^3 - 3t_0^2 t_f + 3t_0 t_f^2} \tag{5}$$

$$\alpha_1 = \frac{6t_0 t_f S_f}{t_0^3 - t_f^3 - 3t_0^2 t_f + 3t_0 t_f^2} \tag{6}$$

$$\alpha_2 = \frac{-3\left(t_f + t_0\right) S_f}{t_0^3 - t_f^3 - 3t_0^2 t_f + 3t_0 t_f^2} \tag{7}$$

$$\alpha_3 = \frac{2S_f}{t_0^3 - t_f^3 - 3t_0^2 t_f + 3t_0 t_f^2} \tag{8}$$

Where: in equ. (2) to equ. (8), S(t) is the displacement function, $\dot{S}(t)$ is the derivateive of function of S(t), t_0, t_f are the start time and the end time, a_0,a_1,a_2,a_3 are the coefficients of the cubic polynomial function.

For parabolic transition linear interpolation, the time of duration t_a at start and end is the same, the value of acceleration is the same but the direction is inverse. Assuming that the middle time of the whole process is t_a, the middle displacement is S_h. According to equal acceleration at the transition time, equation (9) can be obtained:

$$\ddot{S}(t_\alpha) = \frac{S_h - S_\alpha}{t_h - t_\alpha} \tag{9}$$

According to,

$$\begin{cases} S(t_\alpha) = S(t_0) + \frac{1}{2}\ddot{S}(t_\alpha)^2 \\ S(t_h) = \frac{1}{2}(S(t_0) + S(t_f)) \end{cases} \tag{10}$$

According to equation (9), (10), equation (11) can be obtained:

$$t_\alpha = \frac{t_f}{2} - \frac{\sqrt{\ddot{S}(t_\alpha)t_f^2 - 4\ddot{S}(t_\alpha)(S(t_f) - S(t_0))}}{2\ddot{S}(t_\alpha)} \tag{11}$$

Where: t_a is the duration time, S_h is the middle displacement, t_h is the middle time, t_f is the end of time.

4. Result and Discussion

In here, in order to analyze the dynamics and trajectory of transportation fork of the robot, the software ADAMS is used. To minimize the complexity, the torque of the output shaft of the motor is not considered. Simulation results are shown in Figure 3, and their comparisons are in Figure 4, and Figure 5.

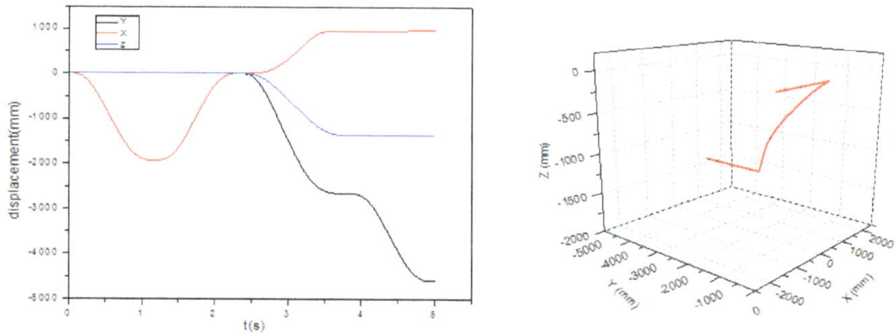

Fig. 3. Kinematics trajectory of transportation fork of Cartesian robot.

According to simulation results, as is shown in Fig. 4, there is little difference between cubic polynomial interpolation and parabolic transition linear interpolation, the velocity of transportation fork change smoothly over time and no mutation. But the Max of velocity of transportation fork by using parabolic transition linear interpolation is greater than by using cubic polynomial interpolation. As shown in Fig. 5, when using cubic polynomial interpolation algorithm, there is the acceleration mutation in the transportation fork, which may cause the decrease of the stability.

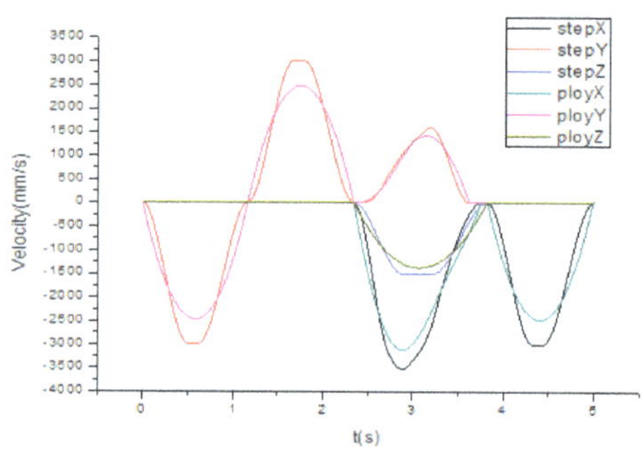

Fig. 4. Comparison curve of velocity.

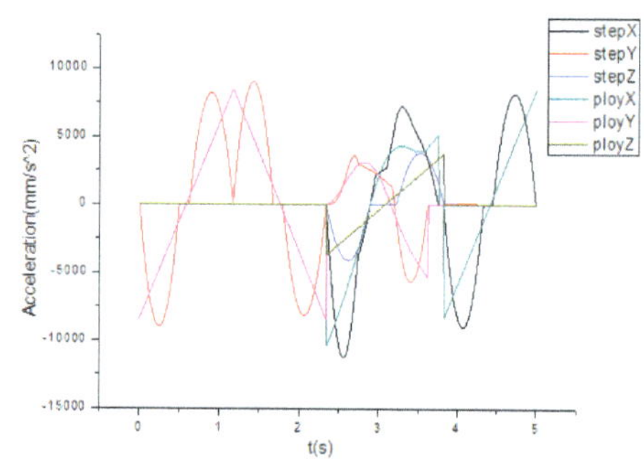

Fig. 5. Comparison curve of acceleration.

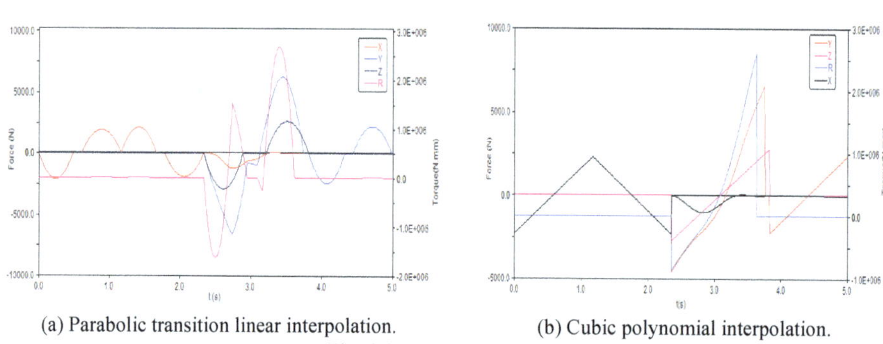

(a) Parabolic transition linear interpolation.

(b) Cubic polynomial interpolation.

Fig. 6. Force curve of joints.

According to Fig. 6, there is the force mutation and torque mutation both on translation and rotation joints, which may lead to severe impact on joints and has a bad influence on the stability of the whole mechanical system. In summary, parabolic transition linear interpolation is more proper than cubic polynomial interpolation for this transmission robot, the movement of the transportation fork is smooth, and stress on the joints is uniform.

5. Conclusion

In this paper, kinematics and dynamics of Cartesian robot were analyzed through ADAMS in case of cubic polynomial interpolation and parabolic transition linear interpolation separately, the result showed that the movement of the transportation fork is smooth, and stress on the joints are uniform when using parabolic transition linear interpolation, which lay a foundation on the vibration analysis of transportation fork. The future main work may focus on improve the stability of vibration of transportation fork and structure optimization.

Acknowledgments

This research work is funded by the National Natural Science Foundation of China (grant no. U1564203) and National Natural Science Foundation of China (grant no. 51405171).

References

1. H. Karbasian and A. E. Tekkaya, A review on hot stamping, *J. Mech. Sci. Technol.* **210**, 2103 (2010)
2. R. Featherstone and D. E. Orin, Robot dynamics: equations and algorithms, in *Proceedings of the 2000 IEEE International Conference on Robotics and Automation*, (San Francisco, USA, 2000).
3. C. Vibet, Symbolic modeling of robot kinematics and dynamics, *Robot. Auton. Syst.* **14**, 301 (1995).
4. Y. Sun, B. Zhu and Y. S. Zhang, The structure design and performance analysis of the transportation fork in hot stamping production line, in *International Conference of ICA3M*, (Shenzhen, China, 2015).
5. Z. Wang and Y.S. Zhang, Optimization of industrial robot's transmission path in hot stamping production line, *Forging and Stamping Technology* **37**, 36 (2012).
6. A. Gasparetto and V. Zanotto, Optimal trajectory planning for industrial robots, *Adv. Eng. Softw.* **41**, 548 (2010).

Research on the Multi-Robots Coordination Conveying Strategy in Hot Stamping Line

W. Fu[1], L. Wang[1], Q. Wang[2], B. Zhu[1] and Y. S. Zhang[1,†]

[1]*State Key Laboratory of Material Processing and Die & Mould Technology, Huazhong University of Science and Technology, Wuhan 430074, P.R. China*

[2]*Dongguan Vision Hot Stamping Technology Co., Ltd. Taozi Industrial Park, Hengli Town, Dongguan City, Guangdong, 523475, China*

[†]*E-mail: zhangys@mail.hust.edu.cn*

In the conveying system of hot stamping line, there exists the static sheet transfer process between the feeding and unloading robot. This paper presents a method called coordination conveying to solve the problem of long conveying time of static transferring. It makes experimental research on the coordinated control strategies of constant and variable speed, analyzes how these parameters, such as overlapping distance, synchronization speed and acceleration impact on the cycle time, and finds the best strategy. The results show that, under the same condition, the constant velocity makes the shortest cycle time, and the variable has the longer synchro time, meaning the abundant time can be used for transfer. There exists the optimal value of overlap distance and synchro speed within the allowable bearing range of the mechanical structure, and the more rapidly the robot conveys, the faster the acceleration is and the shorter the cycle time becomes. With a total conveying distance of 4.9 m as an example, the cycle time can be optimized from 3.2 s to 2.1 s, and reduced by 35 %.

Keywords: Hot stamping line; Feeding/unloading robot; Coordination conveying method.

1. Introduction

The auto's parts, which are made of high strength steel in hot stamping can meet the requirement of lightweight auto, and improve its strength. So it is widely used in auto industry. In hot stamping processes, a high strength steel sheet is heated up in furnace to austenitic state. The hot sheet would be transferred from the furnace to the die for quick stamping and held for several seconds to achieve stable shape and size of parts [1]. The feeding time from the furnace to the die will directly affect the sheet's temperature before forming, and the forming temperature will determine the part's strength. The key link in the conveying process of hot stamping line is to remove the hot sheet form the furnace by the Furnance Unloading Robot (FUR) to the die by the Press Feeding Robot (PFR). The static transfer method is widely used in this process of the current production line. This method increases the conveying time, decreases the sheet temperature quickly, and oxidizes the uncoated sheet severely [2-3]. But moving transfer method can overcome the limitation. This paper presents a quick-run feeding method called coordination conveying. It makes experimental research on the coordinated control strategies of constant and variable speed, analyzes how these parameters such as overlap distance, synchro speed and acceleration impact on the cycle time, and finds the best strategy of the shortest cycle time.

2. Multi-robots Coordination Conveying Method

Coordination conveying is to set an overlap distance in the horizontal conveying direction of the FUR and the PFR. So the two robots achieve the moving transfer by using the overlap segments in the synchro movement and reduce the cycle time. Based on the idea of coordination conveying, the simple model is shown in Fig. 1, and the experimental platform (Fig. 2) is established. The x_1 axis is FUR and the x_2 axis is PFR. The x_2 parallels to the x_1. The PFR's moving space is the vertical area of BC (B'C'). The overlap area is AD (A'D') and the distance is S_c. BC (B'C') is synchro area and the distance is S_S. OE is the total conveying distance. The motion of coordination conveying can be divided into the active motion (x_1) and following motion (x_2). Its process is that the following axis follows the active axis. The x_1 starts first, takes out the hot sheet form furnace, moves forward and regulates the velocity. Then x_2 starts from A' point and regulates velocity. Two axes achieve synchro conditions in the B (B') point. And x_2 starts the synchro motion to follow the x_1.The synchro motion will be removed , x_1 will slow down to stop, and x_2 will sent the hot sheet to the die, if the two axes arrive at C (C') point. The conveying time of BC (B'C') segment is the synchro time. The time must be greater-or-equal to the t_0 which the PFR grabs the sheet from the FUR in the vertical direction to complete the transfer. The aim of coordination conveying is to make the two robots complete the distance of OE conveying in the shortest time [4-5].

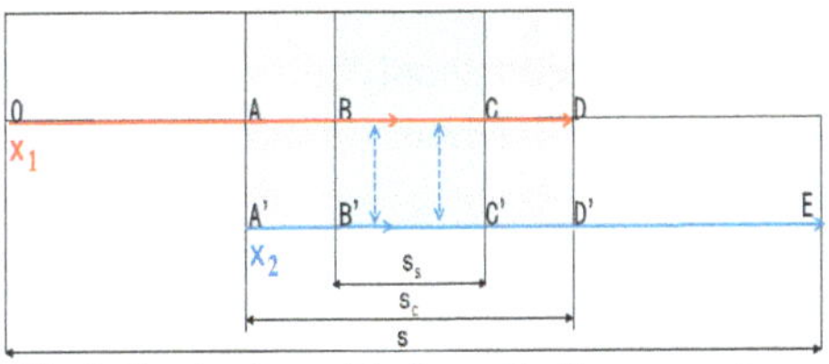

Fig. 1.The simple model of coordination conveying.

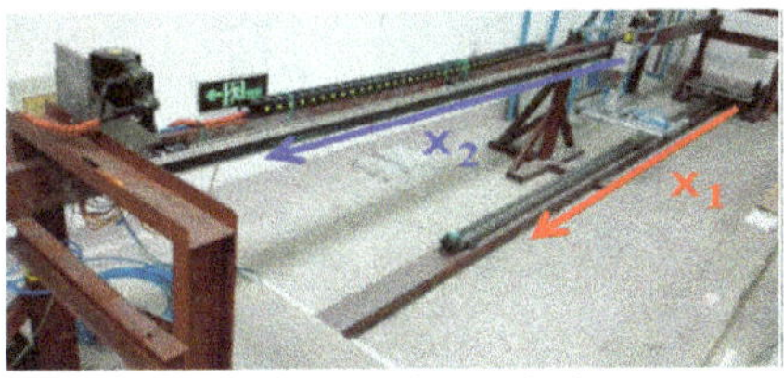

Fig. 2. The experiment platform.

According to the real-time moving performance of following axis, the coordination control strategy is divided into off-line and on-line. It is also divided into the Constant Velocity Coordination Conveying (CVCC) and the Variable Velocity Coordination Conveying (VVCC) according to the velocity state of synchro motion. Off-line coordination conveying is feedforward control. It obtains the optimal value of the operation quantity on the target value through the calculation and repeated experiments. On-line coordination conveying is closed loop feedback control. When the measured value and the target are deviated, the automatic adjustment of the feedback algorithm is achieved. The CVCC refers to that the synchro transfer region is in state of uniform motion. The VVCC refers to that the state is uniform motion and variable motion. The x_1 is slowing down to stop while the x_2 is still following it in this method. The key of achieving CVCC is to keep the two axes' displacement and velocity same. So the two axes achieve the synchro motion in the sheet transfer area and reach a relatively static state of motion to complete the transfer. The adjusted condition is the displacement error of two axes. If the x_2 is behind of x_1, the speed of x_2 will be increased. When the x_2 is ahead of x_1, the speed of x_2 will be decreased. The key of achieving VVCC is the x_2's deceleration ability is stronger than x_1. Based on two closed-loop control method, the adjusted condition is the displacement error and velocity error of two axes. It regulates the x_2 target velocity, making it close or equal to the x_1's actual velocity. If the x_2 is behind of x_1, the x_2's speed will be increased. When the x_2 is ahead of x_1, the x_2's speed will be decreased. By adjusting the velocity, two axes' displacement error in the synchro process will be stabilized within the mechanical structure tolerance [4-6].

3. Experiment and Results

Coordination conveying requires a longer synchro time and smaller and more stable displacement deviation. Its aim is to shorten the cycle time to its maximum under the premise of keeping the sheet transfer completed. The key factor of shortening the tempo is to regulate the overlap distance and the synchro velocity. Increasing the overlap distance properly will shorten cycle time, but the cost of production line will increase. And the synchro velocity is inversely proportional to the overlap distance. Therefore, it is necessary to analyze the influence on the tempo, such as overlap distance, synchro velocity and so on, and find the optimal solution of coordination conveying. In this paper, the synchro velocity is its optimal value. Because of the transfer time t_0 unchanged, if the overlap distance is determined, as long as the synchro time is exactly equal to the transfer time, the velocity will be the optimal value in this case. This paper realizes the two strategies of CVCC and VVCC.

In the CVCC, the relationship between the experimentally measured overlap distance, synchro speed, cycle time, shortening time and acceleration is shown in Fig. 3.

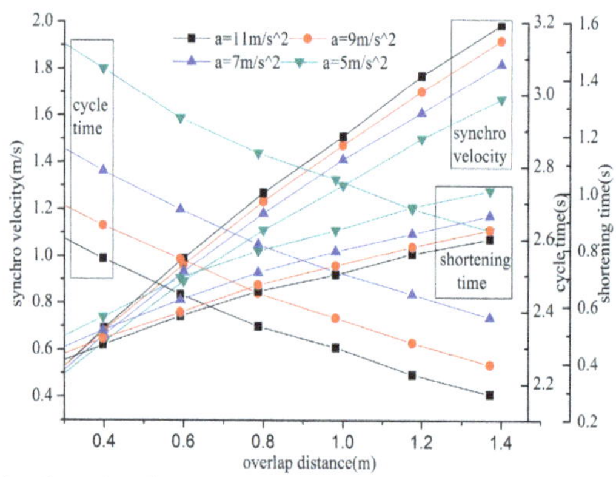

Fig. 3. The relationship of overlap distance, synchro speed, cycle time, shortening time and acceleration.

Under the same overlap distance, as the acceleration increased, the cycle time is reduced and the synchro velocity is increased. When the overlap distance is larger (more than 0.6 m), the optimized tempo by enhancing acceleration is more obvious than increasing the synchro velocity. Clearly observing the figure, when the acceleration is greater than 9 m/s^2, the tempo optimized space decreases continuously. If take into account the cost and other factors, the reasonable acceleration should be below 9 m/s^2.

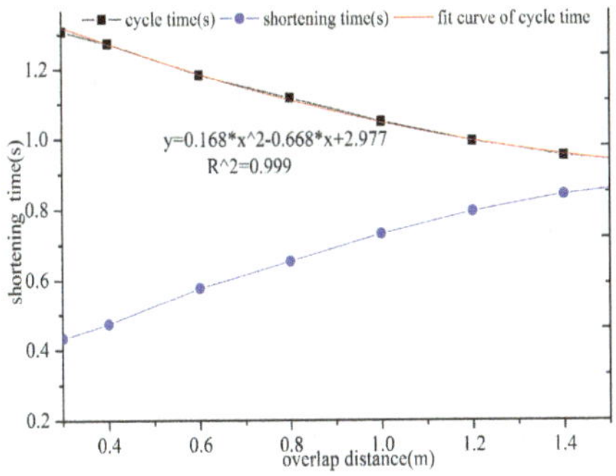

Fig. 4. The analyse of optimized effect.

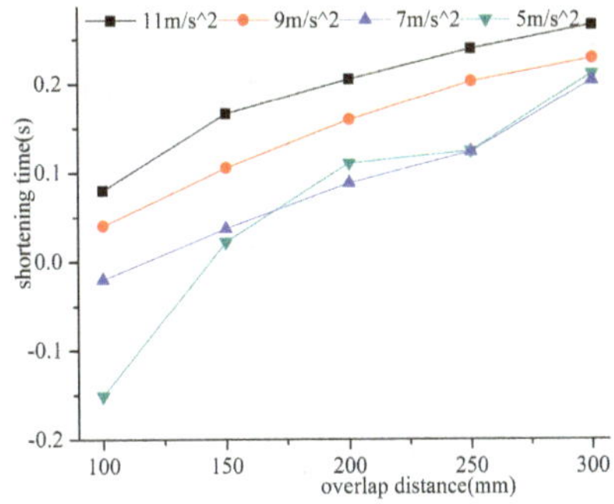

Fig. 5. The optimized effect in limited distance.

Under the same acceleration, with the overlap distance increasing, the synchro velocity is increased, the optimized effect is enhanced, and the cycle time is reduced. However, if overlap distance increases to a certain value, there will be the shortest cycle time. Taking the acceleration of 7 m/s^2 as an example, the cycle time, overlap distance and shortening time are analyzed in the Fig. 4. The fitting curve of cycle time is similar to quadratic function. When the overlap distance is 1.996 m, the shortest cycle time is 2.313s. The overlap distance ranging from 0.3 m to 1.6 m, the optimized effect is better. And the distance ranging from 1.6 m to 1.996 m, the effect is decreased. As the two axes' regulating distance is limited in the limited overlap distance, it's difficult for coordination conveying to shorten the cycle time. The overlap distance ranging from 0.1 m to 0.3 m is analyzed in the Fig. 5. When the overlap distance is less than 100 mm and the acceleration is less than 7 m/s^2, there is no space-optimized by coordination conveying. When the overlap distance is less than 300 mm, the space-optimized of cycle time is less than 0.3 s. If it takes into account the safety and reliability of actual production operating as well as the cost of production line, the reasonable overlap distance should be in the range of 0.3 m to 1.6 m.

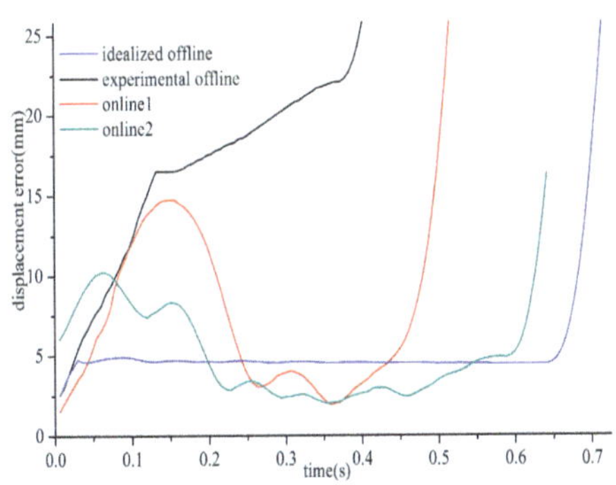

Fig. 6. The displacement error of VVCC.

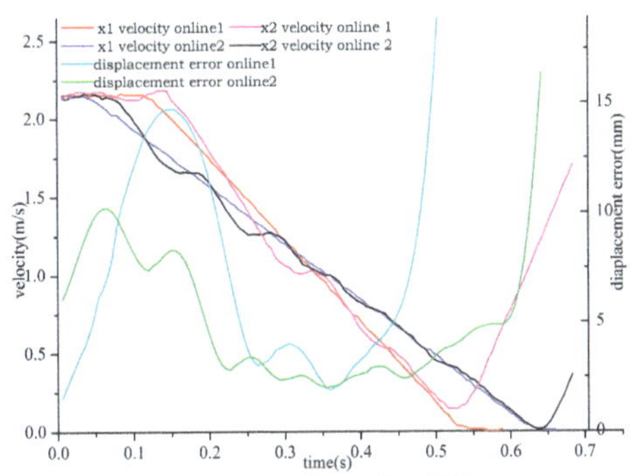

Fig. 7. The effect of VVCC.

Compared with the CVCC, the active axis's speed is dynamic in the VVCC. The following axis needs to capture the active axis's speed in real time, so it can keep pace with the active axis. It's difficult to make the two axes at the same time, same speed and same displacement. According to the data from the off-line condition of two axes independent operation, it is shown in Fig. 6. If the two axes' velocity almost remains same, namely at the same speed and time, in theory, the two axes won't accumulate the displacement error. But in the actual off-line experiment, it is very difficult to ensure that the two axes at the same time and same speed. And the displacement error is accumulating. It is difficult for the two axes to ensure the same position. In the first level closed-loop control strategy, the speed error is gradually reduced and the displacement error (Online1) is improved by adjusting the target speed continuously. Based on it, the second level closed-loop control strategy makes the displacement error as a secondary judgment condition. By adjusting the target speed again, the displacement error (online2) is almost stable in the tolerance of the mechanical structure. So it can realize the effect of VVCC, showing in Fig. 7.

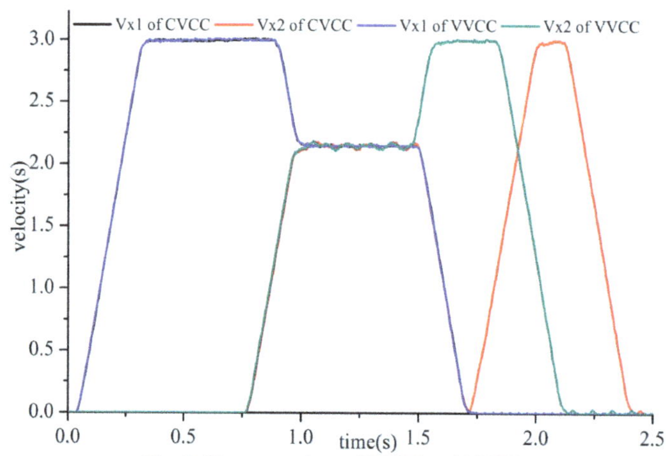

Fig. 8. The comparison of CVCC and VVCC.

Under the same overlap distance, the experiment results of CVCC and VVCC shown in Table 1 and Fig. 8.

Table 1. The comparison of CVCC and VVCC.

Strategy	Cycle time (s)	Shortening time (s)	Synchro time (s)
CVCC	2.100	1.110	0.474
VVCC	2.394	0.816	0.714
No coordination	3.210	0	0

Note: 1. Total conveying distance is 4.9m; overlap distance is 1.4m. 2. Acceleration is 7.92 m/s^2; maximum velocity is 3 m/s; sychro velocity is 2.1 m/s.

Coordination conveying has achieved a significant effect. The CVCC is the most obvious, optimized for 35 % and the VVCC optimized 25 %. But synchro time of the VVCC increased by 51 %. That is, in the same overlap distance, the VVCC provides plenty time for the sheet transfer. Although the VVCC increases cycle time, when the production line layout appears insufficient overlap distance, it can ensure a plenty transfer time. It has more practical significance.

4. Conclusion

This paper studies the control strategy of CVCC and VVCC. It shows that, in the same condition, the CVCC achieves the shortest cycle time, and the VVCC has more synchro time, meaning the abundant time can be used for transfer. There exists the optimal value of overlap distance and synchro velocity within the allowable bearing range of the mechanical structure, and the more rapidly the robot conveys, the faster the acceleration is and the shorter the cycle time becomes. With a total conveying distance of 4.9m as an example, the cycle time can be optimized from 3.2 s to 2.1 s, and the cycle time is reduced by 35 %.

Under the limited overlap distance ranging from 0.1m to 0.3m, this paper did not obtain the desired effect of shortening time by adjusting the coordinated control strategy. The control strategy has no research significance in the limited distance. If the mechanical structure is optimized, the structure design of the clamping claw and the pallet is improved, the mechanical tolerance of the transfer is increased, and the flexible cooperation of the clamping jaw and the pallet can be improved, the problem of coordination conveying in the limited distance will be solved.

Acknowledgements

This research work is funded by the National Natural Science Foundation of China (Grant No. U1564203) and (Grant No. 51405171).

References

1. H. Karbasian and A. E. Tekkaya, A review on hot stamping, *J. Mater. Process Tech.* **210**, 2103 (2010).
2. T. Heller, B. Engl, B. Ehrhardt and J. Esdohr, New high-strength steels-production, properties and applications, *Iron and Steel Society/AIME* **36**, 25 (1998).
3. Z. Wang and Y. S. Zhang, Optimization of industrial robot's transmission path in hot stamping production line, *Forging and Stamping Technology* **37**, 36 (2012).
4. G. Zheng, L. Wang, Q. Wang, Y. Sun, J. Meng and Y. S. Zhang, Research on the cooperative work of multi in hot stamping line, in *International Conference of ICA3M*, (Shenzhen, China, 2015).
5. L. Wang, B. Zhu, Q. Wang, J. Meng and Y. S. Zhang, Production Control and Optimization of Hot Stamping Line, in *5th International Conference on Hot Sheet Metal Forming of High-Performance Steel*, (Toronto, Canada, 2015).
6. T. P. Lu and H. Y. Yi, An agent-based production control framework for multiple-line collaborative manufacturing, *Int. J. Product Res.* **39**, 2155 (2001).

Development and Application of High-Performance Manipulator for Hot Stamping

R. L. He, J. Z. Miao[†] and D. F. Zheng
Jinan Aotto Automation Co. Ltd,
Jinan, 250306, China
[†]*E-mail: miao_jinzhong@aotto.cn*

In the present paper, the development and application of high-performance fast loading manipulator for hot stamping were introduced, based on characteristics and requirements of the hot stamping process, the parameters of high-performance fast manipulator has been proposed, the major components and mechanical hand and arm of fast manipulator have been identified through the manipulator's optimized design, stiffness analysis and reasonable selection, the reasonable matching parameters of fast manipulators and presses for fast closing have been determined by self-developed multi-position analysis control software, which provided the theoretical and experimental basis for system debugging. The developed manipulator achieved the expected effect, which has characters of reliable performance and easy maintenance. It had practical applications in production, and achieved good results.

Keywords: Hot stamping; Fast loading manipulator; Performance.

1. Introduction

The critical process of hot stamping contain blank preparation, unstacking, transformation blank in heating furnace, discharge, transfer die, stamping, pressure maintaining, remove parts, cutting and shot blasting oiled (for no coating sheet parts). Heating process and transfer process after heating of workpiece will directly affects the performance of high-strength steel stamping and subsequent hardening effect, stamping process is accompanied by quench hardening, which will determine the hardening and mechanical properties of parts. In the entire process, furnace, hot press forming, the workpiece transfer apparatus, hot stamping die and laser cutting machine are the main core of the system. The implementation of key process proposed a very high technical requirement for the entire system's adaptability, stability and security. Currently, there are more than 60 hot stamping production lines in China, but 90% of them were imported from abroad [1], the import of the production line not only shows high prices, but also increases the investment in fixed assets and the cost of production for users.

Many key technologies in the import of production lines in were not imported, which will affect the run and maintain of production lines, and the urgent need of corporation among automation companies, equipment manufacturers and research institutes to develop the entire heat molding process and equipment manufacturing technology capability, break the blockade of foreign technology, reduce the completion of the investment and accelerate the development of hot stamping in China. In 2011, Huazhong

University of Science and Technology built the first electric servo press stamping equipment, multi-chamber furnaces for the heating equipment for hot stamping test line in China, based on the technology prototype, the first hot-stamping production line, which is fully digital servo electric press combined multi-chamber furnaces has been built in 2015.

The present paper aimed to develop the loading and unloading manipulator with characteristics of fast, efficient, large stroke, high rigidity, easy maintenance and low fault rate, and match-related software systems to improve the produce take time of entire line of hot stamping production line to ensure quenching quality of components.

2. The Design and Performances of Loading Manipulator

Depend on the requirements of hot stamping technology, the time that fed by manipulator blank from crawling to the mold positioning location should be controlled between 2s to 3s, that means the blank to the mold location is rapid process, so the manipulator should be a fast, efficient and stable accurate requirements. The manipulator in the present paper used X, Y, Z-axis movement to achieve with typical coordinate movement shown in Fig. 1. The manipulator completed the clamp, lift, feed stroke, rall down, unclamp stroke, return stroke in a cycle to achieve the transfer of the blank. The horizontal-axis adopted the multiplier speed structure, which can realize large distances and fast transfer for heated sheet. Because tooling hold blanks from both ends, which is in favor of reduction the press stroke and directly improve the stamping tact. The movements in three directions for manipulator are all driven by a servo motor which movement parameters can be controlled by programming.

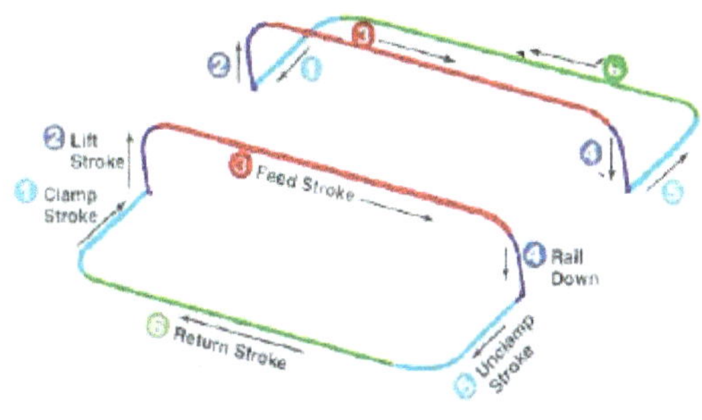

Fig. 1. Three coordinate movement model of fast manipulator.

2.1. *Technical parameters of manipulator*

According to movement model and loading take time and process requirements of hot stamping, the technical parameters were determined as follows:

Feed travel	0～3600 mm	Speed V⩽6 m/s
Unilateral clamp travel	0～1200 mm	Speed V⩽1.2 m/s
Raising travel	0～600 mm	Speed V⩽1.2 m/s
Positioning accuracy	<±0.3 mm	

Maximum load 160 Kg

Empty running tact with loading 15 spm

Number of moving axis 12

Number of servo-actuator X axis (transfer direction) 4

Y axis (clamping direction) 4

Z axis (raising direction) 4

Speed of feeding and rising of manipulator is higher than that of foreign, which are 5 m/s and 0.8 m/s, respectively.

2.2. *Optimization of function and analysis of design for manipulator*

In order to ensure the operation reliability and stability for functions of manipulator, the optimizations of design have been achieved.

(1) High speed conveyer function

To ensure the motion performance of each axis in manipulator, preliminary design of the manipulator core component contains selecting of motor reducer, which are analyzed in a professional design selection software by adopting Siemens Sizer software [2]. The design parameters of manipulator have been input into the software as shown in Fig. 2.

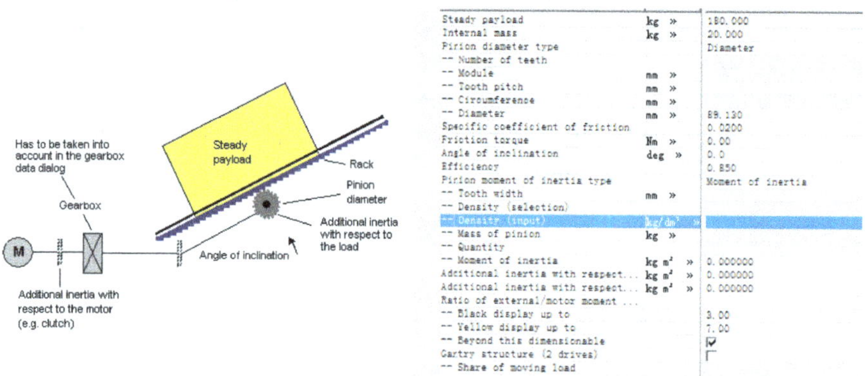

Fig. 2. Results after inputting of mechanical parameters screenshot in software.

The selected results including in the power of each axis manipulator configuration and associated components should be ensure that the manipulator is running under the required speed to meet the take time. And at the same time there is enough room for improvement for the results. The Rockwell's movement analysis software has been adopted to analysis the Y-axis motion parameters [3], the acceleration, displacement and other parameters under different time periods of manipulator which are shown in Fig. 3. It can be noted from Fig. 3 that the correlation results meet the design requirements. The Y-axis power transmission capacity were analyzed by utilization Otto's self-developed multi-position motion control software, the results show that the manipulator can be transmitted work piece with weight of 160 kg, and there is surplus capacity.

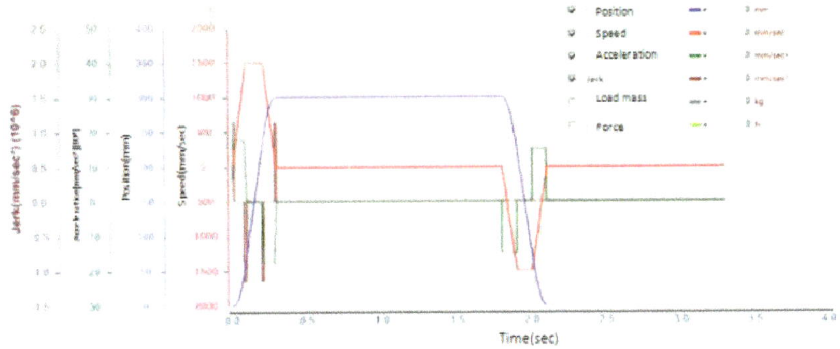

Fig. 3. Acceleration-displacement curves of movement parameter of Y axle.

(2) Analysis of stationarity of manipulator operation

Dynamic analysis of the ability of the Y-axis drive system is shown in Fig. 4. These results show that the original design of the manipulator can meet the requirements of high-speed movement.

In general uniform accelerated motion, although there is not "rigid impact" caused by the mutation rate, but there is "soft impact" caused by the mutation of acceleration of which will bring dithering and noise when the manipulator start and stop. Currently, in debugging process of hot stamping production line, it is very important that each axes motion parameters (speed, acceleration, jerk) of manipulator will be set up to ensure the smooth operation of the manipulator.

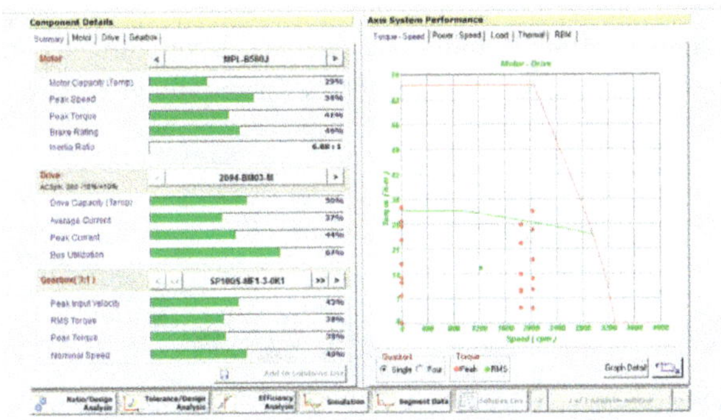

Fig. 4. Powertrain capacity of Y-axis.

The best motion parameters of axles were calculated by developed multi-station calculation procedure according to axles motion distance and time. As the on-line debugging of manipulator, the axles of manipulator were set according to the various parameters calculated by developed software to avoid the flexibility impact of axles during the process of high speed frequently starting and shutting down, and to ensure stable operation. The matching result of manipulator and press motion parameters calculated by multi-station analysis software are shown in Fig. 5.

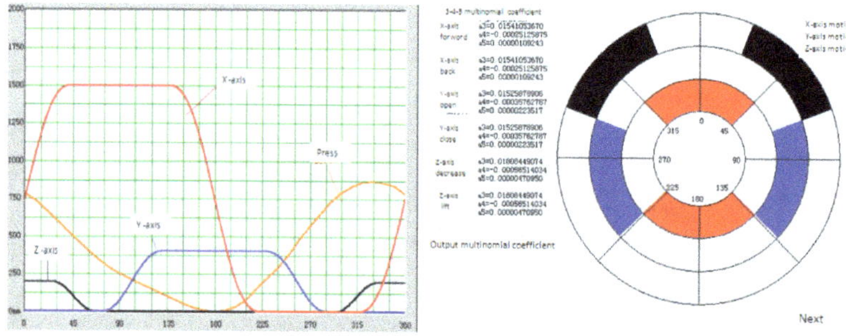

(a) (b)

Fig. 5. (a) match curves of manipulator and press; (b) time sequence colour fig. and polynomial coefficients.

Manipulator chooses double arms structure, which ensured the operation smoothness of parts feeding process, also the feeding speed was increased, the delivery distance was decreased, and the interferential time generated during the press closing process was significantly reduced.

(3) Optimization design, stiffness analysis and material selection of mechanical arm for manipulator

In order to meet the large size sheet hot stamping requirement, the length of carrying arm was 2.5 meters, its motion stroke was 1.8 meters, and 3.6 meters carrying distance were finished by times speed axle.

The main body of mechanical arm was manufactured by hollow Q345 sheet welding, stabilization treatment after welding and machining process, compact overall size of mechanical arm and higher straightness and stiffness were obtained by this processes and design, also guaranteeing the stability and repeated positioning accuracy of manipulator. Optimization design and deformation analysis of manipulator were shown in Fig. 6.

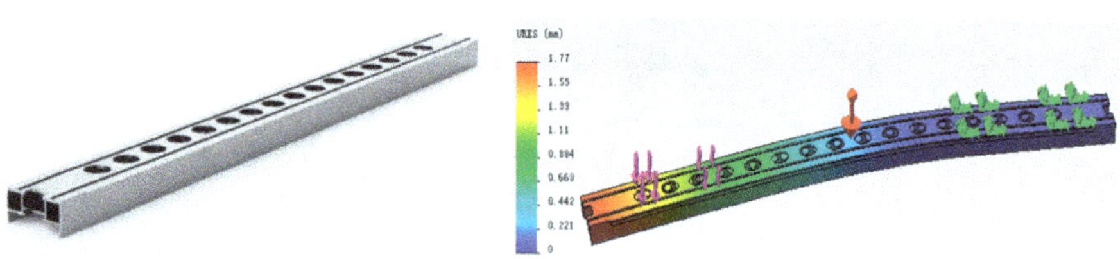

Fig. 6. (a) Manipulator optimization; (b) Arm deformation analysis figure.

In order to verify the suitable design of manipulator, static analysis, stiffness analysis and loading test of manipulators were carried out with 80kg loading on the front end of arm, the deflection of loading position was 1.5 mm, which was far below similar product, the result showed a higher stiffness of manipulator.

(4) Electric control

Controller of manipulator chose Siemens D series, with Profibus-DP / ProfiNet BI, the I / O points and safety related in-out points were directly connected with press control systems and automatic controller by Profibus-DP / ProfiNet.

Servo-motion axle chose Siemens numerical control servo systems, precise actual position and speed linear motion were obtained by servo drive and the stable running, no tremble and precise positioning were also ensured.

The HMI chose hand-held type movement screen with buttons and touch function which can make to convenient for installation, operation, monitoring and data input.

(5) Maintain and use

The hot stamping equipments were run at a relative wicked condition compared with traditional cold stamping process, dust and high temperature have an significant influence on equipments life. The motion components of axles of developed speed manipulator were selected by organ shield, also the gear and rack of axles chose centralization lubrication, regular intervals adding oil to pump should be carried out to execute axles automatic lubrication. Maintain is convenient, comparing with present lubrication station with changing battery and nitrogen tank, the cost was reduced, and the equipments performance stability and using life can be ensured.

3. Development and Application of Simulation Software

As the improvement and application of hot stamping lines, more and more hot stamping parts were used in Body in White, because the various parts size and shape, different press holding time and finite motion space, so coherent motion between mechanical axles and press was a complex question, the manipulator and press should reach a best match up state to ensure the high speed and no interferential motion. Therefore, based on manipulator motion data and press quickly closing coherent analysis, the simulation software was developed, based on different parts, tools and press motion parameters, the best manipulator motion times and state were analyzed by software to ensure the motion match between manipulator and press, which was listed into general control systems as reference data, saving large debugging time.

The 3D figure of developed feeding manipulator was shown in Fig. 7, which has been used in DongFeng TQM hot stamping lines, shown in Fig. 8, obtaining a good application effect.

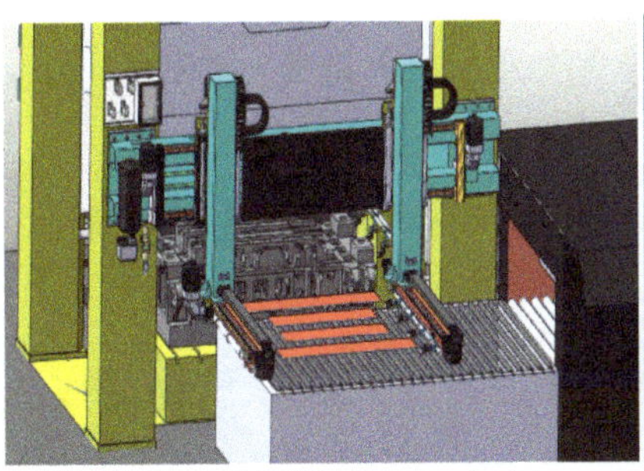

Fig. 7. 3D model of feeding manipulator.

Fig. 8. Application of feeding manipulator.

4. Conclusion

(1) Based on hot stamping processes requirement and feature, through reverse engineering analysis, the work parameters of high speed feeding manipulator was proposed (stroke and speed).
(2) According to function optimization design and analysis of manipulator, the typical and key parts parameters for manipulator and the reasonable matching among parameters of every motion parts were analyzed and calculated by using of selecting software "Sizer" and AB motion analyzer, multi-station calculation procedure to ensure high speed and stable operation of manipulator.
(3) The optimization design, material choosing, heat treatment and deformation simulation analysis of mechanical arm were carried out to meet the requirement of high stiffness and large stroke of arm.
(4) Self-developed simulation software and application make the motion data for motion parts of manipulator and press get reasonable matching, which was used as reference data for system debugging and Bus-Mastering. The final developed manipulator reached the expected effect with reliable performance and convenient maintenance, also application in actual hot stamping lines.

References

1. M. T. Ma, Y. S. Zhang, L. F. Song and Y. L. Wang, Research and Progress of Hot Stamping in China, *Adv. Mater. Res.* **1063**, 151 (2015).
2. Drive Design Tool SIZER for Siemens Drives, 2016.
3. Allen-Bradley, Motion Analyzer, Rockwell Automation software, 2010.

Product Properties

Study on Fatigue Property of Wheel Steel Under Different Processing Methods

Y. C. Ling† and Z. Yan

China Automotive Engineering Research Institute Co., Ltd.,
Automotive Lightweight Engineering Research Center, Chongqing, 401122, China
Chongqing Automotive Lightweight Engineering Research Center,
Chongqing 401122, China
†E-mail:lingyangchao@163.com

The effects of different process routes on fatigue properties of DP600 and FB540 two wheel steels are studied, including the raw material (R), pre drawing (RP), and baking (RPB). With standard tensile specimens as the experimental object, the tensile test and the tension-compression cyclic loading fatigue test have yielded basic parameters of materials, including the material's stress-strain curve and strain life fatigue curve. Simple sample simulation results verify the quasi static analysis method with the feasibility study. After obtaining the basic parameters of the material, according to the basis of the wheel to change wheel shape, the fatigue life of the wheel and the new wheel of modified shape, the quasi-static analysis method for stress and strain on the wheel were undertaken using ANSYS software to study the strain fatigue analysis between the original and new modified wheel. Studying the influence of the bending fatigue life with an original wheel and a diverted wheel, analysis results showed the influence of different material technologies on the material of the original wheel and the diverted wheel. The influence had the same tendency: the wheel modification achieved the goal of losing wheel weight. Finally, experimental verification of the modified wheel, simulation and experimental results show that, simulation results and fatigue life test results were approximate. The crack that appeared in the wheel fatigue experiment had a consistent position with the fatigue life simulation minimum cycle times. The validity of the fatigue strain and fatigue life analysis is verified. This paper provides an effective way to study the influence of material technology on the properties of materials, and it has a certain practical value.

Keywords: New process of hot stamping; High strength steel; B pillar; Side collision performance; Optimization design.

1. Introduction

With the improvement of the level of materials technology, performance materials and material handling processes are closely related. Suitable process can be related to the mechanical properties of the path a greater degree of increase of the material.

Traditional fatigue strength and fatigue life analysis often using only raw material parameters as input parameter analysis, and the impact neglect basic process path on the material properties, into the raw materials used to analyze parameters of fatigue simulation results will bring greater error process. The method is significant performance impact on the material studied.

Wheel is rotating components in direct contact with the ground; the vehicle was in shock load by more complex, automotive wheel process for materials research and structural lightweight design has some practical value [1-3].

2. Simple Specimen Fatigue Test and Simulation

2.1. *Simple specimen fatigue test and simulation*

In this paper, the object wheels include wheel material library DP780 materials, raw materials (DP780r), pre-stretched (DP780rp) and three different pre-stretching material + baking process path (DP780rpb) standard tensile specimen fatigue life test is conducted the test object. Material fatigue tension and compression testing machine for the SANS CMT5305, as shown in Figure 1. Simple sample size is shown in Figure 2.

Fig. 1. Fatigue testing equipment.

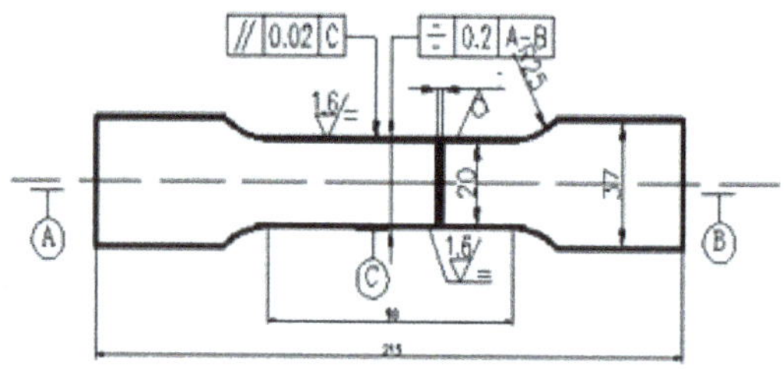

Fig. 2. Material simple sample size.

According to GB / T 15248-2008 conduct strain fatigue tests, the cyclic stress-strain curve DP780r, DP780rp and DP780rpb three processes under the path shown in Figure 3.

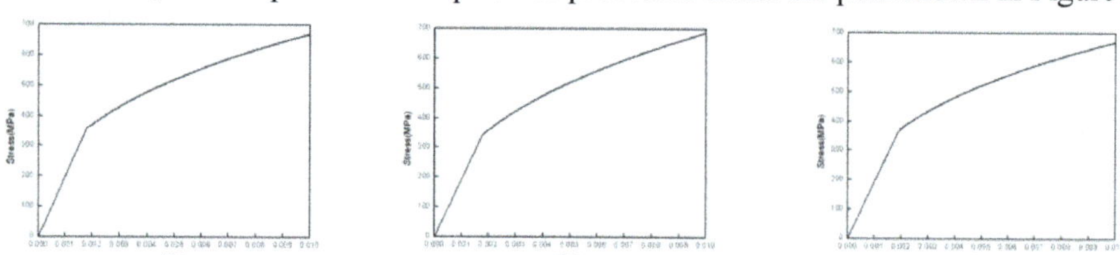

(a) DP780r (b) DP780rp (c) DP780rpb

Fig. 3. DP780 cyclic stress-strain curves under different processes path.

2.2. *Simple sample strain fatigue analysis*

Simple sample strain fatigue analysis using quasi-static strain fatigue analysis method by calculating revolves once each time the quasi-static stress-strain magnitude, edit, load spectrum for a simple sample loop course will be described. To arrive at a simple sample loop cycles throughout the course of the process, get a load cycle under certain operating conditions of the sample, which is the strain cycle fatigue life of the specimen. Fatigue simulation process shown in Figure 4.

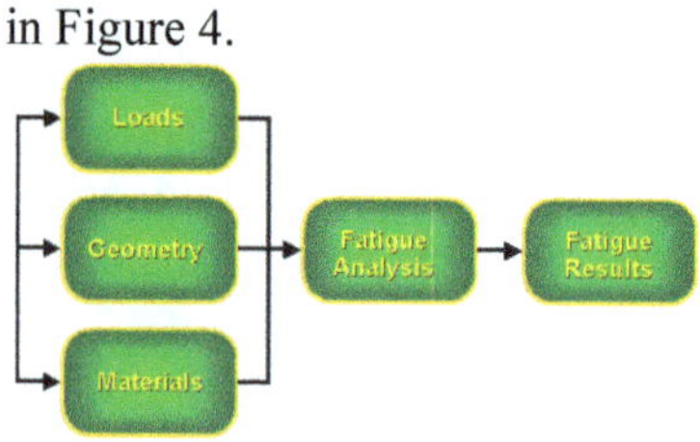

Fig. 4. Fatigue simulation process.

Here is given only DP780r simple specimen tensile and compressive strain overall working conditions and strain analysis results shown in Figure 5.

The simple sample strain calculation result as input load basis, the preparation of cyclic load spectrum, a simple sample different strain fatigue life under load analysis, the conclusion that DP780 materials strain fatigue life curve under different paths.

(a) stretch condition (b) compressed condition

Fig. 5. DP780r strain contours in the quasi-static conditions.

2.3. *Simple sample strain fatigue analysis*

Comparing simple sample test results and simulation of strain fatigue life curve shown in Figure 6.

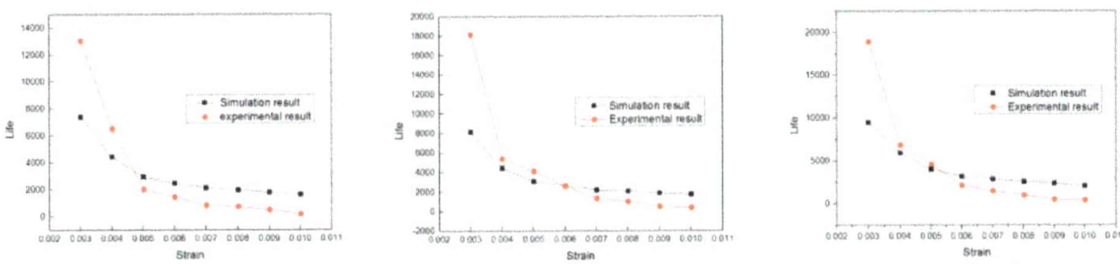

(a) DP780r (b) DP780rp (c) DP780rpb

Fig. 6. DP780 strain-life curve comparison test and simulation under different process paths.

Test and analysis results show that:

1. The technological path has a greater effect on the fatigue properties of the material, different process paths has the different material properties.
2. The resulting strain fatigue life simulation with the experimental curves obtained are basically the same, indicating that the use of quasi-static strain fatigue life analysis methods it is feasible, the method may use engineering analysis.

3. Wheel Fatigue Life Analysis

A section on structural improvements wheels optimizes the use of quasi-static strain fatigue life analysis method for improving the wheel bending fatigue life analysis. Three different wheel materials is a material process path, that DP780r, DP780rp and DP780rpb.

The first step in quasi-static strain field analysis wheels, wheel static load analysis model shown in Figure 7, at the end of the loading lever bending load is applied, the load distribution shown in Figure 8.

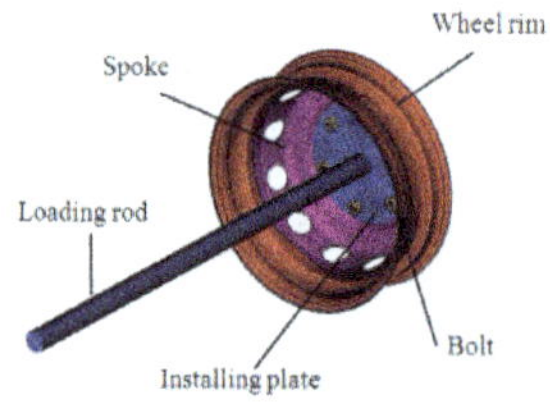

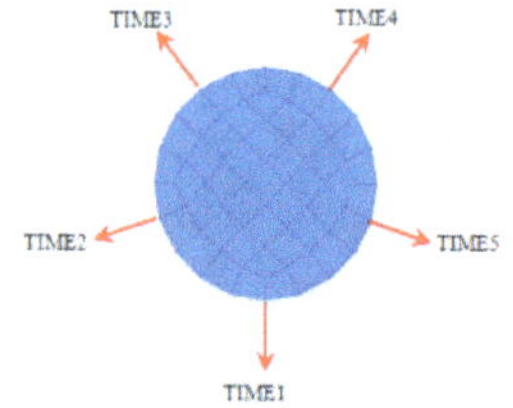

Fig. 7. Wheel static load analysis model. Fig. 8. Wheel static load substep.

They were calculated DP780r, DP780rp, DP780rpb lower at 5 substep of wheels strain field distribution. Only three materials are given here in Figure 9 strain distribution field TIME1 time shown.

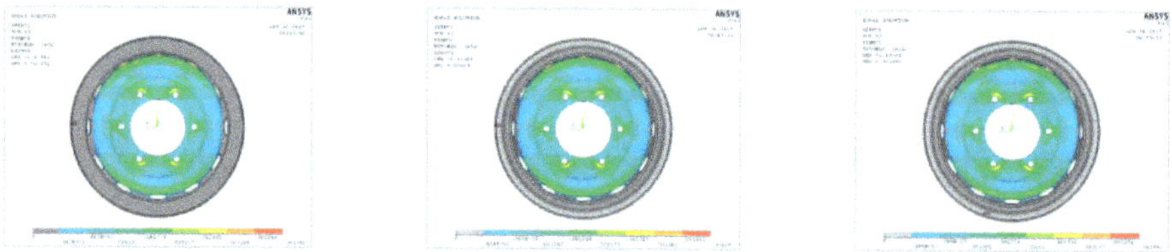

(a) DP780r (b) DP780rp (c) DP780rpb

Fig. 9. DP780 in TIME1 time under different process paths strain distribution.

After the calculation of the three processes under way five substep strain field distribution obtained by editing the simulation wheel rotation load time history fatigue using software ANSYS Design Life strain fatigue module wheel strain fatigue analysis. Three materials derived wheel fatigue life based on contours shown in Figure 10.

The results showed that different process paths material fatigue properties of the material having a greater impact, DP780r cycles is 41358, the shortest fatigue life; DP780rp to 47342 cycles, the fatigue life is longer; DP780rpb cycles is 57621, the longest fatigue life.

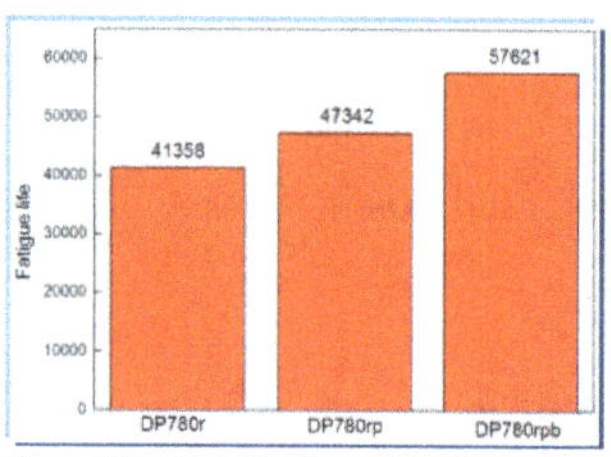

Fig. 10. DP780 fatigue life distribution at different process paths.

4. Wheel Fatigue Life Analysis

Car wheel bending fatigue test machine as shown in Figure 11, DP780r flexural fatigue crack forms on the wheel shown in Figure 12.

The results showed that, DP780r wheel bending fatigue crack position and the minimum life simulation position consistent simulation of fatigue life cycles and test results are basically consistent.

Fig. 11. Wheel bending fatigue test machine.

Fig. 12. DP780r flexural fatigue crack forms.

5. Conclusion

Simple sample tensile and compressive cyclic loading fatigue test results DP780r, cyclic stress-strain curve DP780rp, DP780rpb three different paths craft materials, strain fatigue life curve. Explore the use of quasi-static strain fatigue life simple sample method fatigue life analysis, the results show that this method can reflect the real fatigue properties, to predict fatigue life.

The simple sample material fatigue test results of the cyclic stress-strain curve and strain-life curve as the basic material parameters of the wheel bending fatigue life analysis and the results showed that three different technology paths, fatigue life from small to large order of DP780r <DP80rp <DP780rpb. A process material has great influence on the fatigue properties of the material, pre-stretched and baked treatments could improve the fatigue properties of the material.

The use of raw materials wheel bending fatigue test, the test results show that the fatigue cracking location and Analysis cycles minimum position is consistent, the number of cycles or less.

Acknowledgment

This research work is funded by the fund project: State key laboratory of automobile lightweight construction unit open issue (Serial number: 20130203).

References

1. M. Ma and B. Wu, *Dual steel-physical and mechanical metallurgy*, 2nd edition. (Metallurgical industry Press, 2009).
2. F. Hou, Research on US side impact regulations based on computer simulation of the car door impact bars, 47 (2005).
3. S. Zhao and Z. Wang, *Design of fatigue resistance-method and data*. (Mechanical Industry Press, 1997).

The Application of High-Strength Steel for Threshold Beam

J. W. Li[†], T. Q. Fan, Q. J. Zhao and F. Wang

China Automotive Engineering Research Institute Co., Ltd.
Automotive Lightweight Engineering Research Center,
Chongqing, 401122, China
Chongqing Automotive Lightweight Engineering Research Center
Chongqing 401122, China
[†]*E-mail: jinwei926@163.com*

At present, more and more high strength steel are applied in the automotive industry due to their promising achievements in light-weight and crashworthiness. To verify the safety attribution of high strength steel components, this paper deals with the performance evaluation of the threshold beam used for high strength steel on its rigidity and crashworthiness. In the three-point static compression test, two threshold beams using different high strength steel are used and their performance are compared to each other. Then, a component simulation model is built, with which the crash performance of the threshold beam system in the completed car impact test can be evaluated. Finally, a white body test is carried out to analyze the threshold of the beam's dynamic performance using different high strength steel.

Keywords: High strength steel; White body test; Threshold beam.

1. Introduction

Faced with the increasing demand of automotive safety and lightweight design, High Strength Steels have gained a widespread popularity. Studies have shown that, in the form of a variety of car collision accident, side impact occupant injury as a result of major accidents. Upon impact, the deformation is small due to the high strength steel plate in the process of collision, and can in the moment of collision; the impact energy dispersion will transfer out, to ensure that the body deformation is small. This can effectively reduce the deformation of the cockpit, to protect the safety of occupants, so more and more high-strength steel material is applied on body safety parts.

As the key part of the body frame, the threshold beam is the key factor that affects the load transfer and the side impact resistance of the whole vehicle. To ensure the vehicle has good resistance to impact performance, increase security, we need the threshold beam has a strong strength and flexural properties. And with the rising demand for automotive lightweight, high-strength steel sill beam on the vehicle body has a strong prospect.

2. Simulation Analysis of the Threshold Beam under Side Impact

Studies have shown that components separated from the vehicle and remain on it, that the crash performance of the parts are not identical, but both trends is still the same.

Therefore, we can use the components tests to evaluate the vehicle' performance. For the deformation characteristics of the threshold beam under side impact [1, 2, 3], the method of the collision test of the threshold beam is designed, as shown in Fig. 1 in accordance with C-NCAP management procedures; the white body is fixed and mounted on the car. The initial speed is 50 km/h.

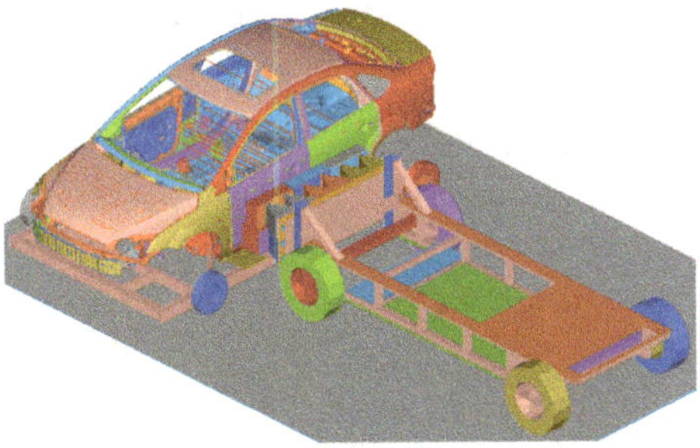

Fig. 1. Side impact finite element model.

After the collision, the intrusion and energy absorption of the threshold beam are used as the evaluation index. The deformation of the threshold beam is shown as follows in Fig. 2. As we can see, the deformation trend of the threshold beam is right, and the maximum deformation of the threshold beam is 34.2 mm. Energy absorption is 4.4 KJ.

Fig. 2. The deformation of the threshold beam.

3. The Side Impact Test Method of the Threshold Beam

In order to verify the validity of the simulation, a white body test is carried out. If the intrusion between the test and simulation are in coincidence, then it can ensure the effectiveness of the body collision model, we can use the white body test to replace the vehicle test to evaluate the performance of the threshold beam. When installed, the BIW installed in the trolley according to the actual installation state, the initial velocity of the moving vehicle is 50 km / h. as shown in Fig. 3.

Fig. 3. The test of the threshold beam.

After the collision of the deformation of the threshold beam shown as in Fig. 4, the maximum deformation of the threshold beam is 25.6 mm.

Fig. 4. Deformation between test and simulation.

As can be seen from the test results, the test results and simulation results are consistent with each other. This can illustrate the effective of the simulation model.

4. Application: Light Weight Research of the Threshold Beam

With the continuous improvement of the lightweight and safety requirements, the high strength steel is widely used in the key structure of the car body. In the existing model, the material of the threshold beam is B410LA, the thickness is 2mm.taking account of the need for lightweight, trying to adopt HC950/1180MS to replace the current material. At the same time, is needed to meet the crash performance and stiffness properties. According to the two kinds of high strength steel material, the thickness and weight are as shown in Table 1.

Table 1. The comparison between two materials.

Material	Thickness (mm)
HC950/1180MS	1.4
B410LA	2.0

4.1. *Research on the threshold beam's stiffness properties*

We can get support reaction force through three-point static compression test. This can seem as a performance evaluation of the threshold beam stiffness index. One ends of the threshold beam are from constraints, and the other of the ends are fully constrained. The center position of the pressure head loading point is located in the sample; the force is always perpendicular to the direction of the two pivot connection. The loading velocity is 0.001m/s, when obvious cracks yield or the loading displacement reaches 130mm, the tests stop. As shown in Fig. 5.

Fig. 5. Three-point static compression test.

The results of the three-point static compression tests of the threshold beam are shown in Fig. 6, two groups of samples without obvious damage, and the deformation are consistent with each other.

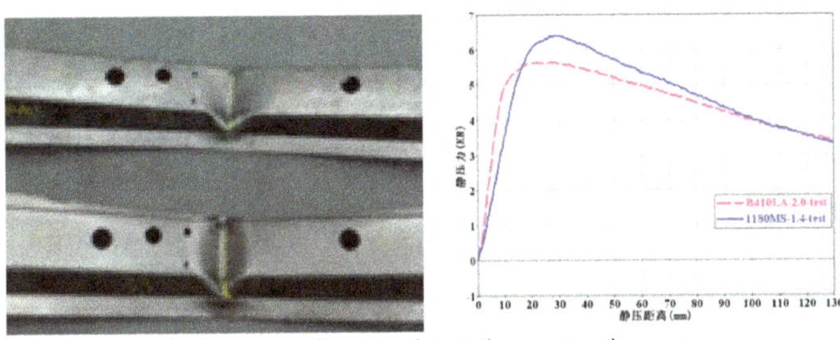

Fig. 6. Result of three-point static compression test.

The peak of the static pressure of the threshold beam are shown in Table 2, as can be see, the peak of the material with HC950/1180MS is higher than with B410LA, have a good stiffness performance.

Table 2. The comparison between two materials.

Material	Thickness (mm)	Weight (Kg)	Peak of pressure	
			Stress (KN)	Displacement (mm)
HC950/1180MS	1.4	4.74	6.39	28.7
B410LA	2.0	6.77	5.66	27.6

4.2. *Study on the side impact performance of the threshold beam*

Based on the above method, the finite element model of the white body is established, and the finite element simulation analysis of the threshold of the two materials is made. After the end of the simulation, the deformation of the threshold beam as shown in Figure 7 below.

Obviously, the deformation of the threshold beam with two different materials are basically the same, the large deformation occurred in the middle position.

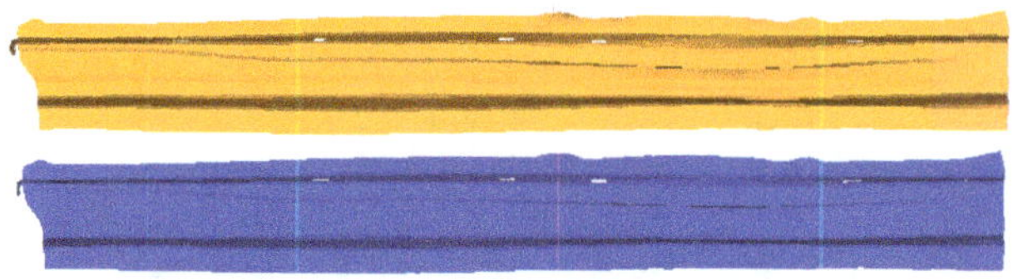

Fig. 7. Deformation between test and simulation.

After the collision, the trolley acceleration curve as shown below: As can be seen from Fig. 8, the acceleration trend of the threshold beam are consistent with each other, the peak of the acceleration is also similar. The acceleration of the vehicle that the threshold beam used B410LA is 32.7 g, the acceleration used HC950/1180MS is 33.0g.

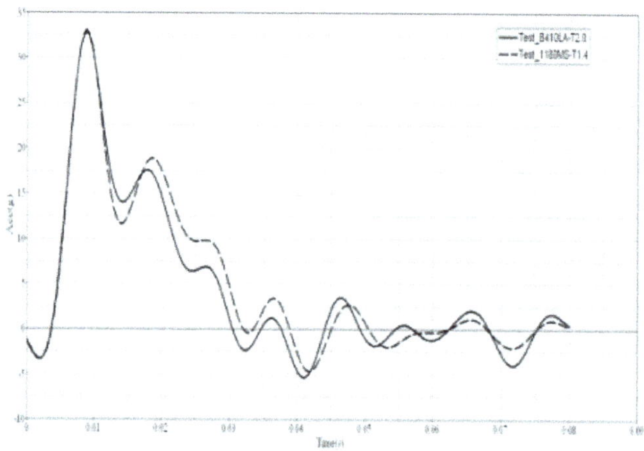

Fig. 8. Acceleration curves.

The maximum displacement of the trolley that used B410LA is 32.1 mm, while that used HC 950/1180 MS is 32.7 mm. the stability deformation displacement with two kinds of materials are shown as follows, for the threshold beam that used B410LA, the displacement at the left end is 11.2 mm, at the middle position is 6.2mm, at the right position is 12.4 mm, for the threshold beam that used HC950/1180MS, the displacement at the left end is 8.7 mm, at the middle position is 5.5mm, at the right position is 10.7mm. As shown in Fig. 9 and Fig. 10.

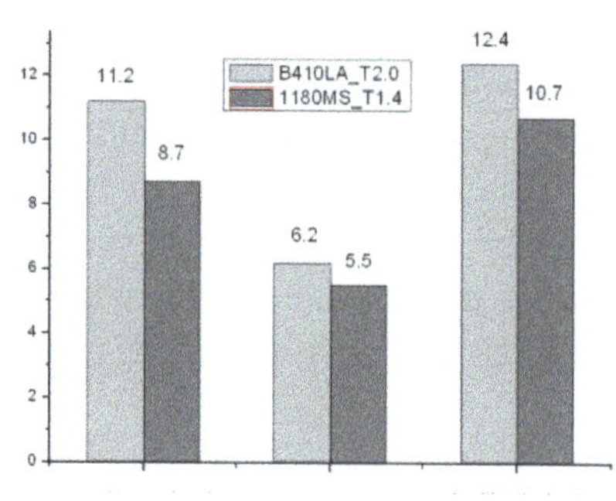

Fig. 9. Acceleration curves.

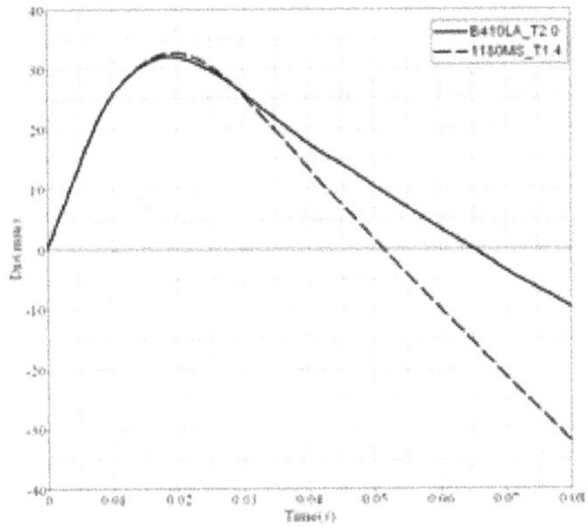

Fig. 10. Displacement of threshold beam.

As can be seen, the trolley's displacement after the collision with two different materials' are similar to each other, but, the threshold beam's deformation is not similar, the threshold beam's displacement that used HC950/1180M are less than that used B410LA.

5. Conclusion

The threshold beam plays an important role in the side impact. In this paper, a design method for threshold beam is carried out based on C-NCAP, which is carried out the evaluation method and evaluation index of the threshold beam, provided an important reference for light design of the threshold beam.

(1) In our study, a method of combing simulation analysis and tests is used to carry the light design of the threshold beam, put forward the evaluation index of the threshold beam.

(2) The weight of the threshold beam reduced 2.07kg, account for 30.6% of the threshold beam, this can provide that the high strength steel is useful in the lightweight research.

Acknowledgment

This research work is funded by the State key laboratory of automobile lightweight construction unit open issue (Serial number: 20130203).

References

1. L. Li, R. D. Liu, X. Xu, X. Wang, K. Q. Wang and J. Y. Guo, Research on non-isothermal warmed B pillar impact on the automotive side impact performance, *Shanghai Metal* **1**, 6 (2014).
2. H. Fei, Research on US side impact regulations based on computer simulation of the car door impact bar, **1**, 47 (2005).
3. H. J. Long, Experimental study on the side impact collision of B-pillar, *Automobile and Shipbuilding* **1**, 6 (2009).

The Study on Spot Welding Technology and Evaluation Method of the Spot Welding Quality for Hot Stamping Quenched Steel

Y. Feng[†], J. Zhou and M. T. Ma

China Automotive Engineering Research Institute Co. Ltd.
Automotive Lightweight Engineering Research Center,
Chongqing, 401122, China
[†]*E-mail: fengyi@caeri.com.cn*

In this paper, the spot welding technology of 22MnMoB hot stamping quenched steel was systematically studied. Through a series of experiments, the optimal spot welding technological parameter has been obtained. The interfacial fracture tendency of spot welds of hot stamping quenched steel behavior was more obvious than that of mild steel. The test results showed that the traditional prediction equation of critical nugget diameter for the changing of fracture mode was not in accordance with test results. A new kind of sphere model which considering the structural complexity of spot weld of hot stamping quenched steel was created to obtain the critical fracture diameter value. The fracture test results of five kinds of hot stamping quenched steel showed that the model has a perfect accuracy which is more accurate and helpful to actual engineering.

Keywords: Resistance spot welding; Hot stamping quenched steel; Critical nugget diameter; Fracture mode.

1. Introduction

As the automotive industry rapidly evolves, the demands for safety and emission become more and more strict, indicating the advent of the lightweight era for automotive. The lightweight of vehicles with firm safety becomes the widespread and significant goal. For meeting this requirement, the development and application of ultra-high strength steel becomes a hot area. Now, hot stamping quenched steel sheets have been already widely applied as many automotive components such as reinforced sheet of door, front bumper and B-pillar, to resist impact and collision, profiting from its high strength ($\sigma_b \geqslant$ 1500MPa) after quenching [1-3]. It is can be foreseen that hot stamping steel sheet becomes one of the most advanced materials for manufacturing lightweight vehicles, having extensive application future.

The body strength of automotive largely depends on the connection strength of the automotive body. Resistance spot welding (RSW) has been the mostly general and important connection technology on the vehicle body profiting from its high welding efficiency and convenience of automation. It had been proved in the past that the fracture mode and mechanism of hot stamping quenched steel is very different with mild steel or normal high strength steel due to its' high resistance rate, low heat conductance and complicated structural property of nugget. Compared to mild steel, hot stamping quenched steel tends to appear the interfacial fracture (IF) mode under shear load condition and the traditional computational formulas for critical fracture nugget diameter such as 4.0 $t^{0.5}$ or 4.0 $t^{0.5}$ were not suitable for hot stamping quenched steel.

In this paper, the spot welding technology of several kinds of hot stamping quenched steel sheet was studied using power frequency spot welder. A new kind of sphere model of fusion zone, which considered the structural complexity of spot weld of hot stamping quenched steel was created to obtain the critical fracture diameter.

2. Experimental Process

The experimental material was 22MnMoB hot stamping quenched sheets with thickness of 1.0 mm ~ 1.8 mm. Mechanical property of the quenched steel sheet is shown in table 1. Different welding parameters were selected for welding experiments. During optimizing process, one of the parameters was changed while other two parameters are set invariable. Shear and peel tests were carried out to obtain the nugget diameter and fracture mode of spot welds.

Table 1. Mechanical property of 22MnMoB sheet.

Material	Yield strengh/MPa	Tensile strengh/MPa	Fracture elongation/%
22MnMoB	1102	1519	5.4

3. Experimental Results and Discussions

3.1. *Spot welding technology*

The effect of welding current to maximum shearing force and nugget diameter is shown in Fig. 1. As the increase of welding current, the maximum shearing force of welds raises and the limit value of maximum shearing force exceeds 30kN when the current is in the range of 6.0 ~ 7.0kN. The variation of welding nugget diameter with the change of welding technological parameters is absolutely similar to the maximum shearing force.

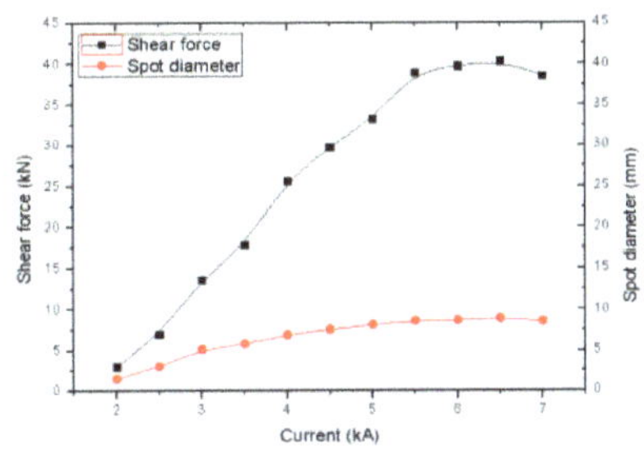

Fig. 1. Shear force and spot diameter vs. welding current.

The micro structure photos of welding area under optimal technological parameter condition are shown in Fig. 2. The Vickers hardness test result is shown in Fig. 3. It is indicated that the hardness of HAZ is lower than other zones.

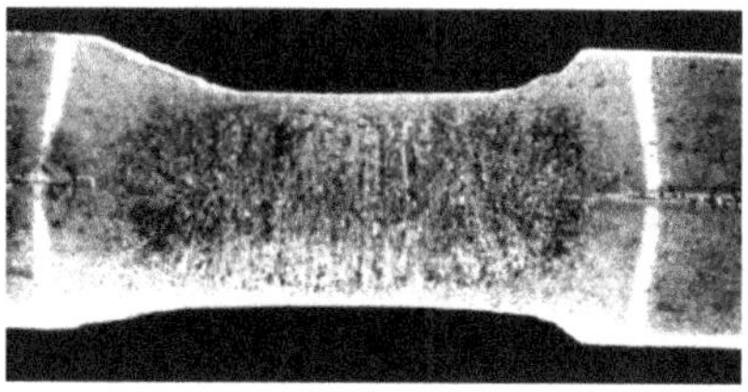

Fig. 2. Structure of welding area.

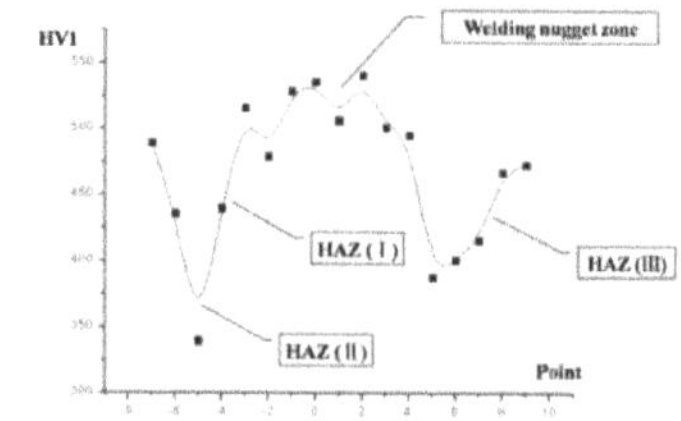

Fig. 3. Vickers hardness of welding area.

3.2. *Failure mode*

As shown in Fig. 4, during shearing tensile test, the joint of 22MnMoB behave two failure modes: nugget pullout and interfacial fracture. According to a lot of study results before, for welding specimens, nugget diameter, remarkably affecting the strength and failure modes of welds, is directly influenced by welding parameters [2-3].

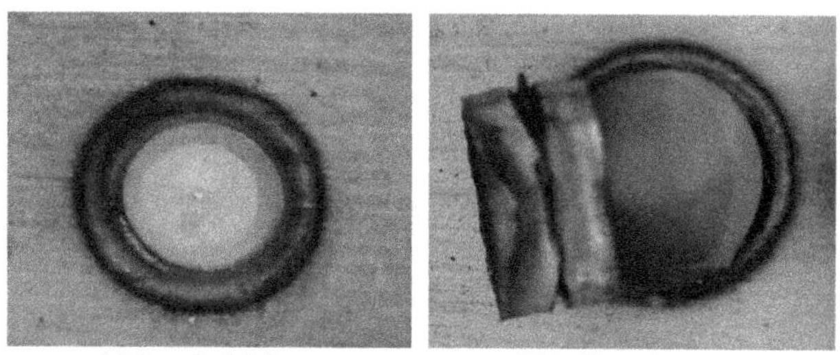

(a) Interfacial fracture. (b) Pullout fracture.

Fig. 4. Failure modes of hot stamping steel sheets.

3.3. *The prediction method of welding quality*

The failure modes and mechanisms of welding joint are very complex, a great deal of researches have proved that there is a critical nugget diameter for a given thickness which is able to be used as the criterion for failure modes of spot welds [3]. A new kind of fracture prediction method based on spot diameter was created for hot stamping quenched steel here and checked on the 1.8mm 22MnMoB sheet. As shown in Fig. 2–3 that the spot zone of hot stamping quenched steel is composed of nugget and heat treat zone (HAZ). In order to achieve the veracity and applicability of fracture prediction method for hot stamping quenched steel, the interface between nugget and HAZ was assumed a sphere considering the structural complexity of spot zone and geometric shape, stress state of spot under a certain load condition (Fig. 5). Here the d (mm) is the nugget diameter, t (mm) is the steel sheet thickness. Moreover, the inner structure of nugget was assumed well-distributed during the load state analysis process.

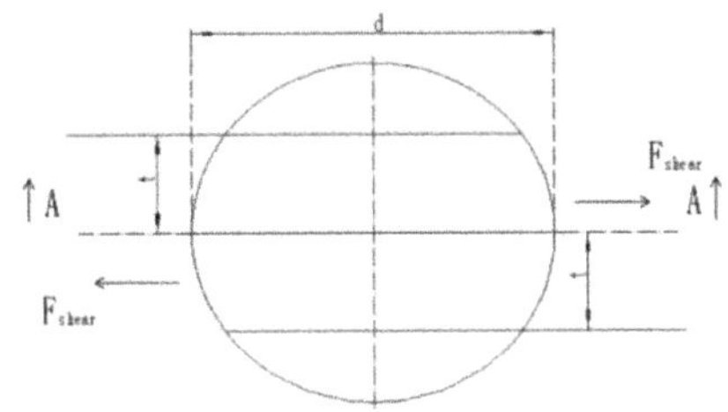

Fig. 5. Nugget stress state model.

As shown in Fig. 6, the shear stress τ on the interface between up and down nugget reaches the critical stress value when the IF mode happens.

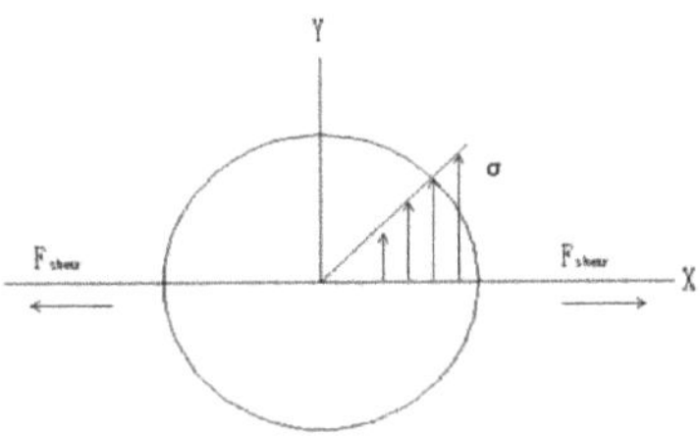

Fig. 6. Stress state distribution (IF).

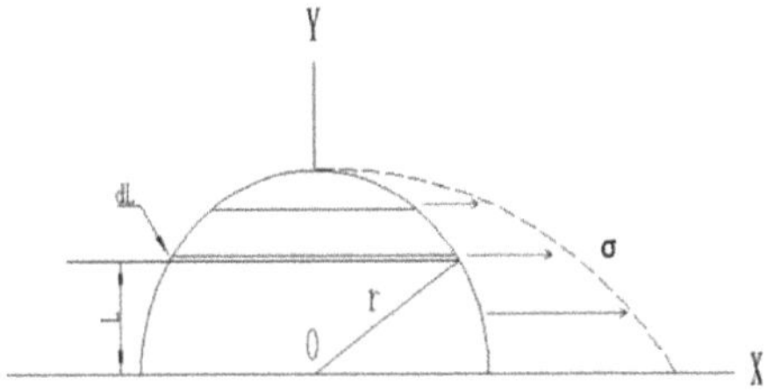

Fig. 7. Stress state distribution (PF).

Based on Fig. 6, the τ on a random position x can be expressed as Eq. 1. The r is nugget diameter value. The differential area on a random position x can be expressed as Eq. 2.

$$\tau_x = \frac{x}{r}\tau_{\max} \tag{1}$$

$$d_s = 2\sqrt{r^2 - x^2}d_x \tag{2}$$

Based on Eq. 1~2, the differential and actual maximize shear force on the whole interface area corresponding to IF mode can be expressed as Eq. 3-4.

$$dF_{IF} = T_x \cdot d_s \tag{3}$$

$$F_{IF} = \frac{2}{3}\tau_{\max}r^2 = \frac{1}{6}\tau_x d^2 \tag{4}$$

As the increasing of welding current, the nugget diameter also increases. The pullout fracture (PF) mode will happen when the tension stress σ at the boundary between nugget and HAZ reaches critical stress value which is as shown in Fig. 7. The maximize tension force on the whole boundary area corresponding to PF mode can be expressed as Eq. 5 through the same basic mechanics analysis process as IF mode.

$$F_{PF} = \frac{\sigma_{\max}}{r}\iint (r^2 - l^2)\cos\beta \cdot d\beta \cdot dl = \frac{\pi}{2}r\sigma_{\max}t - \frac{\pi}{6}\sigma_{\max}\frac{t^3}{r} \tag{5}$$

According to many study results in the past, the final fracture mode of super high strength steel such as hot stamping quenched steel is the result of competition between shear stress τ and tension stress σ. The maximal shear force will equal to tension force (FIF=FPF) when nugget diameter reaches a certain critical value, which can be used to control the spot welding quality. We can obtain the critical nugget diameter value based on Eq. 6 as follows:

$$F_{IF} = \frac{1}{6}\tau_x d^2 = F_{PF} = \frac{\pi}{2} r\sigma_{\max} t - \frac{\pi}{6}\sigma_{\max}\frac{t^3}{r} \tag{6}$$

Here set the *t/d*=x, the Eq. 7 can be translated into another type as follows:

$$\frac{\sigma_{\max}}{\tau_{\max}}x^3 - \frac{3}{4}\frac{\sigma_{\max}}{\tau_{\max}}x + \frac{1}{2\pi} = 0 \tag{7}$$

According to the TRESCA fracture criterion, the ultimate tension stress is twice as much as the value of ultimate shear stress, so the Eq. 7 can be translated to a new type as follows:

$$\frac{(\sigma_{\max})_{HAZ}}{(\sigma_{\max})_{FZ}}x^3 - \frac{3}{4}\frac{(\sigma_{\max})_{HAZ}}{(\sigma_{\max})_{FZ}}x + \frac{1}{4\pi} = 0 \tag{8}$$

Cahoon J. R. had pointed out that there is a kind of proportional relation between ultimate tension stress and Vickers hardness of metal [4]. So above Eq. 8 can further be translated to Eq. 9 as follows:

$$\frac{(H)_{HAZ}}{(H)_{FZ}}x^3 - \frac{3}{4}\frac{(H)_{HAZ}}{(H)_{FZ}}x + \frac{1}{4\pi} = 0 \tag{9}$$

Based on Eq. 9, the critical nugget diameter value can be obtained through testing the Vickers hardness of fusion zone (FZ) and HAZ. As shown in Fig. 3, the Vickers hardness of FZ is about HV 538, the Vickers hardness of HAZ is about HV 321. By bringing the Vickers hardness values of FZ and HAZ into Eq. 9 and solving the computational formula, the x (tc/dc) calculated equals to about 0.189. So through a further conversion process, the dc equals to about 4.97 tc. Here tc is the thickness corresponding to welding indentation which often equals to 0.86 t0 (original thickness of sheet). So finally the dc equals to about 4.30 t0.The further checking tests were carried out on other kinds of hot stamping quenched steel of different thickness values. As shown in Fig. 8, that the theoretical critical diameter value is similar with test value but is unlike to the value obtained by the traditional computational formula such as 4.0 t0.5 or 4.5 t0.5 which proves the reliability of the prediction method for critical fracture nugget diameter in this paper.

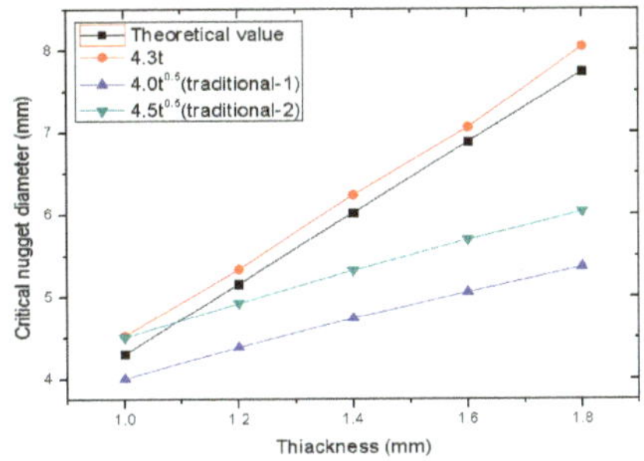

Fig. 8. Fracture test results.

4. Conclusion

In this paper, the spot welding technology of 22MnMoB was studied. Through a series of technology and test experiments, we had obtained the optimal spot welding technological parameter condition. The fracture mode of spot welds of hot stamping quenched steel has a more obvious tendency of interfacial fracture mode (IF) compared to mild steel. The test results showed that the traditional prediction equation of critical diameter of fracture mode was not in accordance with test critical diameter values of hot stamping quenched steel. A new kind of sphere model of fusion zone, which considered the structural complexity of spot weld of hot stamping quenched steel was created to obtain the critical fracture diameter value. The computational formula for critical nugget diameter (d_c =4.30 t_0) has a perfect accuracy which is more accurate and helpful to actual engineering.

Acknowledgment

This work is funded by China Automotive Light-weight technology Research Institute Co. Ltd. (No. 20130202).

References

1. M. Merklein and J. Lechler, Investigation of the thermo-mechanical properties of hot stamping steels, *J. Mater. Process. Tech.* **177**, 452 (2006).
2. Weld button criteria recommended practices for test methods for evaluating the resistance spot welding behavior of automotive sheet steel materials. American National standard, ANSI/AWS/SAE/D8.9-97, Section 5.7:1997.
3. M. Pouranvari and S. Marashi, Critical review of automotive steels spot welding, *Sci. Technol. Weld. Joi.* **18**, 361 (2013).
4. J.R. Cahoon, W.H. Broughton and A.R. Kutzak, The determination of yield strength from hardness measurment, *Metall. Mater. Trans. B* **2**, 1979 (1971).

Comprehensive Performance Evaluation of Hot Forming Steel Based on the Crashworthiness of B-Pillar Reinforcement

Q. Yang[1,†], B. Liu[1], G. Fang[2], Z. C. Qiu[1], B. H. Tan[1], Z. T. Zhu[1], W. Zhang[3] and A. D. Luan[1]

[1]*Chongqing Changan Automobile Co. Ltd, Chongqing, China*

[2]*China Automotive Engineering Research Institute Co. Ltd, Chongqing, China*

[3]*Ma'anshan Iron And Steel Co. Ltd, Ma'anshan, Anhui, China*

[†]*E-mail: yangqin1@changan.com.cn*

Hot forming has been an advanced technology, which fulfills lightweight of automotive body and guarantees crash safety at the same time. At present, hot forming has been an efficient way for automotive manufacturers to improve competence of their products. In this paper, a hot forming B-pillar reinforcement and its original steel sheet are evaluated, including component size, microstructure and mechanical properties. Meanwhile, three-point-bending and side impact testing are conducted for B-pillars manufactured from different raw material suppliers.

Keywords: B-pillar reinforcement; Hot forming steel; Performance evaluation; Three-point bending; Crashworthiness.

1. Introduction

Safety, fuel efficiency and environment have become critical issues in automotive industry, which, in a sense, promote the development of lightweight technology. During past two decades, the ultra-high strength steels (UHSS) has been widely used in car body increasingly, which confirmed to be an efficient way to reduce the weight of car body and guarantees crash safety as well. At present, hot forming is an efficient way for automotive manufacturers to improve competence of their products [1-2].

In a general sense, for crash safety components, as small as possible intrusion must be guaranteed, moreover, as much energy as possible should be absorbed when crash occurs. Especially under the situation of side impact, safety components cracking is not allowed until a certain amount of energy has been absorbed. To evaluate the possibility of cracking, some related performance testing, i.e. three-point-bending experiments can be conducted to reflect the property of components accurately [3-4], which can measure bending strength, obtain static pressure-displacement curves and bending angle. The results measured through three-point-bending experiments can reflect performance of components and materials very well.

2. Material Performance Analysis

The measured material A (22MnB5/1.60mm thickness) is provided by MAANSHAN Iron And Steel CO, LTD. The material B with 1.60mm thickness is provided by compared manufacturer, which has been certificated by vehicle safety test. Chemical composition, mechanical properties and microstructure of material A and B are as shown in Table 1, Table 2 and Fig. 1 respectively.

Table 1. Chemical composition of material A and B (wt.%).

Grade	C	Si	Mn	Cr	Als	B	Ti
A	0.24	0.25	1.31	0.16	/	0.003	0.037
B	0.23	0.24	1.18	0.15	0.047	0.0025	0.028

Table 2. Mechanical properties of material A and B in delivery condition.

Grade	Rp 0.2 (MPa)	Rm (MPa)	A50 (%)
A	448	670	24.67
B	399	584	26.71

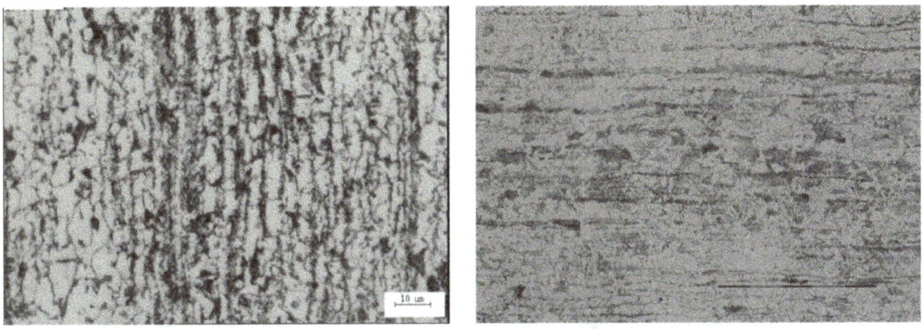

a) Raw material A. b) Raw material B.

Fig. 1. Analysis results of microstructure of material A and B.

Testing results show that both A and B can meet the requirements of composition and mechanical performance as hot forming steels under delivery status. Microstructures of both A and B are ferrite + fine pearlite and bainite.

3. Component Performance Testing

3.1. *Performance testing for the component*

The measured B-pillar reinforcement sample and the corresponding sampling position for mechanical performance testing are shown in Fig. 2, the testing results are summarized in Table 3.

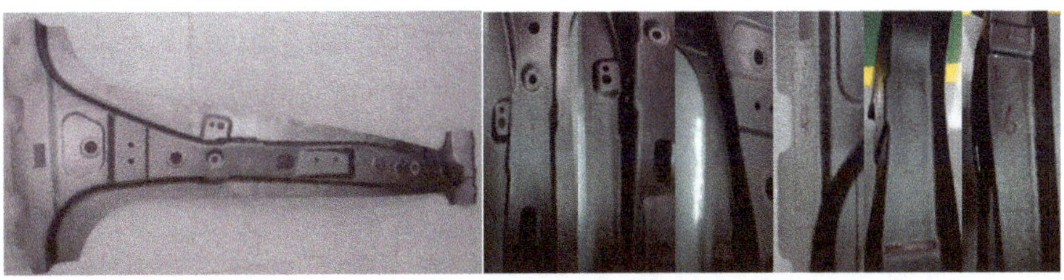

Fig. 2. B-pillar reinforcement sample and sampling position for mechanical properties test.

Table 3. Mechanical properties of B-pillar reinforcement components.

Grade	Number	Rel/MPa	Rm/MPa	A50/%
A	1	1148.57	1614.50	7.11
	2	1126.71	1623.11	7.54
	3	1166.32	1626.57	7.24
	4	1018.44	1567.70	6.70
	5	1124.59	1599.58	7.68
B	1	1097.56	1512.22	7.26
	2	1101.08	1539.97	6.72
	3	1099.65	1536.57	7.22
	4	1069.56	1574.09	6.58
	5	1214.65	1633.52	6.44
	6	1109.94	1578.12	7.56

Sampling position of B-pillar reinforcement and the related analysis results are shown in Fig. 3 and Table 4, respectively.

Fig. 3. Sampling position of B-pillar reinforcement for microstructures analysis.

Table 4. Different region of B pillar microstructure.

	Piece number	microstructure	Micro hardness		
A	Pressing edge position	Martensite	488	493	491
	Side wall	Martensite	467	457	460
	bottom	Martensite	483	460	459
B	Pressing edge position	Martensite	465	460	463
	Side wall	Martensite	466	458	462
	bottom	Martensite	476	469	470

From the results of mechanical performance and microstructure, it can be indicate that tensile strength of hot stamping components is over 1500 MPa and the main microstructure is martensite. Moreover, the micro hardness (HV) surpasses 450.

3.2. *Component three point bending test*

As a testing method, three-point-bending test cannot only measure the bending strength of the tested components stably, but also can yield the static pressure-displacement curve.

The testing method is Simple and intuitive for test results, thus, it is useful to examine the performance of individual component.

During three-point-bending test, the testing state of B-pillar reinforcement is designed as its actual assembly state in the vehicle. The upper and lower ends of the B-pillar reinforcement are fixed by tooling. Static pressure is loaded in the middle of B-pillar reinforcement through a pressure head. The loading location on B-pillar reinforcement is where the door hinge mounted and the force direction is perpendicular to the center plane along the longitudinal direction of the vehicle body. The rigid pressure head used in this test is a radius of curvature, R, 152mm and its compression speed is 2mm/s. Test equipment and schematic diagram are shown in Fig. 4.

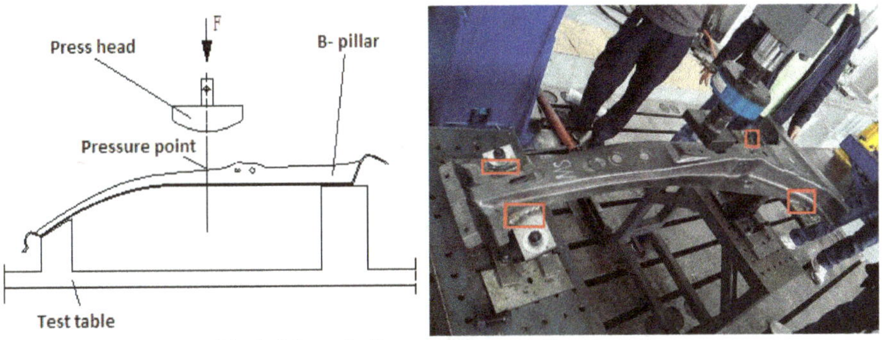

Fig. 4. Schematic diagram of three point bending test.

Peak value of static pressure in B-pillar reinforcement during three-point-bending test is shown in Table 5.

Table 5. Peak value of static pressure in B-pillar reinforcement.

	Peak value (KN)	Displacement under peak pressure (mm)
A	22.5	48.0
B	22.9	50.0

As shown in Table 5, peak values of static pressure for B-pillar reinforcement made of A and B are 22.5 KN and 22.9 KN respectively. Besides, the corresponding static pressure distances are 48mm and 50 mm respectively. The results indicate that material A and B have similar static mechanical properties.

3.3. *Crashworthiness testing*

Side impact testing scheme is shown in Fig. 5. The tested B-pillar reinforcement sample is fixed on a rigid wall through some tooling. The fixing state is in accord with its assembly status in the whole vehicle, and the relative fixing location on the rigid wall is determined by R point of the vehicle. The velocity of the moving sled is 14 km / h ± 0.5 Km / h.

Fig. 5. Fixing status of B pillar in side impact testing.

According to packaging of the vehicle, 4 sensors are placed in B-pillar reinforcement corresponding with the location of dummy's head, chest, abdomen, and pelvis respectively to capture the intrusion. A high-speed camera is located in the left front of the tested B pillar to record the deformation process. The measured items are presented in Table 6.

Table 6. Measured items of B pillar during side impact.

Order	Measured items	Sub items	Remarks
1	Intrusion acceleration	Location of dummy's head	Y direction sensor
		Location of dummy's chest	Y direction sensor
		Location of dummy's abdomen	Y direction sensor
		Location of dummy's pelvis	Y direction sensor
2	high-speed camera	located in the left front of B pillar	one

Table 7 lists the testing conditions. The weight and velocity of the moving sled, the height of B pillar located satisfy the design standards. Table 8 depicts the Peak intrusion of dummy in different location.

Table 7. Testing conditions.

Testing conditions	test order		Standards
	Material A	Material B	
Weight of sled	952	952	950±20
Impact velocity (km / h)	14.19	14.17	14±0.5
Height of B pillar located (mm)	314	315	315±5

Table 8. Peak intrusion of dummy in different location.

Measured items	Testing order		Differences between A and B
	1	2	
	Material A	Material B	
Peak intrusion of dummy's head (mm)	33.1	33.6	+0.5
Peak intrusion of dummy's chest (mm)	74.5	76.9	+2.4
Peak intrusion of dummy's abdomen (mm)	85.1	88.1	+3.0
Peak intrusion of dummy's pelvis	73.9	77.0	+3.1

The results show that peak intrusion tested with material B is greater than that of material A, which means that Material A has better crashworthiness than material B.

4. Conclusion

1) The tensile strength and the yield strength of the hot forming B pillar are over 1500Mpa, and 1000Mpa respectively, and with the elongation rate between 6% and 8%.
2) Energy-absorption performance of materials can be reflected through three-point bending test. In a general sense, the bigger the bending angle is, the more energy is absorbed. Hence, German OEMS such as Mercedes Benz and BMW, have specific demands on energy-absorption of hot forming component, i.e. the bending angle must be over 65 degree during three-point bending test.

Acknowledgments

The authors sincerely thank the funding from Ministry of Industry and Information Technology National industry transfer type to upgrade basic engineering practice project, the National Key Technology R&DProgram (No. 2011BAG03B01) and Chongqing Research Program of Foundation and Advanced Technology program (No: cstc2013jcyjC600001).

References

1. H. Z. Lu, Z. W. Wan, M. T. Ma, Y. L. Cheng and G. M. Chen, Multi-Objective evaluation regulation study of automotive lightweight, in *Proceedings of the 34th World Automotive Congress FISTIA*, (Beijing, China, 2012).
2. G. Fang, J. P. Zhang and Q. S. Jin, Study on mechanical properties and failure modes of hot forming steel under different strain rates, in *the 2nd International Conference on Advanced High Strength Steel and Press Hardening*, (Changsha, China, 2015).
3. M. Abspoel, B. M. Neelis and P. Liempt, Constitutive behavior under hot stamping conditions, *J. Mater. Process. Tech.* **228**, 34 (2016).
4. H. Z. Lu, W. J. Wang, A. M. Guo and J. Bian, Opimize the hot stamping technology base on vehicle safety, *Aut. Tech. Mater.* **10**, 8 (2013).

Crashworthiness Evaluation Method Research for Vehicle Body and its Typical Security Component

T.Q. Fan[†], F. Wang, J. W. Li, Y. Li and Q. J. Zhao

China Automotive Engineering Research Institute Co., Ltd., Chongqing, China, 401122
Chongqing Automotive Lightweight Engineering Research Center, Chongqing, China, 401122
[†]*E-mail: fantiqiang@caeri.com.cn*

This paper focuses on vehicle body and its typical safety consideration, in the form of vehicle crashworthiness. The study undertaken by the us aims to investigate the construction the Body-In-White (BIW), in terms of the safety, crashworthiness with the desire to deploy hot forming steel to reduce the weight of the vehicle. From our study, through simulation analysis in the assessment of crashworthiness determined by BIW and its safety features, we have demonstrated that hot forming components could deliver the safety requirement to protect the passengers without the need for heavy weight steel structure for BIW.

Keywords: Body-In-White (BIW); Security components; Lightweight; Crashworthiness.

1. Introduction

In the event of impact, the vehicle body absorbs most of the energy, and the choice of Hot forming steels [1] for the construction of Body-In-White in the form of front panel, side panel, B pillar, and bumper bean have been under scrutiny; the core consideration is the choice of materials; could the light-weighted hot forming steel withstand the impact as compare with its conventional heavy weight counterpart. Is it safe for the passengers within the capsule [2], has been attracting attention amongst engineers and researchers alike in the automotive industry.

In according to the NCAP (New Car Assessment Program) [3] assessment requirements, which China NCAP (C-NCAP) closely follow. The first big milestone was in 2012 when C-NCAP's test protocols took a significant step closer to matching European requirements:

- There was a general increase in the thresholds for dummies' injury scores
- The main frontal offset impact speed increased from 56 to 64km/h
- C-NCAP added a rear occupant evaluation element
- The organization already introduced a full-width frontal impact against a rigid barrier at 50km/h in 2006

In this article, we will present the test results on front impact, side impact, B pillar impact, Bumper bean impact and Threshold beam impacts tests.

2. Vehicle Crashworthiness Parameters

The vehicle safety performance test evaluation includes: Frontal impact (including offset impact), side impact and side impact pole test. Based on the current domestic research requirement, in here, we use C-NCAP frontal impact and side impact, ENCAP side impact pole test for our study. The vehicle models used in the study are shown in Figures 1~3 below. Contains the A0 and A-class car, respectively.

Fig. 1. A vehicle simulation model.

Fig. 2. B vehicle simulation model.

Fig. 3. C vehicle simulation model.

2.1. *Important assembly*

Body-In-White (BIW) in which a car body's sheet metal components have been welded together, includes include front bumper and B pillar, the threshold beam are all likely candidates for absorbing the energy of the impact and the impact force transfer as shown in Figure 4, and Figure 5, impact force transfer in car body, through the main structure component, eventually dissipate. Absorb most of the energy of the impact; ensuring the structure safety of crew capsule. Typical structure safety where the passengers are will be safely protected. In recent years, following the trend of the light weighting design, the hot forming steels are used to deliver the safety to protect the crew capsule on the one hand, and reducing weight of the vehicle on the others.

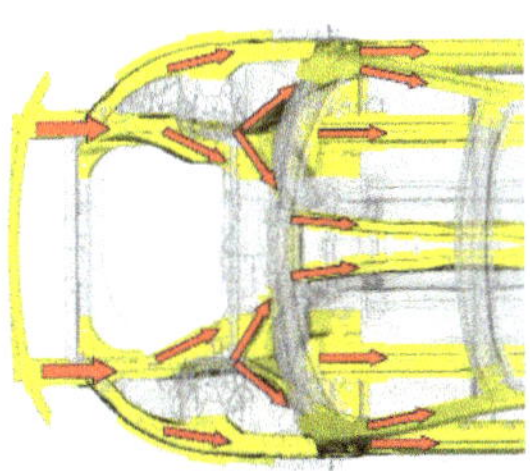

Fig. 4. Front impact force transfer.

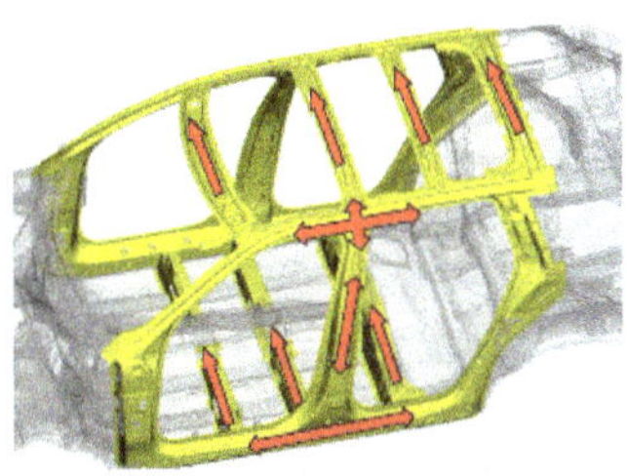

Fig. 5. Side impact force transfer.

2.2. *The performance parameters of the frontal impact*

In the event of crash encountered at the front of the vehicle, the front engine cabin compartment will be the main energy absorption components. The parameters under examination are body deformation and energy absorption of the BIW parts, the bumper beam, longitudinal beam, as shown in table 1.

Table 1 Parts in the frontal crash energy absorption table.

Parts	A vehicle (KJ)	B vehicle (KJ)	C vehicle (KJ)
BIW (with closure)	65.8	93.7	106.2
Bumper beam	3.6	42.4	37.3
Longitudinal beam	32.5	18	25.8

2.3. *The performance parameters of the side impact*

In event of side impact, the side structure of BIW will bear the brunt, resulted in deformation to absorb the impact. BIW main performance is associated with dummy position B pillar and the door intrusion and intrusion speed, as well as the energy absorption of important part, as shown in Figure 6, and Figure 7.

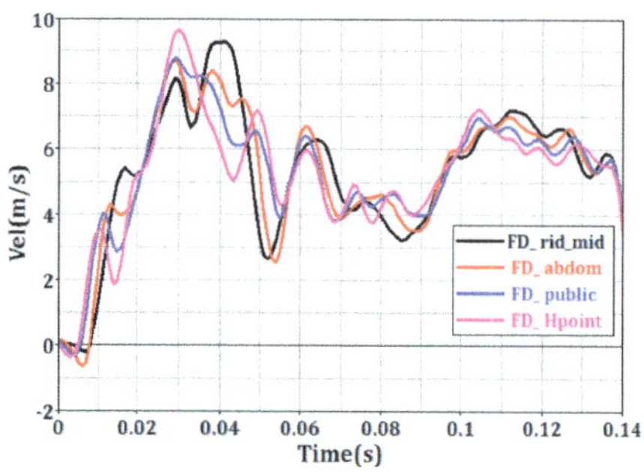

Fig. 6. Side impact intrusion speed.

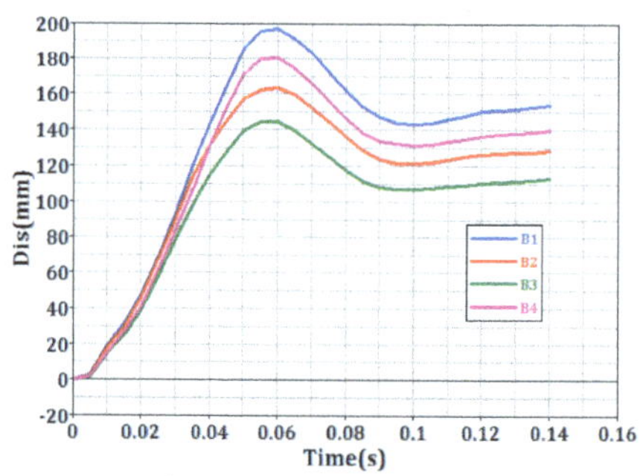

Fig. 7. Side impact intrusion.

Various models of vehicle side impact major parts of the body and energy absorption are shown in table 2 below.

Table 2 Parts in the side crash energy absorption table.

Parts	A vehicle (KJ)	B vehicle (KJ)	C vehicle (KJ)
BIW (with closure)	16.5	14.5	16.3
BIW (without closure)	10.5	9.6	13.3
B pillar	1.5	1.1	3.3
The left front door	3.4	3.1	1.9
The left rear door	2.6	2.0	1.1
The threshold	1.7	2.4	2.3

2.4. *The performance parameters of the side pole impact*

In event of the side pole impact, the threshold beam is the main component bearing the brunt of impact; it is, by far, the biggest deformation and energy absorption of the threshold beam.

3. BIW Crashworthiness Evaluation Method Research

Based on the vehicle impact safety analysis, undertake tests on the vehicle front and side impact form to include vehicle body contains BIW and closure (with door, hood and hatchback), but excluding the engine and transmission parts.

The test for front impact are conducted with dragging the BIW towards a rigid wall as shown in Figure 8, where the deformation and damage at the front cabin parts of the vehicle are recorded and assessed.

As for test on the side impact, we adopted C-NCAP side impact Standard by moving 950 kg deformation barrier at a certain speed towards the side of the vehicle as shown in Figure 9.

The data recorded suggests that: at the speed of 50km/h, the impact on the vehicle body, which includes left, right, and front longitudinal beam, as well as the bumper beam energy absorption to be fine and consistent, tests after tests, event on the speed of vehicle moving at 50km/h, the BIW survive to live another day.

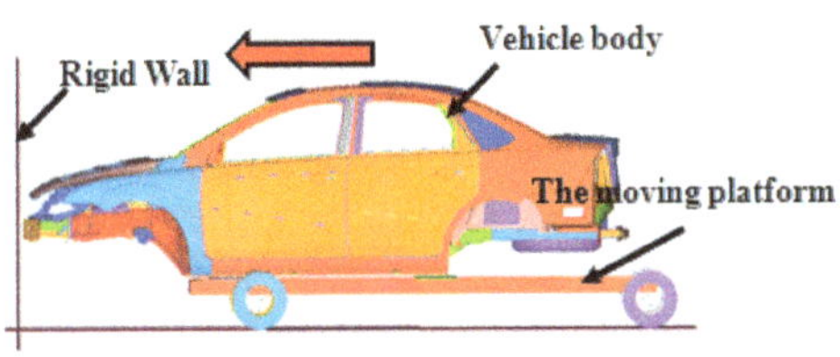

Fig. 8. Vehicle body front impact simulation model.

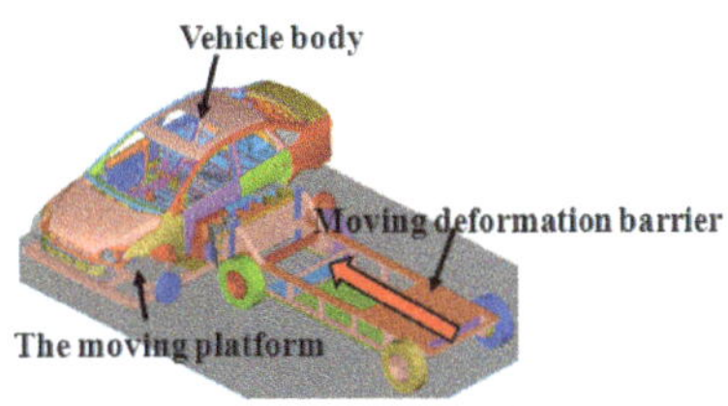

Fig. 9. Vehicle body side impact simulation model.

4. Typical Parts Crashworthiness Evaluation Method

According to the results recorded for crashworthiness at each important parts of the vehicle under test; including bumper beam and B pillar and the threshold beam, for deformation, energy absorption and force transmission, etc., The BIW built from light-weighted hot forming steels, shown remarkable resistance of holding various parts of together in one piece after the impact.

1) Bumper Impact Test on impact energy absorption, deformation and acceleration, to evaluate bumper beam on impact performance: the bumper beam assembled and fixed onto a rigid wall, hit by a 1000 kg moving barrier, at the speed of 22 km/h as shown in Figure 10.
2) Threshold Beam Impact Tests on impact energy absorption, deformation, to evaluate the design of threshold beam on impact performance: the threshold

beam is installed on a 1000 kg moving barrier, crashing toward a rigid pillar, at a speed of 14 km/h. as shown in Figure 11.

3) B Pillar Impact Test on impact energy absorption and the deformation, to evaluate the design of B pillar on impact performance: Adopting C-NCAP side impact Standard, B pillar installed on a rigid wall, to be crashed by adopt 950 kg moving barrier at the speed of 12 km/h to study the deformation, as shown in Figure 12.

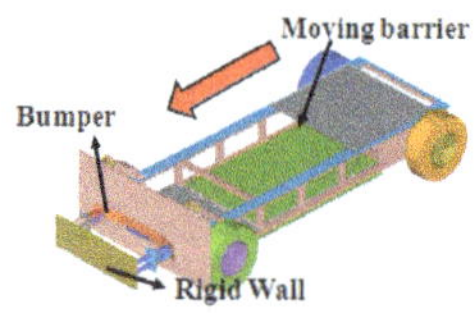

Fig. 10. Bumper beam impact model.

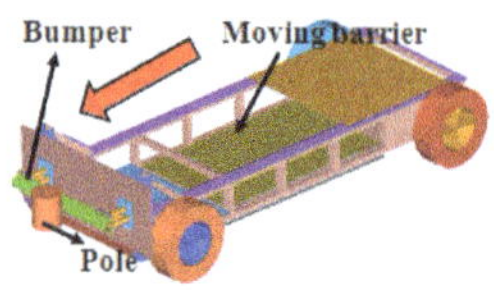

Fig. 11. Threshold beam impact model.

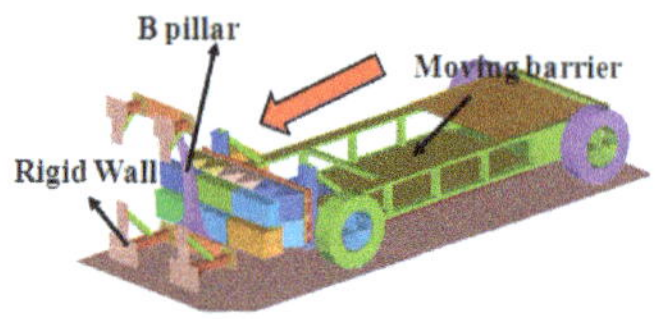

Fig. 12. B pillar impact model.

5. Validation Test

According to the research conducted, including the assessment of crashes at the front, at the side, and the bumper in the events of impact on Body In White or **BIW**, in which a car body's sheet metal components have been welded together, could withstand the impact on a BIW constructed from hot forming steels. The tests undertaken on front and side impact test, the bumper beam, frontal crashes test, B pillar side impact test and the threshold side pole impact test are shown as follows in Figures 13 to 17 as evidences to verify our claims the suitability of hot forming steels for Body In White or **BIW**.

Fig. 13. Vehicle body front impact test.

Fig. 14. Vehicle body side impact test.

Fig. 15. B pillar impact test.

Fig. 16. Bumper bean impact test.

Fig. 17. Threshold beam impact test.

All tests are conducted in according to the NCAP (New Car Assessment Program) [3] assessment requirements, which include front impact, side impact, B pillar impact, Bumper bean impact and Threshold beam impacts tests. BIW constructed out of hot forming components pass all the tests with flying colors.

6. Conclusion

Safety consideration of vehicle body is one of the key criteria in car design. In this paper, through the simulation of crashworthiness determined by BIW and safety analysis in the assessment, we have showed that the use of hot forming components could deliver the safety requirement without the need for heavy weight steel structure.

Acknowledgment

The research work is funded by State key laboratory of automobile lightweight construction unit open issue (Serial number: 20130203).

References

1. M.T. Ma, H.L. Yi, H.Z. Lu and X.M. Wan, On the light weighting of automobile, *Eng. Sci.* **11**, 24 (2009).
2. W.G. Zhang and Z.H. Zhong, Computer simulation of the crashworthiness of the energy absorbing front end of a passenger car, *Automot. Eng.* **24**, 6 (2002).
3. Y.G. Liu and Z.X. Liu, Status and development tendency of the test and assessment technique of the NCAP in different countries, *J. Automot. Saf. Energy* **4**, 16 (2013).

The Study of Vehicle Body Side-Impact Performance

C. Y. Lou[a], Q. Yang[b], B. H. Tan[c], F. Liu[d], Q. Zhang[e] and L. Tang

Chongqing Changan Auto R&D Center, Chongqing, P.R. China
E-mail: [a]loucy@changan.com.cn, [b]yangqin1@changan.com.cn, [c]tanbh1@changan.com.cn, [d]liufeng2@changan.com.cn, [e]zhangqiang4@changan.com.cn

The car body serves as main security structure when collision occurs, which has played an important role in protecting the security of the crew compartment as well as establishing the first line of collision. Only a door and B column exist between the vehicle crew and impactors at the small distance when side impact occurs, which gets closer to the crew compartment without enough space for collision energy absorption, giving rise to great threat for passengers safety. At present, the accumulation of researching on side-impact safety is so less at home that passengers suffer a lot from injury in case of side impact. In order to improve the vehicle safety performance, it is necessary to study the safety of the car side impact.

In this paper, taking one certain car body structure as the research object, which was used to carry out the collision simulation of lateral rigidity column, the analysis of the car body deformation state and passengers injury index, provided some references for the vehicle side-impact research in the future.

Keywords: Side impact analysis; B column and threshold beam; Car-body structure.

1. Introduction

Side impact experiment, according to the C-NCAP program, installed the deformation honey-comb aluminum in the mobile car front end, keeping the vertical relationship between the driving direction of mobile barrier and test vehicle. Besides, the center line of mobile barrier was aimed at the R point of test vehicle and the impact velocity was 50+1Km/h (the test speed shall be not less than 50 Km/h). The distance between the mid vertical plane of mobile barrier and the cross-sectional vertical plane of test vehicle through the collision side in the R point of the front seat should be within ±25mm [1]. A dummy of the EuroSIDII type was placed in driver position in order to measure the driver position being hurt. As shown in Fig. 1.

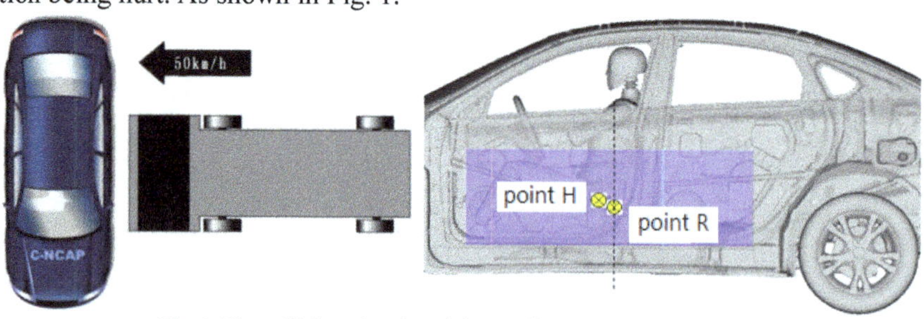

Fig. 1. The collision virtual model according to EuroSIDII design.

[b] Corresponding author.

The main forms of side impact damage include: side impact of the front anti-collision beam and B pillar crushing deformation, leading to the damage of occupant cephalothorax; Plastic inner decoration can occupant abdominal injury; side impact of the front anti-collision beam door panel provides vertical impact load for pelvis, leading to the sacrum fracture, etc. [1, 2]. In order to protect and reduce the occupant injury in side impact, car body structure need to strengthen devices correspondingly.

2. Mechanical Analysis of Side-impact Process

Mechanical analysis of side impact. Due to side impact occurs when mobile barrier keeps vertical relationship with car-body direction, the mainly burden parts of which are B column and the front and back doors, and the car doors are in closed state under impact with B column. As a result, good doors strength and B column strength can prevent structure deformation into car compartment injuring passengers as long as collision occurs.

The transfer of lateral impact force. In car body structure, the decoration of essential parts has important influence on car body stiffness, strength and driving comfort. In general, car body pillar not only plays a role in supporting function but also transfers the force when it comes to collision [2]. When the car side is hit, the front and back doors and B column bear the influence of the transfer force from lateral, causing the trend of car inward deformation. The front and back doors set up anti-collision beam, avoiding that the deformation of car outside door plane is passed to interior, reducing the car door deformation and providing safe space for the crew compartment. The side-impact force of B column receiving, scattering into the structure of top beam and the threshold beam and so on through lap structure, forms a closed frame with A column, the top beam, the threshold beam and C column, effectively transferring the impact of collision, avoiding excessive deformation in part, ensuring the integrity of their lateral structure, so as to make sure the safety of the crew compartment.

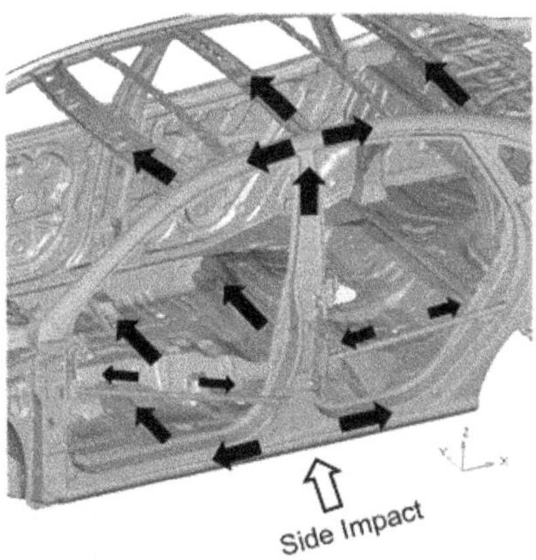

Fig. 2. The transitive graph of force path.

3. The Simulation Analysis of Side-impact Process

3.1. *The purpose of analysis*

Referring to C-NCAP management rules, 50Km/h side impact of one certain car body was analyzed from the aspects of acceleration, displacement and deformation forms; Through this analysis, achieving the optimization of passive safety performance of one certain car and the optimization of car-body structure and materials. The transitive force path is shown in Fig. 2.

In order to guarantee the five-star overall safety goal of the vehicle CNCAP, decomposing that the target value of car-body side impact is of 90 mm or less for maximum invasion and is of 7m/s or less for the largest acceleration of invasion.

3.2. *Establishing of the car-body model*

Simulation model is established according to the car structure crash of one certain car, and the finite element model of one certain car is established by material performance parameters and so on, a total of 1496388 nodes, 1556749 units [3]. The finite element model includes: white body, closed parts, chassis system, cooling system, power system, mobile deformation barrier, rigid ground parts, etc. Curb weight: 1199kg, the center of mass: 1052,-1.7,321,R point coordinates:1270,-325,348,H point coordinates:1180, -325,350. The speed of pallet truck according to C-NCAP test program is set to 50Km/h=13.88mm/ms, the average height difference is 31mm in overlapping between the lower edge of pallet barrier and the top of threshold [3].

3.3. *The result analysis of side-impact stimulation*

The head, ribs and legs of dummy, door panel and H point are taken as the observation point. In the process of collision, an initial collision velocity is set to Y negative 13.88m/s, the velocity of X axis and z axis is zero, so only needing to consider the invasion change of the Y direction. By the computer simulation analysis, each part of the deformation and acceleration statistics is shown as follow table1.

Table 1. Deformation and acceleration statistics.

Input point	Intrusion measurement	Invasion of acceleration
Dummy head corresponding to the output point of B column	71	6.8
Beltline corresponding to the output point of B column	103	6.9
Upper ribs of dummy corresponding to the output point of B column	106	6.9
Under ribs of dummy corresponding to the output point of B column	105	6.9
Dummy abdomen corresponding to the output point of B column	105	7.0
H point of dummy corresponding to the output point of B column	97	7.1
Upper ribs of dummy corresponding to the output point of door inside plane	98	6.9
Upper ribs of dummy corresponding to the output point of door inside plane	105	6.9
Dummy abdomen corresponding to the output point of door inside plane	105	7.0
H point of dummy corresponding to the output point of door inside plane	113	7.0

By statistics, the side-impact model has seven output points, the invasion of which is more than target value, and the maximum invasion of door inside panel is 113mm on H output point. At the same time, the deformation is more than 100mm at the position of under ribs and abdomen. The deformation is so large that it will impact the crew compartment space when accident occurs and affect the door unlocked and the staff's escape from crew compartment. The maximum invasion of model B column is 106mm (as shown in Fig. 3), occurred in the upper ribs. The dummy H point of the biggest intrusion displacement is corresponded to the position of door plank (as shown in Fig. 4), and relative displacement within 0~50ms rapidly increase to the peak of 113mm, then it decreases few due to the springback after the deformation of B column. By the end of collision, the permanent deformation of the relative displacement is 67mm, and the amount of springback is 66mm.

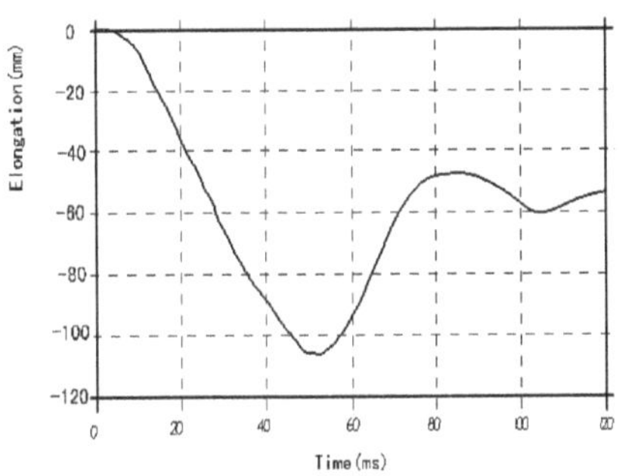

Fig. 3. Upper ribs of dummy corresponding to the invasion in B column position.

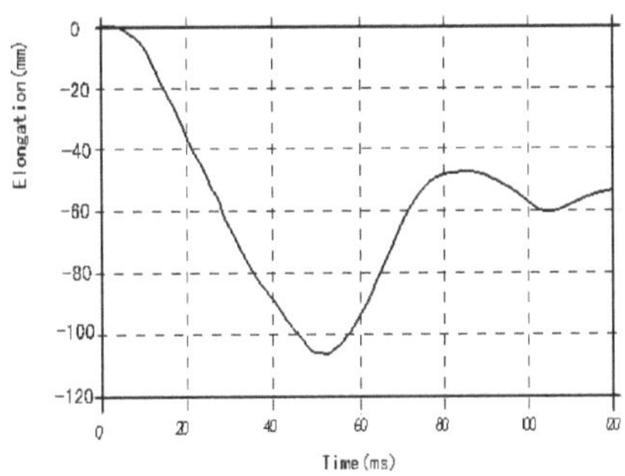

Fig. 4. H point of dummy corresponding to the invasion in door plank position.

The corresponding acceleration value of H point of dummy corresponding to the B column output point is beyond the target value. The acceleration change after the car crash depends on the closing speed of two cars, and the average trend of acceleration is inversely proportional to the quality. The acceleration curve of dummy abdomen corresponding to the output point of B column is as shown in the figure. The maximum change of relative speed in Y direction is 7.1m/s, as shown in Fig. 5.

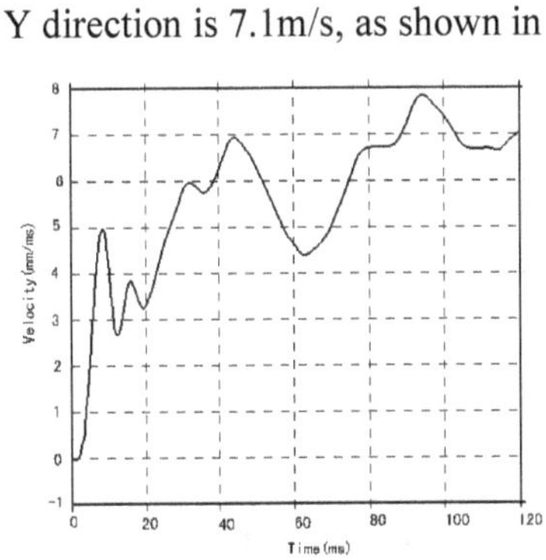

Fig. 5. Dummy abdomen corresponding to the invasion speed of B column in Y direction.

4. Structure Optimization

According to the results of car side impact, it is found that the doors, B column, floor, the threshold beam and the roof have had the lager deformation in the process of side impact, and the door deformation invades the crew compartment. To analyze the relative displacement change of the bottom and central of the B column and the relative speed variation, it is obvious that the invasion can't meet the requirement of design when model lies in B column except the invasion of dummy head corresponding. Although the intrusion velocity meets the requirement, it is necessary to promote the car-body safety performance and carry out partial optimization for structures due to a large amount of invasion and risks being. Front door anti-collision beam, B column and threshold beam are the main power transmission path. Therefore, in the optimization of body structure, the door anti-collision beam, B column and threshold beam are taken into consideration for improving its strength.

Improve measures are shown as follows:

(1) To optimize the lap structure of sheet metal pieces, and to make A column, threshold beam, B column, top beam and C column form complete power transmission path so as to avoid partial deformation and prevent the security space of crew compartment from invading.

(2) To increase the strength of threshold beam, and materials are promoted from HC550/980DP to HC950/1180MS. At the same time, to modify the threshold beam section and increase the cavity size; to increase the B column strength and optimize the forming process, using B1500HS to enhance partial strength.

To verify the structure, the results show that the maximum invasion of model side-impact car is 94mm, which is measured in rib point of B column. The maximum invasion velocity is 6.9m/s when dummy abdomen is corresponded to the output point of B column, which satisfies the requirement of design.

5. Conclusion

Through the analysis of vehicle body side-impact performance, it provides advice for the optimization and improvement of car-body structure. The improvement measures are verified to be effective and feasible, thus providing reference for the design and improvement of subsequent models.

Acknowledgments

The authors sincerely thank the financial supports from Ministry of Industry and Information Technology National industry transfer type to upgrade basic engineering practice project, the National Key Technology R&D Program (No. 2011BAG03B01) and Chongqing Research Program of Foundation and Advanced Technology program (No: cstc2013jcyjC600001).

References

1. H.Y. Wang and W.M. Gao, The passive safety of car structure, *Car and Accessory* **3**, 28 (2001).
2. X.C. Zhu, The investigation on the impact safety of vehicle by application of computer simulation technique, *The World Automob.* **35**, 15 (1999).
3. X.C. Zhu, The current situation and the trend of development of the evaluated system of vehicles' impact safety, Tong Ji University (2006).

The Study of Mechanical Properties and Deformation Mechanism of Spot Weld of Advance High Strength Steel Under Combined Load Condition

Y. Feng†, J. Zhou, M. T. Ma and Y. Zhao

China Automotive Engineering Research Institute Co., Ltd.,
Automotive Lightweight Engineering Research Center, Chongqing, 401122, China
†E-mail: fengyi@caeri.com.cn

In this paper, through strengthening plate method and a special kind of fixture, the failure force values, fracture modes of spot welds of several kinds of advance high strength steels (AHSS) were obtained. Based on the test data, an elliptical failure criterion was used to describe the fracture action of spot weld through function fitting. The minimum failure force of spot weld always appears at 30° load angle which was in accordance with the theoretical value. Compared to other kinds of conventional failure models, the improved elliptical failure criterion which was referenced in this paper is more accurate and helpful to actual engineering.

Keywords: Spot weld; Combined load condition; Failure criterion.

1. Introduction

As one of the most important welding methods, resistance spot welding (RSW) is widely used in automobile industry. Normally, spot welds will be under a very complicated stress condition in an actual crash process. When an external force acts on a component, failure of the spot weld is likely to occur prior to the base metal due to the extremely high stress concentration at the interface between spot weld and matrix [1-3]. The final failure mode of spot weld is a result of combined stress condition which contains many kinds of load forms such as shear load, axial load, etc. Until now, most study works are limited to some simple load conditions [4-6] such as pure shear or axial load which often results in an evident difference between study result and actual case especially in FE analysis area. So it is very necessary to study the mechanical properties, deformation mechanism and failure mode of RWS under combined condition to meet engineering requirement.

In this paper, the failure tests of spot welds of DC01with thickness of 1.5mm, DC04 with thickness of 1.5mm, DC06 with thickness of 1.5mm, DP590 with thickness of 1.5mm, DP780 with thickness of 1.5mm, and DP980 with thickness of 1.5mm, TRIP590 with thickness of 1.5mm, TRIP780 1.5mm, TRIP980 with thickness of 1.5mm were carried out under different combined load conditions to develop corresponding failure criterions. The failure criterions were obtained based on an improved elliptical formula which had been created by J. H. Song, etc [1]. To ensure the effect of experiments, a series of methods were used to decrease the rotation of samples during the failure test process. Failure tests of spot welds were finished under seven different load angles range from 0° to 90°. The tensile strength data were obtained to propose the failure criterion by elliptical function fitting.

2. Experimental Method

As shown in Fig. 1(a), the structure of fixture was designed referring to Arcan-type fixture [2]. A strengthening-board method was used during the test process in order to avoid above problem. To shear tension samples, two boards were used to restrict the deformation of matrix around the nugget as shown in Fig. 1(b). As shown in Fig. 1(c), all samples were welded with one 2.0 mm sheet before formal resistance spot welding process in order to increase the rigidity of samples to prevent excessive bending in the test.

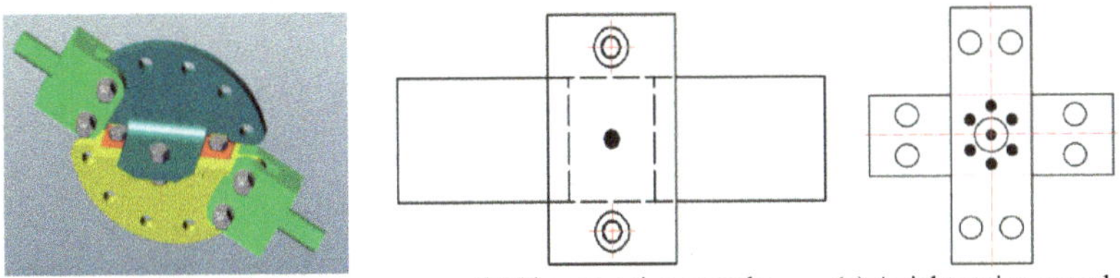

(a) Test fixture (b) Shear tension sample (c) Axial tension sample

Fig. 1. Structure of fixture and samples.

The welding experiments were finished on an advanced intermediate frequency welding machine. The choosing of welding parameters was based on the conventional criterion of resistance spot welding and the welding parameter standards of some famous automobile corporations. The test process stopped till the samples were separated into two components. To increase the reliability of test data, five samples were tested under each angle condition.

3. Experimental Results and Discussions

3.1. *Results*

Table 1 shows the failure force values which were obtained under different load angle conditions. It can be noted that the results are similar to J. H. Song [1, 8] Markiewicz E, Langrand B, etc [9], but unlike to Lee, etc [7]. The total variation tendency of failure force is that as the load angle increasing at earlier stage, the failure force decreases until the minimum force value (30° angle). Then as the load angle increasing subsequently, the failure force increases accordingly and this rule is suited to all kinds of steels. But according to the result of Lee, etc [7], from pure shear to pure axial load condition, the failure force will decrease at all times.

To meet the safety performance requirement and design goals, especially in the FE area, the accurate failure criterion of spot weld should be considered. In this paper, the model which was proposed by J. H. Song, etc [1] was adopted. Based on the model, the key point is to create the failure contour of spot weld which is based on the test data under different combined load conditions. The key parameters in this test are shear failure load ($F_s = F_0 \sin\theta$), axial failure load ($F_n = F_0 \sin\theta$) of the spot weld and the so-called shape coefficient β.

Table 1. Failure force under different load angles.

Material	Failure force (kN)						
	Angle (°)						
	90	75	60	45	30	15	0
DP590	16.24	12.00	9.58	8.31	7.83	8.13	8.92
DP780	19.07	13.98	11.21	9.82	9.05	9.15	9.77
DP980	23.98	17.07	14.03	12.33	11.42	11.49	12.01
TRIP590	16.66	11.98	9.67	8.42	7.90	8.04	8.84
TRIP780	19.63	14.22	11.49	10.12	9.38	9.46	9.99
TRIP980	20.98	16.11	13.23	11.53	10.44	10.49	10.59
DC01	6.37	5.09	4.57	4.22	4.00	4.27	5.09
DC04	6.28	5.00	4.44	4.11	3.99	4.17	4.88
DC06	6.17	4.91	4.26	3.98	3.90	4.05	4.70

3.2. *Failure criterions*

The failure criterion was translated from the elliptic failure model which had been advanced by Lee [7] and can be expressed as follows:

$$\left(\frac{f_s}{F_s}\right)^2+\left(\frac{f_n}{F_n}\right)^2=1 \tag{1}$$

Where f_s is the shear force, f_n is the shear force.

Fig. 2 describes the actual stress distribution around the nugget, in order to create a kind of more simple and logical model, the singularities which are caused by the welding gap between nugget and matrix material are ignored here. The stress state is assumed as a linger distribution around the nugget. Here for simplicity the lower half of the nugget is considered and idealized as a circular cylinder. The average stress through thickness direction substitutes the actual non-linear distributed stress. When the shear load and axial load act on the nugget, the assumed stress distribution state is shown in Fig. 3.

Based on above hypothesis, when the axial load acts on the center point of the nugget, f_n must meet the equilibrium condition (Eq. (2)) that f_n should be equal to the shear force through the thickness direction.

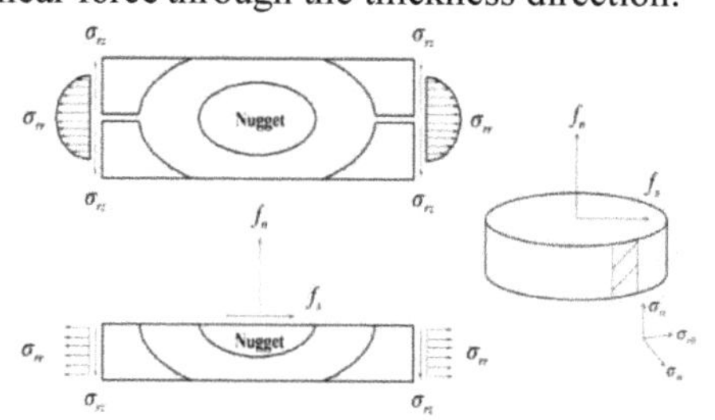

Fig. 2. Stress distribution around the nugget.

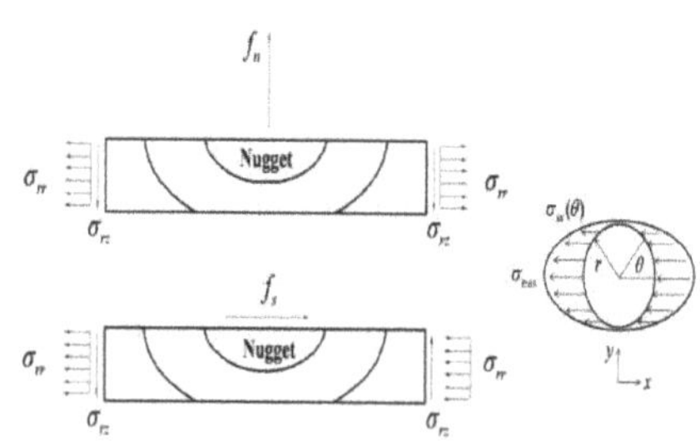

Fig. 3. Assumed stress distribution state.

$$\int_0^{2\pi} \sigma_{rz} rtd\theta + f_n = 0 \tag{2}$$

Where r is the radius of nugget. According to Eq. (2) and plate theory, the σ_{rz} and σ_{rr} can be expressed as Eq. (3)~(4).

$$\sigma_{rz} = -\frac{f_n}{2\pi rt} \tag{3}$$

$$\sigma_{rr} = k\frac{f_n}{rt^2} \tag{4}$$

Where k is the parameter which depends on the sample dimension and test condition.

Equally, the shear load f_s also must meet equilibrium condition Eq. (5)~(6) that f_s should be equal to the average tensile load through the thickness direction around the nugget circumference.

$$\int_0^{2\pi} \sigma_{xx}(\theta) rt\cos\theta d\theta + f_s = 0 \tag{5}$$

$$\sigma_{xx} = -\frac{f_s}{\pi rt}\cos\theta \tag{6}$$

The average shear load σ_{xz} which is balanced with the moment due to the f_s and the tensile load due to the average tensile σ_{xx} is shown as Eq. (7)~(8).

$$\int_0^{2\pi} \sigma_{xz}\cos^2(\theta) r^2 td\theta - f_s\frac{t}{2} = 0 \tag{7}$$

$$\sigma_{xz} = \frac{f_s}{2\pi r^2} \tag{8}$$

Based on above theoretical analysis and calculation, the parameter σ_{xx} and σ_{xz} all can be transformed into cylindrical coordinates which are shown as Eq. (9)~(13).

$$\sigma_{rr} = \sigma_{xx}(\theta)\cos^2\theta = -\frac{f_s}{\pi rt}\cos^3\theta \tag{9}$$

$$\sigma_{r\theta} = -\sigma_{xx}(\theta)\sin\theta\cos\theta = \frac{f_s}{\pi rt}\sin\theta\cos^2\theta \tag{10}$$

$$\sigma_{\theta\theta} = \sigma_{xx}(\theta)\sin^2\theta = -\frac{f_s}{\pi rt}\sin^2\theta\cos\theta \tag{11}$$

$$\sigma_{rz} = \sigma_{xz}\cos\theta = \frac{f_s}{2\pi r^2}\cos\theta \tag{12}$$

$$\sigma_{\theta z} = \sigma_{xz}(-\sin\theta) = -\frac{f_s}{2\pi r^2}\sin\theta \tag{13}$$

At last, the final stress field distribution around the nugget under different combined load condition can be expressed with the cylindrical coordinate system as Eq. (14)~(18):

$$\sigma_{rr} = -\frac{f_s}{\pi r t}\cos^3\theta + k\frac{f_n}{rt^2} \tag{14}$$

$$\sigma_{r\theta} = \frac{f_s}{\pi r t}\sin\theta\cos^2\theta \tag{15}$$

$$\sigma_{\theta\theta} = -\frac{f_s}{\pi r t}\sin^2\theta\cos\theta \tag{16}$$

$$\sigma_{rz} = \frac{f_s}{2\pi r^2}\cos\theta - \frac{f_n}{2\pi r t} \tag{17}$$

$$\sigma_{\theta z} = -\frac{f_s}{2\pi r^2}\sin\theta \tag{18}$$

Based on the hypothetical theory in this paper, during tensile process, the point of 180°(θ) can be considered as the weakest point. So the final failure criterion of spot weld is obtained as Eq. (19).

$$((\frac{f_s}{F_s})^2 + \beta(\frac{f_s}{F_s})(\frac{f_n}{F_n}) + (\frac{f_n}{F_n})^2 = 1 \tag{19}$$

Where F_s and F_n are the maximum shear and axial failure forces of nugget. The coefficient β which is mentioned before can be obtained by elliptical fitting in Origin software. The β of different kinds of steels are shown in table 2.

Table 2. Coefficient β of different materials.

Material	F_s	F_n	β
DC01	6.37	5.09	1.38
DC04	6.28	4.88	1.39
DC06	6.17	4.70	1.41
DP590	16.24	8.92	1.35
DP780	19.07	9.77	1.33
DP980	22.13	11.01	1.30
TRIP590	18.99	10.17	1.37
TRIP780	19.55	10.27	1.34
TRIP980	19.44	9.59	1.23

According to table 2, the final failure criterions of different kinds of steels in the paper are shown as follows (table 3).

Table 3. Failure criterions of different steels.

Materials (1.5mm)	Failure criterion
DC01	$\left(\frac{f_s}{6.37}\right)^2+1.38\left(\frac{f_s}{6.37}\right)\left(\frac{f_n}{5.09}\right)+\left(\frac{f_n}{5.09}\right)^2=1$
DC04	$\left(\frac{f_s}{6.28}\right)^2+1.39\left(\frac{f_s}{6.28}\right)\left(\frac{f_n}{4.88}\right)+\left(\frac{f_n}{4.88}\right)^2=1$
DC06	$\left(\frac{f_s}{6.17}\right)^2+1.41\left(\frac{f_s}{6.17}\right)\left(\frac{f_n}{4.70}\right)+\left(\frac{f_n}{4.70}\right)^2=1$
DP590	$\left(\frac{f_s}{16.24}\right)^2+1.35\left(\frac{f_s}{16.24}\right)\left(\frac{f_n}{8.92}\right)+\left(\frac{f_n}{8.92}\right)^2=1$
DP780	$\left(\frac{f_s}{19.07}\right)^2+1.33\left(\frac{f_s}{19.07}\right)\left(\frac{f_n}{9.77}\right)+\left(\frac{f_n}{9.77}\right)^2=1$
DP980	$\left(\frac{f_s}{23.98}\right)^2+1.30\left(\frac{f_s}{23.98}\right)\left(\frac{f_n}{12.01}\right)+\left(\frac{f_n}{12.01}\right)^2=1$
TRIP590	$\left(\frac{f_s}{16.66}\right)^2+1.37\left(\frac{f_s}{16.66}\right)\left(\frac{f_n}{8.84}\right)+\left(\frac{f_n}{8.84}\right)^2=1$
TRIP780	$\left(\frac{f_s}{19.63}\right)^2+1.34\left(\frac{f_s}{19.63}\right)\left(\frac{f_n}{9.99}\right)+\left(\frac{f_n}{9.99}\right)^2=1$
TRIP980	$\left(\frac{f_s}{20.98}\right)^2+1.23\left(\frac{f_s}{20.98}\right)\left(\frac{f_n}{10.59}\right)+\left(\frac{f_n}{10.59}\right)^2=1$

Above model (Eq. (19)) can be applied to forecast the failure force under different kinds of load conditions and it will be more helpful to FE analysis process in automobile manufacture area. The f_s - f_n curve represents the critical failure condition during test. As shown in Fig. 4, when the parameter point is located below the f_s- f_n curve, the spot could be considered at a relatively safe state.

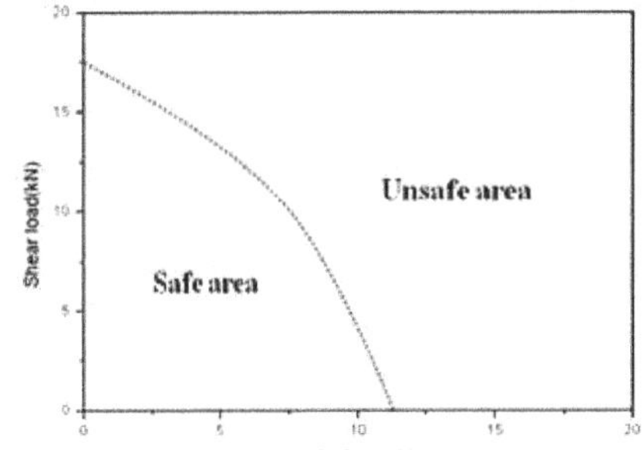

Fig. 4. The partition of safe and unsafe areas.

4. Conclusion

The deformation action, failure criterion of spot weld of AHSS under different load conditions were studied through failure test in this paper, based on test results, an elliptical failure criterion was used to describe the fracture action of spot weld through function fitting. The minimum failure force of spot weld of AHSS always appears at 30° load angle which is almost in accordance with the theoretical value. Compared to other some kinds of conventional failure model, the improved elliptical failure criterion referenced is more accurate and helpful for actual engineering.

Acknowledgements

This work was carried out with the financial supports of Chongqing Basic and Frontier Research Program for Distinguished Young Scientists Project (cstc2013jcyjjq60001) and Chongqing Training Program for Science and Technology Talents (cstc2013kjrc-qnrc60003) and Chongqing Basic and Frontier Research Program (cstc2013jcyjC60001) and Chongqing Automotive Collaborative Innovation Center. These supports are greatly acknowledged.

References

1. J. H. Song and H. Huh, Failure characterization of spot welds under combined axial–shear loading conditions, *Int. J. Mech. Sci.* **53**, 513 (2011).
2. S. H. Lin, J. Pan, S. R. Wu, T. Tyan and P. Wung, Failure loads of spot welds under combined opening and shear static loading conditions, *International Journal of Solids and Structures* **39**, 19 (2002).
3. X. Yang, Y. Xia and Q. Zhou, A simplified FE model for pull-out failure of spot welds, *Engineering Fracture Mechanics* **77**, 1224 (2010).
4. Y. Chao, Ultimate strength and failure mechanism of resistant spot weld subjected to tensile, shear, or combined tensile/shear loads, *ASME Journal of Engineering Materials and Technology* **125**, 125 (2003).
5. P. Wung, A force-based failure criterion for spot weld design, *Experimental Mechanics* **41**,107 (2001).
6. P. Wung, T. Walsh, A. Ourchane, W. Stewart and M. Jie, Failure of spot welds under in-plane static loading, *Experimental Mechanics* **41**,100 (2001).
7. Y. L. Lee, T. J. Wehner, M. W. Lu, T. W. Morrissett and E. Pakalnins, Ultimate strength of resistance spot welds subjected to combined tension and shear, *Journal of Testing and Evaluation* **26**, 213 (1998).
8. J. H. Song, H. Huh, J. Ha and J. H. Lim, Prediction of failure characteristics of spot welds of DP and trip steels with an equivalent strength failure model, *International Journal of Automotive Technology* **14**, 67 (2013).
9. B. Langrand and E. Markiewicz, Strain-rate dependence in spot welds: non-linear behavior and failure in pure and combined modes I/II, *International Journal of Impact Engineering* **37**, 792 (2010).

Author Index

www.ingramcontent.com/pod-product-compliance
Lightning Source LLC
LaVergne TN
LVHW060618300926
849162LV00008B/1652
* 9 7 8 9 8 1 3 2 0 7 2 9 5 *